IET ELECTROMAGNETIC WAVES SERIES 31

Series Editors: Professor P.J.B. Clarricoats
Professor Y. Rahmat-Samii
Professor J.R. Wait

Ionospheric Radio

Other volumes in this series:

Volume 1 **Geometrical theory of diffraction for electromagnetic waves, 3rd edition** G.L. James
Volume 10 **Aperture antennas and diffraction theory** E.V. Jull
Volume 11 **Adaptive array principles** J.E. Hudson
Volume 12 **Microstrip antenna theory and design** J.R. James, P.S. Hall and C. Wood
Volume 15 **The handbook of antenna design, volume 1** A.W. Rudge, K. Milne, A.D. Oliver and P. Knight (Editors)
Volume 16 **The handbook of antenna design, volume 2** A.W. Rudge, K. Milne, A.D. Oliver and P. Knight (Editors)
Volume 18 **Corrugated horns for microwave antennas** P.J.B. Clarricoats and A.D. Oliver
Volume 19 **Microwave antenna theory and design** S. Silver (Editor)
Volume 21 **Waveguide handbook** N. Marcuvitz
Volume 23 **Ferrites at microwave frequencies** A.J. Baden Fuller
Volume 24 **Propagation of short radio waves** D.E. Kerr (Editor)
Volume 25 **Principles of microwave circuits** C.G. Montgomery, R.H. Dicke and E.M. Purcell (Editors)
Volume 26 **Spherical near-field antenna measurements** J.E. Hansen (Editor)
Volume 28 **Handbook of microstrip antennas, 2 volumes** J.R. James and P.S. Hall (Editors)
Volume 31 **Ionospheric radio** K. Davies
Volume 32 **Electromagnetic waveguides: theory and applications** S.F. Mahmoud
Volume 33 **Radio direction finding and superresolution, 2nd edition** P.J.D. Gething
Volume 34 **Electrodynamic theory of superconductors** S.A. Zhou
Volume 35 **VHF and UHF antennas** R.A. Burberry
Volume 36 **Propagation, scattering and diffraction of electromagnetic waves** A.S. Ilyinski, G. Ya.Slepyan and A. Ya.Slepyan
Volume 37 **Geometrical theory of diffraction** V.A. Borovikov and B.Ye. Kinber
Volume 38 **Analysis of metallic antenna and scatterers** B.D. Popovic and B.M. Kolundzija
Volume 39 **Microwave horns and feeds** A.D. Olver, P.J.B. Clarricoats, A.A. Kishk and L. Shafai
Volume 41 **Approximate boundary conditions in electromagnetics** T.B.A. Senior and J.L. Volakis
Volume 42 **Spectral theory and excitation of open structures** V.P. Shestopalov and Y. Shestopalov
Volume 43 **Open electromagnetic waveguides** T. Rozzi and M. Mongiardo
Volume 44 **Theory of nonuniform waveguides: the cross-section method** B.Z. Katsenelenbaum, L. Mercader Del Rio, M. Pereyaslavets, M. Sorella Ayza and M.K.A. Thumm
Volume 45 **Parabolic equation methods for electromagnetic wave propagation** M. Levy
Volume 46 **Advanced electromagnetic analysis of passive and active planar structures** T. Rozzi and M. Farinai
Volume 47 **Electromagnetic mixing formulas and applications** A. Sihvola
Volume 48 **Theory and design of microwave filters** I.C. Hunter
Volume 49 **Handbook of ridge waveguides and passive components** J. Helszajn
Volume 50 **Channels, propagation and antennas for mobile communications** R. Vaughan and J. Bach-Anderson
Volume 51 **Asymptotic and hybrid methods in electromagnetics** F. Molinet, I. Andronov and D. Bouche
Volume 52 **Thermal microwave radiation: applications for remote sensing** C. Matzler (Editor)
Volume 53 **Principles of Planar Near-Field Antenna Measurements** S. Gregson, J. McCormick and C. Parini
Volume 502 **Propagation of radiowaves, 2nd edition** L.W. Barclay (Editor)

Ionospheric Radio

Kenneth Davies

The Institution of Engineering and Technology

Published by The Institution of Engineering and Technology, London, United Kingdom

First published 1990
Reprinted 1996, 2008

The Institution of Engineering and Technology
Michael Faraday House
Six Hills Way, Stevenage
Herts, SG1 2AY, United Kingdom

www.theiet.org

British Library Cataloguing in Publication Data
Davies, K.
Ionospheric radio.
1. Radio waves. Propogation of effects on ionosphere
I. Title
621.3841'1

ISBN (10 digit) 0 86341 186 X
ISBN (13 digit) 978-0-86341-186-1

Printed in the UK by Short Run Press Ltd, Exeter
Reprinted in the UK by Lightning Source UK Ltd, Milton Keynes

Preface

This text replaces two earlier publications (Davies 1965, 1969). The older material has been updated, and additional references have been included. Several new topics are treated that were either omitted in the earlier publications or treated in a limited fashion. These include, for example, Earth–space propagation (Chapter 8), medium frequencies (Chapter 11), and ionospheric modification (Chapter 14). Discussions of other topics have been considerably expanded, for example, the discussion of modern ionospheric sounding techniques in Chapter 4.

Although the book is intended primarily for scientists and engineers working in the disciplines that involve effects of the ionosphere, it will also be of value to teachers and students who require introduction to the subject. Questions have been included for each chapter to test the reader's comprehension of the text.

I thank I. Ahl and M. Wiarda for their help in the preparation of the typescript, J. Bankhead and J. Watterson of the Boulder Laboratories Library for help in literature research, and L. Murdock for editing.

K. Davies
Space Environment Laboratory
Boulder, Colorado
May 1989

Acknowledgments

It is impossible to acknowledge everyone who has contributed to this book, but special thanks are due to scientific colleagues who read the typescript of the various chapters and gave valuable advice: Chapter 1, W.L. Flock; Chapter 2, H. Leinbach; Chapter 3, A.K. Paul; Chapter 4, R. Conkright; Chapter 5, E.K. Smith; Chapter 6, T.A. Croft; Chapter 7, R.F. Donnelly; Chapter 8, J.A. Klobuchar; Chapter 9, J.A. Joselyn; Chapter 10, J.A. Ferguson; Chapter 11, R.D. Hunsucker and J.C.H. Wang; Chapter 12, C.M. Rush; Chapter 13, G.R. Sugar; Chapter 14, T.B. Jones. X.-M. Liu, and C. Stubenrauch read the script and made numerous valuable suggestions. These colleagues and many authors and publishers have generously given permission for the reproduction of published figures and/or provided material. Wherever appropriate, these have been acknowledged in the text. I thank E. Hildner and J.R. Wait for their encouragement during the process of publication.

Finally, my deepest appreciation goes to my wife Joyce Davies for proofreading and other chores during the preparation of the typescript.

Contents

Chapter 1

Characteristics of waves and plasmas

1.1 Historical introduction

The ionosphere is defined, for our purposes, as that part of the upper atmosphere where sufficient ionization can exist to affect the propagation of radio waves. The ionosphere can, therefore, be considered as lying between about 50 km and several Earth radii (Fig. 1.1) (see IRE 1950). The peak electron density usually occurs in the F region (140 km and above). The F region is subdivided into the F_1 layer and the F_2 layer, which are produced by solar extreme ultraviolet light (EUV). Below the F region is the E region (90 km to about 140 km), produced by solar soft X-rays, which contains both the normal E layer and patches of sporadic E (E_s). The D region covers the height interval 50 km to 90 km and contains both the D layer and the cosmic ray, or C, layer. Above the F-peak of electron density is the topside and above that is the protonosphere or plasmasphere. The boundaries between these regions are not well defined.

The alphabetic nomenclature of the ionosphere was introduced by Sir Edward Appleton (Silberstein 1959) who used the letters E and F for electric and field. The origin of the term 'ionosphere' was investigated by Gillmor (1976) and accounts of some of the early investigations can be found in the following: Checcacci 1975; Clark 1971; Ratcliffe 1959; Maeda 1986; Mitra 1984; and Snyder and Bragaw 1986; and the references listed therein.

It is worth noting that ionospheric physicists use the term 'ionosphere' to distinguish the ionized part of the upper atmosphere from the background neutral gas (Banks and Kockarts 1973). The structures and peak densities in the ionosphere vary greatly with time (sunspot cycle, seasonally and diurnally), with geographical location (polar caps, auroral zones, equatorial regions), and with certain solar disturbances.

1.2 The radio spectrum

The radio spectrum affected, to some extent, by the ionosphere ranges from extremely low frequencies up to super high frequencies. The frequency ranges of these bands are given in Table 1.1 together with their primary propagation characteristics and uses. In general, on frequencies below 30 MHz the ionosphere is essential to users whereas on frequencies much above 50 MHz, the ionosphere is a source of annoyance as, for example, in Earth–space communications. Ionospheric radio propagation is characterized by amplitude fading, ray penetration, absorption, etc., making it unreliable for certain uses. Hence, in recent years, there has been a shift away from the lower frequency bands to the upper frequency bands on which undesirable ionosphere effects are less pronounced. Nevertheless, high frequencies are still used extensively for long distance broadcasting and point-to-point communication, especially when one terminal is mobile, e.g. ship-to-shore.

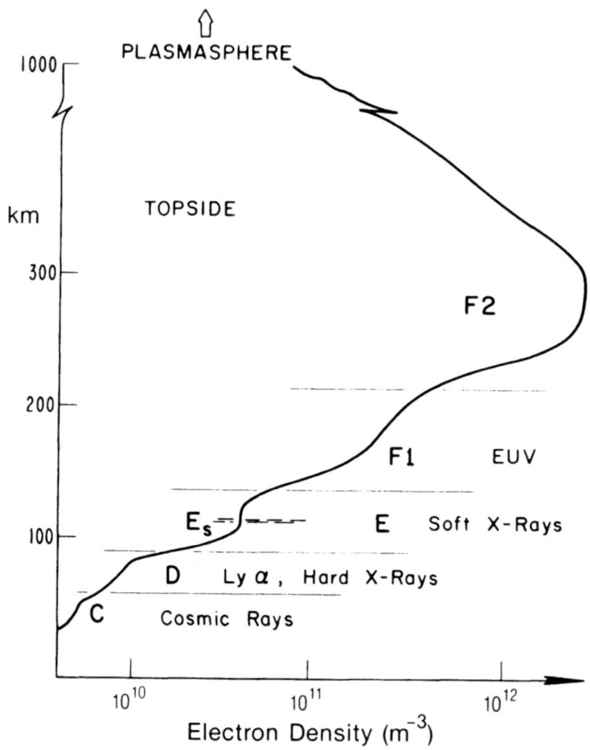

Fig. 1.1 *Ionosphere structure on a summer day in a middle latitude, and the main bands of solar and cosmic ionizing radiations*

The mathematical theory of radio propagation in the ionosphere in the presence of the Earth's magnetic field was developed in the 1920s by Appleton and his co-workers (Gillmor 1982; Ratcliffe 1959), and the formula for the refractive index is usually associated with his name, e.g. Appleton or Appleton–

Table 1.1 *The radio spectrum*

Name	Frequency range	Primary propagation modes	Primary uses
Extremely low frequency (ELF)	< 3 kHz	Earth–ionosphere waveguide penetrates sea water	Land-to-submarine communications
Very low frequency (VLF)	3 – 30 kHz	Waveguide (between ground and lower ionosphere), Ground wave	Navigation, Communication, Standard frequency and time
Low frequency (LF)	30 – 300 kHz	Waveguide, Ground wave	Maritime, Loran C, Broadcasting
Medium frequency (MF)	300 – 3000 kHz	E-region reflection (night), Ground wave	Maritime, Aeronautical, International distress, AM broadcasting, Maritime and land mobile
High frequency (HF)	3 – 30 MHz	Reflection from E and F regions	Maritime and aeronautical fixed services, Broadcasting (amateurs, citizens)
Very high frequency (VHF)	30 – 300 MHz	Line of sight, Scatter from ionosphere	Television, FM broadcasting, Public safety, Aeronautical
Ultra high frequency (UHF)	300 – 3000 MHz	Line of sight (affected by ionospheric irregularities)	Space communications, Television, Radar, Broadcasting, Navigation (fixed, mobile)
Super high frequency (SHF)	3000 – 30 000 MHz	Line of sight (tropospheric, affected by ionospheric irregularities)	Space communications, Television, Radar, Broadcasting, Navigation (fixed, mobile)

Hartree formula (Budden 1985) and, in Germany, the Appleton–Lassen formula (Rawer and Suchy 1976). The Appleton formula has been generalized to include effects of heavy ions (Stix 1962) and the dependence of electron collision frequency on electron energy (e.g. Sen and Wyller 1960).

1.3 Units and symbols

In this book we use the International System (SI) of units based on the meter, kilogram, and second together with the coulomb as the unit of electrical charge (see Goldman and Bell 1986). Since many plasma and atmosphere physicists use the centimeter-gram-second system, some care is necessary when comparing expressions in this book with other texts.

The following symbols and, where appropriate, values will be used unless stated otherwise:

1.3.1 Fundamental quantities

c = free space speed of electromagnetic waves = $2{\cdot}998 \times 10^{8}$ meters/second
e = charge on an electron = $-1{\cdot}602 \times 10^{-19}$ coulomb
h = Planck's constant = $6{\cdot}6256 \times 10^{-34}$ joule seconds
K = Boltzmann's constant = $1{\cdot}381 \times 10^{-23}$ joule/kelvin
m = rest mass of an electron = $9{\cdot}107 \times 10^{-31}$ kilogram
ε_0 = permittivity of free space = $8{\cdot}854 \times 10^{-12}$ farad/meter
μ_0 = permeability of free space = $12{\cdot}57 \times 10^{-7}$ henry/meter

1.3.2 Electromagnetic symbols

$\boldsymbol{B}$ = magnetic flux density (teslas = 10^4 gauss = 10^9 gammas)
$\boldsymbol{B}_0$ = ambient magnetic flux (teslas)
$\boldsymbol{D}$ = $\varepsilon \boldsymbol{E}$ dielectric displacement (coulombs/meter^2)
$\boldsymbol{E}$ = electric field (volts/meter)
$\boldsymbol{H}$ = magnetic field (amperes/meter)
V = magnetic potential
$\boldsymbol{J}$ = current density (amperes/meter^2)
$\boldsymbol{P}$ = volume polarization
ε = permittivity (farads/meter)
μ = magnetic permeability
ρ = volume charge density
σ = surface charge density (coulombs/meter^2)
σ = electrical conductivity (mhos)

1.3.3 Wave symbols

α = angle between $\boldsymbol{k}$ and $\boldsymbol{S}$
κ = $\omega\chi/c$ absorption per unit distance (nepers/meter)

λ = wavelength (meters)
μ = $c/v = ck/\omega$ refractive index (real)
μ' = c/v_g group refractive index
ϕ = phase
ϕ = angle between wave normal and vertical
ω = $2\pi f$ = angular wave frequency (radians/second)
$\Delta\omega$ = angular Doppler frequency shift
Ω = Faraday rotation
Å = angstrom = 0·1 nm = 10^{-10} meter
f = wave frequency (hertz)
Δf = Doppler frequency shift
$\boldsymbol{k}$ = propagation vector (radians/meter)
L = $\int \kappa\, ds$ total absorption (nepers or decibels)
n = $\mu - i\chi$ complex refractive index
P = $\int \mu\, ds$ phase path
P' = $\int \mu'\, ds$ group path
q = $\mu \cos\phi$
R = wave polarization
s = real path
$\boldsymbol{S}$ = energy flow vector
$\boldsymbol{u}$ = packet velocity
v = speed
v_p = phase speed
v_g = group speed

1.3.4 *Magneto-ionic symbols*

f_H = electron gyrofrequency (hertz)
f_N = plasma frequency (hertz)
N, N_e = electron number density (per meter^3)
X = $(\omega_N/\omega)^2$
Y = (ω_H/ω)
Y_L = $Y\cos\theta = f_L/f$
Y_T = $Y\sin\theta = f_T/f$
Z = ν/ω
θ = angle between the direction of propagation $\boldsymbol{k}$ and the external magnetic field $\boldsymbol{B}_0$
ν = collision frequency (per second)
ν_m = collision frequency for monoenergetic electrons with most probable energy (per second)
ψ = forward tilt of the $\boldsymbol{E}$ field in direction of propagation
ϕ = forward tilt of the $\boldsymbol{P}$ vector in direction of propagation
ω_c = $\omega_H \sin^2\theta/2\cos\theta$ critical angular collision frequency
ω_N = $2\pi f_N$ angular plasma frequency (radians/second)
ω_H = $2\pi f_H$ angular gyro- or cyclotron frequency (radians/second)

1.3.5 Plasma symbols

D = diffusion coefficient
D = Debye length
H_{pl} = plasma scale height
K = kelvin, unit of temperature
L = electron loss rate
M = rest mass of an ion
q = rate of production of ion pairs
T = temperature (kelvins)
T_e, T_i = electron and ion temperatures (kelvins)
T_g = gas temperature (kelvins)
w = vertical component of neutral wind
α = pitch angle
α = recombination rate
β = electron–ion linear loss rate (per second)
Δ, δ = fractional energy loss on collision
ε = ratio of electron kinetic energy to the thermal kinetic energy
σ = cross section (absorption, ionization, collision, etc.)
ρ = neutral mass density (kilograms/meter^3)
τ = optical depth (meters)

1.3.6 Atmosphere symbols

C = speed of sound (meters/second)
g = acceleration due to gravity (meters/second^2)
H_p = pressure scale height
p = pressure (newtons/meter^2)
z = reduced height in units of H_p
χ = solar zenith angle
ρ = atmospheric mass density (kilograms/meter^3)
ω_a = angular acoustic cutoff frequency
ω_g = angular buoyancy frequency for an isothermal atmosphere

1.3.7 Geographic, geomagnetic, and solar symbols

ϕ = geographic latitude
Φ = geomagnetic latitude
λ = geographic longitude
Λ = geomagnetic longitude
F = magnitude of the geomagnetic field
H = horizontal component of the geomagnetic field
I = magnetic inclination or dip
Z = vertical component of the geomagnetic field
a = Earth radius (6371·2 km)
x = modified magnetic dip ($= I/\sqrt{\cos\phi}$)

A_p = geomagnetic daily planetary amplitude
AE = auroral electrojet index
$F_{10\cdot7}$ = 10·7 cm radio noise flux (10^{-22} W m^{-2} Hz^{-1})
I_{F_2} = ionospheric F_2 index
K_p = 3-hour planetary magnetic index
R = sunspot number

1.3.8 Ionospheric symbols

f_c = critical or penetration frequency
f_o = critical frequency of ordinary wave
f_x = critical frequency of extraordinary wave
f_z = critical frequency of *z*-wave
f_{min} = lowest frequency seen on a vertical ionogram
h = real height (meters)
h' = group, equivalent, or virtual height (meters)
h_m = height of maximum electron density (meters)
N_m = maximum electron density (electrons/meter^3)
N_p = columnar electron content of the plasmasphere (electrons/meter^2)
N_F = Faraday content (electrons/meter^2)
N_T = total electron content (electrons/meter^2)
τ = slab thickness (meters)

1.3.9 Propagation symbols

f_o = frequency of wave propagating obliquely
f_v = equivalent vertical frequency
f_m = maximum frequency reflected with oblique propagation
p_t, p_r = radiated and received power flux (watts/meter^2)
t = time (seconds)
B = bandwidth (hertz)
D = ground range (kilometers)
F = coupling factor between wave and an antenna
M = MUF factor
M = modulation depth
MUF = maximum usable frequency
Q = mode availability
R = reflection coefficient
T = transmission coefficient
Δ = angle of elevation
ϕ_c = critical angle for occurrence of spitze
ϕ_0 = angle between ray and vertical at entry into a layer
ρ = reflection coefficient
ρ = circuit reliability.

1.4 Some properties of propagation media

1.4.1 Homogeneous and inhomogeneous media

Consider a volume as illustrated in Fig. 1.2*a* in which we take test samples near points A, B, C, . . . , etc. If a property of the medium is the same at A, B, C, . . . , etc. and at all other points, the medium is said to be homogeneous (or uniform) in that property. The medium is inhomogeneous when the property differs from one sample to another. It is important to remember that a medium can be homogeneous in one characteristic and inhomogeneous in another. For example, in an isothermal atmosphere under gravity, the temperature is homogeneous but the density is inhomogeneous. Provided that the spatial changes are sufficiently small an inhomogeneous medium can be treated as a locally uniform medium.

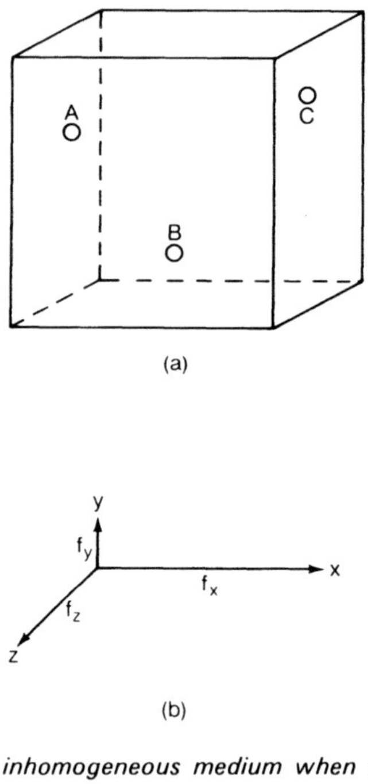

Fig. 1.2 (*a*) *Concept of an inhomogeneous medium when a property of the medium at different points (A, B, C, . . . , etc.) varies*
(*b*) *Concept of an anisotropic medium when a property of the medium at a given point varies with direction (x, y, z)*

1.4.2 Isotropy and anisotropy

Consider the situation, illustrated in Fig. 1.2*b*, at a point in a homogeneous medium. We measure a property f in different directions (f_x, f_y and f_z). If the property is the same in all directions, the medium is said to be isotropic. On the other hand, if the value of the property depends on direction, the medium is said to be anisotropic. Anisotropy is generally related to a preferred direction such as that of gravity or the Earth's magnetic field.

1.5 Properties of waves

1.5.1 Plane waves

A sinusoidal wave of infinite wavefront and infinite duration in a homogeneous medium can be specified by the peak amplitude E_m and the phase ϕ relative to a spatial coordinate system. Thus, in general, the instantaneous amplitude E can be expressed as

$$E = E_m \exp i\phi = E_m \cos\phi + iE_m \sin\phi \tag{1.1}$$

where $i = \sqrt{-1}$. The phase ϕ is a function of time and of space. The phase depends on two parameters:

(*a*) ω, the change of phase (in radians) per unit time
(*b*) k, the change of phase per unit distance.

Figure 1.3 shows a section of an infinite plane wave propagating in the direction of the vector $\boldsymbol{k}$. The phase ϕ at P relative to that at O is the same as that at Q, viz.,

$$\phi = -ks\cos\theta = -\boldsymbol{k}\cdot\boldsymbol{s} \tag{1.2}$$

where $\boldsymbol{S}$ is the position vector of P. The negative sign is an arbitrary convention.

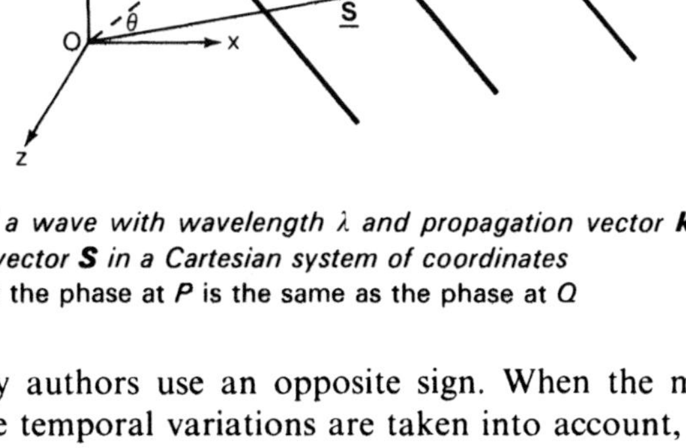

Fig. 1.3 *Phase of a wave with wavelength λ and propagation vector **k** at a point P, with position vector **S** in a Cartesian system of coordinates*
Note that the phase at *P* is the same as the phase at *Q*

Note that many authors use an opposite sign. When the medium is slowly varying, and the temporal variations are taken into account, we have

$$\phi = \omega t - \int \boldsymbol{k}\cdot d\boldsymbol{s} \tag{1.3}$$

The angular frequency ω (in radians per second) is related to the frequency f (in hertz) by

$$f = \omega/2\pi \tag{1.4}$$

and the propagation constant k is related to the wavelength λ, measured in the direction of $\boldsymbol{k}$, by

$$k = \frac{2\pi}{\lambda} = \frac{\omega}{v} = n\frac{\omega}{c} \tag{1.5}$$

where $v = \lambda f$ is the speed of a wavefront, c is a reference speed (e.g. free space speed), and n is the refractive index. Note that, in general, v, λ, k, and n are all properties of the medium.

Equation (1.1) is a solution of the wave equation

$$\nabla^2 E \equiv \frac{\partial^2 E}{\partial x^2} + \frac{\partial^2 E}{\partial y^2} + \frac{\partial^2 E}{\partial z^2} = \frac{1}{v^2}\frac{\partial^2 E}{\partial t^2} = \left(\frac{k}{\omega}\right)^2 \frac{\partial^2 E}{\partial t^2} \tag{1.6}$$

From (1.3) we have

$$\boldsymbol{k} = -\operatorname{grad}\phi \tag{1.7}$$

which is the eiconal equation (Budden 1985).

1.5.2 Propagation in a dispersive medium

Consider the propagation of a modulated wave in a homogeneous medium. The process of modulation creates a spectrum of wavelets with slightly different frequencies and slightly different propagation factors, or wavenumbers. In Fig. 1.4 we see the simplest case in which a modulated wave is made from the interference of two waves of equal amplitudes E_1 and E_2 with angular frequencies ω and $\omega + \delta\omega$ and wavenumbers k and $k + \delta k$ respectively. Following Stratton (1941) we have

$$\psi_1 = \cos(\omega t - kx) \tag{1.8a}$$

and

$$\psi_2 = \cos[(\omega + \delta\omega)t - (k + \delta k)x] \tag{1.8b}$$

which when superimposed give

$$\begin{aligned}\psi = \psi_1 + \psi_2 &= 2\cos\tfrac{1}{2}(t\delta\omega - x\delta k) \\ &\quad \times \cos\left[\left(\omega + \frac{\delta\omega}{2}\right)t - \left(k + \frac{\delta k}{2}\right)x\right] \\ &\approx 2\cos\tfrac{1}{2}(t\delta\omega - x\delta k)\cos(\omega t - kx)\end{aligned} \tag{1.9}$$

The first cosine term gives the modulation envelope whereas the second cosine term gives the carrier signal. It follows from (1.9) that the carrier signal travels with the phase speed v_p given by

$$v_p = \frac{\omega}{k} \tag{1.10}$$

whereas the envelope travels at the group speed v_g given by

$$v_g = \lim_{\delta k \to 0} \frac{\delta\omega}{\delta k} = \frac{d\omega}{dk} \tag{1.11}$$

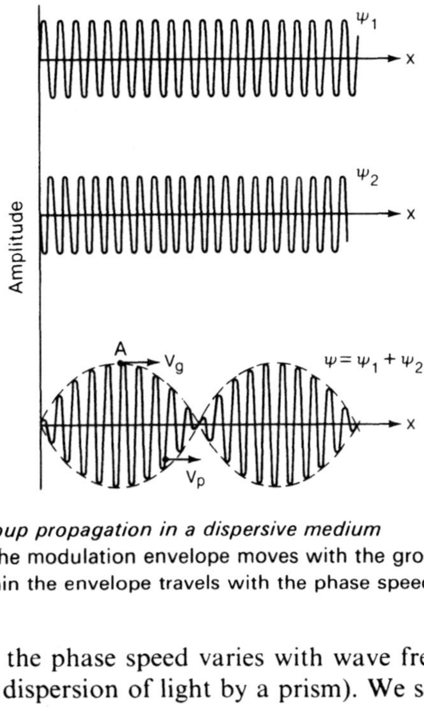

Fig. 1.4 *Concept of group propagation in a dispersive medium*
A point *A* on the modulation envelope moves with the group speed v_g whereas the waveform within the envelope travels with the phase speed v_p

A medium in which the phase speed varies with wave frequency is said to be dispersive (from the dispersion of light by a prism). We see that, except in the case of a linear dependence of ω on k, the envelope will travel at a different speed from the individual component wavelets. In the case where v_p is greater than v_g, a wavelet will appear at the back of a modulation envelope and work its way through the front. Were we to tag an individual wave crest we should see it start at the rear of an envelope with a small amplitude, which would increase to a maximum and then decrease as the crest approached the front. Thus, in order to define the group speed, it is essential to be able to track some identifiable feature, e.g. point A in Fig. 1.4; thus the modulation envelope must not suffer appreciable distortion.

It is possible to create a pulse, a single envelope, by the addition of a suitable spectrum of individual wavelets (called Fourier composition), and (1.13) gives the speed of the pulse (Jordan and Balmain 1968). In general, a dispersive medium is characterized by a resonant frequency (or frequencies).

1.5.3 Propagation in an anisotropic medium

As mentioned in Sec. 1.4.2, certain media have characteristic directions and, for a monochromatic wave, the phase speed can be a function of the angle θ of propagation between the propagation vector $\boldsymbol{k}$ and the characteristic direction. For illustration consider the case of a wave front emanating from a point source in a homogeneous, lossless, and anisotropic medium as sketched in Fig. 1.5*a*. After a time t, a noncircular wavefront will be as shown. Since the medium is homogeneous there will be no refraction and the energy will travel out radially. On the other hand, the propagation vector $\boldsymbol{k}$ at P is normal to the phase surface and we have the interesting situation in which the energy propagation vector $\boldsymbol{S}$ has a different direction from that of phase propagation $\boldsymbol{k}$ (Ratcliffe 1972). Figure 1.5*b* illustrates an equivalent view in which a composite wave is made up of monochromatic plane wavelets with slightly different directions of propagation. Although the wavelets are propagating close to the direction of the principal $\boldsymbol{k}$ vector, the wavelets reinforce at O and C and the direction $\boldsymbol{S}$ of energy flow is along OC. From the geometries of either Fig. 1.5*a* or Fig. 1.5*b*, it can be shown that

$$\tan\alpha = \frac{1}{k}\frac{dk}{d\theta} = \frac{1}{\mu}\frac{d\mu}{d\theta} \tag{1.12}$$

where μ is the real refractive index.

From (1.11) and (1.12) it follows that the vector wave-packet velocity $\boldsymbol{u}$ is given by

$$\boldsymbol{u} = \frac{d\omega}{d\boldsymbol{k}} = \hat{x}\frac{\partial\omega}{\partial k_x} + \hat{y}\frac{\partial\omega}{\partial k_y} + \hat{z}\frac{\partial\omega}{\partial k_z} = \frac{\partial\omega}{\partial k}\left(\hat{k} + \frac{\hat{\theta}}{k}\frac{\partial k}{\partial\theta}\right) \tag{1.13}$$

where $\hat{\boldsymbol{x}}$, $\hat{\boldsymbol{y}}$, and $\hat{\boldsymbol{z}}$ are unit vectors in the x, y, and z directions, respectively; θ is the angle between $\boldsymbol{k}$ and an arbitrary direction; and $\hat{k}$ and $\hat{\theta}$ are unit vectors in the $\boldsymbol{k}$ direction and perpendicular to $\boldsymbol{k}$, respectively.

1.5.4 Wave packets

In practice there is no monochromatic wave with an infinite wavefront. Wave signals start and end in time and they are confined laterally. Thus real wave signals are composed of wave packets, which can be represented by a suitable frequency spectrum and a suitable angular spectrum of component wavelets. The frequency spectrum confines the packet in the direction of propagation by interfering destructively in front of and behind the packet and constructively within. In the case of a suitable angular spectrum of the wavelets, there is destructive interference to the right and left of the packet and there is constructive interference within the packet. The shape of the packet is determined by its frequency spectrum and angular spectrum.

It is important to distinguish between the group speed v_g and the packet velocity $\boldsymbol{u}$. By definition the term group speed, or group velocity, usually denotes the component velocity of a wave packet in the direction of wave propagation (i.e. in the direction of $\boldsymbol{k}$) and is given by (1.11) whereas the packet velocity $\boldsymbol{u}$ is in the direction of energy flow $\boldsymbol{S}$ and is given by (1.13). It can be thought of as the motion, in space, of the center of the packet (Ratcliffe 1972).

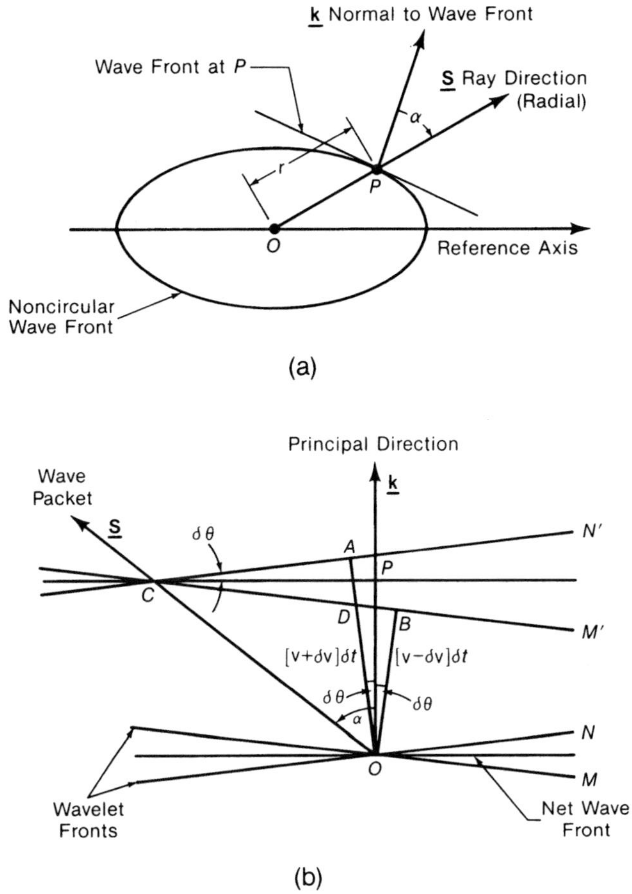

Fig. 1.5 (*a*) *The difference between the direction of wave propagation* $\boldsymbol{k}$ *and the (ray) direction of energy flow* $\boldsymbol{S}$ *in an anisotropic medium*
The ray direction makes an angle α with the wave normal $\boldsymbol{k}$
(*b*) *Propagation of a wave packet produced by the constructive interference between plane wavelets traveling in slightly different directions in an anisotropic medium*
The wavelets move essentially in the direction OP whereas the wave packet moves along OC

The group refractive index μ' is defined by analogy with the phase refractive index μ, viz.,

$$\mu' = \frac{c}{v_g} = c\,\frac{dk}{d\omega} = \frac{d}{d\omega}(\omega\mu) = \frac{d}{df}(f\mu) \tag{1.14}$$

It is important to note that, whereas the refraction of the wave normal is given by the phase refractive index (Snell's Law), there is no such relationship between the direction of energy flow and the group refractive index.

1.5.5 Phase paths and packet paths

Let us consider the path of a pulse (packet) of waves in a dispersive, anisotropic, and slowly varying inhomogeneous medium between a start point S and an end point E (Fig. 1.6). The actual path in space of an identifiable point on the wave packet (Fig. 1.6*a*) is called the ray path. It is convenient to define certain properties of the wave at a point P on the path. The phase characteristics, such as wavelength λ and phase speed v_p, are defined in the direction of $\boldsymbol{k}$, which is normal to the phase fronts. In the ray direction, which is tangent to the ray path at P, the wavelength λ_r, the ray (phase) speed v_r, and the packet speed u are given by (Fig. 1.6*b*)

$$\lambda_r = \lambda \sec\alpha \tag{1.15a}$$

$$v_r = v_p \sec\alpha \tag{1.15b}$$

$$u = v_g \sec\alpha \tag{1.15c}$$

The phase difference ϕ between S and E is given by

$$\phi = \int_S^E k\, ds \cos\alpha = \int_S^E \boldsymbol{k}\cdot d\boldsymbol{s} \tag{1.16}$$

where $d\boldsymbol{s}$ is an element of the ray path at P. The phase path P is defined as the distance, in free space, required to produce the given phase change ϕ; that is,

$$P = \frac{c}{\omega}\int_S^E \boldsymbol{k}\cdot d\boldsymbol{s} = \int_S^E \mu \cos\alpha\, ds \tag{1.17}$$

The phase transit time T_p, or time for a wave front to travel from S to E, is

$$T_p = \frac{P}{c} = \frac{\phi}{\omega} \tag{1.18}$$

The time of flight T_g of a packet traveling from S to E is

$$T_g = \int_S^E \frac{ds}{u} = \frac{d}{d\omega}\int_S^E k\cos\alpha\, ds = \frac{d\phi}{d\omega} \tag{1.19}$$

Thus the time of flight of a wave packet is the rate of change of phase with frequency. The group path P' is

$$P' = cT_g = \int_S^E \frac{c}{v_g}\cos\alpha\, ds = \int_S^E \mu' \cos\alpha\, ds \tag{1.20}$$

where μ' is as defined in (1.14).

P' and P are related by

$$P' = \int_S^E \frac{c}{v_g} \cos\alpha \, ds = c\int_S^E \frac{dk}{d\omega} \cos\alpha \, ds = \frac{d}{d\omega}(\omega P) \tag{1.21}$$

and

$$\mu' = \frac{d}{d\omega}(\omega\mu) \tag{1.22}$$

$$T_g = \frac{d}{d\omega}(\omega T_p) \tag{1.23}$$

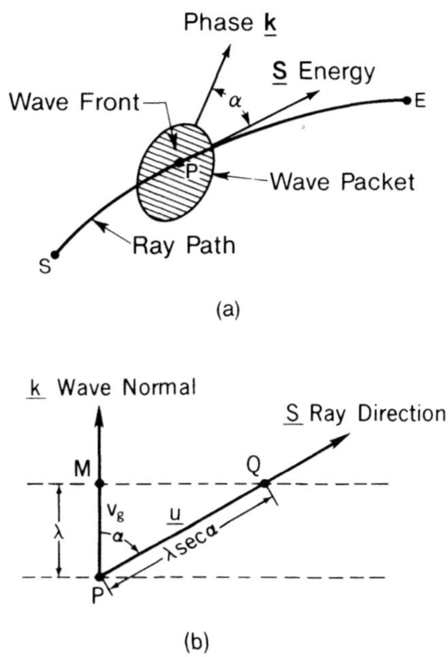

Fig. 1.6 (*a*) *Wave packet moving along a ray SE in an anisotropic medium*
The energy flows along the ray path, SE, while the wave normal at a point P is inclined at an angle α to the tangent to the ray path at P
(*b*) *Relationships between wave characteristics measured in the direction of phase propagation (such as wavelength λ, group speed v_g) and the corresponding quantities measured in the direction of energy propagation (e.g. wavelength $\lambda \sec\alpha$, packet velocity* ***u***)

It must be remembered that group paths and phase paths do not represent real paths but rather times of transit. The only real path is the ray path S given by

$$S = \int_S^E ds \tag{1.24}$$

When the phase path P is fixed in time, the number of wavefronts reaching E per second is the same as the number of wavefronts leaving S per second. When the phase path changes, e.g. increases, there is a larger number of fronts along the path; thus fewer per second reach E, and the received frequency is Doppler shifted downward by $\Delta\omega$. When the rate of change $\dot{P}$ of P is small compared with the free space speed c, we have

$$\Delta\omega = -\frac{2\pi}{\lambda_0}\dot{P} \tag{1.25}$$

In terms of linear frequency

$$\Delta f = -\frac{1}{2\pi}\dot{\phi} = -\frac{1}{\lambda_0}\dot{P} = -f\dot{P}/c \tag{1.26}$$

1.6 Electromagnetic waves

Radio waves are electromagnetic waves, and their propagation in a medium must satisfy two sets of conditions: (i) Maxwell's equations, which relate electric and magnetic fields, and (ii) the response of the medium to the wave fields – the constitutive relationships. For the purposes of our discussions we relate the electric field $\boldsymbol{E}$, the displacement field $\boldsymbol{D}$, the volume polarization $\boldsymbol{P}$, the magnetic intensity $\boldsymbol{H}$, and the magnetic flux density $\boldsymbol{B}$. The constitutive relations are

$$\boldsymbol{D} = \varepsilon\boldsymbol{E} = \varepsilon_0\boldsymbol{E} + \boldsymbol{P} \tag{1.27a}$$

$$\boldsymbol{B} = \mu_0\boldsymbol{H} \tag{1.27b}$$

where ε is the dielectric permittivity of the medium (generally a tensor), ε_0 is the permittivity of free space ($8{\cdot}85 \times 10^{-12}\,\mathrm{F\,m^{-1}}$), and μ_0 ($=4\pi \times 10^{-7}\,\mathrm{H\,m^{-1}}$) expresses the fact that the magnetic properties of an ionized gas are essentially those of free space. Since $\boldsymbol{E}$ and $\boldsymbol{P}$, and hence $\boldsymbol{D}$, are not necessarily in the same direction, (1.27a) expresses the anisotropy of the medium.

Maxwell's equations can be expressed in integral and differential forms. The integral forms are

$$\int_A \varepsilon\boldsymbol{E}\cdot d\boldsymbol{A} = \int_v \rho\, dv = Q \tag{1.28a}$$

$$\int_A \boldsymbol{B}\cdot d\boldsymbol{A} = 0 \tag{1.28b}$$

$$\int_s \boldsymbol{H}\cdot d\boldsymbol{s} = \int_A \boldsymbol{J}\cdot d\boldsymbol{A} + \frac{d}{dt}\int_A \boldsymbol{D}\cdot d\boldsymbol{A} \tag{1.28c}$$

$$\int_s \boldsymbol{E}\cdot d\boldsymbol{s} = -\frac{d}{dt}\int_A \boldsymbol{B}\cdot d\boldsymbol{A} \tag{1.28d}$$

where A represents a surface area, v is a closed volume containing an electric charge of density ρ, $\boldsymbol{J}$ is an electric current density, and s is a closed path.

The corresponding differential forms are

$$\nabla \cdot \varepsilon \boldsymbol{E} \equiv \operatorname{div} \varepsilon \boldsymbol{E} = \rho \tag{1.29a}$$

$$\nabla \cdot \boldsymbol{B} \equiv \operatorname{div} \mu_0 \boldsymbol{H} = 0 \tag{1.29b}$$

$$\nabla \times \boldsymbol{H} \equiv \operatorname{curl} \boldsymbol{H} = \boldsymbol{J} + \dot{\boldsymbol{D}} \tag{1.29c}$$

$$\nabla \times \boldsymbol{E} \equiv \operatorname{curl} \boldsymbol{E} = -\dot{\boldsymbol{B}} = -\mu_0 \dot{\boldsymbol{H}} \tag{1.29d}$$

where the point represents the time differential.

Let us consider conditions in free space where the polarization $\boldsymbol{P}$, the charge density ρ, and the current $\boldsymbol{J}$ are all zero. We take the curl of (1.29*c*) and (1.29*d*) and use the relationship

$$\operatorname{curl} \operatorname{curl} \boldsymbol{E} = \operatorname{grad} \operatorname{div} \boldsymbol{E} - \operatorname{div} \operatorname{grad} \boldsymbol{E} \tag{1.30}$$

Because there is no free charge, substitution of $\rho = 0$ in (1.29*a*) gives

$$\operatorname{div} \boldsymbol{E} = 0 \tag{1.31a}$$

Also, from (1.29*b*),

$$\operatorname{div} \boldsymbol{H} = 0 \tag{1.31b}$$

Hence

$$\operatorname{div} \operatorname{grad} \boldsymbol{E} = \mu_0 \operatorname{curl} \dot{\boldsymbol{H}} = \varepsilon_0 \mu_0 \ddot{\boldsymbol{E}} \tag{1.32}$$

and

$$\operatorname{div} \operatorname{grad} \boldsymbol{H} = -\varepsilon_0 \operatorname{curl} \dot{\boldsymbol{E}} = \varepsilon_0 \mu_0 \ddot{\boldsymbol{H}} \tag{1.33}$$

Now

$$\operatorname{div} \operatorname{grad} \boldsymbol{E} = \frac{\partial^2 E}{\partial x^2} + \frac{\partial^2 E}{\partial y^2} + \frac{\partial^2 E}{\partial z^2} \tag{1.34}$$

(CRC 1961). Thus (1.32) and (1.33) are the same as the wave equation (1.6) in which the phase velocity

$$v_p = c = 1/(\varepsilon_0 \mu_0)^{1/2} \tag{1.35}$$

Substituting the appropriate numerical values we find $c \approx 3 \times 10^8\ \mathrm{m\,s^{-1}}$, which is essentially the free-space speed of light. This conclusion was first drawn by James Clerk Maxwell (1873) and he postulated that visible light is an electromagnetic wave. In free space an electromagnetic wave is a transverse wave in the sense that the electric and magnetic fields of the wave are perpendicular to the direction of propagation; they are also perpendicular to each other.

In a homogeneous isotropic medium, $\boldsymbol{P}$ is parallel to $\boldsymbol{E}$, and for an infinite transverse plane wave traveling in the x-direction, for which $\partial/\partial y = \partial/\partial z = 0$, eqns. (1.29) lead to

$$\frac{\partial^2 H_y}{\partial x^2} = \varepsilon \mu_0 \ddot{H}_y \tag{1.36}$$

so, by analogy with (1.35), the phase speed is

$$v_p = 1/(\varepsilon\mu_0)^{1/2} \tag{1.37}$$

and the refractive index μ is given by

$$\mu^2 = \frac{c^2}{v_p^2} = \frac{\varepsilon}{\varepsilon_0} = 1 + \frac{1}{\varepsilon_0}\frac{P}{E} \tag{1.38}$$

So far we have said little or nothing about the nature of the refractive index and/or the wavenumber k. In certain media the wavenumber k can be complex; that is

$$\boldsymbol{k} = \boldsymbol{k}_r - i\boldsymbol{\kappa} \tag{1.39}$$

Combining (1.1), (1.3), and (1.39) yields

$$E = E_m \exp\left(-\int \boldsymbol{\kappa}\cdot d\boldsymbol{s}\right) \exp i\left(\omega t - \int \boldsymbol{k}_r \cdot d\boldsymbol{s}\right) \tag{1.40}$$

which is a wave with amplitude decreasing exponentially with distance. The factor κ is the attenuation per unit distance and is defined in the direction of the wavenumber $\boldsymbol{k}$. Some authors work in the time domain, and they express the attenuation in terms of the imaginary part of a complex frequency.

We now consider plane electromagnetic waves in an electrically neutral anisotropic medium, that is, one in which the volume polarization vector $\boldsymbol{P}$ is in a different direction from the driving electric field $\boldsymbol{E}$. Let the planes of constant phase be parallel to the y–z plane and let the direction of propagation be in the x direction so that

$$\frac{\partial}{\partial x} = -ik \tag{1.41a}$$

$$\frac{\partial}{\partial t} = i\omega \tag{1.41b}$$

$$\frac{\partial}{\partial y} = \frac{\partial}{\partial z} = 0 \tag{1.41c}$$

Writing out (1.27) and (1.29*a* and 1.29*b*) with conditions (1.41) we have for the x-components (setting $\rho = 0$)

$$D_x = \varepsilon_0 E_x + P_x = 0 \tag{1.42a}$$

$$H_x = 0 \tag{1.42b}$$

which tells us that, in general, the electric field of the wave has a component in the direction of propagation. However, the magnetic field lies in the plane of the wavefront. Further, putting $J = 0$ we have from (1.27), (1.29*c*), (1.29*d*), and (1.41),

$$i\omega D_y = ikH_z \tag{1.43a}$$

$$i\omega D_z = -ikH_y \tag{1.43b}$$

$$i\mu_0\omega H_y = ikE_z \tag{1.44a}$$

$$i\mu_0\omega H_z = -ikE_y \tag{1.44b}$$

From (1.43) and (1.44) we define the wave polarization R:

$$R = -\frac{H_y}{H_z} = \frac{E_z}{E_y} = \frac{D_z}{D_y} = \frac{P_z}{P_y} = \frac{z}{y} \tag{1.45}$$

where z and y are the spatial displacements of the charges (electrons) producing the volume polarizations. From (1.27), (1.43*a*), and (1.44*b*) we have

$$\frac{k^2}{\varepsilon_0\mu_0\omega^2} = \frac{c^2k^2}{\omega^2} = \mu^2 = \frac{D_y}{\varepsilon_0 E_y} = \frac{D_z}{\varepsilon_0 E_z} \tag{1.46a}$$

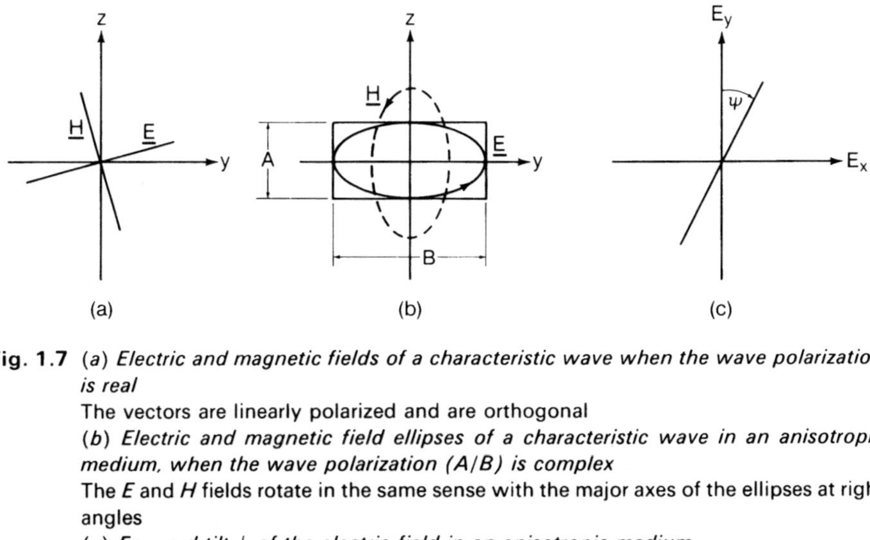

Fig. 1.7 (*a*) *Electric and magnetic fields of a characteristic wave when the wave polarization is real*
The vectors are linearly polarized and are orthogonal
(*b*) *Electric and magnetic field ellipses of a characteristic wave in an anisotropic medium, when the wave polarization (A/B) is complex*
The *E* and *H* fields rotate in the same sense with the major axes of the ellipses at right angles
(*c*) *Forward tilt ψ of the electric field in an anisotropic medium*
The wave is not a transverse wave as far as the electric field is concerned

that is,

$$\mu^2 = 1 + \frac{P_y}{\varepsilon_0 E_y} = 1 + \frac{P_z}{\varepsilon_0 E_z} \tag{1.46b}$$

Note that the refractive index is expressed in terms of the ratios of the components of the $\boldsymbol{E}$ and $\boldsymbol{P}$ vectors and not, as in (1.38), in terms of the total magnitudes. Note also that μ^2 is not given by the ratio of the x components $P_x/\varepsilon_0 E_x$, which is -1, from (1.42a). It is important to keep in mind that the concept of refractive index applies only to characteristic waves, or normal modes, which are waves for which R is constant for propagation in a given direction in a homogeneous medium. When R is real, the wave is linearly polarized in the wavefront with the electric vector making an angle $\tan^{-1}(R)$ with the y-axis. When R is complex, the wave is elliptically polarized and the ends of the $\boldsymbol{E}$ and $\boldsymbol{H}$ vectors describe ellipses in the y–z plane as seen in Fig. 1.7. The $\boldsymbol{E}$ ellipse can be enclosed in a rectangle with sides parallel to the y- and z-axes, and the ratio of the lengths of the sides is $R(=B/A)$. More generally, the $\boldsymbol{E}$ polarization ellipse is tilted forward in the direction of propagation by an angle ψ given by

$$\tan\psi = E_x/E_y \tag{1.47}$$

and the $\boldsymbol{P}$ ellipse is tilted forward by an angle ϕ, where

$$\tan\phi = P_x/P_y = +E_x/(1-\mu^2)E_y = \tan\psi/(1-\mu^2) \tag{1.48}$$

From (1.42) it follows that there are no forward tilts for the $\boldsymbol{H}$ and $\boldsymbol{D}$ vectors (Allis *et al.* 1963).

1.7 Plasma properties

The ionosphere is usually considered a cold weakly ionized gas (plasma) with the Earth's magnetic field superposed (magnetoplasma). To all intents and purposes the plasma contains equal numbers of positive and negative charges and, hence, is electrically neutral. For our purposes, we confine our attention to plasmas consisting of equal numbers of electrons (with charge $-e$ and mass m) and single-charged positive ions (with charge $+e$ and mass M). In a plasma, electrons can oscillate about the heavy ions and spiral around the external magnetic field $\boldsymbol{B}_0$.

1.7.1 Plasma frequencies

Consider the forces acting on an electron in a slab of plasma in which the electron gas is displaced a small distance x with respect to the gas of heavy positive ions (Fig. 1.8). The surface charge density $\sigma = Nex$ and the electric field

$$E = -Nex/\varepsilon_0 \tag{1.49}$$

Hence, the equation of motion of an electron is

$$\ddot{x} = \frac{e}{m}E = -\frac{Ne^2}{\varepsilon_0 m}x \tag{1.50}$$

which represents a simple harmonic motion with angular plasma frequency ω_N

given by

$$\omega_N^2 = 4\pi^2 f_N^2 = \frac{Ne^2}{\varepsilon_0 m} \tag{1.51}$$

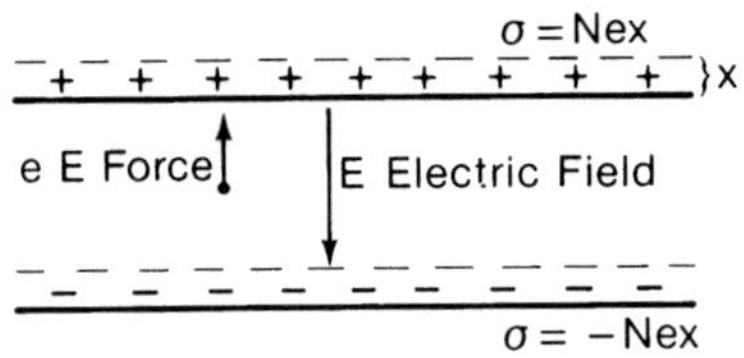

Fig. 1.8 *Force on a charged particle in an electric field **E** produced by polarization of a plane slab of electrically neutral plasma*

where f_N is the plasma frequency. Note that the plasma frequency is independent of the sign of the charge.

By analogy with (1.51), a single-charged ion plasma frequency ω_{Ni} is given by

$$\omega_{Ni}^2 = 4\pi^2 f_{Ni}^2 = \frac{N_i e^2}{\varepsilon_0 M} \tag{1.52}$$

Therefore,

$$f_{Ni} = f_N (m/M)^{1/2} \tag{1.53}$$

1.7.2 Gyrofrequencies

When an ion with charge e is injected into free space containing an imposed uniform magnetic field $\boldsymbol{B}_0$ it describes a helical path. The motion parallel to the magnetic field is unaffected by the field whereas there is a Lorentz force $ev_\perp B_0$ resulting from the perpendicular component of the motion where $v_\perp$ is the ion velocity perpendicular to the magnetic field. The projection of the ion motion onto a plane perpendicular to $\boldsymbol{B}_0$ is a circle of radius r_H with an angular speed ω_H called the angular cyclotron frequency or angular gyromagnetic frequency. Equating the magnetic and centrifugal forces we have

$$ev_\perp B_0 = m r_H \omega_H^2$$

and

$$v_\perp = \omega_H r_H$$

$$\omega_H = 2\pi f_H = \frac{|e|}{m} B_0 \approx 1{\cdot}76 \times 10^{11} B_0 \tag{1.54}$$

In the case of electrons, f_H is the electron gyrofrequency sometimes called the Larmor frequency. For a single-charged ion of mass M the ion gyrofrequency f_{Hi} is given by

$$f_{Hi} = \frac{2\pi}{\omega_{Hi}} = \frac{|e|}{2\pi M} B_0 \tag{1.55}$$

hence

$$f_{Hi} = \frac{m}{M} f_H \tag{1.56}$$

Table 1.2 *Some estimated gyrofrequencies and plasma frequencies in the ionosphere*

Ion	Gyrofrequency (Hz)	Plasma frequency (Hz)	r_H	v ($m\,s^{-1}$)
Electron	$1{\cdot}40 \times 10^6$	10×10^6	2 cm	$2{\cdot}13 \times 10^5$
Proton	762	$23{\cdot}3 \times 10^4$	1 m	5×10^3
O^+	47·7	$5{\cdot}83 \times 10^4$	4·1 m	$1{\cdot}2 \times 10^3$
NO^+	25·4	$4{\cdot}26 \times 10^4$	5·7 m	909
O_2^+	23·8	$4{\cdot}13 \times 10^4$	5·9 m	879

Given condition:
$N = 1{\cdot}24 \times 10^{12}$ ions m^{-3}
$B_0 = 0{\cdot}5 \times 10^{-4}$ tesla
$T = 1000$ K

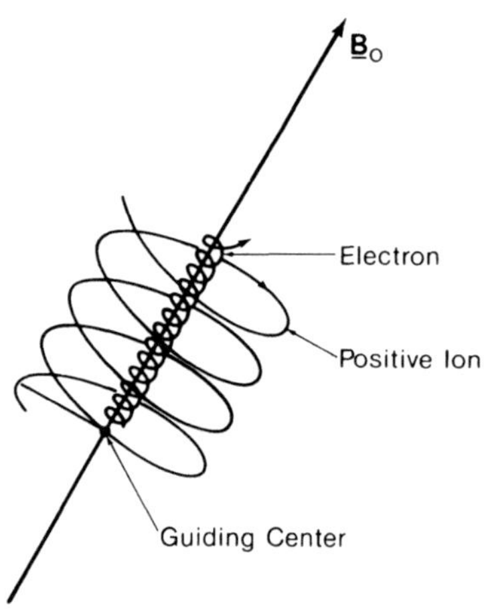

Fig. 1.9 *Motions of electrons and positive ions in the presence of a magnetic field* $\boldsymbol{B}_0$
The light electron moves in a helix of small radius whereas the heavy ion moves in the opposite sense in a helix of larger radius. The guiding center is at the foot of the perpendicular from the ion (or electron) to the magnetic-field line

It is seen that the gyrofrequencies are independent of the speeds of the ions. Because of their thermal motions the electrons and ions are continuously spinning in the Earth's magnetic field. *The sense of rotation of a negatively charged ion is given by the fingers of the right hand when the right thumb points in the direction of the magnetic field.* Positive ions rotate in the opposite sense. Table 1.2 gives an idea of the magnitudes of the gyrofrequencies and plasma frequencies of some ions in the F_2 layer of the ionosphere. Thus, for a given temperature, the gyroradius increases with mass as $(M/m)^{1/2}$. The gyroradius is sometimes called the Larmor radius. We see that electrons are tied much more closely to a magnetic field than are the heavy ions (Fig. 1.9). The foot of the perpendicular from the ion to the magnetic field is called the guiding center of the ion motion.

1.8 Motions of ions in crossed electric and magnetic fields

In crossed electric and magnetic fields (Fig. 1.10), electrons and ions drift in a direction perpendicular to both fields with a velocity $\boldsymbol{V}$ given by

$$\boldsymbol{V} = (\boldsymbol{E} \times \boldsymbol{B})/B^2 \qquad (1.57)$$

This is called the E-crossed-B drift of the guiding center and, in the absence of collisions, it is independent of the magnitude and sign of the charge and of the ion mass (Spitzer 1962).

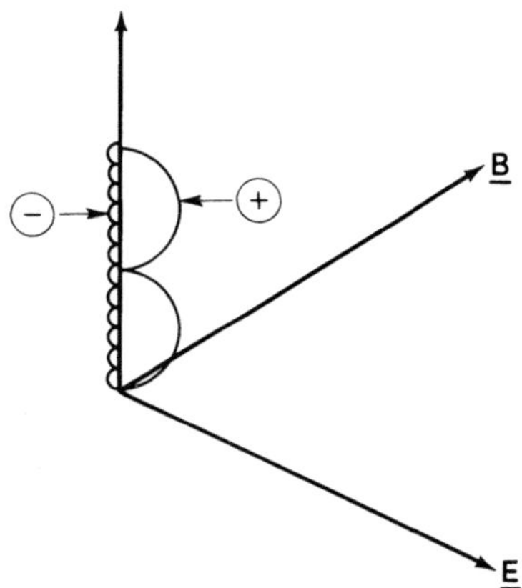

Fig. 1.10 *Motions of light electrons and heavy positive ions in crossed electric and magnetic fields*
Electrons and ions drift in the same direction (perpendicular to the plane containing $\boldsymbol{E}$ and $\boldsymbol{B}$) with the same speed ($\boldsymbol{E} \times \boldsymbol{B}/B^2$)

We see that the response of a magnetoplasma to an imposed electric field is anisotropic because the relationship between an electric field $\boldsymbol{E}$ and the resulting current density $\boldsymbol{J}$ depends on the direction of the magnetic field. This can be represented formally by introducing a conductivity tensor $\boldsymbol{\sigma}$ such that

$$\boldsymbol{J} = \boldsymbol{\sigma}\dot{\boldsymbol{E}} = \dot{\boldsymbol{P}} \qquad (1.58)$$

1.9 Collisions

When a wave traverses a plasma, the ions are set in motion and, in general, the electron speeds will exceed those of the heavy ions and neutrals. The excess speeds are dissipated by re-radiation of electromagnetic waves and by conversion of the ordered motion into thermal energy. In plasmas with high neutral densities, collisions may be a dominant factor in particle motion. In the simplest picture the electron–neutral and electron–ion collisions are elastic collisions that take place when the centers of the particles approach within a distance $r_e + r_a$, where r_e and r_a are the radii of the electrons and neutrals respectively (which are assumed to be spheres). Since $r_e \ll r_a$ the number of collisions per second ν_{ea} between electrons and atoms is given by

$$\nu_{ea} = v_e/l_e = \pi r_a^2 n_a (3KT/m)^{1/2} \tag{1.59}$$

where

$$l_e = (\pi r_a^2 n_a)^{-1} \tag{1.60}$$

is the mean free path of the electrons, K is Boltzmann's constant, T is temperature (kelvins), m is electron mass,

$$v_e = (3KT/m)^{1/2} \tag{1.61}$$

is the thermal speed of electrons, and n_a is the number density of neutral particles.

This simple picture of elastic collisions has to be modified when we consider the microscopic picture of collisions. For example, when an electron approaches a neutral molecule, the latter can be polarized, and there is an electrostatic attraction that will affect the motion of the electron. Similarly, when an electron approaches an ion, the Coulomb force will cause the electron velocity to change without either center's approaching within the critical ballistic distance. Thus, when we think of collisions in an ionized gas, we reduce the gas to an equivalent ballistic ensemble in which one effective collision may be made up of several of the types mentioned above. The important point to remember is that encounters between particles are complicated processes; (1.59) does not accurately represent the dependence of electron collision frequency on electron temperature, molecular density, radius, etc. In fact, laboratory experiments have shown that the collisional cross section for an electron–neutral collision is, generally, a function of the electron energy (see Sec. 3.8).

Collision frequencies of electrons with neutrals ν_{en}, electrons with ions $\nu_{e\pm}$, ions with neutrals $\nu_{\pm n}$, and ions with ions $\nu_{\mp\pm}$ can be approximated by

$$\nu_{en} = 5 \times 10^{-1} n_n T^{1/2} \tag{1.62a}$$

$$\nu_{e\pm} = [34 + 8{\cdot}36 \log (T^{3/2}/n_e^{1/2})] n_\pm T^{-3/2} \tag{1.62b}$$

$$\nu_{\pm n} = 3{\cdot}35 \times 10^{-21} n_n/M^{1/2} \tag{1.62c}$$

$$\nu_{\mp\pm} = 3{\cdot}06 \times 10^{-14} \nu_{e\pm}/M^{1/2} \tag{1.62d}$$

where n_n, $n_\pm$ and n_e are the number densities of neutrals, positive and negative ions, and electrons respectively, M is the mean molecular mass, and T is temperature (Matsushita 1967, p. 382).

1.10 The Debye length

The ability of an ion to affect the motion of another ion is limited by the Debye shielding caused by the surrounding plasma (Sec. 4.7). For example, a heavy positive ion will tend to attract electrons, and repel other positive ions, and it will have a 'cloud' of electrons associated with it. At a certain distance D from the center of the ion there will be no electric field due to the ion. This distance, called the Debye length, is given by

$$D = \left(\frac{\varepsilon_0 KT_e}{e^2 N_e}\right)^{1/2} \approx 69\,(T_e/N_e)^{1/2} = \frac{v_e}{\sqrt{3}\,\omega_N} \qquad (1.63)$$

where N_e is the electron number density. Inserting values appropriate to the F_2 layer, namely $N_e = 10^{11}\,\mathrm{el.\,m^{-3}}$, and $T_e = 1000\,\mathrm{K}$, we have $D = 7 \times 10^{-3}\,\mathrm{m} = 7\,\mathrm{mm}$. The concept of a Debye sphere is valid only if there are sufficient electrons to provide the shielding. Organized plasma motion (e.g. plasma waves) cannot occur on a scale smaller than the Debye length.

1.11 References

ALLIS, W. P., BUCHSBAUM, S. J., and BERS, A. (1963): 'Waves in anisotropic plasmas'. (MIT Press, Cambridge, MA)

BANKS, P. M., and KOCKARTS, G. (1973): 'Aeronomy, A and B' (Academic Press, New York) 2 vols.

BUDDEN, K. G. (1985): 'The propagation of radio waves' (Cambridge University Press, Cambridge, England)

CHECCACCI, P. F. (Ed.) (1975): 'Special Marconi issue', *Radio Sci.*, **10**, p. 654

CLARK, R. W. (1971): 'Sir Edward Appleton' (Pergamon Press, Oxford, England)

CRC, (1961): 'Standard mathematical tables', HODGMAN, C. D. (Ed.) (Chemical Rubber Publishing Co., Cleveland, Ohio)

GILLMOR, C. S. (1976): 'The history of the term "ionosphere"', *Nature*, **262**, p. 347

GILLMOR, C. S. (1982): 'Wilhelm Altair, Edward Appleton and the magneto-ionic theory', *Proc. Amer. Phil. Soc.*, **126**, p. 395

GOLDMAN, D. T., and BELL, R. J. (1986): 'The international system of units (SI).' NBS Special Publication 330 (US Govt. Printing Office, Washington, DC)

IRE (1950): 'Standards on wave propagation, definition of terms', *Proc. IRE*, **38**, p. 1264

JORDAN, E. C., and BALMAIN, K. G. (1968): 'Electromagnetic waves and radiating systems' (Prentice-Hall, Englewood Cliffs, NJ) 2nd edn.

MAEDA, K. (1986): 'Fifty years of the ionosphere', *J. Radio Res. Lab.*, **33**, p. 103

MATSUSHITA, S. (1967): 'Solar quiet and lunar daily variation fields' *in* MATSUSHITA, S., and CAMPBELL, W. H. (Eds.): 'Physics of geomagnetic phenomena: I' (Academic Press, New York, pp. 302–424)

MAXWELL, J. C. (1873): 'A treatise on electricity and magnetism' (Dover Publications, Inc., New York) 2 vols.

MITRA, A. P. (1984): Fifty years of radio science in India. Status report series 2 (Indian National Science Academy, New Delhi)

RATCLIFFE, J. A. (1959): 'The magneto-ionic theory' (Cambridge University Press, Cambridge)

RATCLIFFE, J. A. (1972): 'An introduction to the ionosphere and magnetosphere' (Cambridge University Press, Cambridge)

RAWER, K., and SUCHY, K. (1976): 'Remarks concerning the dispersion equation of electromagnetic waves in a magnetized cold plasma', *J. Atmos. Terr Phys.*, **38**, p. 395

SEN, H. K., and WYLLER, A. A. (1960): 'On the generalizations of the Appleton–Hartree magnetoionic formulas', *J. Geophys. Res.*, **65**, p. 3931

SILBERSTEIN, R. (1959): 'The origin of the current nomenclature of the ionospheric layers', *J. Atmos. Terr. Phys.*, **13**, p. 382

SNYDER, W. E., and BRAGAW, C. L. (1986): 'Achievement in radio'. NBS Special Publication 555, US Govt. Printing Office, Washington, DC).

SPITZER, L. (1962): 'Physics of fully ionized gases' (Wiley Interscience, New York) 2nd edn.

STIX, T. H. (1962): 'The theory of plasma waves' (McGraw-Hill, New York)

STRATTON, J. A. (1941): 'Electromagnetic theory' (McGraw-Hill, New York)

Chapter 2

Solar-terrestrial relationships

2.1 The solar-terrestrial system

The Earth's upper atmosphere is ionized by radiations, both electromagnetic and corpuscular, from the Sun. The electromagnetic radiations, such as radio, infrared, visible, ultraviolet, extreme ultraviolet, and X-rays, travel directly between Sun and Earth at the free space speed with a transit time of about 8·3 minutes. The region between and including Sun and Earth has these components: the Sun, the interplanetary medium, the Earth's magnetosphere, the Earth's neutral atmosphere, and the Earth's ionosphere as depicted in Fig. 2.1. Understanding the features of the solar-terrestrial system discussed in this Chapter is essential to an understanding of the spatial and temporal structure of the ionosphere.

2.2 The Sun

2.2.1 The quiet Sun

2.2.1.1 General features

From an astrophysical point of view the Sun is an average star with a radius of 696 000 km and surface (photospheric) temperature of about 5800 K. The Sun consists of the solar interior, the visible surface or photosphere, the chromosphere or lower solar atmosphere, and the corona or outer solar atmosphere. The composition of the Sun is approximately as follows: hydrogen 92·1%; helium 7·8%; oxygen 0·06%; carbon 0·03%. Although most other elements known on Earth have been detected on the Sun, they amount to less than about 0·01%. It must be realized that the Sun is a gaseous body and has no surface like that of Earth. The reason for the apparent sharp edge of the visible photosphere is the rapid decrease, with increasing distance from the center, of the solar density. When we look directly at the photosphere, at the center of the solar disk we see a deep hot part ($\approx$ 6390 K) whereas near the limb we see a higher and cooler part (≈ 5000 K) which gives rise to limb darkening. Temperature falls off above the photosphere and reaches a minimum of about 4200 K at a height of some 5000 km, above which level the temperature climbs until it reaches 10^6 K and more in the outer corona.

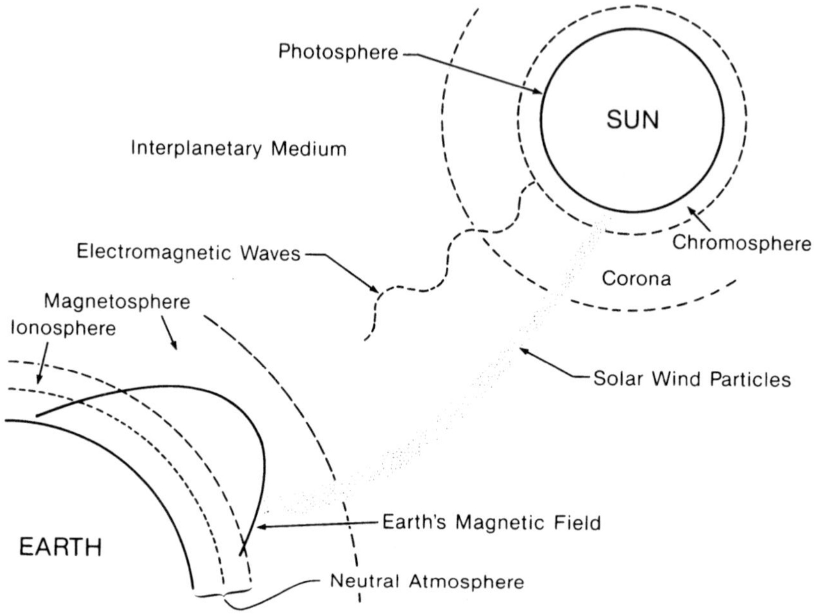

Fig. 2.1 *Solar-terrestrial system and solar ionizing radiations*

Viewed with the human eye the Sun appears as a fairly uniform disk, radiating as a black body at a temperature of about 6000 K. Superimposed on the black body continuum are certain lines, some of the more conspicuous being Hα, 6563 Å (656·3 nm); Lyman α, 1216 Å (121·6 nm); HeI, 10 830 Å (1083·0 nm); HeII, 304 Å (30·4 nm); and singly ionized Ca IIK, 3934 Å (39·3 nm). The flux of visible solar energy reaching the Earth's surface varies little from day to day or from year to year ($< 0{\cdot}5\%$). However, solar emissions at shorter wavelengths, in the extreme ultraviolet and X-ray wavelengths, vary by orders of magnitude depending on sunspot number and solar activity.

2.2.1.2 Sunspots

Sunspots, first noticed by Theophrastus around 325 BC, are one of the more notable phenomena on the solar surface. These are dark spots (Fig. 2.2), which appear and disappear with time. The life of a sunspot is highly variable; some spots last for a few days only whereas a few survive for four or five solar rotations (of about 27 days each). Sunspots appear dark because the surface temperature is low (≈ 3000 K) compared with the 6000 K of the ambient photosphere. Sunspots tend to group together. A group may contain a single spot or dozens of spots. The spots rotate with the Sun, that is, from east to west. The axis of rotation of the Sun is approximately perpendicular (83°) to the plane of the Earth's orbit, the ecliptic plane, whereas the Earth's rotation axis is inclined at $23\frac{1}{2}°$ to the perpendicular to the ecliptic plane. Sun and Earth rotate in the same direction. Whereas the Earth rotates equally at all latitudes, the Sun rotates faster the lower the heliolatitude.

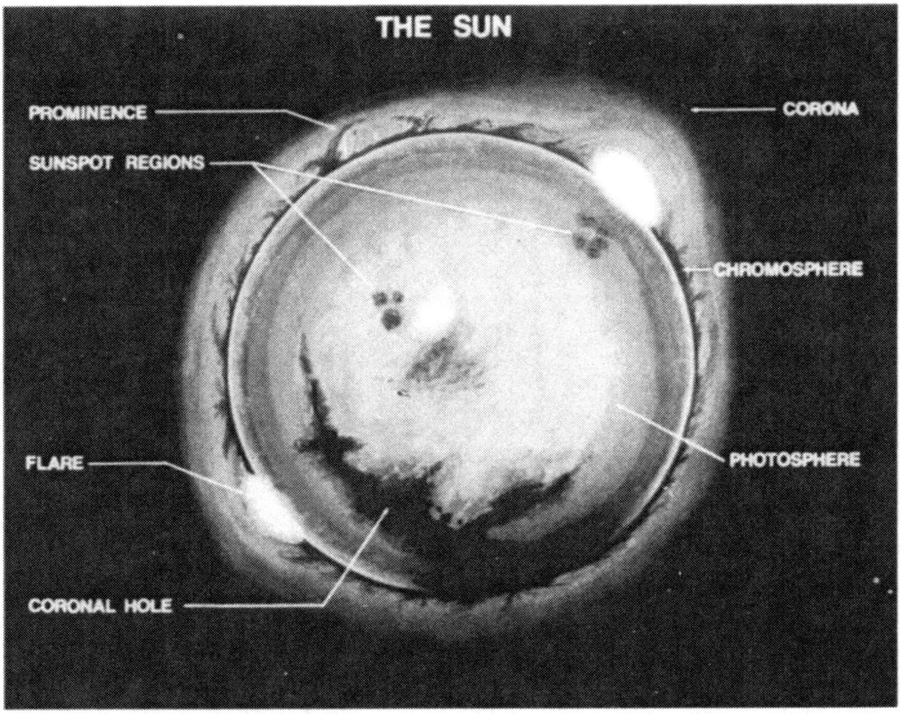

Fig. 2.2 *Solar features (observed at various times on different wavelengths) that can have important consequences on the ionosphere (NASA photograph MSFC–5/80–STE 3826)*

The most notable feature of sunspots is the approximately 11-year periodicity in occurrence. The occurrence of sunspots is measured by the Wolf, or Zurich, sunspot number R given by

$$R = k(10g + s) \tag{2.1}$$

where g is the number of sunspot groups and s is the number of observed individual spots. The factor k is a correction factor, approximately unity, which takes into account equipment and observer characteristics. The sunspot number is weighted heavily in favor of groups. It is sometimes difficult to tell whether a large complex of spots should be counted as one group or two, so that the sunspot number's value as an index of a specific solar phenomenon is questionable. Nevertheless, it is valuable as a long-term index because data are available over several hundred years (Fig. 2.3). Sunspot numbers are available daily and

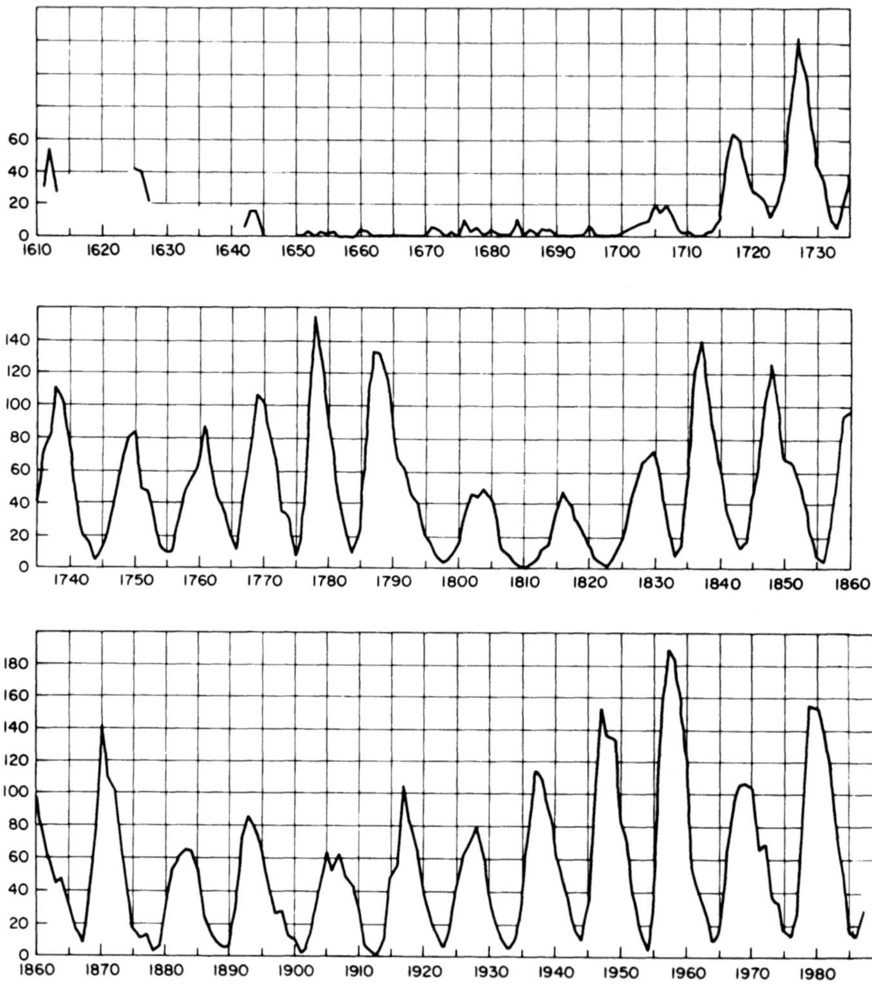

Fig. 2.3 *Yearly sunspot numbers from 1600 to 1980 (Courtesy of J. A. Eddy)*
The period between about 1640 and 1710 is called the Maunder minimum

as monthy and yearly averages of the international sunspot number R_i and the American relative sunspot number R_a (see *Solar-Geophysical Data*). The average 11-year cycle of sunspot numbers is not related to the lifetimes of individual spots, which are relatively short-lived features ($\approx$ days). As some sunspots appear, others disappear. Spots tend to occur at different solar latitudes. The spots, after a sunspot minimum, occur at latitudes of 20° to 30° north and south whereas at sunspot maximum spots occur at $\pm 15°$ and, as the sunspot number declines, spots occur in latitudes of 5° to 10°.

Sunspots are characterized by strong magnetic fields which may approach 0·4 tesla (4000 G). Sunspot groups tend to be bipolar, having both positive and negative magnetic polarities. The leading spots in sunspot groups have one polarity for almost all northern hemisphere groups, and the opposite polarity for almost all southern hemisphere groups. The polarities are reversed from one 11-year cycle to the next, indicating a 22-year polarity cycle. The magnetic configuration on the Sun varies in periods ranging from days to years. Figure 2.4 shows a McIntosh Hα synoptic chart that illustrates the varying regions of magnetic polarity for Carrington rotation 1744 (that is, the 1744th solar rotation since 1853). Magnetic fields dominate the behavior of the solar plasma; for example, magnetic structures determine when and where various features (e.g. flares, filaments, prominences) appear. The position of a feature on the Sun is usually given by its latitude with respect to the solar equator and its central meridian distance (CMD) which is the longitude measured from the central meridian as seen from Earth. The month and decimal day on which the feature crosses the central meridian is called the central meridian passage (CMP).

The observed times between sunspot maxima have varied between 7 and 15 years with an average of 11·1 years. Furthermore, the amplitude of the yearly means fluctuated between about 40 and 200 between about the years 1700 and 1980. As seen in Fig. 2.3, there were few sunspots during the 70-year interval between 1645 and 1715. This interval is called the Maunder Minimum (Eddy 1979; Eddy 1983; Noyes 1982). Spectrum analyses of sunspot numbers show the presence of periods of about 80 years, 22 years, and 11 years, and there is usually a pronounced $\sim$27-day periodicity associated with solar rotation. The most widely used index in ionospheric work is the 12-month smoothed relative sunspot number defined by

$$R_{12} = \left[\sum_{n-5}^{n+5} (R_k) + 0{\cdot}5(R_{n+6} + R_{n-6}) \right] \Big/ 12 \qquad (2.2)$$

in which R_k is the mean value of R for a single month k, and R_{12} is the smoothed index for the month represented by $k = n$.

In addition to the relatively dark sunspots on the photosphere, bright areas often appear in the chromosphere in the light of ionized calcium and hydrogen. These areas, called plages, are comparatively long-lived regions. The chromosphere is best observed during an eclipse when it appears as a red ($H\alpha$) rim above which protrude prominences, like giant 'flames' rising above the Sun's surface (Fig. 2.2). Individual spike-like jets in the chromosphere are called spicules. The chromosphere can be photographed in light of different colors with a spectroheliograph (Fig. 2.5). Dark filaments can be seen which extend over tens or hundreds of thousands of kilometers over the disk. Filaments are prominences

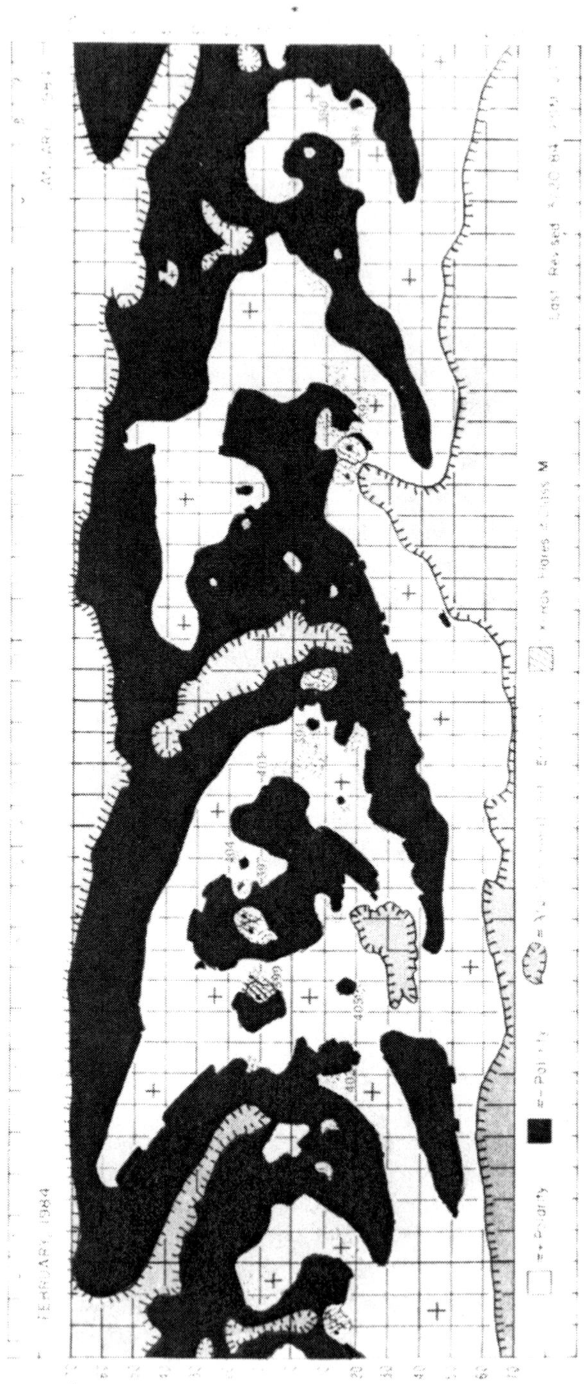

Fig. 2.4 *A solar* Hα *synoptic chart for Carrington rotation 1744 (January 6 – February 7, 1984), showing the large-scale magnetic structures deduced from visible wavelengths (Courtesy of P. A. McIntosh)*
The + sign indicates outward field and the – sign indicates inward field

that appear against the disk and they are believed to portray magnetic structures. Figure 2.5*c* shows also the appearance of the Sun in X-rays. Here a far different Sun is seen from that seen in visible light. X-rays reveal the hot plasma in the corona viewed against the dark cooler photosphere. The dark area running down from the pole is called a coronal hole and is thought to indicate a region where the magnetic field is open to the interplanetary field and ionized particles can escape into the solar wind.

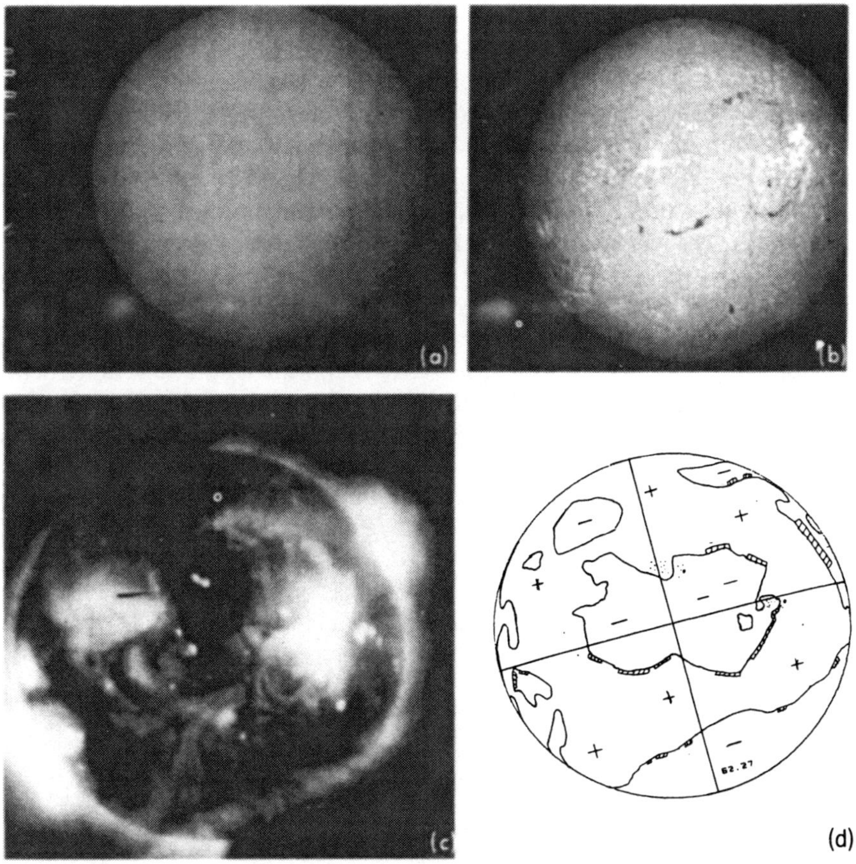

Fig. 2.5 *The Sun seen (a) in white light, (b) in 6563 Å Hα light, and (c) in soft X-rays; (d) the Sun's magnetic field (Courtesy of W. Wagner)*
The + sign indicates an outward-pointing field and the − sign indicates an inward-pointing field

2.2.1.3. Ionizing radiations

Solar radiations are responsible for most of the ionization in the ionosphere. The D region is produced by Lyman α (1216 Å) and X-rays (< 10 Å) whereas the E region is formed by X-rays in the range $30 < \lambda < 100$ Å, by Lyman β (1025·7 Å), and by ultraviolet radiation in the Lyman continuum (910 to 980 Å) (Banks and Kockarts 1973). Solar X-ray fluxes are measured in units of watts per square meter. The letters A, B, C, M, and X stand for powers of 10. The letter B denotes 10^{-7}, C denotes 10^{-6}, M denotes 10^{-5}, and X denotes any power greater than 10^{-4}. Digits from 1·0 to 9·9 act as multipliers; for example, C3·2 stands for a flux of $3{\cdot}2 \times 10^{-6}$ W/m^2. The primary source of ionization in the F region appears to be the Lyman continuum and the band 350 to 200 Å, which includes the HeII (304 Å) resonance line.

2.2.1.4 Solar radio emissions

The Sun is a source of radio noise that is produced by plasma oscillations, cyclotron emissions, and random collisions of electrons with heavy particles. The noise from the quiet Sun is of the last type, and it has a minimum base level that is the thermal emission from the quiet Sun. The daily full-disk radio noise flux on a wavelength of 10·7 cm (2800 MHz), measured with a reflector of 1·8 m diameter near Ottawa at about 1700 UT, is one of the most commonly used and the most available indicators of the general level of solar activity. On a long-term basis it is highly correlated with many ionosphere and upper atmosphere parameters because it is well correlated with X-ray, extreme ultraviolet, and ultraviolet fluxes. Since it is unrealistic to expect a single radio measurement of a full-disk emission at one wavelength and only one time of day to represent the ionizing effect, it is amazing that the 10·7 cm flux correlates so well with upper atmosphere phenomena. The radio noise flux is measured in units of 10^{-22} W m^{-2} Hz^{-1} and varies from a minimum near 65 (at the minimum of the sunspot cycle) to a maximum of about 200. Daily and monthly-mean values of the solar flux on 2800 MHz, adjusted to 1 AU (the mean Earth-Sun distance) are published in *Solar-Geophysical Data.* This publication of the National Geophysical Data Center in Boulder, Colo., includes daily solar flux values, from Sagamore Hill, Mass., adjusted to 1 AU on 15 400, 8800, 4995, 2695, 1415, 609, 410, and 242 MHz.

The long-term variation in the 10·7 cm radio flux is shown in Fig. 2.6, which also shows the monthly high and low values of the daily radio flux. The dashed curve approximately separates the flux associated with centers of activity from a quiet Sun.

2.2.2. The active Sun

2.2.2.1. Flares

A solar flare is a burst of 'light' occurring in the chromosphere near a sunspot (Fig. 2.2). It is most easily observed in red Hα light (6563 Å) although, in some rare cases, flares have been seen in white light. Flares are frequent around the peaks of sunspot cycles. They are observed over a wide band of wavelengths from radio to X-rays. The optical flare appears as a sudden brightening in an

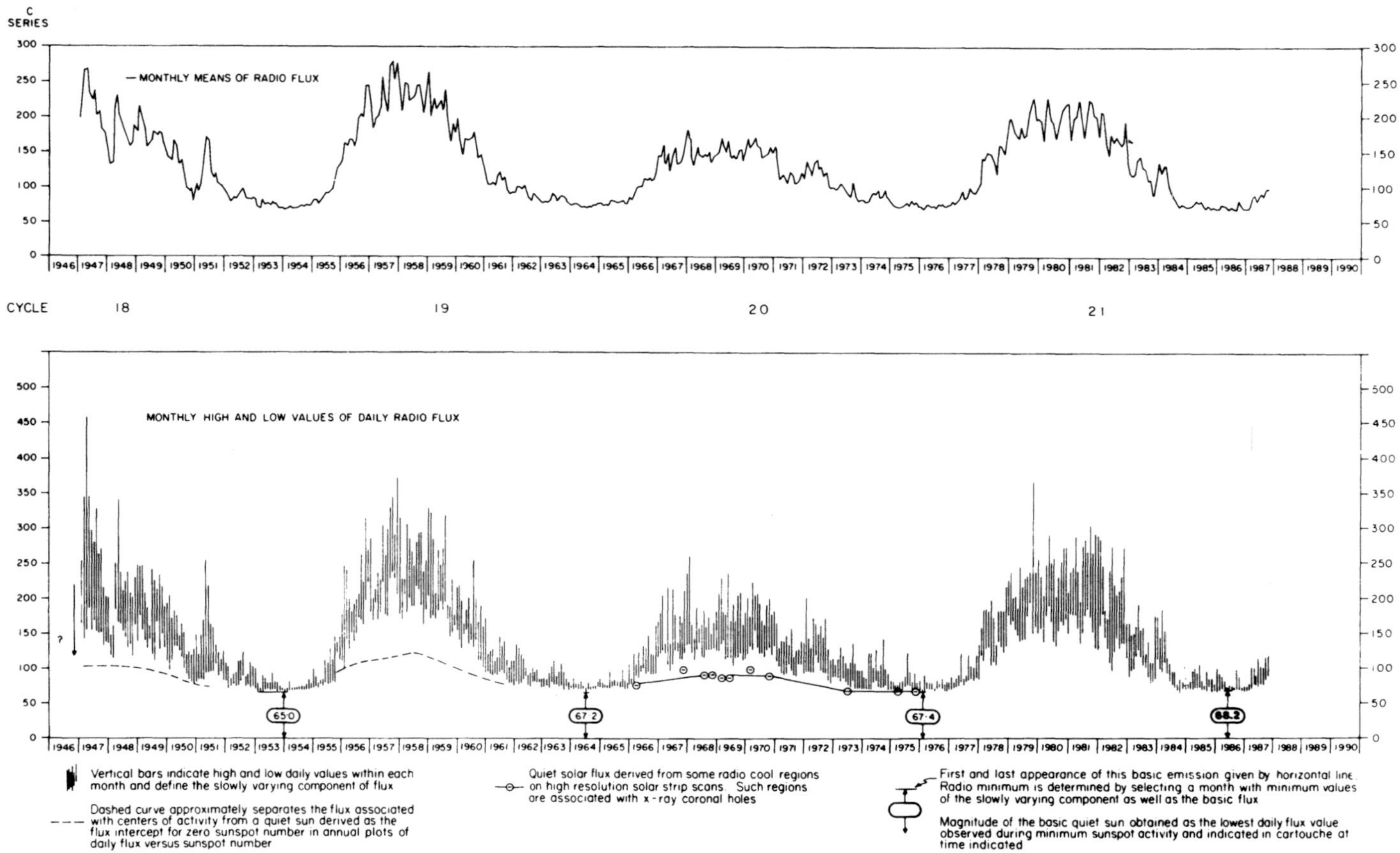

Fig. 2.6 *Long-term variation of 10·7 cm solar radio noise flux (From Solar-Geophysical Data)*
Top: Monthly means
Bottom: Monthly high and low values of daily radio flux

irregular patch within an active area. Its brightness increases rapidly as the area expands to a maximum, in a few minutes, and decays. Flare durations vary from about 3 min up to 2 h. Some large flares are characterized by an 'explosive phase'; these are the flares most likely to produce geophysical responses (Evans 1972). Data on the sizes of flares, their start times, durations, etc., obtained from a worldwide network of 31 flare-patrol observatories, are given in *Solar-Geophysical Data*. Flares are ranked in importance in several ways; for example, the average optical importance is a measure of flare size and brilliance. The first character ranks the area on a 5-level scale:

S (Subflare): $\leqslant 2{\cdot}0$ hemispheric square degrees
Importance 1: 2·1–5·1 square degrees
Importance 2: 5·2–12·4 square degrees
Importance 3: 12·5–24·7 square degrees
Importance 4: $\geqslant 24{\cdot}8$ square degrees

The second character grades the plage brilliance on a 3-level scale: f = faint, n = normal, and b = bright. Under this classification scheme, an Sf event is the smallest and 4b is the largest and brightest.

X-ray flares, in the wavelength range 1–8 Å are classified as C, M, and X depending on the power flux level (Table 2.1). A C3·2 burst, for example, indicates an X-ray burst with a maximum flux of $3{\cdot}2 \times 10^{-6}\,\mathrm{W\,m^{-2}}$. There is a close statistical relationship between the number of flares per solar rotation (N_F) and the corresponding mean sunspot $\bar{R}$, given by

$$N_F = C(\bar{R} - 10) \tag{2.3}$$

where C is a proportionality factor. The mean values of C for the cycles having peaks in 1937 and 1947 are 1·98 and 1·47 respectively.

Table 2.1 *Classification of X-ray flares, 1–8 Å (0·1–0·8 nm)*

Class	Peak flux ($\mathrm{W\,m^{-2}}$)
C	$10^{-6} \leqslant \phi < 10^{-5}$
M	$10^{-5} \leqslant \phi < 10^{-4}$
X	$10^{-4} \leqslant \phi$

2.2.2.2 *Radio emissions*

Radio bursts are generally associated with flares and originate from all levels of the solar atmosphere between the lower chromosphere (millimeter and centimeter [300–10 000 MHz] waves) and the outer corona to heights of several solar radii (meter [30–300 MHz] and decameter [10–30 MHz] waves) (Kundu 1965).

Bursts of centimeter waves are relatively simple: a sharp rise is followed by a slower decline. There are three types: the impulsive burst (few minutes duration), the post-burst (up to tens of minutes), and the gradual rise and fall (up to tens of minutes). Decimeter bursts are more complicated; frequencies drift from high to low, or low to high, at a fast rate ($\approx 100\,\mathrm{MHz\,s^{-1}}$).

Bursts of meter and decameter waves are very complicated. They are illustrated on the basis of their spectra in Fig. 2.7. A Type I or noise storm can last from several hours to several days. It consists of a slowly varying broadband enhancement of the continuous solar radiation. Superimposed on the slowly varying component are short-lived bursts that are mostly on frequencies below about 300 MHz and that can last from a fraction of a second to about 1 min. Type I bursts extend over a narrow frequency range, from 5 to 30 MHz for example, and the frequency of a single burst may drift with time to higher or to lower frequencies. The radiation is strongly circularly polarized, and its onset may occur in association with a flare or an active region and may involve enhancement of the order of a million times the quiet values (Kundu 1965).

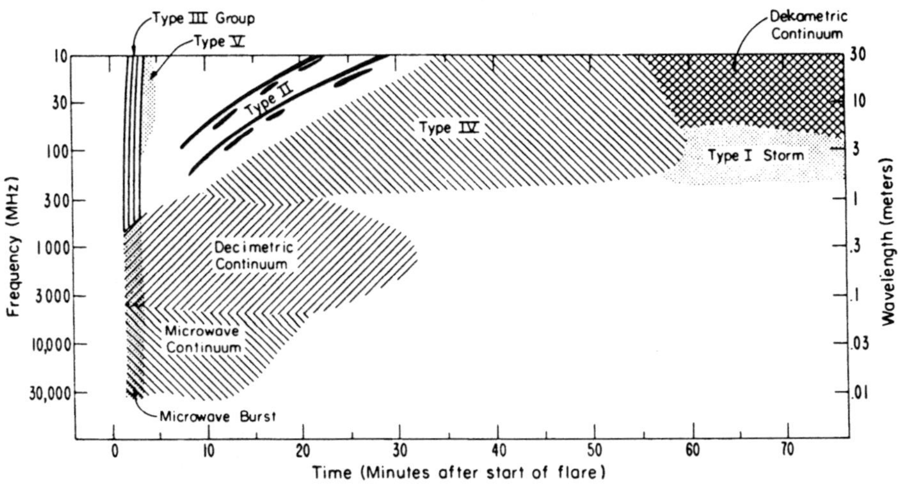

Fig. 2.7 *Types of meter and decameter radiowave bursts (From Solar-Geophysical Data 1987)*
I: Storm burst
II: Slow drift burst
III: Fast drift burst
IV: Prolonged continuum
V: Brief continuum
Also shown are microwave burst, microwave continuum, decimetric continuum, and decametric continuum

Type II (slow) and Type III (fast) bursts are intense events of minutes and seconds duration, respectively. Their spectra drift toward lower frequencies at rates of about $1\,\mathrm{MHz\,s^{-1}}$ and $20\,\mathrm{MHz\,s^{-1}}$, respectively. These bursts are due to plasma oscillations, at the local plasma frequencies (see Sec. 1.7), excited by a source moving outward through the corona in which the electron density decreases with radial distance. The exciting sources travel outward through the corona with speeds of about 1000 km s^{-1} (Type II) and 100 000 km s^{-1} (Type III).

Type IV bursts are long-lived (10 min to 1 h) continuum emissions occurring over a wide range of wavelengths from centimeters to decameters. Sources of Type IV meter waves have large sizes and speeds of several hundred km s^{-1} (Kundu 1965, Chapter 11). Type V is a brief (few seconds) continuum enhancement normally following a Type III burst and is nearly always limited to meter waves. Further information on solar radio wave spectral observations is given in *Solar-Geophysical Data* (No. 515, Explanation of data reports, July 1987).

2.3 The interplanetary medium

2.3.1. The solar wind

A simple picture of the solar corona has two kinds of regions: (*a*) those with closed magnetic field lines where plasma is confined, and (*b*) those with open magnetic field lines from which the coronal plasma expands into interplanetary space and forms the solar wind. A typical wind speed is 400 km s^{-1}, which can increase to about 1000 km s^{-1} during disturbed periods. The largest scale structures in the interplanetary medium are those caused by the rotation of the Sun. When particles are emitted radially they will appear, to a nonrotating observer above a solar pole, to follow spiral Archimedean paths. A typical density is 5×10^6 protons m^{-3} with an equal density of electrons to ensure electrical neutrality. Fluctuations of density of an order of magnitude occur over times of minutes to hours, which implies that there are irregularities of 10^5 km and larger within the solar wind. The mean proton energy is about 0·5 keV, and the mean electron energy is about 0·25 eV (Hargreaves 1979; Noyes 1982). The solar wind speed is small near the Sun but increases with radial distance and becomes supersonic (exceeds the sound speed of about 200 km s^{-1}) at a critical distance of about 5 solar radii above the photosphere. On the average, the transit time between Sun and Earth is about $4\frac{1}{2}$ days during which time the Sun rotates through about 60°. Coronal holes, regions of open magnetic fields, are important sources of enhanced solar wind.

2.3.2 The interplanetary magnetic field

The interplanetary magnetic field (IMF) has a spiral form (Fig. 2.8). The field can be directed either toward or away from the Sun, giving rise to a sector structure. There is an even number of sectors, half having the magnetic field pointing inward and half pointing outward. The number of sectors varies with sunspot cycle and disturbance. Plasma emitted from solar flares, coronal mass ejections, and eruptive prominences usually travels faster than the ambient solar wind, and this produces a shock front as the fast plasma overtakes the ambient plasma.

2.4 The Earth's magnetic field

2.4.1 The dipole approximation

To a first approximation, the Earth is a sphere uniformly magnetized in the direction of the dipole axis. The centered dipole axis cuts the Earth's surface

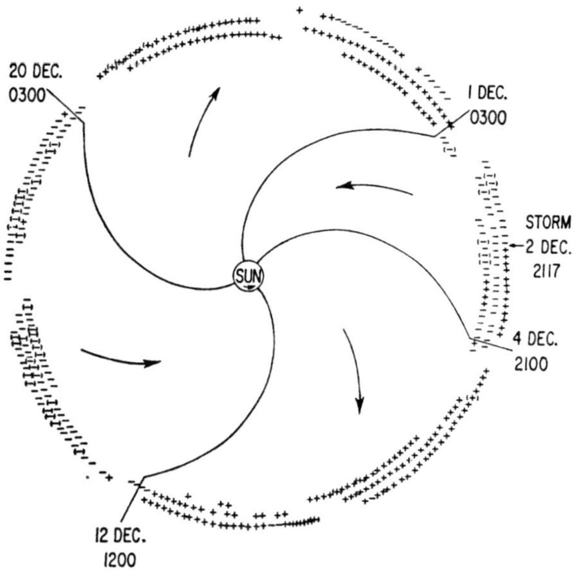

Fig. 2.8 *Spiral structure of the interplanetary magnetic field, showing sectors in which the interplanetary magnetic field points inward and outward (From Wilcox 1968 © Reidel Publishing Co.; reprinted with permission)*

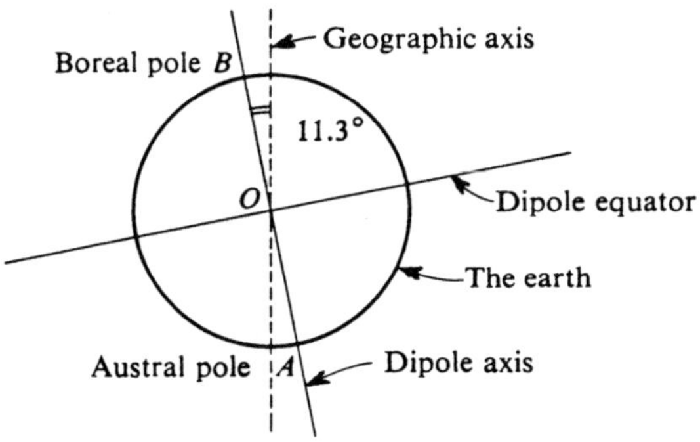

Fig. 2.9 *Earth's dipole field showing the north (B) and south (A) dipole poles and the dipole equator*

(Fig. 2.9) at two points A and B called the south (austral) and north (boreal) dipole poles. The dipole axis is inclined at about 11° to the axis of rotation. The positions A and B change because of the secular variation of the Earth's magnetic field. The plane through the Earth's center perpendicular to BA is called the dipole equator. Dipole latitude Φ is reckoned relative to this equator. The semicircles joining B and A are called the dipole meridians; that passing through the south geographic pole is chosen as the zero of dipole longitude Λ. The relationships between the dipole coordinates (Φ, Λ) and the corresponding geographic coordinates (ϕ, λ) at a point P are given by

$$\sin \Phi = \sin \phi \sin \phi_0 + \cos \phi \cos \phi_0 \cos (\lambda - \lambda_0) \tag{2.4a}$$

$$\sin \Lambda = \frac{\cos \phi \sin (\lambda - \lambda_0)}{\cos \Phi} \tag{2.4b}$$

where ϕ_0 and λ_0 are the geographic latitude and longitude of the north dipole pole. In the International Geomagnetic Reference Field (IGRF) (Barraclough 1987) the values for January 1988 were $\phi_0 = 79{\cdot}1°$N and $\lambda_0 = 289{\cdot}1°$E (Peddie 1988, private communication).

The dipole magnetic potential V at a point P whose vector position relative to the dipole center is $\boldsymbol{r}$, is given by

$$V = \frac{\boldsymbol{M} \cdot \boldsymbol{r}}{r^3} = -\frac{M \sin \Phi}{r^2} \tag{2.5}$$

where $\boldsymbol{M}$ is the vector magnetic moment.

The radial (vertical) component Z of the field, reckoned positive when inward, is given by

$$Z = -\frac{\partial V}{\partial r} = -\frac{2M \sin \Phi}{r^3} \tag{2.6}$$

and the tangential field H, which lies in the magnetic meridian through P (directed northward), is given by

$$H = \frac{1}{r}\frac{\partial V}{\partial \Phi} = -\frac{M \cos \Phi}{r^3} \tag{2.7}$$

The magnetic dip, or inclination I, measured positive when directed below the horizontal, is given by

$$\tan I = Z/H = 2 \tan \Phi \tag{2.8}$$

which is independent of the radial distance.

The total field F is obtained from

$$F^2 = H^2 + Z^2 \tag{2.9}$$

From (2.6) and (2.7) it is seen that $Z = 0$ at the Equator (horizontal field) and $H = 0$ at the poles (vertical field). Let B_{00} represent the value of the Earth's field B_0 at the Equator on the Earth's surface of radius a.

$$H = B_{00}(a/r)^3 \cos \Phi \tag{2.10a}$$

$$Z = -2B_{00}(a/r)^3 \sin \Phi \tag{2.10b}$$

$$F = B_{00}(a/r)^3(1 + 3 \sin^2\Phi)^{1/2} \tag{2.10c}$$

At a given radial distance the field strength at the poles is twice that at the Equator. For the Earth's field $B_{00} \approx 0{\cdot}31$ gauss $= 0{\cdot}31 \times 10^{-4}$ tesla.

The equation of a line of force is obtained from

$$\tan I = 2 \tan \Phi = \frac{1}{r}\frac{dr}{d\Phi}$$

which gives, on integration,

$$r = aL \cos^2\Phi \tag{2.11}$$

where L is the distance, measured in Earth radii, at which the line of force intersects the equatorial plane. Surfaces of constant L or L shells are important in magnetospheric studies (Hess 1968) because the guiding centers (Sec. 1.7.2) tend to follow these surfaces.

Along a single field line

$$B(L, \Phi) = B_{00}\left(\frac{1}{L}\right)^3 \frac{(1 + 3 \sin^2\Phi)^{1/2}}{\cos^6\Phi} \tag{2.12}$$

The field line cuts the Earth's surface at a latitude

$$\Phi = \cos^{-1}(1/\sqrt{L}) \tag{2.13}$$

called the invariant latitude. Because of the divergence of the field lines, the cross section A of a tube of flux at a latitude Φ is

$$A = A_0 \frac{\cos^6\Phi_0}{\cos^6\Phi}\left(\frac{1 + 3 \sin^2\Phi}{1 + 3 \sin^2\Phi_0}\right)^{1/2} \tag{2.14}$$

where A_0 is the cross section on the ground at a dipole latitude Φ_0, since the flux AB = constant. At the equatorial plane,

$$A = A_0L^3(1 + 3 \sin^2\Phi)^{1/2} = A_0L^3(4 - 3/L)^{1/2} \tag{2.15}$$

The length l of a field line between a point in one hemisphere and its magnetic conjugate point in the other hemisphere is

$$l = \frac{a}{\sqrt{3}}\left[CG + L \ln (C + G) - \frac{L}{2} \ln L\right] \tag{2.16}$$

where $C^2 = 3(L - 1)$ and $G^2 = 4L - 3$.

2.4.2 The real field

The magnetic intensity $\boldsymbol{F}$ at any point P on the Earth's surface can be specified by its downward vertical component Z and its vector horizontal component $\boldsymbol{H}$, or by $\boldsymbol{H}$ and the angle I by which $\boldsymbol{F}$ dips below the horizontal. On the Earth, the direction of $\boldsymbol{H}$ is specified by the angle D between $\boldsymbol{H}$ and the geographic north; D is called the magnetic (or compass) declination and is reckoned positive if eastward. The northward and eastward components of H are denoted by X, Y, respectively, and related as follows:

$$\tan D = \frac{Y}{X}, \quad X = H \cos D, \quad Y = H \sin D, \; H^2 = X^2 + Y^2$$

The seven quantities F, Z, H, I, D, X, and Y are called the magnetic elements, and any set of three independent elements serves to specify $\boldsymbol{F}$, that is, F, I, D; H, I, D; H, Z, D; X, Y, Z; D, I, Z; X, Y, F; and D, H, F. The elements F, H, Z, and I are called *intrinsic* because their only reference to direction is to the natural direction characteristics of P, namely, the vertical. The other three elements, D, X, Y, are called *relative* because they are defined relative to the geographical (or rotational) axis NS, which has no necessary relation to the geomagnetic field.

Magnetic surveys on land, ships, and spacecraft have made possible construction of numerical models of the Earth's magnetic field. Figures 2.10 and 2.11 are examples of models that are useful for ionospheric radio propagation purposes. For the Earth-centered dipole model the field on the Earth's surface at the magnetic equator (a = 6371·2 km) is 30 854 nT. The dipole that better fits the real field is the eccentric dipole. The axis of this dipole is parallel to the Earth-centered dipole but displaced by 510 km (adjusted to 1988·0, IGRF–1985 model). The north geomagnetic or dipole pole is near 79·1°N, 289°E in northwestern Greenland; the dip pole, where the field is vertical, is over northern Canada. The magnetic dip poles $P_N(+90)$ and $P_S(-90)$ are not at opposite ends of a diameter. The antipode of P_S is about 2450 km from P_N, and the line $P_N P_S$ misses the Earth's center by about 1000 km. The positions of the dip poles in the IGRF–1985 (Personal communication, N.W. Peddie 1988) are P_S = (64·994°S, 139·347°E) and P_N = (77·609°N, 256·71°E). The moment of the Earth's magnetic dipole has been decreasing since at least 1835; this decrease is continuing at a rate of about −28 nT per year (Peddie and Zunde 1987). Since 1965, the North Magnetic Pole has migrated northwestward at a rate of about 9 km per year, while the South Magnetic pole has migrated northward at a rate of about 6 km per year.

In geomagnetism the field is usually measured in nanoteslas. The older units, the gauss (G) and the gamma (γ), are related to the tesla (the SI unit) by

$$1\gamma = 10^{-5}\,\mathrm{G} = 10^{-9}\,\mathrm{T} = 1\,\mathrm{nT}$$

The distribution of the Earth's magnetic potential V (from which the vector field can be derived) can be expressed by a mathematical process called spherical

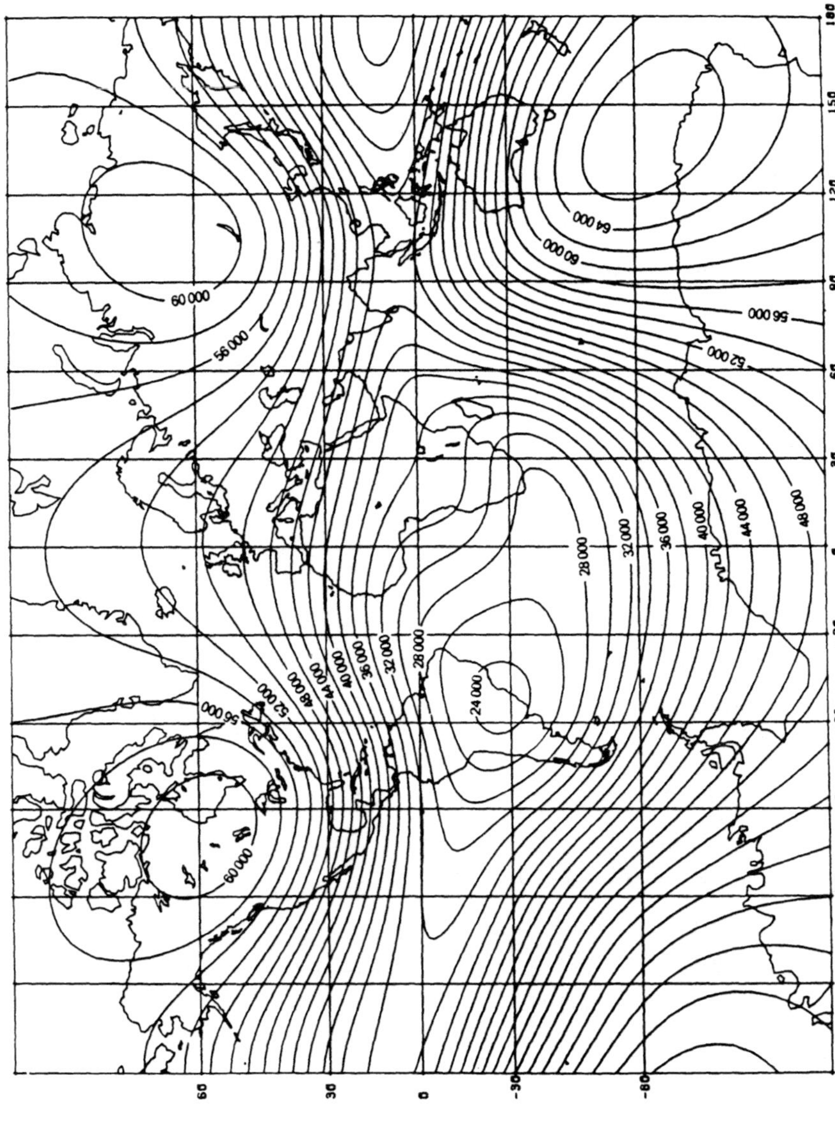

Fig. 2.10 *Numerical model of the Earth's magnetic total intensity (F), from the International Geomagnetic Reference Field – 1985 (From Barraclough, 1987 © Elsevier Science Publishers; reprinted with permission)*
Units are nanoteslas

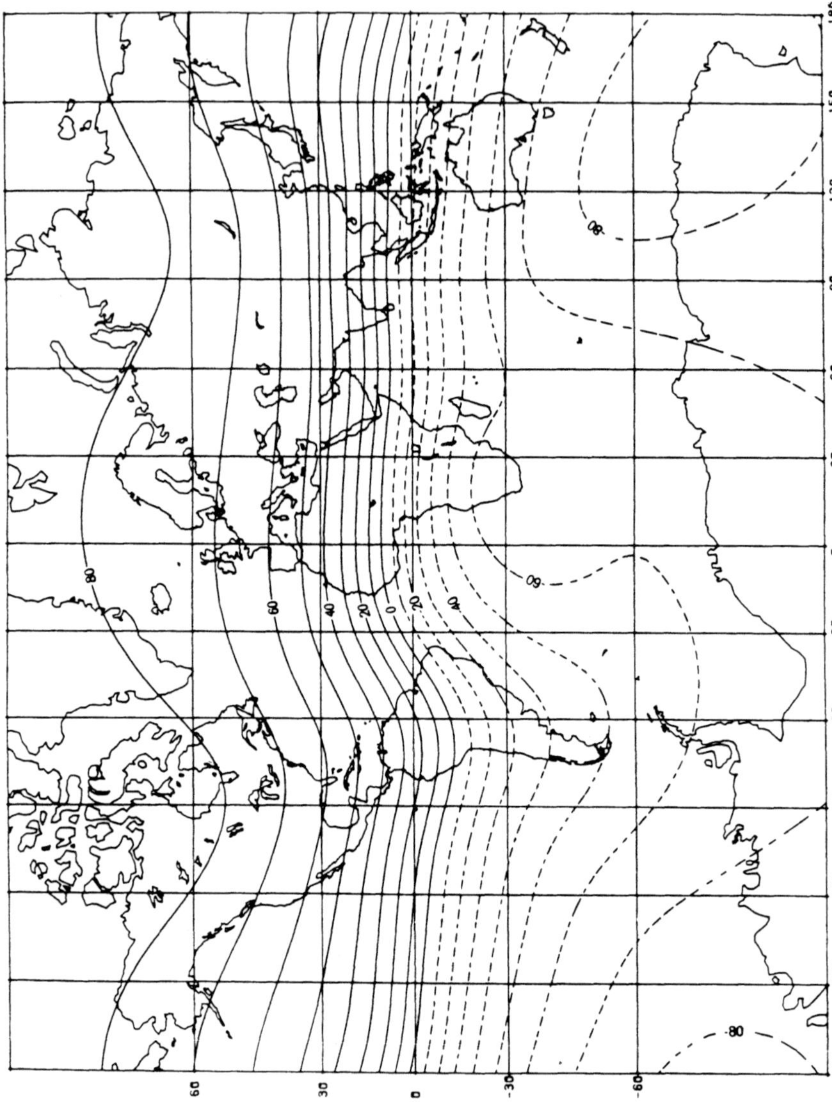

Fig. 2.11 *Numerical model of the Earth's magnetic dip, from the International Geomagnetic Reference Field – 1985 (From Barraclough, 1987 © Elsevier Science Publishers; reprinted with permission)*
Units are degrees of inclination

harmonic analysis (Vestine 1967). The potential V can be expressed as a sum of orthogonal functions of the type

$$V = a \sum_{n=0}^{\infty} \sum_{m=0}^{n} \left(\frac{a}{r}\right)^{n+1} P_n^m(\cos\theta)(g_n^m \cos m\lambda + h_n^m \sin m\lambda) \qquad (2.17)$$

where θ is the north co-latitude (i.e. angular distance from the North Pole) and λ is the east longitude. P_n^m (cos θ) are the spherical harmonic Schmidt functions, and the coefficients g_n^m and h_n^m are called the Gauss coefficients. Tables of coefficients for the IGRF–1985 are given by Barraclough (1987). The IGRF–1985 consists of two parts: (*a*) the 1985·0 geomagnetic main field at the start of 1985 and (*b*) the secular variation.

2.4.3 The magnetosphere

The solar wind flows around the Earth's magnetic field, producing a geomagnetic 'cavity' (Fig. 2.12). The effect is somewhat similar to the wake produced

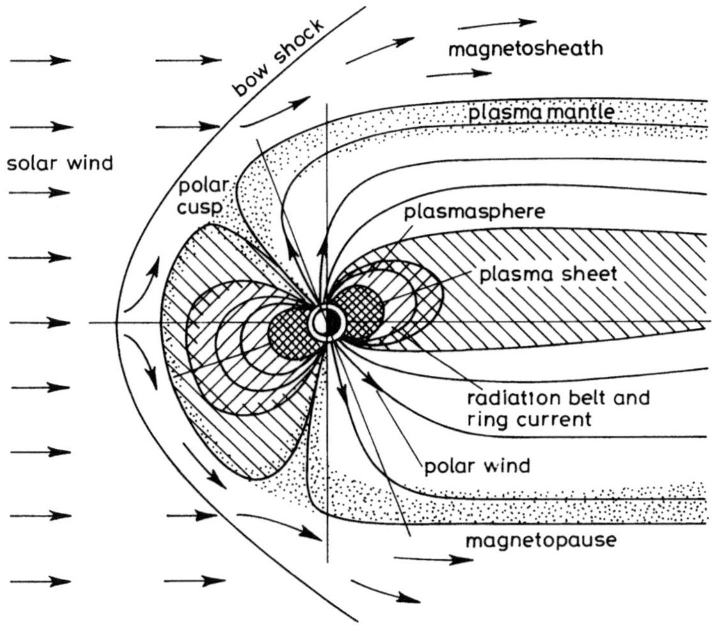

Fig. 2.12 *Magnetosphere formed by the interaction of the solar wind with the Earth's magnetic field*
The magnetic field is compressed on the dayside and drawn out on the nightside

by a stationary object in a river. A bow shock is established upstream (≈ 10 Earth radii) from the Earth, and this compresses the magnetic field on the dayside and stretches the field out (the tail) on the nightside. Within the shock front is the magnetosheath, which contains particles having essentially the same energies as those in the solar wind (energy density $\approx 2 \times 10^6\,\mathrm{eV\,m^{-3}}$). Within the magnetosheath is the magnetosphere bounded by the magnetopause. Most of the solar wind particles deflected by the bow shock continue downwind; however, a small fraction of the magnetosheath plasma interacts with the magnetopause and penetrates into the magnetosphere. Within the magnetosphere, charged particle motion is dominated by forces due to the Earth's magnetic and electric fields. The geomagnetic tail extends some hundreds of thousands of kilometers. Between the lines on the dayside and those on the nightside are two polar clefts or cusps, through which magnetosheath particles have direct access to the ionosphere. The structure of the magnetosphere is essentially fixed in the Sun–Earth coordinate system. The lower part of the magnetosphere, which rotates with the Earth and which contains cold plasma, is called the plasmasphere.

The magnetospheric tail consists, roughly speaking, of oppositely directed field lines separated by a neutral sheet of nearly zero magnetic field. Surrounding the neutral sheet is a plasma of hot particles with energies of 1 to 10 kiloelectronvolts (keV), a density of 10^4 to $10^6\,\mathrm{m^{-3}}$, and a bulk speed of a few hundreds of kilometers per second. Throughout the inner magnetosphere are trapped particles (the radiation belts and ring current). Some of these particles originate in the ionosphere and some come from the solar wind. The trapping of particles arises from the increase of magnetic field strength as the Earth is approached. Consider an ion gyrating about a line of force (e.g. Fig. 1.9); the perpendicular from the ion to that field line, at a given instant of time, gives the 'guiding center' of the particle motions. When the field is constant, the speed $v_\parallel$ of the guiding center is fixed. When the field strength increases, the radius of gyration (see Sec. 1.7) decreases and the perpendicular speed $v_\perp$ of gyration increases. Since the particle motion is adiabatic (no energy change) the parallel speed must decrease. The pitch angle $\alpha\,[=\sin^{-1}(v_\perp/v)]$ is related to the magnetic ic field strength B (Lyons and Williams 1984, eqn. 2.19) by

$$\frac{\sin^2\alpha}{B} = \text{a constant} \tag{2.18}$$

Hence, as B increases, α will increase to 90° at which point $v_\parallel = 0$ and the particle 'mirrors' where

$$B_m = \frac{B_{00}}{\sin^2\alpha_0} \tag{2.19}$$

where B_{00} and α_0 are the values of B_0 and α at the equatorial plane. The bouncing of ions between two hemispheres is illustrated in Fig. 2.13. In the Earth's real field there are zonal motions of the ions caused by zonal gradients of the field. When the mirror height is below about 100 km the particle is absorbed and may ionize the atmosphere. In general, there will be a distribution of pitch angles at the Equator and of particle energies, so particles will be mirroring over a range of altitudes. In middle latitudes, the electron bounce period is of the order of 0·1 to 1·0 second; proton bounce periods at the same energies are about 43 times greater (Lyons and Williams 1984, Table 2.2).

2.4.4 Magnetic variations

2.4.4.1 The quiet field

Owing to electric currents flowing above the Earth, the observed magnetic field at a point on the Earth varies diurnally, seasonally, and with solar activity. The net variation at the Earth's surface is the sum of two parts: (i) that caused by external currents and (ii) that produced by induced Earth currents. Days on which the magnetic field varies smoothly with time are called quiet or q days; other days are called disturbed or d days (Matsushita 1967*a*).

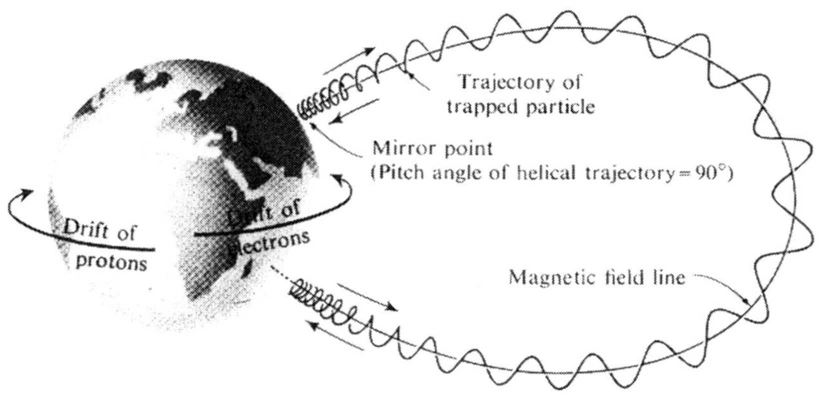

Fig. 2.13 *Motion of a charged particle in a dipole field (From Hess, 1968 © Blaisdell Publishing Co.; reprinted with permission)*
For low energy particles the motion can be divided into three components: (1) spiraling along field lines, (2) bouncing between hemispheres, and (3) drifting in longitude around the Earth, with electrons drifting east and protons west

On q days the magnetic variations proceed mainly according to local solar time, but they also contain a small part that varies with lunar time. The variations are called the solar daily (S_q) and lunar daily (L) magnetic variations. The variations are caused by currents flowing above the Earth's surface, mostly in the E layer where the electrical conductivity maximizes. The S_q currents are stronger by day than by night, stronger in summer than in winter, and about 50% stronger at sunspot maximum than at sunspot minimum. Of particular interest, from the point of view of the ionosphere, is the concentration of enhanced currents in a narrow strip along the magnetic equator that is called the 'equatorial electrojet'.

Even during quiet periods, the daily variations of the geomagnetic components occasionally show marked day-to-day changes of amplitude and phase, which indicate day-to-day variations of the current strength and shifts in the positions of the equivalent-current vortices in the upper atmosphere (Matsushita 1967*a*).

2.4.4.2 *Magnetic disturbances*

During a magnetic disturbance, additional currents circulate in the ionosphere. They are superposed on the S_q and L currents. Unlike the S_q and L currents, they are strong in high latitudes and are often stronger over the night than over the day hemisphere. They are denoted by DP (D = disturbance, P = polar). DP currents are especially concentrated along the auroral zones (i.e. the region of overhead visible aurora) and, hence, the DP currents there are called the auroral electrojet. A strong auroral electrojet may carry a current of the order of 1 million amperes. The return current flows mainly over the polar cap, but some currents flow between the north and south auroral zones. The DP currents are associated with increased coupling of the solar wind to the magnetosphere.

The 'average' magnetic storm is represented in Fig. 2.14. It has an initial positive phase, a main negative phase, and a recovery phase according to the storm time (that is, time reckoned from the beginning of the storm). The storm has Universal Time (UT) and Local Time (LT) components. The UT part is called the storm time variation D_{st}, and the LT part is called the disturbance daily variation.

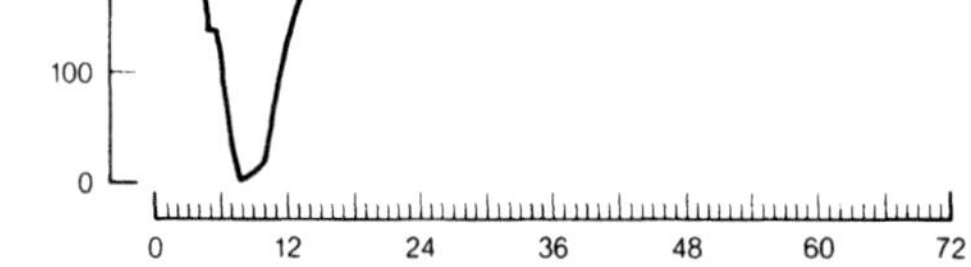

Fig. 2.14 *Magnetic storm variation consisting of a sudden commencement, an initial (positive) phase, a main (negative) phase, and a recovery phase*

A geomagnetic storm begins either gradually or with an abrupt change called a storm sudden commencement (SSC) caused by a solar wind shock hitting the magnetopause and compressing the magnetic field. Short-lived disturbances are called sudden impulses. Roughly speaking, the main phase begins when the horizontal component at the Equator, $D_{st}(H)$, decreases below the prestorm value, and ends when $D_{st}(H)$ reaches its maximum negative value. The initial and recovery phases are simply before and after the main phase. Often these phases are hard to identify (Matsushita 1967*b*). Quasi-sinusoidal oscillations of the magnetic field, in the range of about 0·2 s to about 10 min, are called pulsations (Jacobs 1970); they are thought to be manifestations of hydromagnetic waves in the magnetosphere.

Some of the solar wind is trapped by the geomagnetic field and the trapped particles spiral around the lines of force, between the Northern and Southern Hemispheres; they also drift around the Earth under the influence of the magnetic field. The total motion is equivalent to an electric current in the form of a ring around the Earth. The net effect at the Earth is to decrease the horizontal field, and in many storms the decrease soon overpowers the initial increase. A storm appears to have substorms, or bays. A substorm is an event, or series of events, in the magnetosphere in which energy is suddenly dissipated in association with auroral breakup. This explosive dissipation represents the 'unloading' of energy that was previously stored in the magnetotail, in the form of ring current enhancement, Joule heating, and auroral particle precipitation. The coupling of the solar wind to the magnetosphere depends most strongly on the direction of the interplanetary magnetic field relative to the Earth's field, and on the velocity of the solar wind.

A short-lived disturbance in the magnetic field is called a substorm or magnetic bay (Rostoker 1972). These are particularly prominent near the auroral zones where strong disturbance currents flow. A single isolated substorm lasts less than about 1 h, although a series of substorms may blend together to form a substorm of long duration.

2.5 Magnetic indices

The degree of magnetic disturbance during each Greenwich day is indicated by a variety of indices (Parkinson 1983, Rostoker 1972) adopted internationally. Magnetic observatories throughout the world assign local indices from which the world indices are determined. These include, for example, the K and A indices, and the auroral activity index AE.

2.5.1 The K indices

The K index is a 3 h range designed to measure the irregular variations associated with magnetic field disturbances. Each observatory assigns an integer from 0 to 9 to each of eight 3 h UT intervals (0000–0300, 0300–0600, . . . , 2100–2400). A permanent scale is adopted for a given observatory but the scale differs from observatory to observatory. The amplitude range R (i.e. the highest value minus the lowest, measured in gammas) is used to define a particular K. The range R is obtained for each element D, H, and Z, or X, Y, and Z. The largest of the three values, that is, the most disturbed element, is taken as the basis for K. In assessing K, the daily solar variation (S_q), the daily lunar variation (L), solar-flare effects, and the after-effect of a disturbance are eliminated (Lincoln 1967, Parkinson 1983). The K scale for a standard middle-latitude station (Niemegk: 52°04′N, 12°40′E) is given in Table 2.2. Different R ranges correspond to different K's at various observatories to ensure that a very disturbed day at one (middle-latitude) observatory has essentially the same K as that for an auroral observatory. For example, the values of R for $K = 9$, vary from 1500 nT at Godhaven to 300 nT at Honolulu. The K scale is quasi-logarithmic.

Table 2.2 *Relationship between K indices (K) and amplitude ranges (R) (nT)*

$K =$ 0		1		2		3		4		5		6		7		8		9
$R =$	5		10		20		40		70		120		200		330		500	

The planetary 3 h index K_p is designed to give a global measure of geomagnetic activity. K_p indices are obtained from local variations for 12 stations in geomagnetic latitudes of 63° to 48° in both the Northern and the Southern Hemisphere. K_p is an index in 28 grades from 0o, 0+, 1−, 1o, 1+, 2−, . . . , 9−, 9o. The K_p index is not particularly satisfactory for representing disturbances within the polar cap.

2.5.2 *The A indices*

For some applications a daily index is desirable. The K sum is not suitable because of its quasi-logarithmic nature. The A_p or 'daily equivalent planetary amplitude' is derived from K_p by use of a_p given in Table 2.3. The daily A_p index is the average of the eight a_p values.

Table 2.3 *Equivalent range a_p for given K_p*

K_p	0o	0+	1−	1o	1+	2−	2o	2+	3−	3o	3+	4−	4o	4+
a_p	0	2	3	4	5	6	7	9	12	15	18	22	27	32
K_p	5−	5o	5+	6−	6o	6+	7−	7o	7+	8−	8o	8+	9−	9o
a_p	39	48	56	67	80	94	111	132	154	179	207	236	300	400

2.5.3 *Auroral electrojet index AE*

This index is a measure of the currents flowing in the auroral zone; it is derived from data obtained at stations distributed in longitude near the auroral zone. Daily H magnetograms from the stations are superposed, and the upper and lower envelopes define AU and AL respectively. Hence, by definition

$$\mathrm{AE} = \mathrm{AU} - \mathrm{AL} \tag{2.20}$$

AO is defined as the average of AU and AL:

$$\mathrm{AO} = (\mathrm{AU} + \mathrm{AL})/2 \tag{2.21}$$

2.5.4 Equatorial D_{st}

The aforementioned indices tend to emphasize the high-latitude magnetic disturbances. The hourly D_{st} index measures the strength of the ring current, using magnetic field variations near the dipole equator, averaged over local time. Hourly D_{st} data, together with a variety of other magnetic indices, are given in *Solar-Geophysical Data*. The best single source of indices is Bulletin No. 12 of the International Union of Geodesy and Geophysics (IUGG) International Association of Geomagnetism and Aeronomy (IAGA). Other sources of geomagnetic data include *Journal of Geophysical Research* and *Solar Geophysical Data*.

2.5.5 Selected quiet and disturbed days

For any given month the days are ranked by using the following numbers, giving each equal weight: (i) the sum of the eight K_p values; (ii) the sum of the squares of the eight K_p values, and (iii) the greatest K_p value. The average of the three numbers is taken for each day. The five days with the highest averages are the five disturbed days. The ten days with the lowest averages are the ten quiet days that include the five quietest days.

2.5.6 Bartels' 'musical scales'

Bartels' format for displaying K_p values illustrates the sequence graphically (Fig. 2.15). The data are arranged in rows so as to show the 27-day recurrence tendency of magnetic disturbance, arising from solar rotation. The diagrams are issued monthly in *Solar-Geophysical Data* along with the tables of geomagnetic activity indices and are helpful in making 27-day recurrence forecasts.

2.6 The auroral zones

The auroral zones may be defined as those regions of the Earth where visible aurora occurs overhead. Aurora is the result of excitation of atmospheric atoms and molecules by energetic ions and electrons released from the magnetosphere. Auroral activity can be regarded as a branch of magnetosphere–ionosphere–atmosphere coupling. By some mechanism(s) the ions trapped in the magnetosphere are energized so that they penetrate into sufficiently dense regions of the atmosphere to collide with and excite various atmospheric constituents. The auroral zones are relatively narrow rings situated between magnetic latitudes of about 64° and about 70°. A satellite (*Dynamics Explorer*) image of the quiet auroral oval as taken in the atomic oxygen lines at 1304 Å (130·4 nm) and 1356 Å (135·6 nm) at an altitude of about 3·5 Earth radii is shown in Fig. 2.16. The bright hemisphere is sunlit and there are two bright portions of the auroral oval, one at local midnight and one at local noon; the latter is caused by precipitation of particles through the polar cusp. Lines of equal frequency of occurrence of overhead visible aurora are called isochasms. The maximum

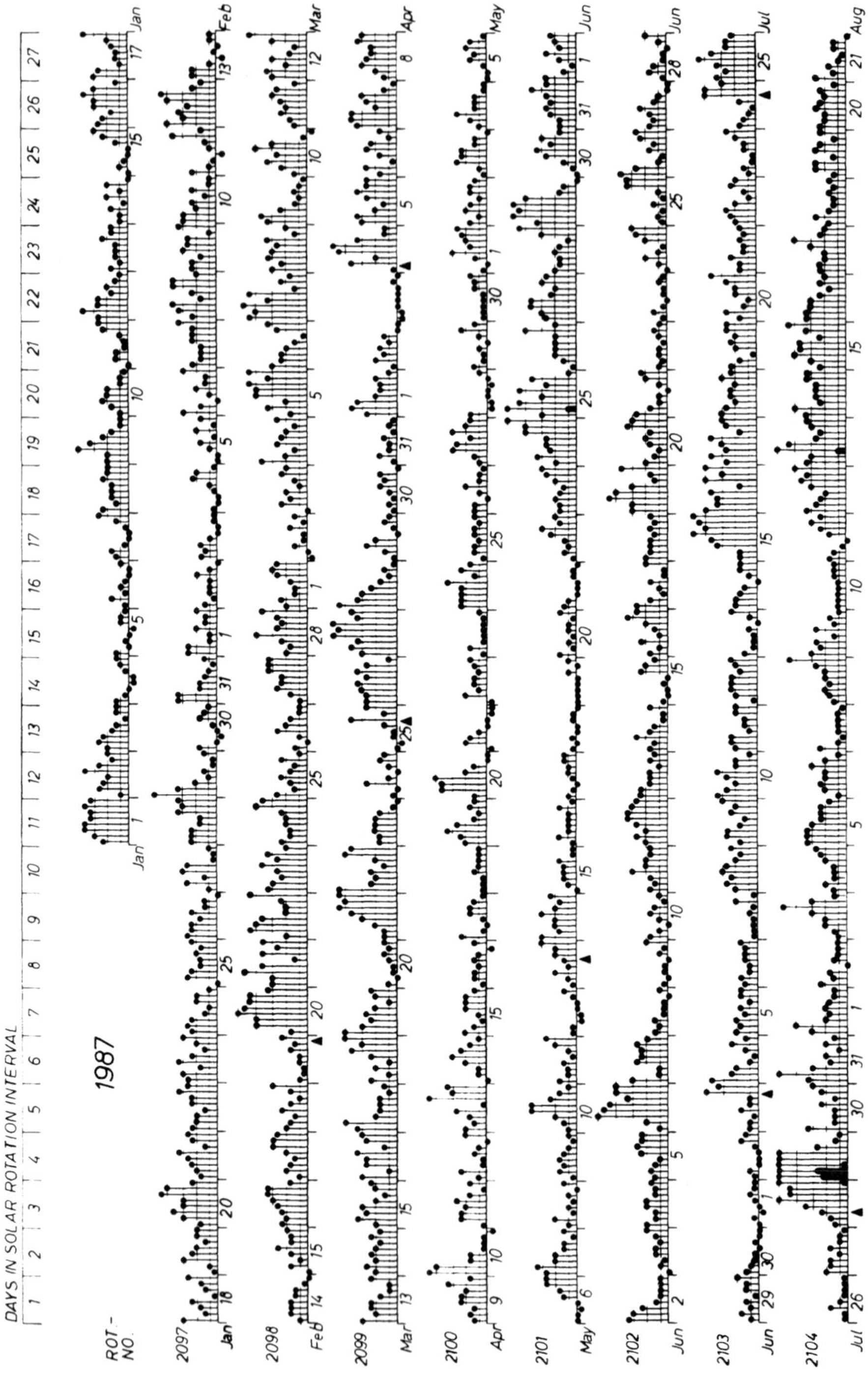
DAYS IN SOLAR ROTATION INTERVAL
1987
ROT-NO.
2097
2098
2099
2100
2101
2102
2103
2104

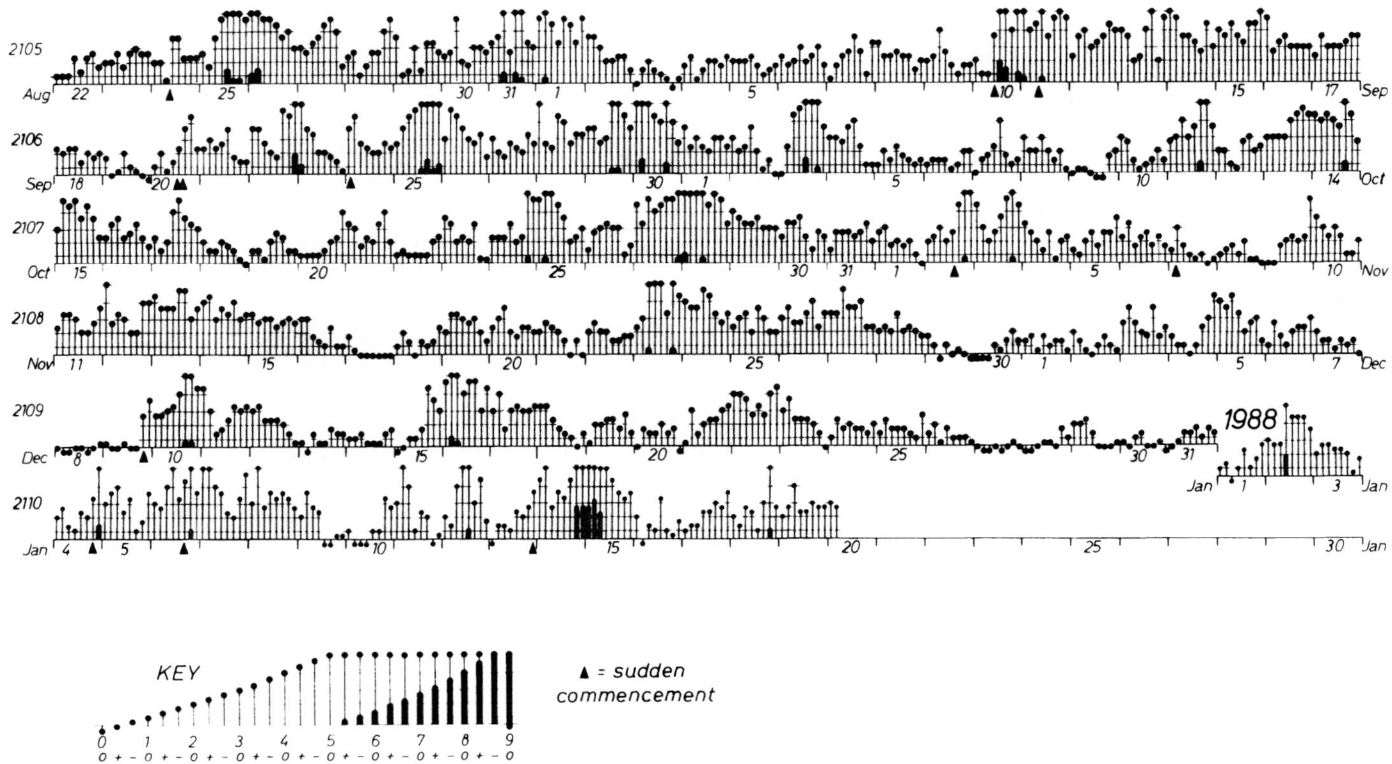

Fig. 2.15 *Bartels' 'musical scales' (By permission of the Institut für Geophysik, University of Göttingen)*
Planetary 3-h-range indices (K_p) are arranged in 27-day rows to show recurring pattern of magnetic disturbance (data for Jan. 1, 1987, to Jan. 20, 1988)

isochasm is located at about 67° geomagnetic latitude. The northern zone is elongated (oval). The geographical areas enclosed by the auroral rings are called the polar caps. For comprehensive treatments of the aurora, see for example Chamberlain (1961) and Hultqvist (1967).

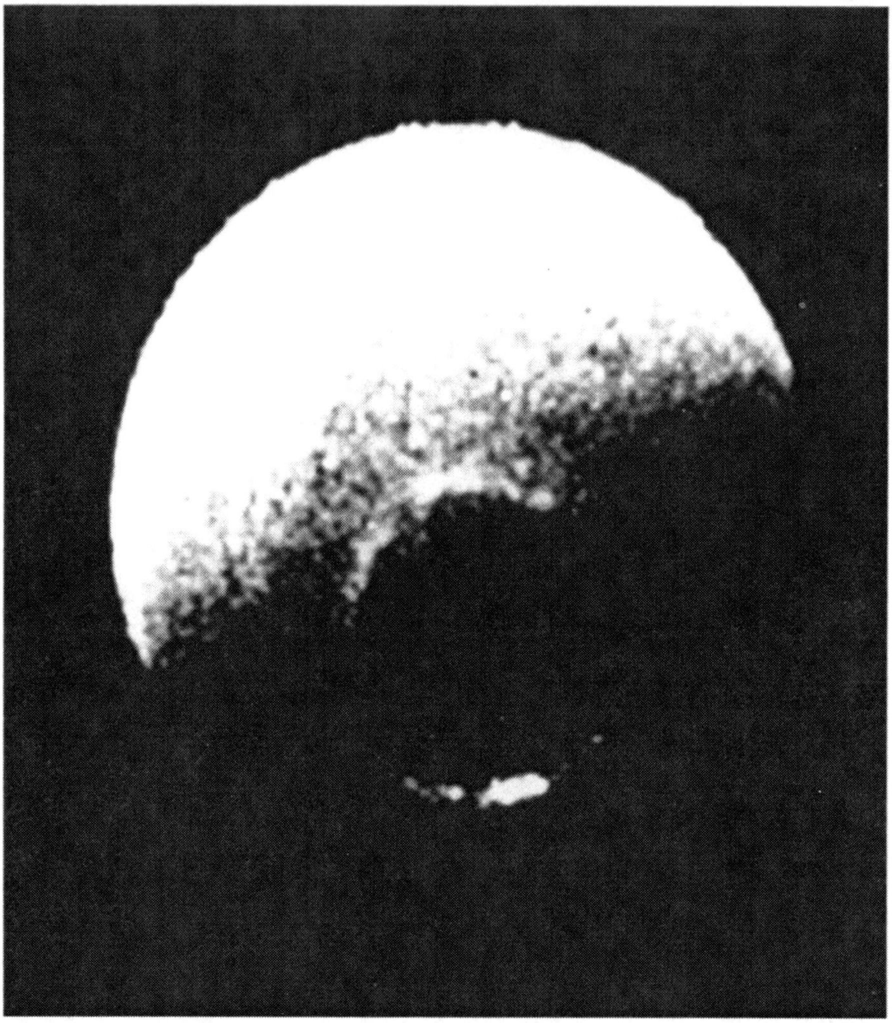

Fig. 2.16 *Dynamics Explorer image of the quiescent auroral oval taken at 1304 Å and 1356 Å at an altitude of about 3·5 Earth radii (Courtesy of L. A. Frank)*
The bright portions of the auroral oval appear at local midnight and local noon

The aurora emits over a wide range of wavelengths from radio to X-rays. In addition to visible aurora, there are infrared aurora, ultraviolet aurora, and X-ray aurora. Visible auroras occur in a wide variety of structures. These range from small-scale sharply defined narrow rays, which are aligned along magnetic field lines, to large-scale structures (such as arcs and rayed bands) that are oriented in the magnetic east–west direction. On the equatorial side an aurora may remain quiet for hours and then may brighten and become active. In general, the intensity and position of the auroral zone is related to magnetic disturbance, the latitude of the zone decreasing with increasing magnetic disturbance. The more important lines in auroral spectra include oxygen 5577 Å (green) and 6300 Å (red), and molecular bands from singly ionized N_2 molecules.

Because of the separation of charges as plasma flows around the Earth's magnetic field, there is a dawn-to-dusk electric potential across the magnetosphere of about 60 000 to 100 000 V. In the morning this potential drop causes field-aligned currents to flow from the magnetosphere on the poleward side of the auroral oval, south across the oval, and back to the magnetosphere at the equatorial edge. On the dusk side the currents are in the opposite direction. Figure 2.17 gives a schematic view of some of these high-latitude processes that are of special interest to ionospheric scientists and radio communicators. The most active and brilliant auroras are located in the midnight sector of the auroral zone. During disturbed periods these displays undergo large-scale expansions and contractions called auroral substorms.

2.7 The plasmasphere

The plasmasphere may be thought of as that part of the (inner) magnetosphere in which the plasma rotates with the Earth. In the plasmasphere, above an altitude of 1000 km say, the neutral atmosphere densities are so small that they have little effect on the charged particle motions determined by electric and magnetic fields. The positive ions are predominantly protons (H^+) and, hence, the plasmasphere is practically synonymous with the protonosphere. The neutral atmosphere in this region is often referred to as the exosphere. To a rough approximation, the plasmasphere is enclosed within an L shell of about 6 called the plasmapause. Equation (2.13) shows that this corresponds to an invariant latitude of 66°, which is near the equatorial side of the auroral oval. In the plasmasphere, there is essentially no ion production; the ionization is produced in the ionsophere during the day and diffuses upward. The protons are produced by charge exchange between oxygen ions and hydrogen:

$$O^+ + H \rightleftarrows H^+ + O \tag{2.22}$$

The ion density at 1000 km, in middle latitudes, is of the order of 10^{10} m^{-3}. At higher altitudes it decays out to the plasmapause at which there is an abrupt drop of 1 to 2 orders of magnitude. Within the plasmapause the ion temperature is about 1000 K, and the plasma is said to be cold, whereas outside the plasmapause the energy is of the order of 1000 eV ($\approx 10^7$ K). During nighttime, when the ionospheric plasma densities decrease, plasma diffuses down from the plasmasphere into the ionosphere.

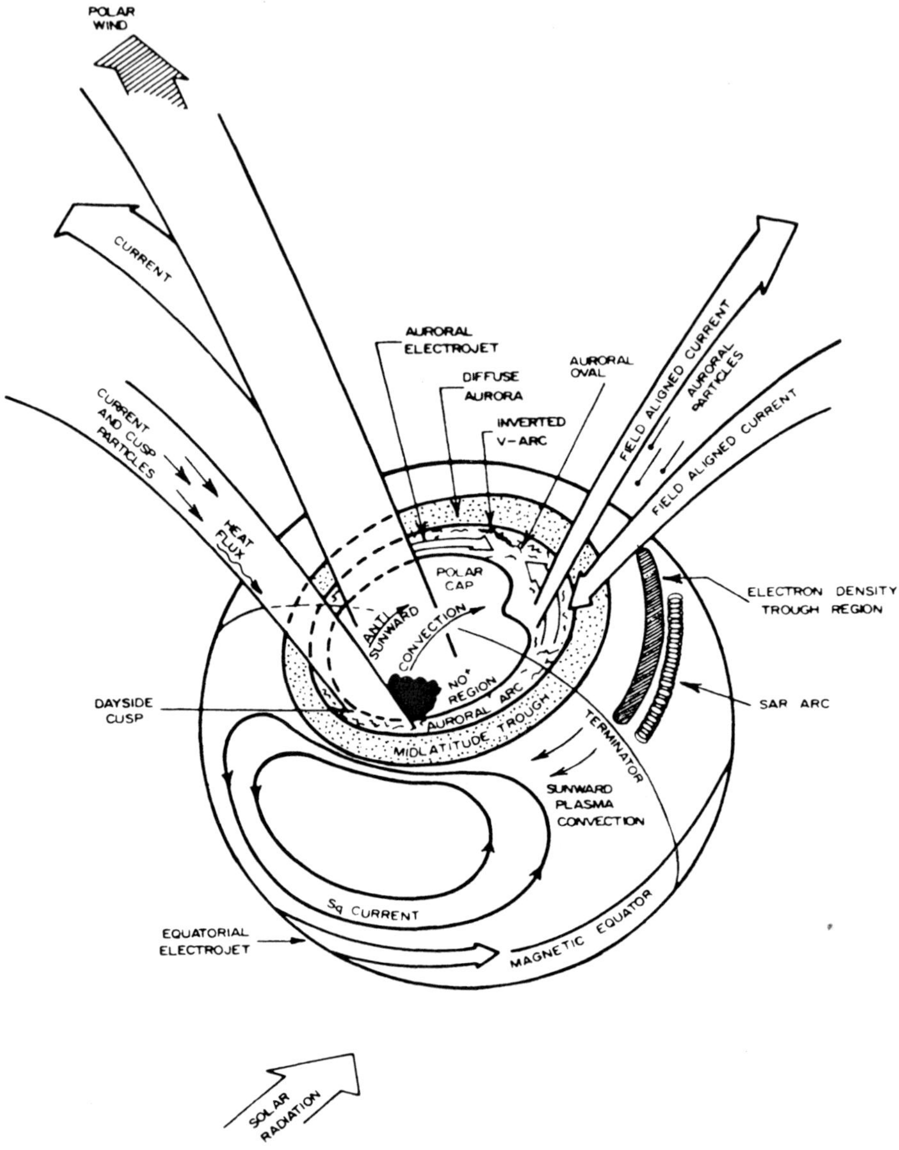

Fig. 2.17 *High-latitude processes of interest to radio communicators*
Electric currents flow from the magnetosphere into the auroral ionosphere along magnetic field lines, then along the auroral zone and back into the magnetosphere. Plasma is convected, from day to darkness, across the polar cap by dawn-to-dusk electric fields in the magnetosphere. The polar wind results from plasma escaping along magnetic field lines. On the nightside auroral ionization is produced by the precipitation of energetic auroral particles from the magnetosphere. Also of interest are the electron-density trough, the auroral electrojet, and the dayside cusp

2.8 The Earth's atmosphere

2.8.1 Nomenclature

The neutral atmosphere can be subdivided according to several properties. For example, in Fig. 2.18 the division is by temperature and dynamics. Below about 100 km, in the turbosphere the composition is maintained essentially constant by mixing, whereas above that altitude, in the diffusosphere or heterosphere, the composition is determined by diffusive separation. Molecular O_2 and N_2 dominate at the lower heights, giving way to atomic oxygen and eventually to helium and hydrogen in the exosphere. The most common subdivision is by temperature: in the troposphere the temperature decreases with height up to the tropopause, the average altitude of which varies from about 18 km at the Equator to about 8 km at the geographical poles. The lapse rate in the troposphere is about

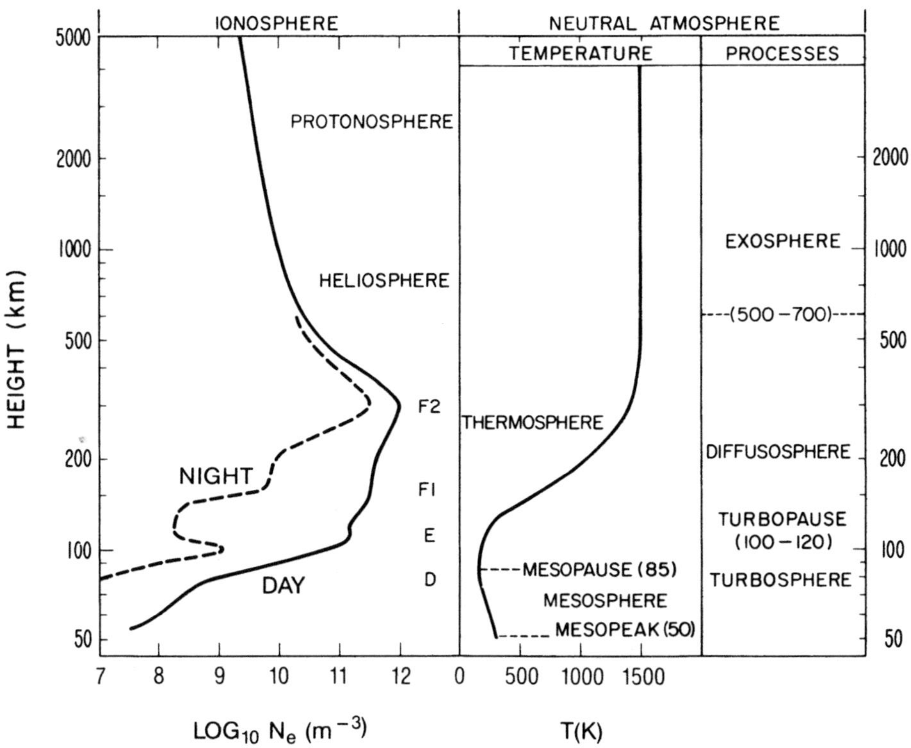

Fig. 2.18 *Atmosphere division by temperature and dynamics*

$-6\,K\,km^{-1}$, which leads to a tropopause temperature of about 200 K. In the stratosphere the temperature increases to about 270 K near 50 km (i.e. at the mesopeak), above which it falls again in the middle atmosphere, or mesosphere, reaching a mesopause minimum near 85 km above which the temperature steadily increases up through the thermosphere.

2.8.2 Pressure, density, and temperature variations

The relationship between pressure p, mass density ρ, and temperature T, at any height h, is given by the barometric equation

$$p = p_0 \exp\left(-\int_{h_0}^{h} \frac{dh}{H_p}\right) \tag{2.23}$$

and

$$\rho T = \rho_0 T_0 \exp\left(-\int_{h_0}^{h} \frac{dh}{H_p}\right) \tag{2.24}$$

where p_0, ρ_0, and T_0 are the values of p, ρ, and T at the reference height h_0 and

$$H_p = \frac{KT}{\bar{m}g} \tag{2.25}$$

is the pressure scale height, which is a function of the mean molecular mass $\bar{m}$, the acceleration due to gravity g, and temperature T, all of which are height dependent. The scale height varies from about 8 km at the Earth's surface to several hundreds of kilometers in the upper ionosphere.

In an isothermal atmosphere

$$p = p_0 \exp\left(-\frac{h - h_0}{H}\right) \tag{2.26}$$

$$\rho = \rho_0 \exp\left(-\frac{h - h_0}{H}\right) \tag{2.27}$$

From the perfect gas law ($p = nKT$) and $\rho = \bar{m}n$ we have, in a temperature-, density-, and composition-varying atmosphere,

$$-\frac{1}{H_p} = \frac{1}{p}\frac{dp}{dh} = \frac{1}{T}\frac{dT}{dh} + \frac{1}{n}\frac{dn}{dh} = \frac{1}{T}\frac{dT}{dh} + \frac{1}{\rho}\frac{d\rho}{dh} - \frac{1}{\bar{m}}\frac{d\bar{m}}{dh} \tag{2.28}$$

The following formulas are useful in estimating the scale height H in kilometers at any altitude h:

$$H = 0{\cdot}85\left(1 + \frac{h}{a}\right)^2 \frac{T}{M} \tag{2.29}$$

where M is in gram mols, and a (≈ 6370 km) is the mean Earth radius. At 550 km, $H = T/M$, which is correct to within 10% between 200 km and 900 km.

It is often convenient to use a reduced height z defined by

$$z = \int_{h_0}^{h} \frac{dh}{H_p} \tag{2.30}$$

which, for an isothermal atmosphere, becomes

$$z = \frac{h - h_0}{H_p} \tag{2.31}$$

Above 120 km the temperature increases asymptotically to a value T_∞ called the exospheric temperature.

The temperature at a height h can be approximated by curves of the form

$$T = T_\infty - (T_\infty - T_{120}) \exp[-s(h - 120)] \tag{2.32}$$

where T_{120} is the temperature at 120 km and s is a parameter that depends on T_∞ (CIRA 1972). The exospheric temperatures can vary with time, especially over the sunspot cycle, and with geomagnetic activity. The nighttime minimum of the global exospheric temperature T_c (when $K_p = 0$) is related to the 10·7 cm radio noise flux $F_{10\cdot 7}$ by

$$T_c = 379 + 3{\cdot}24\bar{F}_{10\cdot 7} + 1{\cdot}3(F_{10\cdot 7} - \bar{F}_{10\cdot 7}) \tag{2.33}$$

where $\bar{F}$ is a long-term average (CIRA 1972). The dependence on geomagnetic activity is

$$\Delta T_\infty = 28K_p + 0{\cdot}03 \exp(K_p) \tag{2.34}$$

where K_p is the 3 hour index. These equations express the fact that the exospheric temperature increases with solar flux (sunspot number) and with magnetic disturbance.

2.8.3 *Chemical composition*

Below the mesopause ($\approx$ 85 km) turbulent mixing keeps the composition of the atmosphere essentially homogeneous. The percentages of the number densities of the major species (molecular nitrogen, molecular oxygen, and atomic argon) at sea level are as follows:

N_2	O_2	Ar	Sum	$\bar{M}$
78·08	20·95	0·93	99·96	28·96

where $\bar{M}$ is the mean molecular weight.

Above the mesopause, molecular oxygen is dissociated by solar UV with wavelengths shorter than 1759 Å (175·9 nm). This process is so rapid that turbulence cannot keep O_2 and O mixed and, therefore, the number density of O increases rapidly with height to a peak near 100 km. The fraction of O also increases until, above about 150 km, the atmosphere becomes mostly atomic oxygen. Above about 100 km (the turbopause) each constituent obeys a separate barometric equation with its own 'partial scale height'

$$H_i = KT/m_i g \tag{2.35}$$

where i denotes the ith constituent (VanZandt 1967). The structure of the diffusosphere is sensitive to the height of the turbopause, which probably varies diurnally, seasonally, and with solar activity. The height variations of the number densities of the main atmospheric constituents according to the mean COSPAR International Reference Atmosphere (CIRA) is given in Fig. 2.19. The height variations of molecular temperature, mass, density, pressure, number density, and pressure scale height are given in the CIRA 1972 tables for altitudes up to 2000 km.

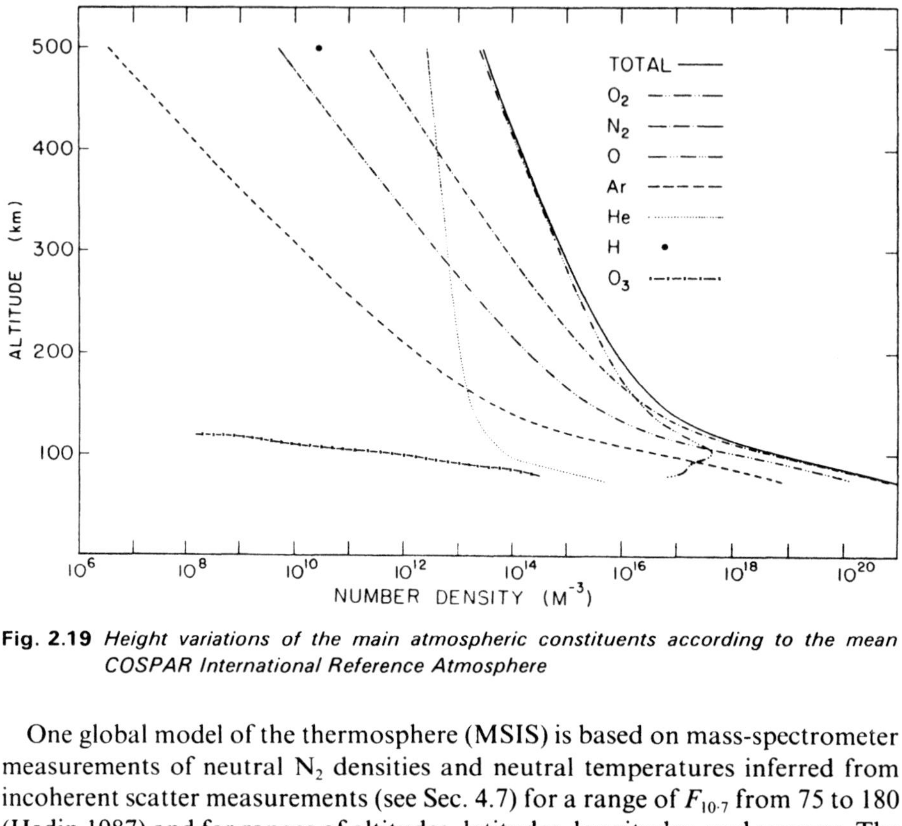

Fig. 2.19 *Height variations of the main atmospheric constituents according to the mean COSPAR International Reference Atmosphere*

One global model of the thermosphere (MSIS) is based on mass-spectrometer measurements of neutral N_2 densities and neutral temperatures inferred from incoherent scatter measurements (see Sec. 4.7) for a range of $F_{10.7}$ from 75 to 180 (Hedin 1987) and for ranges of altitudes, latitudes, longitudes, and seasons. The model contains O, He, Ar, O_2, and H densities starting at a base height of 120 km. This model is used extensively for calculating production and loss of ionization in the ionosphere. Nevertheless, it must be realized that any model gives a simplified picture of a rather complicated phenomenon and cannot be expected to accurately represent conditions at a specific location at a given time.

2.9 Formation of ionized layers

2.9.1 Production

Here we consider the production of ionization in a horizontally stratified layer by the absorption of a parallel, monochromatic beam of solar ionizing radiation incident at an angle χ with the vertical. The formula for the rate of production of ion pairs, in a multi-component atmosphere, by a spectrum of solar radiations has been given elsewhere (e.g. VanZandt 1967). For the simplified case of a single solar wavelength ionizing a one-component isothermal gas, the rate of production is given by the Chapman function:

$$q(z, \chi) = q_0 \exp\left[1 - z - \sec \chi \exp(-z)\right] \tag{2.36}$$

in which

$$z = (h - h_0)/H_p \tag{2.37a}$$

where H_p is the pressure scale height (2·25), and

$$q_0 = \phi(\infty)\eta/H_p\varepsilon \tag{2.37b}$$

is the rate of ion production at the level $z = 0$. Here $\phi(\infty)$ is the solar flux density outside the atmosphere in photons per unit area, η is the number of ion pairs produced by one photon, and $\varepsilon = 2{\cdot}718$. The reference height h_0 is that at which $N\sigma H_p = 1$, where N is the number density of neutrals, σ is the absorption cross section, and the height variation of g is assumed zero. In terms of the optical depth τ of the atmosphere down to the height h, we have

$$q = q_0 \tau \exp(1 - \tau \sec \chi) \tag{2.38a}$$

where

$$\tau = \int_h^\infty N\sigma \, dh = \exp(-z) \tag{2.38b}$$

The reference height h_0 is the height of maximum ion production when the Sun is overhead ($\chi = 0°$).

The variations of q/q_0 with z, for various zenith angles χ, are given in Fig. 2.20 from which we see that the gradient of production is steeper on the lower side than on the upper side of the peak and that, at sufficiently high levels, the production rate (and electron density) is essentially independent of χ.

The height of maximum production z_m is obtained by differentiating (2.36), which yields

$$z_m = \ln \sec \chi \tag{2.39a}$$

That is,

$$h_m = h_0 + H_p \ln \sec \chi \tag{2.39b}$$

The use of a logarithmic scale for q makes the shapes of the q curves identical. The peak of production is

$$q_m = q_0 \cos \chi \tag{2.40}$$

Changing the intensity of the solar flux changes the peak value of q but does not affect the height of maximum production.

The quiet D region is produced by the photoionization of nitric oxide (NO), a trace constituent, by the Lyman α radiation that penetrates to the D region by

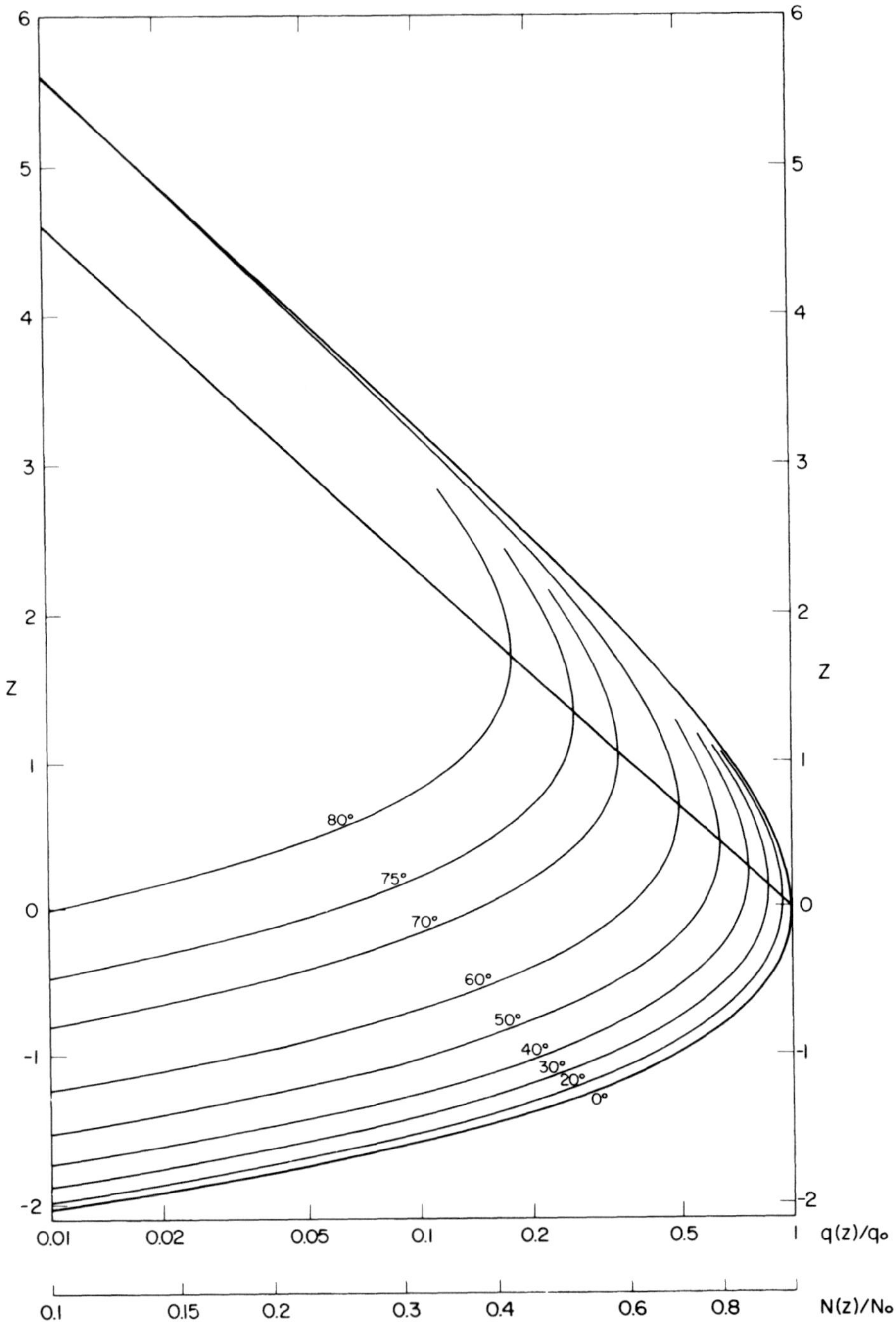

Fig. 2.20 *Normalized rate of photoionization $q(z)/q_0$ and electron density $N(z)/N_0$ according to Chapman theory*

virtue of a window in the absorption cross section of O_2. During sudden ionospheric disturbances there may be great enhancements of solar X-rays in the range 1 to 10 Å (0·1–1·0 nm), which also ionize the D region. In the E region the ionizing radiations lie in the wavelength range from about 8 Å (0·8 nm) to about 140 Å (14 nm) (soft X-rays) and the wavelength range from 912 Å (91·2 nm) to 1027 Å (102·7 nm), which can ionize O_2. In the F layer, wavelengths between about 140 Å (14 nm) and 911 Å (91·1 nm) are important in ionizing atomic oxygen with a peak absorption near 175 km. There are no peaks in absorption of solar radiation in the F_2 layer. The F_2-layer peak is caused by diffusion rather than by recombination (VanZandt 1967).

2.9.2 Ion loss and layer shapes

Electrons are not normally lost by simple recombination with a positive atomic ion. First, the recombination rate is about 10^5 times greater with molecular ions than with atomic ions. Hence, the atomic (oxygen) ions must be converted to molecular ions before the recombination takes place. Secondly, the original molecular ions are converted into another species of molecular ion. Thirdly, in the lower D region at night, electrons can attach themselves to neutral molecules to form negative ions, which can recombine with positive ions.

The following reactions are important in the daytime D region and the E and F regions:

$$O_2^+ + e \rightarrow O + O + 6{\cdot}96\,\text{eV} \tag{2.41a}$$

$$N_2^+ + e \rightarrow N + N + 5{\cdot}82\,\text{eV} \tag{2.41b}$$

$$NO^+ + e \rightarrow N + O + 2{\cdot}76\,\text{eV} \tag{2.41c}$$

$$O^+ + O_2 \rightarrow O_2^+ + O + 1{\cdot}53\,\text{eV} \tag{2.41d}$$

$$O^+ + N_2 \rightarrow NO^+ + N + 1{\cdot}09\,\text{eV} \tag{2.41e}$$

Reactions (*a*), (*b*), and (*c*) are called dissociative recombination reactions; (*d*) and (*e*) are called charge-transfer or atom-ion exchange reactions or charge-exchange reactions. Other reactions in the ionosphere include

$$O_2^+ + N_2 \rightarrow NO^+ + NO + 0{\cdot}87\,\text{eV} \tag{2.42a}$$

$$N_2^+ + O \rightarrow NO^+ + N + 3{\cdot}05\,\text{eV} \tag{2.42b}$$

$$N_2^+ + O_2 \rightarrow O_2^+ + N_2 + 3{\cdot}50\,\text{eV} \tag{2.42c}$$

In the E and F_1 layers, the reactions are fast. In the E region, where almost all the ions are molecular, the rate of recombination of electrons equals the rate of recombination of the molecular ions (i.e. αN_e^2, where α is the mean dissociative recombination coefficient for the molecular ions). Thus the rate of increase of electron density N_e is given by the continuity equation

$$\frac{dN_e}{dt} = q - \alpha N_e^2 \tag{2.43}$$

In quasi-equilibrium (i.e. dN_e/dt small), substitution in (2.36) gives

$$N_e = N_{e0} \exp \tfrac{1}{2} [1 - z - \sec \chi \exp(-z)] \qquad (2.44a)$$

and the maximum electron density is

$$N_m(\chi) = N_{e0} \cos^{1/2} \chi \qquad (2.44b)$$

Equation (2.44*a*) is sometimes called the alpha-Chapman layer, and the variation of electron density $N(z)/N_0$ with reduced height is shown in Fig. 2.20.

At high levels in the ionosphere where photoionization of O is dominant the loss rate depends linearly on N_e. Thus,

$$\frac{dN_e}{dt} = q - \beta N_e \qquad (2.45)$$

At these heights, the lifetimes of the atomic ions are long, and the rate of loss of electrons is determined by charge exchange. At intermediate heights the equation of continuity for electrons is

$$\frac{dN_e}{dt} = q - \frac{\beta \alpha N_e^2}{\beta + \alpha N_e} \qquad (2.46)$$

When $\beta \gg \alpha N_e$, (2.46) reduces to (2.43), whereas when $\beta \ll \alpha N_e$ we obtain (2.45).

The linear loss coefficient β is proportional to molecular density. Therefore, variations in molecular densities (e.g. those caused by heating in the dynamo region) can have a marked effect on electron densities in the F_2 layer.

Assuming quasi-equilibrium and a linear loss with the Chapman production function (2.36) and assuming that β depends directly on molecular density we obtain, for the electron density distribution with height,

$$N_e = N_{e0} \exp [1 - \sec \chi \exp(-z)] \qquad (2.47)$$

This profile has no maximum except at very great heights where there is an insufficient number of molecules that can be ionized.

The solar zenith angle χ can be determined from

$$\cos \chi = \sin \phi \sin \delta + \cos \phi \cos \delta \cos h \qquad (2.48a)$$

where ϕ is geographic latitude, δ is the solar declination, and h is the local hour angle of the Sun measured westward from apparent noon (i.e. mean noon corrected for the equation of time E and difference in time corresponding to the longitude difference between the station and the standard time meridian used at the location). The equation of time E (solar time − mean time) is given to within 2 minutes by

$$E = 7{\cdot}75 \sin [0{\cdot}9856(Y - 3{\cdot}0)] - 9{\cdot}94 \sin [1{\cdot}971(Y - 80{\cdot}7)] \qquad (2.48b)$$

where Y is the day number starting on 1 January. For monthly averages the declination δ (in degrees) is given to a sufficient accuracy by

$$\delta = 23{\cdot}44 \sin [0{\cdot}9856\,(Y - 80{\cdot}7)] \tag{2.48c}$$

For large zenith angles ($> 80°$), the effect of curvature of the Earth is important and sec χ must be replaced by the Chapman function $Ch(R + z, \chi)$, where $R = (a + h)/H_p$. Tables of this function have been published by Wilkes (1954).

Although most of the assumptions underlying the simple Chapman layer do not apply in the upper atmosphere (isothermal, single species, single ionizing radiation), nevertheless the Chapman theory has been an invaluable guide in pointing the way to methods of data analysis and presentation and providing a useful reference (Kelley 1989).

2.9.3 Transport of plasma

Electron production q and loss L can account for the morphologies of the D, E, and F_1 layers. However, in the F_2 layer, movement of ionization is important, and the equation of continuity takes the form

$$\frac{\partial N}{\partial t} = q - L - \operatorname{div}(N\boldsymbol{v}) \tag{2.49}$$

where $\boldsymbol{v}$ is the mean velocity. The div term represents the net loss due to transport. Transport can include, for example, diffusion, neutral winds, and vertical electromagnetic drift (Kato 1980).

2.9.3.1 Plasma diffusion

Vertical diffusion speed w is given by

$$w = -D(h) \sin^2 I \left(\frac{1}{N_e}\frac{dN_e}{dh} + \frac{Mg}{2KT} \right) \tag{2.50}$$

where M is the ion mass, $D(h)$ is the ambipolar diffusion coefficient (the ions and electrons diffuse together), and I is the magnetic dip. Because $D(h)$ is inversely proportional to neutral density, diffusion increases rapidly with increase of altitude. Since $I = 0°$ at the dip equator, vertical diffusion vanishes there because diffusion across the magnetic field is inhibited. Equation (2.50) shows that, other considerations being equal, vertical diffusion maximizes at the dip poles where $I = 90°$. This shows that diffusion maximizes along the magnetic field; such diffusion is called field-aligned diffusion. Even when diffusion is included, the shape of the F_2 layer is approximately that of a Chapman layer.

Diffusion separation in a plasma differs from that in a neutral gas because of the electrostatic field produced by charge separation. The temperatures of photo-electrons T_e can be higher than ion temperatures T_i. In a plasma the heavy ions will tend to sink to lower levels (under gravity) and the electrons will tend to rise. Under equilibrium conditions,

$$\frac{1}{N_e}\frac{d}{dh}(N_e K T_e) = -m_e g + eE \tag{2.51a}$$

$$\frac{1}{N_i}\frac{d}{dh}(N_i K T_i) = -M_i g - eE \tag{2.51b}$$

where N_e, N_i are the number densities of electrons and ions respectively, assumed equal, g is the gravitational acceleration, and the ion mass M_i is very much greater than the electron mass m_e. Adding (2.51*a*) and (2.51*b*) gives

$$\frac{1}{N_e}\frac{dN_e}{dh} = -\frac{M_i g}{K(T_e + T_i)} = \frac{1}{H_{pl}} \tag{2.52}$$

provided that the temperatures are independent of height. The plasma scale height

$$H_{pl} = \frac{K(T_e + T_i)}{gM_i}; \tag{2.53}$$

thus, in thermal equilibrium ($T_e = T_i$)

$$H_{pl} = \frac{2KT}{M_i g} \tag{2.54}$$

which is essentially twice that of the neutral gas. Thus the ions and electrons diffuse together, a process called ambipolar diffusion.

Substracting (2.51*b*) from (2.51*a*) and neglecting m_e relative to M_i gives the electric field *e*. In thermal equilibrium

$$E = gM_i/2e \tag{2.55}$$

which for an electron-proton gas is of the order of $0{\cdot}05\,\mu\mathrm{V\,m^{-1}}$ and about $0{\cdot}82\,\mu\mathrm{V\,m^{-1}}$ for an electron-O^+ gas.

2.9.3.2 Neutral winds

Under quiet conditions the temperature of the upper atmosphere is greatest near the sub-solar point and is lowest on the nightside. This establishes a pressure difference that drives horizontal neutral winds from the dayside toward the nightside (Fig. 2.21). By neutral-ion collisions the wind motion is transferred to the ions and, hence, by Coulomb attraction to the electrons. In the lower D, E, and F_1 layers the collisions are sufficiently numerous that the plasma motion is determined by the neutral motions. In the F_2 layer, however, the collision frequency is relatively low and the horizontal neutral wind cannot move the ions across the Earth's magnetic field. The ions adhere to the magnetic field like rings on a bar; they can slide along the field easily but cannot move across it. The effects of a horizontal wind on plasma motion are shown in Fig. 2.22. Figure 2.22*a* shows that, for a middle-latitude location, an equatorward wind

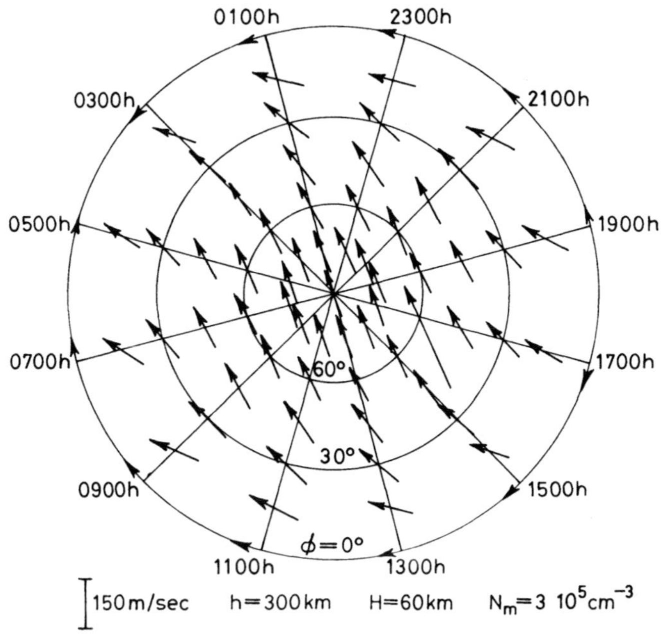

Fig. 2.21 *Atmospheric wind system in the Northern Hemisphere (From Kohl and King 1967 © Pergamon Press Inc.; reprinted with permission)*
Calculated for an altitude of 300 km and for a peak electron density of 3×10^{11} m^{-3}

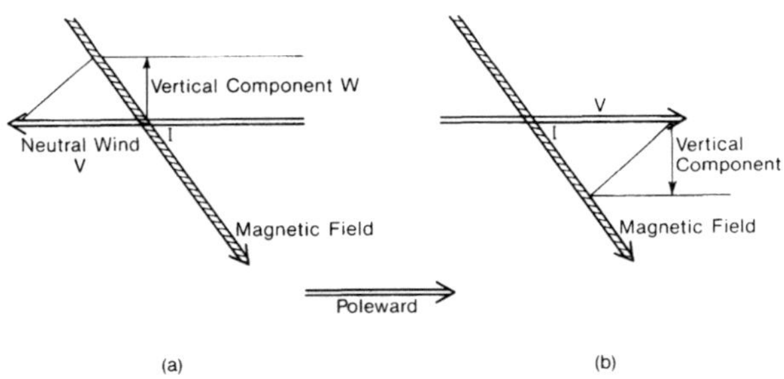

Fig. 2.22 *Plasma motion induced by a horizontal neutral wind*
(*a*) Equatorward wind
(*b*) Poleward wind

lifts plasma, whereas Fig. 2.22*b* shows that a poleward wind lowers plasma. Because the loss rate decreases with height the equatorward wind will increase the electron densities relative to the no-wind densities, whereas the poleward wind decreases the relative electron densities. Since the vertical component *w* of the plasma velocity is

$$w = V \sin I \cos I \tag{2.56}$$

it is seen that there is no vertical motion at the Equator ($I = 0°$) or at the poles ($I = 90°$). For a given wind speed the vertical component maximizes at a dip of 45°.

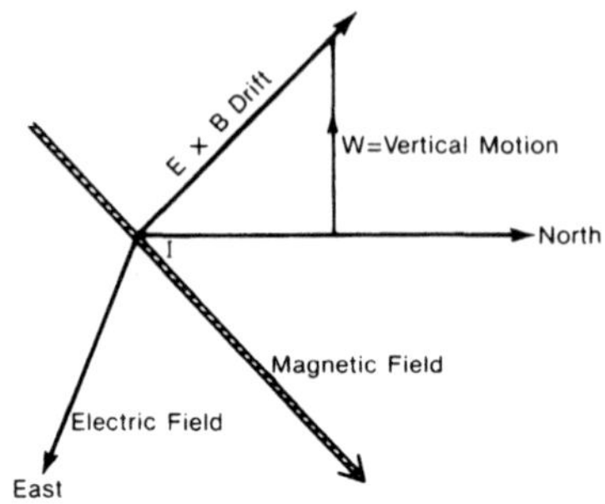

Fig. 2.23 *Electromagnetic drift caused by a west-to-east electric field*

2.9.3.3 Electromagnetic drift

Electromagnetic ($\boldsymbol{E} \times \boldsymbol{B}$) drift is important in the F_2 layer, since it is the main mechanism that can cause plasma to move across magnetic field lines. Figure 2.23 illustrates the vertical drift in a magnetic meridian in the Northern Hemisphere produced by a west-to-east electric field. The magnitude of w is given by

$$w = \frac{E}{B_0} \cos I \tag{2.57}$$

which, for a given electric field strength, is zero at the poles and E/B_0 at the dip equator. Electric fields are particularly important near the dip equator, where vertical motions are important (contrast with diffusion in Sec. 2.9.3.1), and in the auroral zones and polar caps, where horizontal motions are important (see Hall and Barclay 1989, Ch. 5).

2.10 References

BANKS, P. M., and KOCKARTS, G. (1973): 'Aeronomy. A and B' (Academic Press, New York) (2 vols.)

BARRACLOUGH, D. R. (1987): 'International geomagnetic reference field: The fourth generation', *Phys. Earth Planet. Inter.*, **48**, p. 279

CHAMBERLAIN, J. W. (1961): 'Physics of the aurora and airglow' (Academic Press, New York)

CIRA (1972): 'COSPAR International Reference Atmosphere 1972' (Akademie-Verlag, Berlin)

EDDY, J. A. (1979): 'A new sun'. NASA-SP402, National Aeronautics and Space Administration, US Govt. Printing Office, Washington, DC

EDDY, J. A. (1983): 'The Maunder minimum: A reappraisal', *Solar Phys.*, **89**, p. 195

EVANS, J. W. (1972): 'Introductory review of solar activity' *in* McINTOSH, P. S., and DRYER, M. (Eds.): 'Solar activity, observations and predictions' (MIT Press, Cambridge, MA)

HALL, M. P. M., and BARCLAY, L. W. (1989): 'Radiowave propagation' (Peter Peregrinus, Ltd., London)

HARGREAVES, J. K. (1979): 'The upper atmosphere and solar-terrestrial relations' (Van Nostrand Reinhold Co., New York)

HEDIN, A. E. (1987): 'MSIS-86 thermospheric model', *J. Geophys. Res.*, **92**, p. 4649

HESS, W. N. (1968): 'The radiation belt and the magnetosphere' (Blaisdell Publishing Company, Waltham, MA)

HULTQVIST, B. (1967): 'Aurora' *in* MATSUSHITA, S., and CAMPBELL, W. H. (Eds.): 'Physics of geomagnetic phenomena II' (Academic Press, New York) p. 664

JACOBS, J. A. (1970): 'Geomagnetic micropulsations' (Springer-Verlag, Berlin)

KATO, S. (1980): 'Dynamics of the upper atmosphere' (Reidel Publishing Co., Dordrecht, Holland)

KELLEY, M. C. (1989): 'The Earth's ionosphere' (Academic Press, Inc., San Diego, CA)

KOHL, H., and KING, J. W. (1967): 'Atmospheric winds between 100 and 700 km and their effects on the ionosphere', *J. Atmos. Terr. Phys.*, **29**, p. 1045

KUNDU, M. R. (1965): 'Solar radio astronomy' (Interscience Publishers, New York)

LINCOLN, J. V. (1967): 'Geomagnetic indices' *in* MATSUSHITA, S., and CAMPBELL, W. H. (Eds.): 'Physics of geomagnetic phenomena I' (Academic Press, New York) p. 67

LYONS, L. R., and WILLIAMS, D. J. (1984): 'Quantitative aspects of magnetospheric physics' (Reidel Publishing Co., Dordrecht, Holland)

MATSUSHITA, S. (1967*a*): 'Solar quiet and lunar daily variation fields' *in* MATSUSHITA, S., and CAMPBELL, W. H. (Eds.): 'Physics of geomagnetic phenomena I' (Academic Press, New York) p. 302

MATSUSHITA, S. (1967*b*): 'Geomagnetic disturbances and storms' *in* MATSUSHITA, S., and CAMPBELL, W. H. (Eds.): 'Physics of geomagnetic phenomena, II' (Academic Press, New York) pp. 793–819

NOYES, R. W. (1982): 'The Sun' (Harvard University Press, Cambridge, MA)

PARKINSON, W. D. (1983): 'Introduction to geomagnetism' (Scottish Academic Press Ltd., Edinburgh)

PEDDIE, N. W., and ZUNDE, A. K. (1987): 'A model of the geomagnetic secular variation for 1980–83', *Phys. Earth Planet. Inter.*, **48**, p. 324

ROSTOKER, G. (1972): 'Polar magnetic substorms', *Rev. Geophys. Space Phys.*, **10**, p. 157

Solar-Geophysical Data. National Geophysical Data Center, Boulder, CO 80303, issued monthly

Solar-Geophysical Data (1987): Explanation of Data Reports, No. 515 (Supplement), July, National Geophysical Data Center, Boulder, CO 80303

VANZANDT, T. E. (1967): 'The neutral atmosphere and the quiet ionosphere' *in* MATSUSHITA, S., and CAMPBELL, W. H. (Eds.): 'Physics of geomagnetic phenomena I' (Academic Press, New York) p. 509

VESTINE, E. H. (1967): 'Main geomagnetic field' *in* MATSUSHITA, S., and CAMPBELL, W. H. (Eds.): 'Physics of geomagnetic phenomena I' (Academic Press, New York) p. 181.

WILCOX, J. M. (1968): 'The interplanetary magnetic field, solar origin and terrestrial effects', *Space Sci. Rev.*, **8**, p. 258.

WILKES, M. V. (1954): 'A table of Chapman's grazing incidence integral Ch (x, χ)', *Proc. Phys. Soc.*, **67B**, p. 304

Chapter 3

Magnetoionic theory

3.1 The Appleton formula

To interpret radio signals transmitted through, and reflected from, the ionosphere it is necessary to have some understanding of the radio refractive index of the ionosphere. The formula for the refractive index is usually associated with Sir Edward Appleton and is known by several titles such as the Appleton-Hartree formula (Budden 1985) or the Appleton-Lassen formula (Rawer and Suchy 1976); here we follow Ratcliffe (1959) and call it the Appleton formula.

To obtain the refractive index for plane waves in a homogeneous plasma consisting of electrons and heavy positive ions in the presence of a uniform magnetic field $\boldsymbol{B}_0$, and of electron collision frequency ν, we make use of the results of Sec. 1.6, viz., eqns. (1.45) and (1.46). To obtain the constitutive relations of the medium for radio waves of small amplitudes we assume that the collision frequency is independent of electron energy, that the positive ions are infinitely massive and, therefore, do not move under the influence of the sinusoidally varying electric field. The idea of refractive index has meaning only for characteristic waves, or normal modes, that retain their identity as they propagate through a uniform (homogeneous) medium. For our present purposes we assume the following properties:

Wave properties
(i) Simple harmonic progressive waves with small amplitude
(ii) Steady-state solution
(iii) Plane waves with fixed polarization
Properties of the medium
(iv) Electrically neutral
(v) Charges distributed with statistical uniformity, with number density N, so there are no resultant space charges
(vi) Uniform external magnetic field
(vii) Electron collisions independent of electron energy

(viii) Thermal motions of the electrons unimportant; such a medium is called a cold plasma

(ix) Magnetic properties of free space

We consider a plane electromagnetic wave traveling in the x-direction of an orthogonal system of axes (Fig. 3.1) in which the external magnetic field lies in the x–y plane and makes an angle θ with the direction of propagation. The equation of motion of an electron is given by

$$m\ddot{\boldsymbol{r}} = e\boldsymbol{E} + (e\dot{\boldsymbol{r}} \times \boldsymbol{B}) - m\nu\dot{\boldsymbol{r}} \tag{3.1}$$

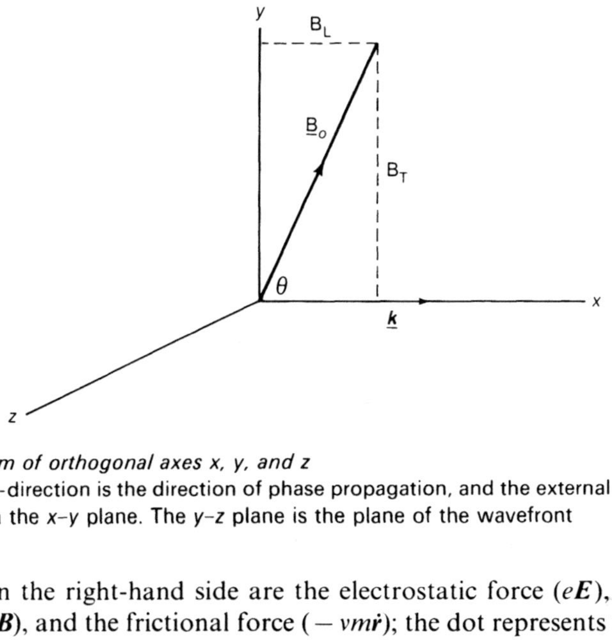

Fig. 3.1 *System of orthogonal axes x, y, and z*
The x-direction is the direction of phase propagation, and the external magnetic field lies in the x–y plane. The y–z plane is the plane of the wavefront

The terms on the right-hand side are the electrostatic force ($e\boldsymbol{E}$), the Lorentz force ($e\dot{\boldsymbol{r}} \times \boldsymbol{B}$), and the frictional force ($-\nu m\dot{\boldsymbol{r}}$); the dot represents the time rate of change, and $\boldsymbol{r}$ is the position vector. Note that the collision frequency ν is based on the assumption that all the ordered momentum, obtained from the wave, is destroyed on collision. It is, therefore, an effective collision frequency and may or may not bear any relationship to a collision frequency defined in terms of gas kinetic theory.

We have, for the x, y, and z components of $\boldsymbol{r}$,

$$m\ddot{x} = eE_x - e\dot{z}B_T - m\nu\dot{x} \tag{3.2a}$$

$$m\ddot{y} = eE_y + e\dot{z}B_L - m\nu\dot{y} \tag{3.2b}$$

$$m\ddot{z} = eE_z + e\dot{x}B_T - e\dot{y}B_L - m\nu\dot{z} \tag{3.2c}$$

We transform these equations as follows:

(i) Multiply throughout by the charge density Ne

(ii) Replace $\dot{x}$ by $i\omega x$, $\ddot{x}$ by $-\omega^2 x$, etc.

(iii) Put $Nex = P_x$ the x-component of the volume polarization; similarly, use P_y and P_z for Ney and Nez, respectively.

With some rearrangement we obtain the constitutive relations:

$$\varepsilon_0 X E_x = -P_x(1 - iZ) + iP_z Y_T \tag{3.3a}$$

$$\varepsilon_0 X E_y = -P_y(1 - iZ) - iP_z Y_L \tag{3.3b}$$

$$\varepsilon_0 X E_z = -P_z(1 - iZ) - iP_z Y_T + iP_y Y_L \tag{3.3c}$$

where $X = Ne^2/\varepsilon_0 m\omega^2$, $Y_L = eB_L/m\omega$, $Y_T = eB_T/m\omega$, and $Z = \nu/\omega$.

The subscripts T and L refer to the transverse and longitudinal components of the imposed magnetic field with reference to the direction of the wave normal (i.e. phase propagation).

Using (1.42*a*) and (3.3*a*) we obtain

$$\frac{P_x}{P_z} = \frac{x}{z} = \frac{iY_T}{1 - X - iZ} \tag{3.4}$$

in which x and z are components of the vector $\boldsymbol{r}$. From (1.45), $P_y E_z = P_z E_y$, so with (3.3*b*) and (3.3*c*) we have, since $R = P_z/P_y$,

$$R^2 - \frac{i(Y_T^2/Y_L)R}{1 - X - iZ} + 1 = 0 \tag{3.5}$$

There are, in general, two solutions of this quadratic, which are given by

$$R = \frac{i}{2Y_L}\left\{\frac{Y_T^2}{(1 - X - iZ)} \mp \left[\frac{Y_T^4}{(1 - X - iZ)^2} + 4Y_L^2\right]^{1/2}\right\} \tag{3.6}$$

For the propagation of high-frequency radio waves through the E and F regions of the ionosphere, Z is usually very small and can be neglected. For instance, with a frequency of 10 MHz at the 100 km level ($\nu = 1{\cdot}75 \times 10^3$) we have $Z = 3 \times 10^{-4}$. Under these conditions R is given by

$$R = \frac{i}{2Y_L}\left\{\frac{Y_T^2}{1 - X} \mp \left[\frac{Y_T^4}{(1 - X)^2} + 4Y_L^2\right]^{1/2}\right\} \tag{3.7}$$

This formula tells us that in general there are two, and only two, characteristic waves that can propagate in this particular medium.

From (1.46*b*), (3.3*b*), and (3.6) we obtain for the Appleton formula

$$n^2 = (\mu - i\chi)^2$$

$$= 1 - \frac{X}{1 - iZ - \dfrac{Y_T^2}{2(1 - X - iZ)} \pm \left[\dfrac{Y_T^4}{4(1 - X - iZ)^2} + Y_L^2\right]^{1/2}} \tag{3.8}$$

When collisions are negligible (e.g. in the E and F regions) $Z \approx 0$ and

$$\mu^2 = 1 - \frac{2X(1 - X)}{2(1 - X) - Y_T^2 \pm [Y_T^4 + 4(1 - X)^2 Y_L^2]^{1/2}} \tag{3.9}$$

When the magnetic field is negligible (i.e. $Y \ll 1$)

$$n^2 = (\mu - i\chi)^2 = 1 - \frac{X}{1 - iZ} \tag{3.10a}$$

$$\text{When } \nu^2 \ll \omega^2,\ \kappa \approx \omega\chi/c \approx \frac{\omega_N^2 \nu}{2\mu\omega^2 c} \tag{3.10b}$$

$$\text{When } \nu^2 \gg \omega^2,\ \kappa \approx \frac{\omega_N^2}{2\mu\nu c} \tag{3.10c}$$

When both collisions and magnetic field effects are negligible

$$\mu^2 = 1 - X = 1 - \left(\frac{f_N}{f}\right)^2 = 1 - k\frac{N}{f^2} \tag{3.11}$$

where $k = 80{\cdot}5$, N is in electrons per cubic meter, and f is in hertz.

Alternative forms of this equation, including effects of heavy ions, are given in books on plasma waves (e.g. Stix 1962, Allis *et al.* 1963, and Ginzburg 1970).

3.2 Some properties of the Appleton formula

For the sake of discussion we start with (3.11). First, we see that the square of the refractive index is less than unity and, since μ must be real for the wave to propagate, μ must lie between 0 and 1. Secondly, we see that, for a given frequency, the refractive index decreases with increase of electron density and that, for a given electron density, the refractive index decreases with decrease of frequency. For small amplitudes the refractive index is independent of the wave amplitude. Furthermore, the refractive index is independent of the sign of the ion charge and varies inversely as the ion mass, which is why, for high frequencies, heavy ions can usually be neglected when compared with electrons.

Consider a radio wave incident at an angle ϕ_0 on a plane refracting layer. As the wave penetrates into the layer the electron density increases and the wave normal (the direction of the $\boldsymbol{k}$ vector) changes according to Snell's law:

$$\mu \sin \phi = \mu_0 \sin \phi_0 \tag{3.12}$$

where ϕ is the angle between the wave normal and the perpendicular to the surfaces of constant refractive index μ, and μ_0 and ϕ_0 are the corresponding values at the base of the layer; that is, $\mu_0 = 1$. At reflection, $\phi = 90°$ so

$$\mu_r = \sin \phi_0 \tag{3.13}$$

Hence, with vertical propagation for which $\phi_0 = 0$ we have

$$\mu_r = 0 \tag{3.14}$$

Thus, from (3.11) and (3.14) vertical reflection occurs when the plasma frequency f_N equals the wave frequency f. Hence, when the wave frequency exceeds the maximum plasma frequency (or critical frequency) of the layer, the wave will penetrate the layer.

To find the reflection conditions for vertical propagation when the external magnetic field is present, we set $\mu = 0$ in (3.9). The positive sign gives

$$X_r = 1 \tag{3.15}$$

and the negative sign gives

$$X_r = 1 - Y \tag{3.16a}$$

or

$$X_r = 1 + Y \tag{3.16b}$$

The wave with the positive sign in (3.9) is usually called the 'ordinary' wave since it is reflected at the same height as it would be in the absence of the magnetic field. The waves with the negative sign in (3.9) are called extraordinary waves. Equations (3.16*a*) and (3.16*b*) show that their height of reflection of the extraordinary wave is independent of the angle θ and, hence, of the magnetic dip. Note that, in general, on frequencies greater than the electron gyrofrequency ($Y < 1$) the extraordinary wave does not normally penetrate to the $X = 1 + Y$ level. Ambiguities that arise in the association of these terms with the respective signs are discussed by Ratcliffe (1959).

Two particularly interesting cases are those in which the propagation is (*a*) parallel to the magnetic field ($\theta = 0°$) and (*b*) perpendicular to the magnetic field ($\theta = 90°$).

3.2.1 With parallel propagation

With parallel (or longitudinal) propagation (e.g. vertical propagation at a magnetic pole) we have

$$\mu^2 = 1 - \frac{X}{1 \pm Y} \tag{3.17}$$

where the plus sign refers to the ordinary wave and the minus sign refers to the extraordinary wave. Thus, reflection with vertical propagation occurs where

$$X_r = 1 \pm Y \tag{3.18}$$

This implies that there is no reflection of the ordinary wave at the $X = 1$ level. Actually there is a reflection level at $X = 1$ (given by an infinitesimal value of θ), and the reflection criteria given by (3.18) both refer to extraordinary waves.

For waves originating at the Earth's surface in high latitudes, it is possible for energy to be 'coupled' from the ordinary wave (at $X = 1$) into the extraordinary wave and, in this way, to reach the $X = 1 + Y$ level. This composite signal is called the z-wave.

3.2.2 With perpendicular propagation

With perpendicular, or transverse, propagation ($\theta = 90°$) we have for the ordinary wave

$$\mu^2 = 1 - X \tag{3.19a}$$

and for the extraordinary wave

$$\mu^2 = 1 - \frac{X(1 - X)}{1 - X - Y^2} \tag{3.19b}$$

Thus, *for this case only*, the propagation of the ordinary wave is independent of the magnetic field whereas the extraordinary-wave refractive index depends on the field. With vertical propagation the reflection conditions are $X_{ro} = 1$ and $X_{rx} = 1 \pm Y$. On frequencies less than the electron gyrofrequency $Y > 1$ and, in this case, the extraordinary wave is reflected at the level $X_{rx} = 1 + Y$ only.

Curves depicting the variations of μ^2 (or μ) with X are called *dispersion* curves and have been discussed extensively in the literature (e.g. Ratcliffe 1959, Davies 1969, Budden 1985). Dispersion curves for $Y = \frac{1}{2}$ and $Y = 2$ are shown in Fig. 3.2 for $\theta \approx 45°$, together with the limiting cases of longitudinal ($\theta = 0$) and transverse ($\theta = 90°$) propagation. Note that the curves are not drawn to scale but are intended to illustrate the general shapes. A few general features should be noted. With $\theta = 0°$ the refractive index of the waves changes rapidly with X, near $X = 1$. There are branches of the curves that do not start at $X = 0$ and, therefore, may not normally apply to waves emitted from a transmitter on the ground. Also the refractive index can exceed unity. With $Y = \frac{1}{2}$, the right-hand branch of the curve for the extraordinary wave goes to infinity at

$$X = \frac{1 - Y^2}{1 - Y_L^2} \tag{3.20}$$

In plasma physics this is called a resonance. As the level of resonance is approached, both the phase and group speeds approach zero. This level has been called the fourth reflection condition but it is not a real reflection condition (e.g. see Budden 1985 for further discussion of this subject). For all values of θ, the ordinary-wave curves pass through $\mu = 0$ at $X = 1$ and through $\mu = 1$ at $X = 0$. Likewise, for the extraordinary wave, with $Y < 1$ the curves pass through $\mu = 0$ at $X = 1 \mp Y$ for all θ. With $Y > 1$ the extraordinary-wave branch with the minus sign disappears and the extraordinary-wave branch with the + sign passes through $\mu = 0$ at $X = 1 + Y$ for all θ. Furthermore, $\mu^2 > 1$ for $0 < X < 1$.

For certain ranges of X, μ^2 is less than zero and the refractive index is

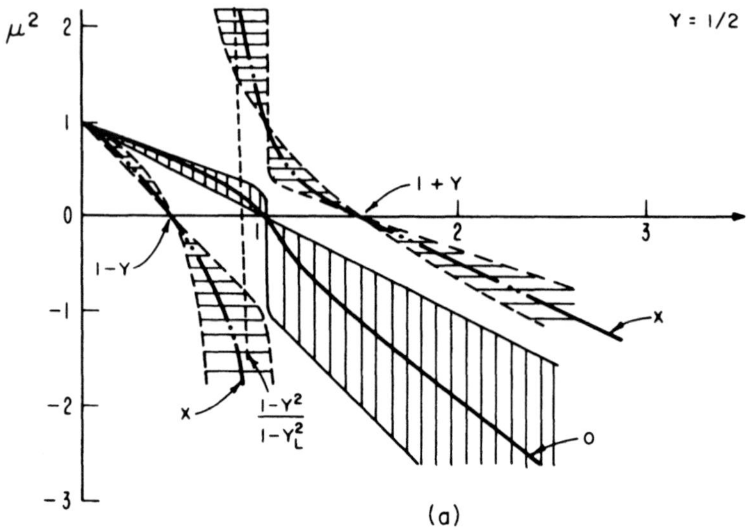

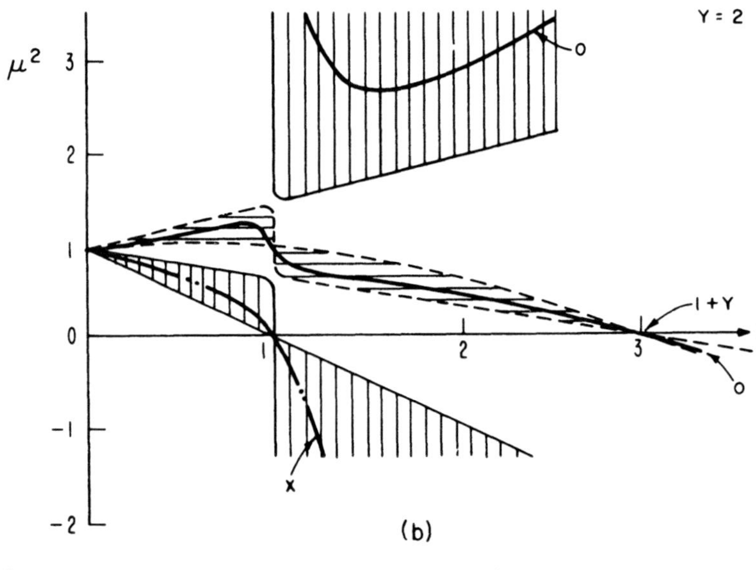

Fig. 3.2 *Dispersion curves for an angle of propagation of about 45° (with no collisions)*
The vertical hatching represents the range for the + sign, and the horizontal hatching gives the range for the − sign
o = ordinary wave
x = extraordinary wave

imaginary — at heights above the height of vertical reflection — and the wave is evanescent. In an evanescent wave the phase is the same at all heights h, the amplitude decays exponentially at the rate $\exp(-|\mu|\omega h/c)$ where $|\mu|$ is the modulus of the imaginary refractive index, and the E and H fields are 90° out of phase. When a wave is turned on, some energy from the wave goes into the establishment of the evanescent wave, and when the wave is turned off energy is restored from the evanescent wave to the reflected wave.

3.3 Approximate formulas

For purposes of discussion it is often convenient to use approximations to the Appleton equation that usually involve the relative magnitudes of the terms under the square root. These approximations are called quasi-transverse (QT) and quasi-longitudinal (QL). The usual rules for determining which approximation (if either) holds are

$$\text{QT: } Y_T^4 \gg 4(1 - X)^2 Y_L^2 \tag{3.21a}$$

$$\text{QL: } Y_T^4 \ll 4(1 - X)^2 Y_L^2 \tag{3.21b}$$

These approximations have been discussed extensively (including collisions) by Budden (1985), Kelso (1964), Ratcliffe (1959), and Davies (1969).

It is important to note that these conditions do not depend solely on the value of θ. Thus the quasi-transverse approximation holds both for $\theta \approx 90°$ with nearly all values of X and for the ordinary wave with $X \approx 1$, when $Y \ll 1$, at nearly all values of θ. Similarly, the QL approximation is good for the extraordinary wave near $X = 1 \pm Y$ for a large range of θ. Near the level of vertical reflection of the *o* wave the following is a good approximation:

$$\frac{X}{1 - \mu^2} = 1 + (1 - X)\cot^2\theta \tag{3.22}$$

which reduces to (3.19*a*) at $X = 1$ and $\theta = 90°$. On very high frequencies the QL approximation is particularly useful, viz.

$$\mu^2 = 1 - \frac{X}{1 \pm Y_L} \tag{3.23}$$

Care should be exercised in the use of these formulas because the approximations may depend on how the values of μ^2 approach the transverse and longitudinal values (Davies 1969, Sec. 5.5). They should not be used near reflection levels with vertical propagation.

3.4 Polarization

3.4.1 Polarization equations

The wave polarization R (eqn. 1.45) gives the ratios of the sides of the rectangle, in the y–z plane, enclosing the ellipses traced out by the tips of the projections of the electric and magnetic fields. In the case where collisions are negligible, the polarizations are obtained from (3.7):

$$R_{\mp} = \frac{i}{2Y_L}\left\{\frac{Y_T^2}{1-X} \mp \left[\frac{Y_T^4}{(1-X)^2} + 4Y_L^2\right]^{1/2}\right\} \tag{3.24a}$$

$$R_{\mp} = \frac{i}{2Y_L(1-X)}\{Y_T^2 \mp [Y_T^4 + 4(1-X)^2 Y_L^2]^{1/2}\} \tag{3.24b}$$

$$R_{\mp} = \frac{i}{Y_L}\left(1 - \frac{X}{1-\mu_\pm^2}\right) \tag{3.24c}$$

From (3.24) it follows that

$$R_+ R_- = 1 \tag{3.25}$$

where the $-$ and the $+$ signs go respectively with the $+$ and $-$ signs in the Appleton equation.

3.4.2 Polarization ellipses

From (1.45) and (3.25) we deduce the following;

(i) The ellipses of the ***D***, ***E***, ***P*** vectors, projected onto the y–z plane, are similar and rotate in the same sense. Note that the D ellipse lies in the y–z plane whereas the ellipses of the ***E*** and ***P*** vectors are, in general, tilted in the x-direction (Fig. 3.3). For the ordinary wave, the major axis lies in the y-direction, and for the extraordinary wave the major axis lies in the z-direction.
(ii) The ***H*** and ***B*** ellipses lie in the y–z plane and are similar to the ***D*** ellipse, but are rotated through 90°, and the sense of rotation is the same.
(iii) The sense of rotation of the extraordinary wave is opposite to that of the ordinary wave (Fig. 3.3*a*), and the major axes of the ordinary and extraordinary ellipses are perpendicular.

The angle ϕ, between the y axis and the plane of the ***P*** ellipse (Fig. 3.3*c*), is obtained from (1.45) and (3.4):

$$\tan\phi = \frac{P_x}{P_y} = \frac{x}{y} = i\,\frac{Y_T R}{1-X} \tag{3.26a}$$

or

$$\cot\phi = \frac{Y_T}{2Y_L} \pm \{(Y_T/2Y_L)^2 + [(1-X)/Y_T]^2\}^{1/2} \tag{3.26b}$$

where $+$ and $-$ signs refer respectively to the ordinary and extraordinary waves. Since ϕ is real, the particle motion in the x–y plane is linear.

The forward tilt ψ of the ***E*** vector is given by

$$\tan\psi = \frac{E_x}{E_y} = \frac{x}{y}(1-\mu_\pm^2) = i\,\frac{(1-\mu_\pm^2)}{1-X}RY_T \tag{3.27}$$

where x and y are the components of the displacement ***r*** of the electrons from their equilibrium positions (Fig. 3.3*d*).

From (3.4), in the absence of collisions and for propagation at an angle θ, we have the electron motion in the x–z plane:

$$\frac{P_x}{P_z} = \frac{x}{z} = \frac{iY_T}{1 - X} \tag{3.28}$$

Thus, the $\boldsymbol{P}$ and $\boldsymbol{r}$ ellipses in the x–z plane are similar for both ordinary and extraordinary waves as sketched in Fig. 3.3*b*. On the other hand, the projection of the $\boldsymbol{E}$ ellipse is given by

$$\frac{E_x}{E_z} = \frac{E_x}{E_y}\frac{E_y}{E_z} = \frac{1}{R}\tan\psi = i\frac{(1 - \mu_\mp^2)Y_T}{1 - X} \tag{3.29}$$

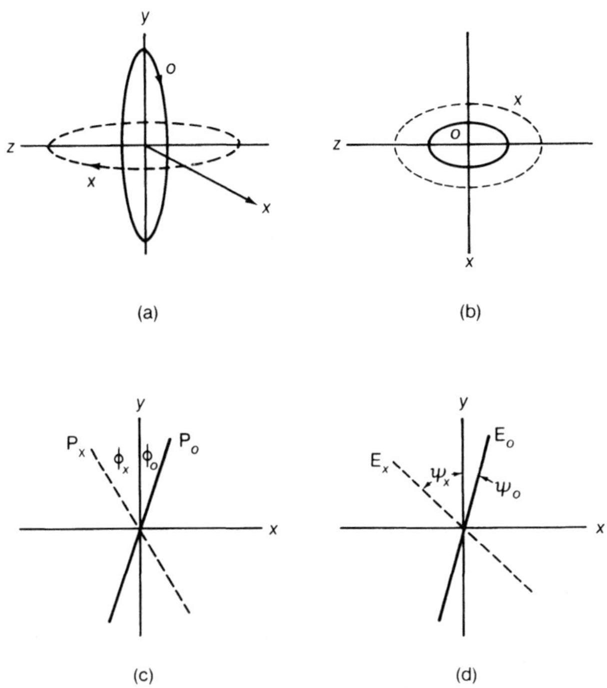

Fig. 3.3 *Projections onto the y–z, x–z, and x–y planes of electron motions and electric fields* Note that the polarizations P are proportional to the electron displacements and they are tilted out of the wavefront as are the electric fields

which shows that the polarization and displacement ellipses of the *o*-wave differ from the corresponding ellipses of the *x*-wave.

It is seen that the $\boldsymbol{P}$ (or $\boldsymbol{r}$) ellipse and the $\boldsymbol{E}$ ellipse are, in general, tilted forward at different angles to the y-direction, although the projections onto the y–z plane are similar. Furthermore, the forward tilts of the ordinary and extraordinary ellipses are different. From (3.26*b*) it is seen that the signs of ϕ for the two characteristic waves are opposite, since the square root term is larger than $Y_T/2Y_L$. Hence, if the ellipse of the *o*-wave is tilted forward, the ellipse of the *x*-wave is tilted backward (Fig. 3.3*c*).

From (3.27) and (3.26) we have

$$\tan \psi_{\pm} = (1 - \mu_{\pm}^2) \tan \phi_{\pm} \tag{3.30}$$

A useful rule for remembering the senses of rotation of the vectors of the characteristic waves is as follows: When the thumb points in the direction of the magnetic field $\boldsymbol{B}_0$, the rotation of the extraordinary-wave vectors is given by the fingers of the right hand; the rotation of the ordinary-wave vectors is given by the fingers of the left hand.

It will be seen from (3.24) that, at a given point in the plasma, the polarization depends only on the conditions at the point and does not depend on the manner in which the wave reached the point. In this respect, polarization is analogous to refraction as given by Snell's law.

3.4.3 Polarization with vertical propagation

Of particular interest are the polarizations at reflection with vertical propagation:

For the *o*-wave $\mu_+ = 0$, $X = 1$, and $R_- = 0$; thus $z = 0$ and the electron motion is in the x–y plane. From (3.24*b*) and (3.26*a*) we have

$$\tan \phi = \frac{x}{y} = \frac{Y_T}{2Y_L(1 - X)^2} \{Y_T^2 - [Y_T^4 + 4Y_L^2(1 - X)^2]^{1/2}\} \tag{3.31}$$

Expanding the square root term and taking the first two terms we have $\tan \phi = \cot \theta$, which implies that the electron motion is parallel to the imposed magnetic field. Also, since $\mu_+ = 0$, $k = 0$ and, from (1.44), $H = 0$ and therefore the oscillation is electric. From (3.30) it follows that $\psi = \phi$ and hence the electric field is parallel to the magnetic field as shown in Fig. 3.4*a*.

For the extraordinary wave $\mu_- = 0$ and $X = 1 \mp Y$; hence, from (3.24*c*), $R_+ = \pm i \sec \theta$, and $\phi = -\theta$, which tells us that the projections of the electron motion onto the x–y and y–z planes are ellipses and that the projection onto the x–y plane is a straight line perpendicular to the magnetic field (Fig. 3.4*b*). Thus, the true electron motion is a circle perpendicular to the magnetic field. Again, since $\mu = 0$, $k = 0$, it follows from (1.44) that the oscillations are electric.

3.4.4 Transverse and longitudinal propagation

Consider, further, the cases of propagation transverse to and parallel to the magnetic field. With transverse propagation, $\theta = 90°$, $Y_T = Y$, and $Y_L = 0$, and for the ordinary wave $R_- = z/y = 0$. From (3.4) with $Z = 0$ we have

$$\frac{x}{z} = \frac{iY}{1 - X} \tag{3.32}$$

Now y is finite since it represents an electron displacement and, since $z/y = 0$, for $X \neq 1$ it follows that $z = 0$ and $x = 0$. This means that, for transverse propagation of the ordinary wave, the electron motion is in the y-direction which is parallel to the magnetic field. Hence, a linear antenna oriented in the north–south direction at the magnetic equator produces an ordinary wave only.

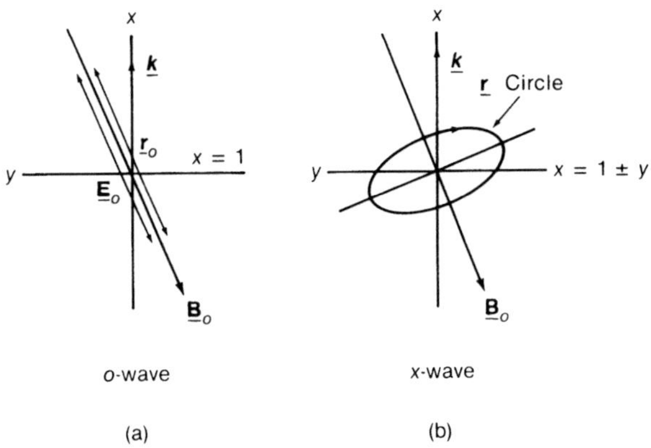

Fig. 3.4 *Electron motions and electric fields at the levels of reflection with vertical propagation*
(*a*) The ordinary-wave motions are parallel to the magnetic field
(*b*) The motions caused by the extraordinary wave are circles in a plane perpendicular to the magnetic field

With the extraordinary wave we have $R_+ = x/y \to \infty$, which tells us that there is no component of motion in the y-direction. Together with (3.30) this implies that the electron motions produced by the extraordinary wave are ellipses lying in the x–z plane (i.e. the vertical, or equatorial, plane).

With longitudinal (parallel) propagation, $\theta = 0$ and $R_{\mp} = \mp i$; also, from (3.4), $x = 0$. It follows that the two characteristic waves are circularly polarized in the plane of the wavefront. Since a linear oscillation can be decomposed into two circular motions of equal amplitudes and opposite senses of rotation, it follows that, at the magnetic poles, a linear antenna on the ground will excite equal powers in the ordinary and extraordinary waves.

3.4.5 Effect of collisions

The introduction of collisions causes rotations of the axes of the various ellipses and changes their shapes. The effects are illustrated in Fig. 3.5. The motion in the x–y plane is no longer linear, but elliptical. The relationship $R_+ R_- = 1$ is valid in the presence of collisions so that the two (o and x) ellipses are still similar. All the ellipses in the y–z plane belonging to the ordinary wave are rotated in one direction and all those belonging to the extraordinary wave are rotated in the opposite direction through the same angle. This fact enables us to deduce the ellipse for one characteristic wave from that of the other by reflecting it in a line making an angle of 45° with the magnetic meridian as shown in Fig. 3.5*a*. For the projection of the motion onto the x–z plane, (3.4) shows that both o- and x-ellipses are rotated about the y axis by the same angle and in the same sense.

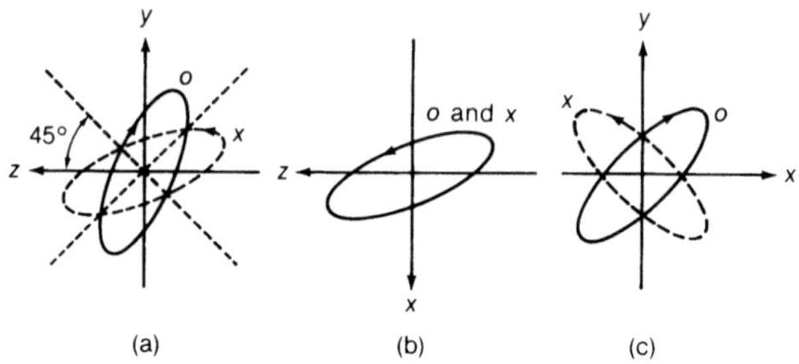

Fig. 3.5 *Rotations of the axes of the polarization ellipses caused by collisions*
(*a*) In the wavefront
(*b*) In the x–z plane
(*c*) In the magnetic-meridian plane

In the case of purely longitudinal propagation ($Y_T = 0$) the polarization is circular. For purely transverse propagation of the ordinary wave $R_- = 0$ and the E-vector is parallel to the magnetic field whereas for transverse propagation of the extraordinary wave (in the x–z plane), the ellipse of electron motion is rotated about the y-direction (i.e., about the magnetic field). The effects of collisions on the $\boldsymbol{E}$ ellipses are more complicated: the rotations about the z-direction, of the o- and x-ellipses, are in opposite senses and the ellipses *fatten* with increase of collision frequency. For a discussion of the effects of collisions on polarization the reader is referred to Ratcliffe (1959) who points out that characteristic waves emerging from the ionosphere in the Northern Hemisphere should all have their major axes in the NE–SW pair of quadrants.

3.4.6 Limiting polarization

When a radio wave enters the ionosphere it is split up into ordinary and extraordinary waves. The extent to which the power in the incident wave is divided between the two characteristic waves is quite complicated and has to be tackled using full-wave theory (Ratcliffe 1959, Budden 1985). The energy distribution depends on the relationship between the polarization of the incident wave and the limiting polarization R_l of the characteristic waves at the bottom of the ionosphere where $X \approx 0$, and hence

$$R_{l\mp} \approx \frac{i}{2Y_L}[Y_T^2 \mp (Y_T^4 + 4Y_L^2)^{1/2}] \tag{3.33}$$

3.5 Coupling

In the foregoing discussions it has been assumed that the ordinary and extraordinary waves propagate independently of each other and that there is no conversion of one wave into another. This state usually holds because the wave characteristics are quite different. Under certain conditions (e.g. around $X = 1$; see Fig. 3.2) the refractive indices, and hence the polarizations of different waves, become almost equal. Under these circumstances an ordinary wave can excite an extraordinary wave and vice versa. Coupling occurs when the polarization changes rapidly with distance as, for example, when a wave passes through the transverse condition ($\theta = 90°$) (Cohen 1964). The coupling factor is given by

$$\Psi = \frac{1}{R^2 - 1}\frac{dR}{dz} \tag{3.34}$$

Substitution for R yields (for fixed θ)

$$\Psi = -\frac{i}{4}\frac{Y_T^2 Y_L}{(1 - X - iZ)^2 Y_L^2 + Y_T^2/4}(X' - iZ') \tag{3.35}$$

where X' and Z' are the spatial gradients of X and Z respectively. Equation (3.35) shows that coupling can be enhanced when there are large gradients of electron density (X') and of collision frequency (Z'). Alternatively, coupling occurs when X and Z are small, but θ changes rapidly, under the QT approximation, when the coupling factor is given by

$$\Psi \approx -\frac{i}{Y}\frac{d\theta}{dz} \tag{3.36}$$

Note that there is essentially no coupling for longitudinal propagation because then the polarizations are different.

3.6 Ray direction in a magnetoplasma

In Sec. 1.5.3 we saw that, in an anisotropic medium, the direction of energy flow $\boldsymbol{S}$ differs from the direction of wave propagation $\boldsymbol{k}$. Now the Appleton formula shows that μ^2, and hence k, depends on the angle θ between the direction of propagation and the imposed magnetic field $\boldsymbol{B}_0$, and therefore the medium is anisotropic. Neglecting collisions, we have from (1.12) and (3.9)

$$\tan\alpha = \frac{1}{2\mu^2}\frac{d\mu^2}{d\theta} = \pm\frac{(\mu^2 - 1)Y_T Y_L}{[Y_T^4 + 4(1 - X)^2 Y_L^2]^{1/2}} \tag{3.37}$$

The $+$ and $-$ signs refer to the corresponding signs in the Appleton equation. Note that α is of opposite sign for the ordinary and extraordinary waves except when the refractive index is greater than unity for one of the characteristic waves and less than unity for the other. Furthermore, the sign of tan α changes as θ goes through 90°. We use the following sign convention for α: α is positive when it is in the same sense as θ, the angle measured from the direction of propagation to the imposed magnetic field. That is, α is positive when measured counter-clockwise in our system of Cartesian coordinates.

Bremmer (1949) summarized the ray behaviour as follows:

(*a*) The wave normal $\boldsymbol{k}$ is parallel to the plane of incidence and satisfies Snell's law.
(*b*) The ray $\boldsymbol{S}$ is situated in the plane through the wave normal and the magnetic field.
(*c*) The angle α measured from the wave normal to the ray is given by (3.37).

By use of these rules, rays have been traced in the ionosphere (e.g. Lawrence and Posakony 1962).

3.7 Effects of heavy ions

In the foregoing discussions the effects of heavy ions have been neglected on account of the large masses. Ratcliffe (1959, Sec. 16.2) pointed out that 'coupling' of the electron and ion motions occurs in the x-direction of propagation. When there is no motion of particles in the x-direction the electrons and ions make separate and independent contributions to n^2 (Ginzburg 1970).

3.8 Generalized magnetoionic theory

3.8.1 Nature of electron collisions

In the derivation of the Appleton formula (Sec. 3.1) we assumed that the electron collision frequency is not dependent on the electron energy and therefore the distribution of electron speeds can be neglected. For medium frequencies in the D and lower E regions, and when the collision frequency is comparable with the angular wave frequency, that assumption is no longer valid. Our simple picture of collisions in a gas is based on elastic encounters between spherical bodies. In a plasma, this picture has to be expanded to consider the effect of a (charged) electron on a neutral molecule and the interaction of an electron with a positive and/or negative ion.

Laboratory measurements have shown that for slow electrons in nitrogen gas, the collision frequency ν for momentum transfer (i.e. friction) is approximately proportional to electron energy; that is,

$$\nu = \nu_m \varepsilon \tag{3.38}$$

Here $\varepsilon = mv^2/2KT$ is the ratio of electron kinetic energy to the thermal kinetic energy of the gas, and ν_m is the collision frequency when $\varepsilon = 1$.

3.8.2 Generalized complex refractive index

In thermal equilibrium it is assumed that the distribution of electron speeds is Maxwellian. However, when a radio wave traverses the plasma, the electric field accelerates the electrons such that the interaction between the wave and the medium is nonlinear. The refractive index can be obtained by solving the Boltzmann transfer equation (e.g. Sen and Wyller 1960, Ginzburg 1970). In the case of longitudinal propagation ($\theta = 0$) (see Sec. 3.3) the complex refractive index is given by

$$n^2 = (\mu - i\chi)^2 = 1 - \frac{X(1 \pm Y)}{Z_m^2}\mathscr{C}_{3/2}\left(\frac{1 \pm Y}{Z_m}\right) - i\frac{5X}{2\mu Z_m}\mathscr{C}_{5/2}\left(\frac{1 \pm Y}{Z_m}\right) \tag{3.39}$$

where $Z_m = \nu_m/\omega$. The + and − signs refer respectively to the ordinary and extraordinary waves, and the functions $\mathscr{C}_{3/2}(x)$ and $\mathscr{C}_{5/2}(x)$ are the so-called semiconductor integrals given by

$$\mathscr{C}_p(x) = \frac{1}{p!}\int_0^\infty \frac{\varepsilon^p \exp(-\varepsilon)\, d\varepsilon}{\varepsilon^2 + x^2} \tag{3.40}$$

Values of $\mathscr{C}_{3/2}(x)$ and $\mathscr{C}_{5/2}(x)$ have been tabulated by Burke and Hara (1963) and by Davies (1969). When $\mu^2 \gg \chi^2$ we have

$$\mu^2 = 1 - \frac{X(1 \pm Y)}{Z_m^2}\mathscr{C}_{3/2}\left(\frac{1 \pm Y}{Z_m}\right) \tag{3.41}$$

and

$$\chi = \frac{c\kappa}{\omega} = \frac{5X}{4\mu Z_m}\mathscr{C}_{5/2}\left(\frac{1 \pm Y}{Z_m}\right) \tag{3.42}$$

In the absence of a magnetic field, and for transverse propagation with the electric field parallel to the magnetic field, the refraction and absorption indices are given by putting $Y = 0$ in (3.41) and (3.42).

In the case of the ordinary wave with quasi-longitudinal propagation, with small θ, Y is replaced by $Y_L = |Y \cos\theta|$. In the case of the extraordinary wave on frequencies between 2 MHz and 4 MHz it may be more accurate to use (3.42) than to replace Y by Y_L (Benson 1964).

3.8.3 Limiting conditions

Case 1: When the wave frequency ω is large compared with the collision frequency, making $x \gg 1$,

$$\mathscr{C}_{3/2}(x) \approx 1/x^2$$

and

$$\mathscr{C}_{5/2}(x) \approx 1/x^2$$

Under these conditions (3.41) gives $\mu^2 = 1 - (\omega_N/\omega)^2$, which is the same as the classical equation (3.11). This is seen in Fig. 3.6. For the absorption index,

$$\chi = \frac{c\kappa}{\omega} = \frac{5\omega_N^2 \nu_m}{4\mu\omega^2} = \frac{\omega_N^2 \bar{\nu}}{2\mu\omega^2} \tag{3.43}$$

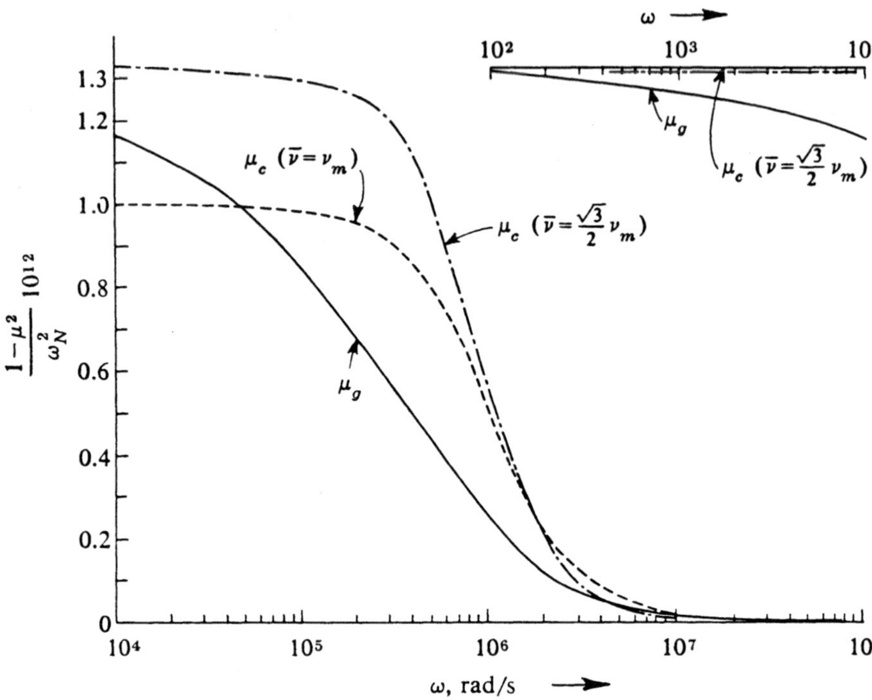

Fig. 3.6 *Comparison of the angular frequency variations of the quantity $(1 - \mu^2)/\omega_N^2$ (From Davies 1969 © Blaisdell Publishing Co.; reprinted with permission)*
μ^2 is obtained from the generalized formula (3.41) with $Y = 0$, and from the classical equation $\mu^2 = 1 - \omega_N^2/(\omega^2 + \bar{\nu}^2)$ [see (3.10*a*)]

This is essentially the same as the classical formula (3.10*b*) with $\nu^2 \ll \omega^2$), provided that we replace ν by

$$\bar{\nu} = 2{\cdot}5\nu_m \tag{3.44}$$

as seen in Fig. 3.7.

Case 2: When the collision frequency is large compared with the wave frequency and thus $\nu_m \gg \omega$, we have $x \approx 0$ and

$$\mathscr{C}_{3/2} = 4/3$$

$$\mathscr{C}_{5/2} = 4/15$$

Thus

$$\mu^2 = 1 - \frac{4\omega_N^2}{3\nu_m^2} = 1 - \frac{\omega_N^2}{\bar{\nu}^2} \tag{3.45}$$

Again, the classical formula can be used, provided that ν is replaced by

$$\bar{\nu} = \sqrt{3}(\nu_m/2) \tag{3.46}$$

as shown in Fig. 3.6.

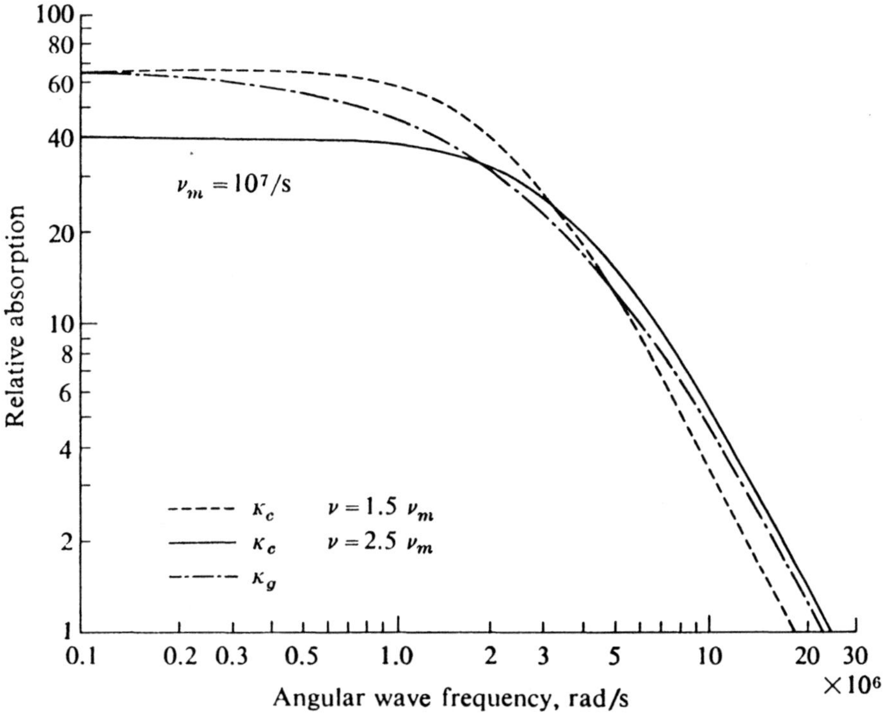

Fig. 3.7 *Comparisons of the variations with frequency of the absorption coefficients (From Davies 1969 © Blaisdell Publishing Co.; reprinted with permission)*

$$\kappa_g = \frac{5}{4}\frac{\omega_N^2}{c\nu_m}\mathscr{C}_{5/2}\left(\frac{\omega}{\nu_m}\right), \text{ and } \kappa_c = \frac{\omega_N^2}{2c}\frac{\bar{\nu}}{\omega^2 + \bar{\nu}^2}$$

For the absorption coefficient

$$\kappa = \frac{\omega_N^2}{3\mu\omega\nu_m} = \frac{\omega_N^2}{2\mu\omega\bar{\nu}} \tag{3.47}$$

Therefore, we can use the classical formula (3.10*c*), provided we replace ν by

$$\bar{\nu} = 3\nu_m/2 \tag{3.48}$$

3.8.4 Frequency dependence of absorption

In the D region the refraction is usually small, so the more important effect of the generalized theory is on absorption. The frequency dependence of absorption is shown in Fig. 3.7. On high frequencies, the classical formula is good, provided that $\bar{\nu} = 2{\cdot}5\nu_m$; on low frequencies it is necessary to use $\bar{\nu} = 1{\cdot}5\nu_m$.

The generalized theory is important in the detailed quantitative interpretation of certain experiments but for most practical radio propagation problems the classical theory suffices, especially when appropriate values of the effective collision frequencies are used.

3.9 References

ALLIS, W. P., BUCHSBAUM, S. J., and BERS, A. (1963): 'Waves in anisotropic plasmas' (MIT Press, Cambridge, MA.)

BENSON, R. F. (1964): 'The quasilongitudinal approximation to the generalized theory of radio wave absorption', *Radio Sci.*, **68D**, p. 191

BREMMER, H. (1949): 'Terrestrial radio waves' (Elsevier Publishing Company, Inc., New York)

BUDDEN, K. G. (1985): 'The propagation of radio waves' (Cambridge University Press, Cambridge)

BURKE, M. J., and HARA, E. H. (1963): 'Tables of the semiconductor integrals $\mathscr{C}_p(x)$ and their approximations for use with the generalized Appleton–Hartree magneto-ionic formulas'. DRTE Report No. 1113, Ottawa

COHEN, M. R. (1964): 'Magnetoionic calculations near the transverse region', *J. Geophys. Res.*, **69**, p. 1935

DAVIES, K. (1969): 'Ionospheric radio waves' (Blaisdell Publishing Co., Waltham, MA)

GINZBURG, V. L. (1970): 'Propagation of electromagnetic waves' (Gordon and Breach, New York)

KELSO, J. M. (1964): 'Radio ray propagation in the ionosphere' (McGraw-Hill, New York)

LAWRENCE, R. S., and POSAKONY, D. J. (1962): 'A digital ray-tracing program for ionospheric research. Space Research II' (North-Holland Publ. Co., Amsterdam)

RATCLIFFE, J. A. (1959): 'The magneto-ionic theory' (Cambridge University Press, Cambridge)

RAWER, K., and SUCHY K. (1976): 'Remarks concerning the dispersion equation of electromagnetic waves in a magnetized cold plasma', *J. Atmos. Terr. Phys.*, **38**, p. 395

SEN, H. K., and WYLLER, A. A. (1960): 'On the generalizations of the Appleton–Hartree magnetoionic formulas', *J. Geophys. Res.*, **65**, p. 3931

STIX, T. H. (1962): 'The theory of plasma waves' (McGraw-Hill, New York)

Chapter 4

Radio soundings of the ionosphere

4.1 Introduction

Most of our knowledge of the ionosphere comes from remote sensing by radio waves. Single-frequency techniques for remotely measuring time of flight, amplitude, phase, polarization, and angles of arrival include reflection by refractive bending and by so-called coherent and incoherent scatter. The sweep-frequency ionosonde technique was developed to probe the vertical structure of the ionosphere. Moving transmitters and/or receivers (on satellites) have been used to investigate the horizontal structure of the ionosphere. A comprehensive discussion of ionospheric sounding technique is given by Hunsucker (1990). We discuss some of these techniques later in this Chapter but first we consider the principles underlying the measurement of wave features.

4.2 Principles of measurement

4.2.1 Time of flight

One of the easiest, and most used, measurements is the time delay of a radio wave. There are two methods of measurement: (i) measurement of the time delay of a pulse and (ii) determination of the stationary phase (see eqn. 1.19). The direct measurement of pulse delay is done by starting a timing device (e.g. cathode ray oscilloscope time base or a digital clock) when the transmitter is triggered and noting the time of the leading edge of the echo pulse(s). Uncertainties arise in identifying which feature on the pulse envelope to use as reference. Even with a relatively 'square' emitted pulse, ionospheric dispersion gives rise to a triangular or trapeziform echo. Two features that have been used are (i) the point of maximum rate of increase obtainable by differentiating the echo pulse, and (ii) the time of peak amplitude. Problems arise because, as the echo fades and changes shape, these features are affected (Paul 1979). In some cases increased accuracy is obtained by using Gaussian-shaped pulses or coded pulses (Devlin *et al.* 1977). However, time accuracy is limited to about 1·3 μs, which corresponds to 0·2 km of virtual height.

The phase approach is based on (1.19), that is, on the phase change $\Delta\phi$ over a small frequency interval Δf. A typical value of Δf is 8 kHz. This frequency interval corresponds to a distance of 37·5 km or a virtual height interval of 18·75 km. Thus, the virtual height must be known to within this interval to avoid an ambiguity. The estimated accuracy of this method is 0·2 km of virtual height, which corresponds to a phase accuracy of 4°.

4.2.2 *Phase*

The concept of phase difference is illustrated in Fig. 4.1*a* where the echo signal

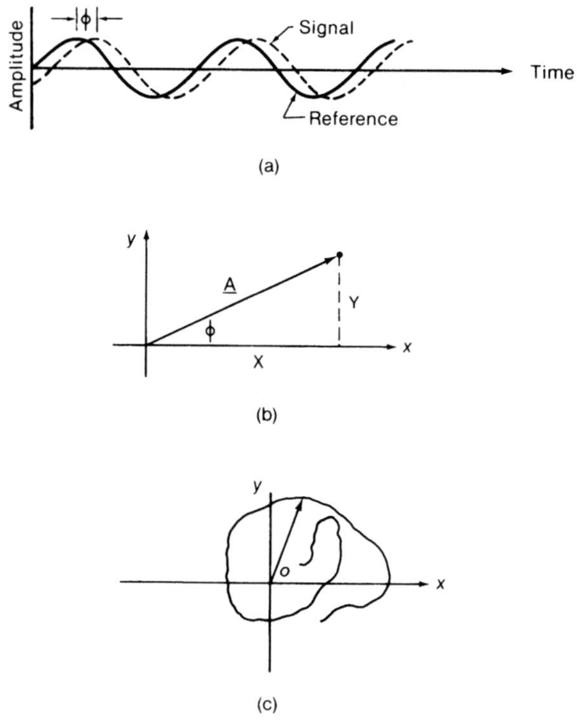

Fig. 4.1 (*a*) *Phase difference between a received sine wave and a reference sine wave*
(*b*) *Concept of complex amplitude*
(*c*) *Locus of the tip of the amplitude vector with changes of phase and amplitude*

is compared with a reference signal of equal amplitude. The time difference between the maxima of the two curves, expressed as an angle, gives the relative phase ϕ. This gives rise to the concept of complex amplitude, depicted as a vector in Fig. 4.1*b*. The echo amplitude is considered to be made up of two orthogonal components: X in phase with the reference signal and Y in quadrature with the reference such that

$$\phi = \tan^{-1}(Y/X) \tag{4.1}$$

Thus, the complex amplitude is

$$A(t) = X(t) + iY(t) \tag{4.2}$$

where $i = \sqrt{-1}$. In general, X and Y vary in complicated fashion as the ionosphere changes and the signal fades so the locus of the tip of the amplitude vector has an irregular locus, somewhat as sketched in Fig. 4.1*c*.

When only the phase difference is changing, the magnitude of the resultant amplitude (A) is given by

$$|A| = (X^2 + Y^2)^{1/2} \tag{4.3}$$

which is independent of time, and the tip of the amplitude vector traces out a circle in time. Since any one cycle is identical with any other cycle, there is an ambiguity in the number of whole rotations of the echo signal compared with the reference signal. In the case of digital data processing, a computer can follow a slowly changing phase from one cycle to another but this is not so when the phase changes rapidly, when there is equipment failure, etc.

Changes of the phase of an ionospheric echo relative to a reference signal are made by mixing the incoming signal with the reference, which produces an interference pattern the modulation of which gives the phase difference. The reference signal can be the transmitter signal, for essentially vertical propagation, or a stable oscillator such as a quartz crystal oscillator or an atomic frequency oscillator. This is called phase-coherent detection.

The total phase is given by

$$\phi = 2n\pi + \tan^{-1}(Y/X) \text{ radians} \tag{4.4}$$

in which n is an integer. The amplitudes X and Y are either displayed on an oscilloscope or recorded digitally. As the phase of the echo signal changes, the output (ϕ versus time) has the appearance of ramps (Fig. 4.2). The integer n is

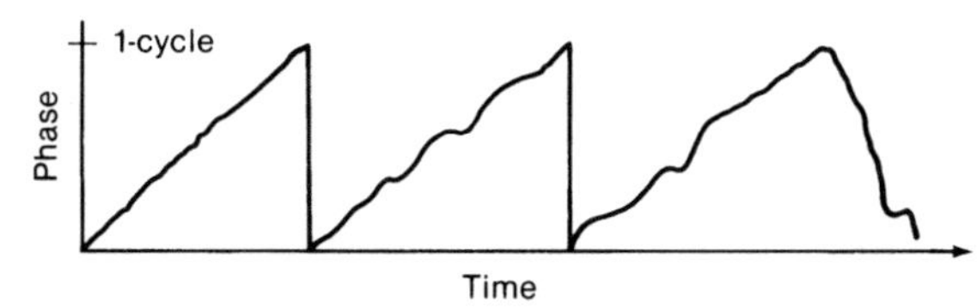

Fig. 4.2 *Variation of phase with time*
Each 'ramp' represents one cycle change

sometimes called the ramp number. The ramp number has to be determined, at some point, by inspection and must thereafter be carefully tracked. This technique has been used extensively on very low frequencies.

4.2.3 Frequency variations

Frequency is the time derivative of phase; that is, the frequency difference Δf between an echo signal and that of a reference signal is given by (see eqn. 1.26)

$$\Delta f = -d\phi/2\pi dt \qquad (4.5)$$

where ϕ is in radians and Δf is in hertz. When the phase ϕ is recorded digitally, Δf can be obtained by numerical differentiation. In general, the echo signal contains components with different Doppler shifts and these can be determined by spectral analysis (e.g., Fast Fourier Transform) of the signals on a computer. The beat frequency can be measured directly with a frequency meter. An alternative method (Watts and Davies 1960) is to record the beat frequency (a few hertz), produced by the mixing of the echo and reference signals, on a slowly moving (e.g. 0·02 ips) magnetic tape. The tape is later played into an audio frequency analyzer at a speed of 30 ips. This results in a frequency multiplication of 1500 and so converts the beat frequency into an audio tone. As the phase path of the ionospheric echo changes, the beat frequency changes; these changes can be recorded and displayed by digital or by analog techniques.

4.2.4 Polarization measurements

Ionospheric echoes usually consist of the superposition of two elliptically polarized (ordinary and extraordinary) waves. At the magnetic dip equator the electric field of the ordinary wave is north–south and the electric field of the extraordinary wave is east–west so that the waves can be separated by the use of two linear antennas oriented north–south and east–west.

In sufficiently high latitudes, and/or on sufficiently high radio frequencies, the polarization ellipses are essentially circles. The two circularly polarized signals can be separated by two orthogonal dipoles oriented in the x and y directions. Consider two oppositely rotating waves of amplitude E_o and E_e incident on the antenna. The outputs E_x and E_y are

$$E_x = E_o \cos \omega t + E_e \cos(-\omega t) = E_o \cos \omega t + E_e \cos \omega t \qquad (4.6a)$$

and

$$E_y = E_o \sin \omega t + E_e \sin(-\omega t) = E_o \sin \omega t - E_e \sin \omega t \qquad (4.6b)$$

We now introduce a delay of 90° into one of the signals (E_x, say) relative to the other signal to obtain

$$E_o = E_x \cos(\omega t - \pi/2) + E_e \cos(\omega t - \pi/2) \qquad (4.7a)$$

$$E_x = -E_o \sin \omega t - E_e \sin \omega t \qquad (4.7b)$$

Subtraction of (4.7b) from (4.6b) gives

$$E_S = 2E_o \sin \omega t, \text{ i.e. the ordinary wave} \qquad (4.8a)$$

Addition of the signals gives

$$E_A = -2E_e \sin \omega t, \text{ i.e. the extraordinary wave} \qquad (4.8b)$$

The addition and subtraction of the signals can be done numerically, when the data are recorded digitally, or electrically.

4.2.5 Angle of arrival

There are three principal methods of measuring angle of arrival: (i) the phase measurement method, (ii) the use of 'musa' or multi-unit steerable antenna, and (iii) the use of two antennas having different radiation patterns in the vertical plane.

The first method is the most common and depends on the fact that when a plane radio wave (of wavelength λ) from a given direction arrives at two receiving antennas, separated by a known distance d, the phase angle ϕ between the induced voltages is

$$\phi = \frac{2\pi d}{\lambda} \cos \Delta \cos \alpha \text{ (in radians)} \tag{4.9}$$

The elevation Δ is the angle measured above the horizontal, and α is the azimuthal angle measured from a line joining the centers of the two antennas. In practice the antennas may be oriented with their centers in the great circle plane passing throught the transmitter and receiver so that $\alpha = 0$. For accurate measurements it is necessary to correct for the variation of α.

The musa system for measuring a vertical angle consists of an antenna array having a very sharp lobe that can be steered up and down in the vertical plane. The vertical angle is determined by the position of the lobe at which the received signal is a maximum.

For the third method, two identical antennas may be at different heights above ground or one can be horizontal and the other vertical. In either case the antenna patterns must be known. One useful application of this method is in normal incidence sounding where dipoles are located at such heights as to give (i) a maximum or (ii) a null in the zenith direction. This arrangement enables a determination to be made of whether echoes are returned from overhead or scattered obliquely. For further information on current techniques of radio direction finding the reader is referred to books by Gething (1978) and Whale (1969).

4.3 Ionosondes

4.3.1 Conventional ionosonde technique

The conventional equipment for measuring the virtual height of the ionosphere is a sweep-frequency pulsed radar device called an ionosonde. The early history of ionosondes and their roles in radio science has been given by Villard (1976). The frequency can range from below 0·1 MHz to 30 MHz or more with a sweep duration from a few (e.g. 10) seconds to a few minutes. A pulse of radio waves is emitted, and a timing device (e.g. the time-base of a cathode ray tube) is triggered. The echo pulse is fed to the Y-plates of a cathode ray tube. The position of the pulse on the time base is a measure of the time of flight of the pulse and, hence, of the virtual height (or equivalent or group height) h' of reflection. The Y-deflection is related to the echo amplitude. This type of presentation is called an A scan (Fig. 4.3a). When the echo pulse is applied to

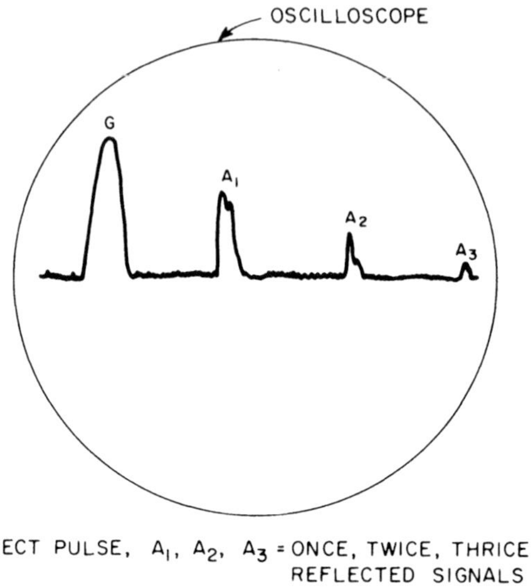

(a) A

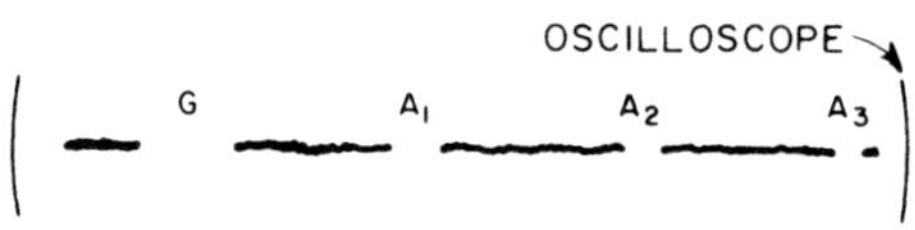

(b) B

Fig. 4.3 (a) *An A scan obtained by applying the echo signal to the Y-plates of an oscilloscope* The position of the echo pulse on the time base gives the time of flight of the pulse (b) *A B scan in which the echo pulses are applied to the grid of an oscilloscope to blank out the time base*

the grid of the oscilloscope so as to blank the time base, the presentation is called a B scan (Fig. 4.3b). As the frequency is increased the pulse penetrates to higher levels in the ionosphere and, in general, the virtual height increases until the signal penetrates at the critical frequency f_c, which is related to the maximum electron density N_{max} by

$$N_{max} = 1{\cdot}24 \times 10^{10} f_c^2 \tag{4.10}$$

where N_{max} is in electrons per cubic meter and f_c is in megahertz.

The virtual (or group) height h' is related to the time of flight t and the group refractive index μ' as follows:

$$h' = \tfrac{1}{2}ct = c\int_0^{h_r} \frac{dh}{u} = \int_0^{h_r} \mu' \, dh \qquad (4.11)$$

where u is the group speed and h_r is the real height of reflection.

A virtual height versus frequency record, called an ionogram, is obtained by photographing a B scan with slowly moving film. The frequency axis is obtained from the motion of the film. Application of short blanking pulses to the time base at certain frequencies provides a frequency calibration. Height calibration is achieved by the application of short blanking pulses from a pulse generator, e.g. pulses every 1/1500 s provide 100 km height marks. Samples of conventional ionograms are shown in Fig. 4.4, for a middle-latitude location on a summer day and night. At night only F-layer echoes are present and the virtual height increases steadily with frequency up to penetration at the ordinary-wave critical frequency f_oF_2. The trace for the extraordinary wave penetrates at a frequency f_xF_2 and shows also retardation on the lower frequencies caused by underlying ionization. This retardation decreases with increase of frequency. The daytime ionogram (Fig. 4.4*a*) shows only ordinary wave traces with reflections from the E, F_1, and F_2 layers. The extraordinary wave has been suppressed. There is marked retardation of the F_2 echo on frequencies just above f_oF_1 and below f_oF_2.

Table 4.1 *Characteristics of a conventional ionosonde*

Frequency range	1·0 to 20 MHz
Sweep duration	15 s to 5 min
Height range	1000 km
Peak power	1 to 10 kW
Frequency resolution	0·1 MHz
Height resolution	1·0 km
Pulse repetition rate	50 per second
Pulse duration	30 μs

Some characteristics of a conventional sounder are listed in Table 4.1. It should be realized that various trade-offs have to be made in a sounder. For example, there is a lower limit to the sweep duration, depending on the height range and frequency range. Consider a linear frequency sweep of 20 MHz, at a pulse repetition rate of 50 pps and a height range of 1000 km; if the frequency change between successive pulses is 0·02 MHz the minimum sweep duration is 6·7 s. The sweep rate cannot be so fast that the receiver bandpass is out of the frequency bandwidth of the echo. The longer the sweep duration the better the signal-to-noise ratio and the higher the quality of the ionogram. However, when the sweep duration is long, the ability to follow temporal variations in the

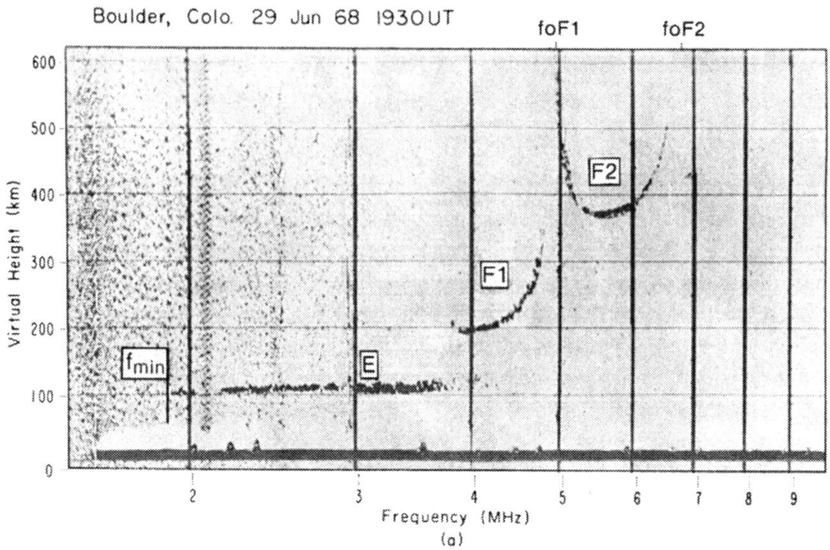

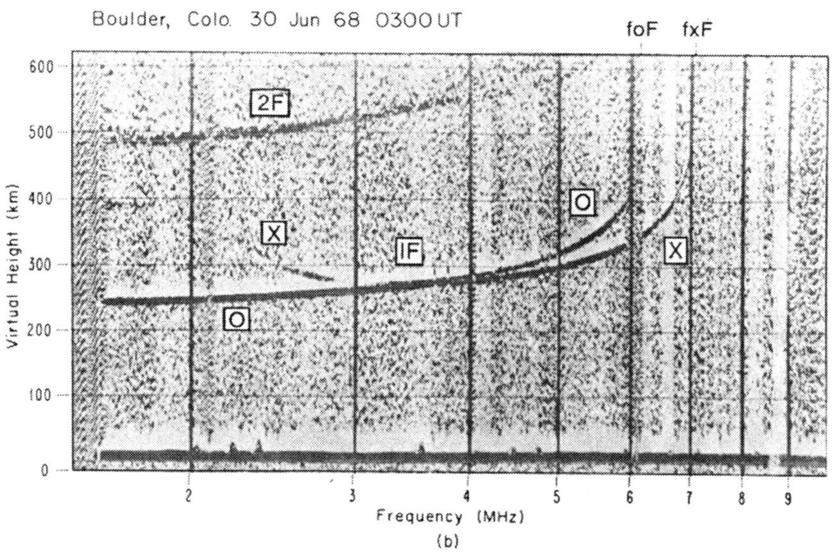

Fig. 4.4 (*a*) *Sample summer daytime ionogram with ordinary wave traces only*
Reflections are from the E layer, the F_1 layer, and F_2 layer at progressively higher heights and on progressively higher frequencies. Notice the group retardation near the critical (penetration) frequencies f_oF_1 and f_oF_2.
(*b*) *Sample ionogram taken on a summer evening in middle latitudes, showing one-hop and two-hop echoes from the F layer*
The signals are split into ordinary waves (O) and extraordinary waves (X) having critical (penetration) frequencies f_oF and f_xF, respectively. The increase of the virtual height at the lower frequency end of the extraordinary trace is caused by retardation in underlying ionization

ionosphere is reduced and the ionospheric characteristics may change appreciably during the sweep. Another disadvantage of a long sweep duration and/or continual sweeping is the interference to other radio systems produced by the relatively wide spectrum of the pulse amplitude modulation.

4.3.2 Some limitations of conventional ionosondes

In spite of the invaluable contributions made by conventional ionosondes to ionospheric knowledge, there are limitations. These limitations revolve around the fact that very few characteristics of the radio echoes are measured. Although some ionosondes have been modified in order to show amplitude (Wakai and Ishizawa 1962), ionosondes in general have ignored direct amplitude information as well as information on direction of arrival, polarization, phase and (Doppler) frequency variations, etc. Furthermore, in recent years digital computer control of ionosonde operation and data acquisition, display, and analysis has become reality, and it is now possible to scale ionograms (especially those in middle latitudes) routinely by computer techniques without human intervention.

4.3.3 Monitoring versus research ionosondes

Ionosondes are used extensively in remote sensing both for monitoring the long-term temporal and spatial variations of the ionosphere and for research into ionospheric physics. The requirements for these two modes of operation are somewhat different (Table 4.2). In a monitoring sounder the ionograms should be relatively insensitive to the peculiar characteristics of the sounder so that ionograms from different sounders can be easily compared and so that equipment deterioration has a relatively minor effect on the measured parameters. Automatic scaling is a desirable feature of a modern sounder because manual scaling (labor) costs may be a large fraction of the total operational costs.

Table 4.2 *Requirements for research and monitoring ionosondes*

Characteristic	Research ionosonde	Monitoring ionosonde
Flexibility	Maximum	Essential minimum
Ease of maintenance	Not essential	Essential
Remote control	Not essential	Desirable
Meeting preset schedules	Not essential	Essential
Low cost	Minor consideration	Desirable
Number of parameters	Maximum	Preset minimum
Portability	Important	Not important
Data continuity	Not essential	Essential

A research instrument needs to be flexible, with the capability of measuring the maximum number of wave parameters, including amplitude, polarization, phase, Doppler, angle of arrival. Cost, reliability, maintenance, availability, and standardization of components are, generally, not of great concern in the case of a research instrument because the operation is intermittent and attended by the experimenter. Automatic scaling and electron density profile inversion are desirable but not essential because an experimenter may be interested in analyzing a few carefully selected ionograms that need expert interpretation.

4.4 Some new sounders

Several sounders have recently been developed that take advantage of modern digital techniques. The techniques involve accurate frequency measurement by synthesis of signals from stable oscillators. The cost of such sounders is reasonable because of the low cost of digital components and peripheral equipment, for example, read-only memories (ROMs) and microprocessors.

4.4.1 The NOAA/HF sounder

This is essentially a research sounder and one of the more sophisticated ionospheric sounders in use. It is a general purpose pulse-amplitude computer-based system for radio measurements of the ionosphere in the range from 100 kHz to 30 MHz. A development of the dynasonde (Wright and Pitteway 1982), it permits full digital signal processing of echo signals. Having a dual-channel receiver, it supplies fast measurements of wave polarization and direction of arrival. With suitable antenna systems this sounder is capable of making many of the possible measurements of echo characteristics. It is intended (Grubb 1979) to provide the following:

(i) A sounder with the minimum hardware restraints on echo characterization
(ii) A capability for real-time data analysis and display and, if necessary, for providing automatic control of data acquisition
(iii) Operation with similar sounders to provide oblique soundings.

These echo measurements are basic:

(*a*) The (complex) vector description of echoes versus range and frequency, to enable the measurement of echo amplitude, phase, pulse group delay, stationary phase ($d\phi/df$) group delay, and Doppler spectra.
(*b*) The direction of arrival, wave polarization, and other irregular structures in the wavefront.
(*c*) The time variations of the above characteristics.

The basic arrangement of the radio frequency (RF) system used for generating the emitted signals and for the reception of echo signals is as follows. An oscillator generates a variable frequency between 40·1 and 70 MHz which is mixed with a signal from a second oscillator that has a fixed frequency of 40 MHz, to produce a frequency in the range 0·1 MHz to 30 MHz. Keying and filtering (pulse shaping) of the emitted signal is performed at the 40 MHz signal stage. This 40 MHz oscillator provides the reference for the complex-amplitude representation of the echoes. The frequency of operation is computer controlled as are the frequency range, sweep time, etc.

The echo signal is amplified and mixed with a signal from the first oscillator to produce an intermediate frequency of 40 MHz which, after amplification, is compared with the reference 40 MHz to obtain the in-phase and quadrature components. The linear receiver has a 30–40 dB operating range. The position of this range is adjusted to prevailing levels between pulses. The overall receiver 3 dB bandwidth is 33 kHz. The virtual height range is 776 km.

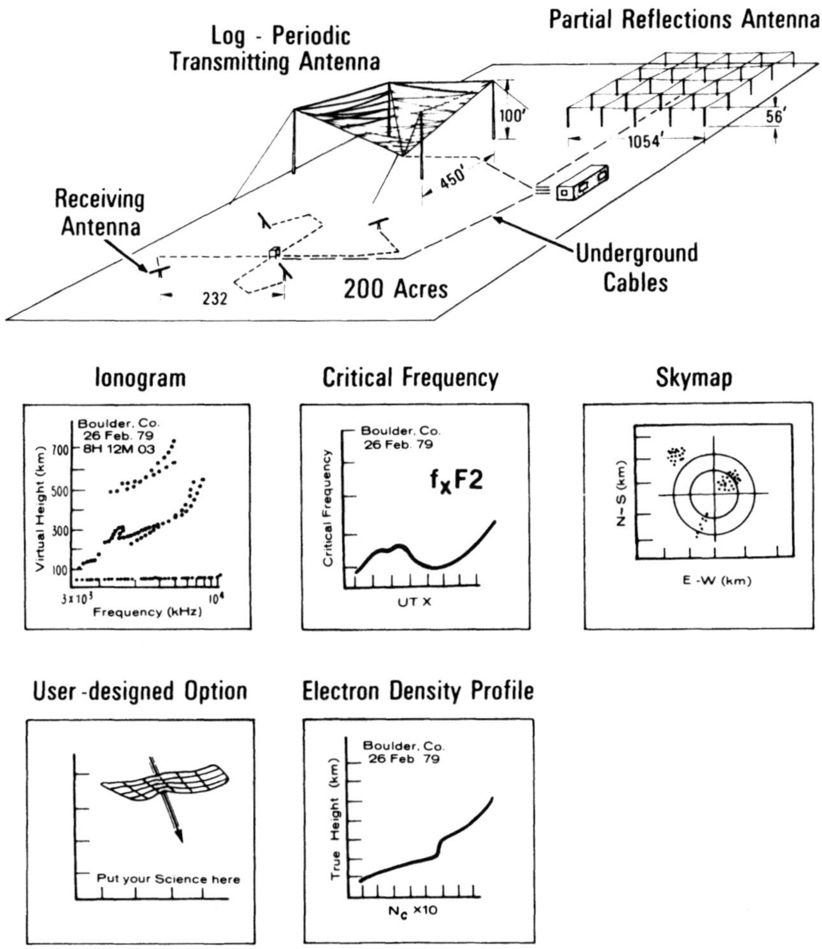

Fig. 4.5 *The field site at Boot Lake, Colorado. The antennas used for transmitting and receiving are shown*
(*Below*): Some sample outputs

The exploitation of a modern sounder requires a more sophisticated antenna system than those used with conventional sounders. For example, Fig. 4.5 shows an antenna arrangement used with a NOAA sounder. The transmitting antenna is an inverted pyramid log-periodic antenna, having a height of 100 ft and a side of 450 ft. It has a relatively constant power output and pattern over the frequency band.

The receiving antennas consist of four short horizontal dipoles at the corners of a square of side 232 ft (70 m) and aligned along the (NS and EW) diagonals, thus providing a pair of crossed dipoles for polarization measurements. The angles of arrival, along the directions of the diagonals, are obtained from the phases of the waves at the antennas (see Sec. 4.2.5).

The sample ionogram in Fig. 4.5 is a graphical representation of the digital ionogram, stored on magnetic tape, which contains all the recorded information. Note that the ionogram is composed of a set of discrete points.

4.4.2 The digisonde

The digisonde (Reinisch 1986) is a pulse amplitude sounder that combines several features required for both monitoring and research purposes. The digisonde measures the following parameters: (i) group travel time or h', (ii) echo amplitude, (iii) phase, (iv) precise Doppler frequency, (v) angle of arrival, (vi) wave polarization, that is, separation of ordinary and extraordinary waves, and (vii) wave-front curvature. Digital pre-processing and formatting are used to reduce the volume of data to manageable size.

The digisonde operates in (i) the ionogram mode, and (ii) the drift mode. In the ionogram mode, the virtual height and amplitude (to $\frac{1}{2}$ dB) are measured over a selected frequency range. The information on angles of arrival and Doppler shifts is limited but the wave polarizations are identified for all frequency-range bins. In the drift mode, only a few frequency-range bins are selected. The digital ionograms can be displayed on a printer (Fig. 4.6). The $h'(f)$ position is represented by the location of a small number giving either the amplitude in 4 and 1 dB increments or the value of one of the other parameters. The operating parameter values can be entered, at a keyboard, to a processing controller (computer). The frequency range is 0·4–40 MHz. The first local oscillator is kept 2 MHz off the transmitted frequency, resulting in a 2 MHz intermediate frequency (IF) which is then converted to an IF of 225 kHz and an overall bandwidth of 15 kHz. For further details on the technical aspects of the digisonde the reader is referred to papers by Bibl and Reinisch (1978) and Reinisch (1986).

Real-time post-processing is performed in a microcomputer, one of the outputs of which is a sky map that gives an immediate picture of the horizontal structure of the ionosphere, that is, the distribution of reflecting centers. A feature of the digisonde is that options are preprogrammed. For example, the ionogram rate can be selected from a set of 12 ionogram sequences, extending from 1 to 120 ionograms per hour. Ionograms can be started manually or automatically by computer. The ordinary and extraordinary waves are separated using signals from crossed-loop antennas. In latitudes for which quasi-longitudinal propagation holds, the signals are added and subtracted by means of a $\pm\pi/4$ phase shifter (see Sec. 4.2.4); at the magnetic equator the loops are used independently. The digisonde is designed for remote control, unmanned operation, and automatic scaling and processing of ionograms (Reinisch 1986).

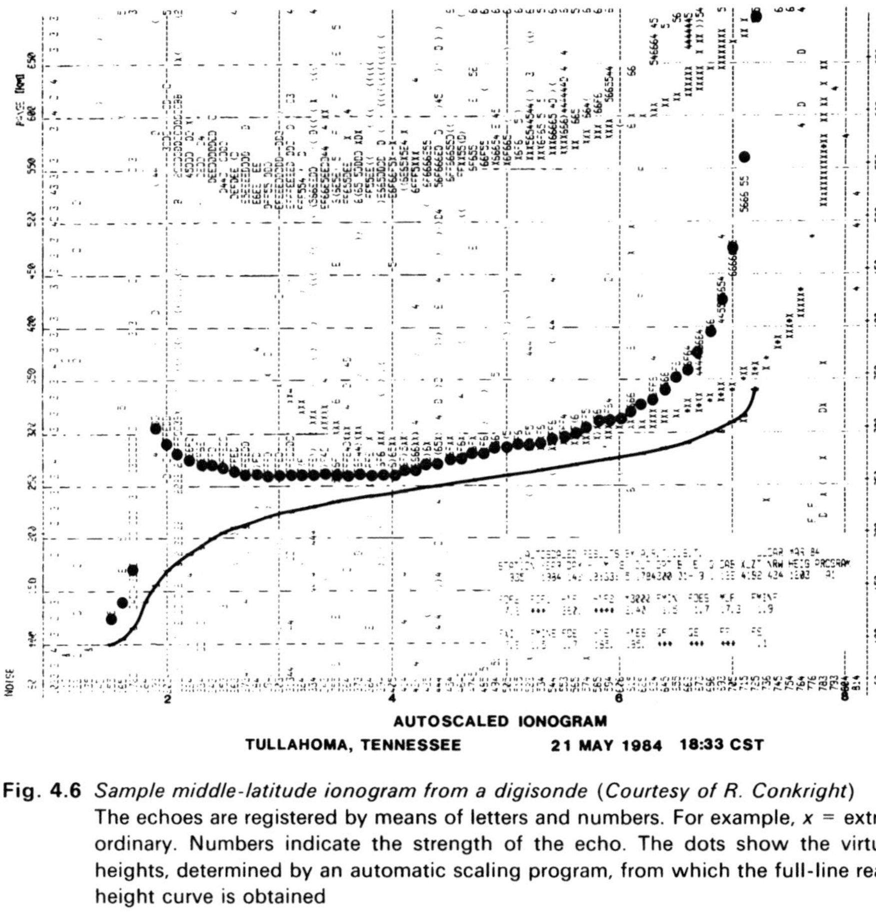

Fig. 4.6 *Sample middle-latitude ionogram from a digisonde (Courtesy of R. Conkright)* The echoes are registered by means of letters and numbers. For example, x = extraordinary. Numbers indicate the strength of the echo. The dots show the virtual heights, determined by an automatic scaling program, from which the full-line real-height curve is obtained

4.4.3 The IPS–42 sounder

A relatively inexpensive modularized sounder that has gained wide acceptance by the monitoring community is that developed at the Ionospheric Prediction Service (IPS) in Sydney, Australia. This is a 5 kW, 40 μs, pulse-amplitude sounder covering the frequency range from 1 MHz to 21·475 MHz in steps of 5 kHz. The ionograms are recorded photographically and/or, with suitable peripheral equipment (Paterson and Bevins 1989; Kelly 1986), digitally recorded and then displayed by a dot matrix printer. The receiver bandwidth is 25 kHz and the logarithmic dynamic range is about 80 dB. The logarithmic frequency range is derived from a stable crystal oscillator. Prearranged programs are incorporated so that ionograms can be generated singly on command, routinely at 15 min, 5 min, and 1 min intervals or continuously every 20 s. An ionogram from an IPS sounder is shown in Fig. 4.7. The IPS sounder can be used as a remote monitor, that is, one in which the monitoring device is separated from the ionosonde. This ionosonde can be controlled (programmed) directly from the monitoring station.

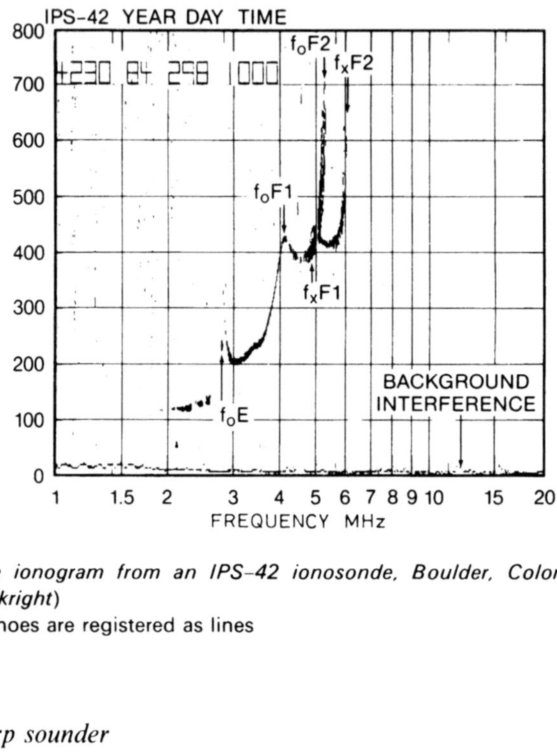

Fig. 4.7 *Sample ionogram from an IPS-42 ionosonde, Boulder, Colorado (Courtesy of R. Conkright)*
The echoes are registered as lines

4.4.4 The chirp sounder

The principle of the chirp, or pulse compression, sounder differs from that of the pulse-amplitude sounders in that the radio signals are continuous waves in which the frequency is modulated (Barry 1971). The principle of operation is that the frequency of the transmitted signal varies (linearly) in time with slope df/dt (Fig. 4.8). Let f be the transmitter frequency when the pulse is emitted. When the echo returns, after a flight time Δt, the transmitter frequency is $f + (df/dt)\Delta t$. When the transmitted frequency is subtracted from the echo frequencies, echoes at various time delays become tones on frequencies proportional to the time delays. Thus, by frequency analysis (Fourier transformation) of the echo signal, the time delays (and hence the virtual heights) can be determined (Poole 1985; Poole and Evans 1985).

Although a chirp sounder has the disadvantage of needing a frequency analyzer and a transmit–receiver switch, it has some advantages, as follows:

(i) Low peak power (10–100 W), hence fewer electronics problems
(ii) Good noise discrimination, since spikey noise tends to be flat in the frequency domain whereas chirp echoes are spike-like
(iii) Less radio interference to other spectrum users
(iv) Easy phase measurement by means of Fourier transformation of the chirp difference signal.

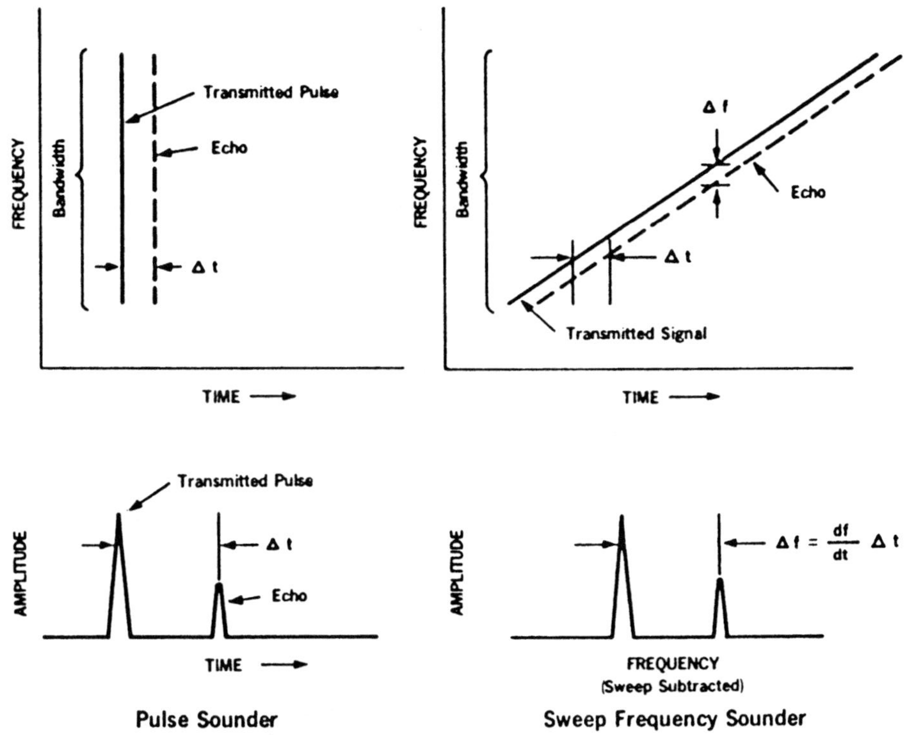

Fig. 4.8 *Frequency–time representation of (left) pulse sounder signals and (right) frequency-modulated continuous-wave (FM–CW) signals of a chirp sounder (From Barry 1971 © IEEE; reprinted with permission)*
The time delay Δt of an FM–CW echo is derived from the difference Δf between the echo frequency and the simultaneous frequency of the transmitter signal

A typical chirp cell of 1 s with a sweep rate of 50 kHz s^{-1} gives a theoretical range resolution of (1/50) kHz = 20 μs = 3 km of group height.

A disadvantage of the chirp sounder is that its ionogram takes longer to complete than a pulse ionogram. For example, with a sweep rate of 50 kHz s^{-1}, a linear sweep from 0·5 to 15·5 MHz takes 5 min whereas a pulse sounding can

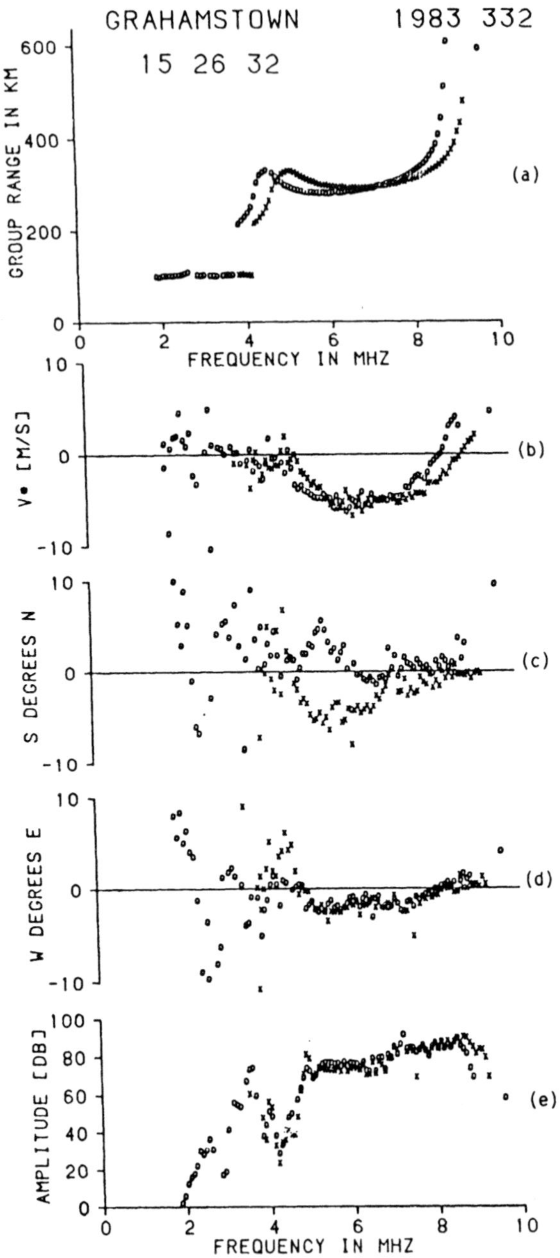

Fig. 4.9 *Sweep frequency ionogram taken with a chirp sounder at Grahamstown, South Africa (Courtesy of A. W. V. Poole)*

The variations of different echo characteristics are functions of wave frequency; *o* = ordinary echo, and *x* = extraordinary echo. (*a*) is the group height; (*b*) is the Doppler velocity; (*c*) and (*d*) are arrival angles in the north-south and east-west directions respectively; (*e*) is the echo amplitude be accomplished in 15 s. Figure 4.9 shows a sample chirp ionogram together with associated measurements such as velocity, angle of arrival, and echo amplitude.

4.5 Modes of operation

In ionospheric soundings it is often necessary to trade frequency resolution for time resolution, depending on the electron-density structure and the dynamics of the ionosphere. For example, when an accurate profile is required, high-frequency resolution is needed whereas for the examination of time changes, frequent measurements at a few frequencies are preferable. Thus, the operation of sounders can be grouped into three modes: the ionogram (I) mode, the kinesonde (K) mode, and the basic or intermediate (B) mode. The principal distinctions between these modes are in the frequency patterns. I-mode provides high-frequency resolution, at the price of rather poor time resolution. K-mode provides high time resolution, but only on a small number (e.g., < 10) of radio frequencies. B-mode is a compromise pattern suitable for certain types of measurement: it retains high sampling-time resolution (in discrete time intervals) and frequency resolution adequate to define an ionogram.

For routine soundings, the ionogram mode is used periodically (e.g. every 15 min or every 60 min) except during special observation periods (e.g. World Days) when soundings at shorter intervals (e.g. 1 min) may be desirable. On the other hand, for observations of transient phenomena (e.g. traveling disturbances) the K-mode is desirable, interspersed with occasional ionograms to keep track of slow changes in the ionosphere.

4.6 Some disadvantages of ionosondes

As discussed above, a modern ionosonde can directly measure all, or most, of the characteristics of a radio echo and, therefore, one might expect it to be an ideal instrument. However, for some purposes (such as continuous monitoring of the ionosphere) simple measurements on a few frequencies have proved very useful. Examples include ionospheric absorption of radio waves, phase and amplitude measurements on very low frequencies, high-frequency Doppler, and total electron content. It is easier to optimize the performance of a single-frequency instrument than one having several frequencies. Vertical incidence sounders often suffer from the following limitations: (i) In certain disturbances echoes are lost because of high D-region absorption; (ii) the F_2-layer trace is lost when f_oF_2 falls below f_oF_1 or f_oE (the G-condition), (iii) when the maximum frequency of the trace exceeds the upper frequency limit of the sounder, or (iv) when the upper and/or lower frequency of the echo falls below the lower frequency limit of the sounder; (v) the sounders provide no direct information on 'valleys' between layers and on the profile above the maximum electron density; (vi) in general, only intermittent operation is possible; (vii) it is uneconomical to tie up an expensive sounder to make simple measurements.

4.7 Incoherent scatter sounders

The ground-based ionosonde technique is essentially restricted to observations of the bottomside of the ionosphere, that is, to heights less than $h_m F_2$. Various methods have been employed to study the topside, nearly all of which involve rockets or satellites. Over the past 30 years or so the incoherent (or Thomson) scatter technique has been developed, which makes possible not only studies of the topside ionosphere from the ground but also the measurement of a variety of other properties of the upper atmosphere (neutral density, temperature, and composition).

Thomson (1906) showed that electrons are capable of scattering electromagnetic waves (X-rays). The scattering cross section (σ) of an electron is independent of the wavelength (λ) of the incident wave and is called the Thomson cross section. Note that $\sigma_e = 4\pi r_e^2 = 10^{-28}\ \mathrm{m}^2$, where r_e is the effective radius of an electron. Gordon (1958) showed that existing radar technology could detect weak scatter from the ionosphere. Gordon predicted also that the random thermal motions of the electrons would produce Gaussian broadening of the spectrum (of the scattered signal) with a center to half power width of 0·71 Δf_e, where Δf_e is the Doppler shift produced by an electron approaching the radar at the mean thermal speed; that is,

$$\Delta f_e = (8KT_e/m_e)^{1/2}/\lambda = 11\,(T_e)^{1/2}/\lambda \text{ (in kilohertz)} \tag{4.12}$$

where K is Boltzmann's constant, and T_e and m_e are the electron temperature and mass, respectively. For a wavelength of 1·5 m and $T_e = 1600\,\mathrm{K}$, $\Delta f_e \approx 300\,\mathrm{kHz}$; the echo spectrum would be broad and, consequently, a very large antenna would be required.

The existence of incoherent scatter was first demonstrated in 1958 by Bowles with a high power (≈ 1 MW) transmitter on a wavelength of 7·5 m and with an antenna of cross-section of about 5 acres. Bowles (1961) observed echoes with a bandwidth of about a factor of 10 less than 0·71 Δf_e. He correctly surmised that the narrow spectrum was due to the coupling between the electrons and the heavy ions that effectively increased the mass m_e in (4.12). Heavy ions affect radio waves when the exploring wavelength is very much longer than the Debye length (eqn. 1.63), which is about 1 cm or less below 1000 km, rising to 6 cm at 2000 km. Thus, to observe the electron-scatter spectrum the radar wavelength has to be so short that the scatter power is insufficient to be of practical use for ionospheric sounding. When the exploring wavelength is much longer than the Debye length, the scattering can no longer be regarded as coming from individual electrons, but instead is best thought of as arising from density fluctuations brought about by longitudinal oscillations of the plasma. Thus, the data processing and interpretation are highly complicated; the interested reader is referred to the review by Evans (1969) and the references therein.

In spite of the highly complicated interpretation of the spectrum of the scattered signals, the incoherent scatter technique provides a means of obtaining the electron density profile. With vertical sounding the power p_r received from a distance R is given by

$$p_r = C\sigma N p_t a / R^2 \tag{4.13}$$

where

- N is electron density
- σ is the effective scattering cross-section of an electron
- a is the antenna aperture (m^2)
- p_t is the radiated power
- C is a calibration that takes into account losses in feeds, effects of side lobes, pulse duration, etc.

The following features of (4.13) are worth noting:

(i) The received power p_r is proportional to R^{-2}, rather than R^{-4} which is the case for a single scatterer. From the divergence of the beam it follows that the scattering volume increases as the square of the range, thus removing a factor of R^{-2}.

(ii) The aperture a appears to the first power only. This arises because the increased aperture (i.e. higher gain) results in a smaller scattering volume.

Because the echoes are weak, considerable data integration is necessary. For a given radar the power output of the receiver is proportional to N/R^2 or, in the case of a vertical radar, to N/h^2. The range R is obtained from the time delay so that the electron density, $N(h)$, profile can be obtained directly. Each electron-density 'point' refers to a bin, or cell, over a specified height range and time interval. In principle, the radars can be calibrated from their known characteristics but it is simpler to calibrate the profile at the maximum electron density $N_m F_2$ obtained from the critical frequency $f_o F_2$ measured directly and accurately by a nearby ionosonde where

$$N_m F_2 = 1{\cdot}24 \times 10^{10} \, (f_o F_2)^2 \text{ (in m}^{-3}\text{)} \tag{4.14}$$

Corrections to the profile due to electron and ion temperature differences are discussed by Evans (1969).

The effective electron cross section σ depends on the Debye length D (see eqn. 1.63) and the ratio T_e/T_i of electron to ion temperature in the following way:

$$\sigma = \frac{\sigma_e}{(1 + \alpha^2)(1 + T_e/T_i + \alpha^2)} \tag{4.15}$$

where $\alpha = 4\pi D/\lambda$. Thus when $T_e = T_i$ and $\lambda \gg D$ (i.e. when $\alpha = 0$), $\sigma = \sigma_e/2$; hence the effective cross section is half the cross section of a single electron.

The echo spectrum is characterized by a double hump, the shape depending on the ratio T_e/T_i (Fig. 4.10). By matching the observed and calculated spectra, the ratio T_e/T_i can be determined. The electron temperature T_e can exceed the ion temperature T_i because of the production of photoelectrons with excess energy.

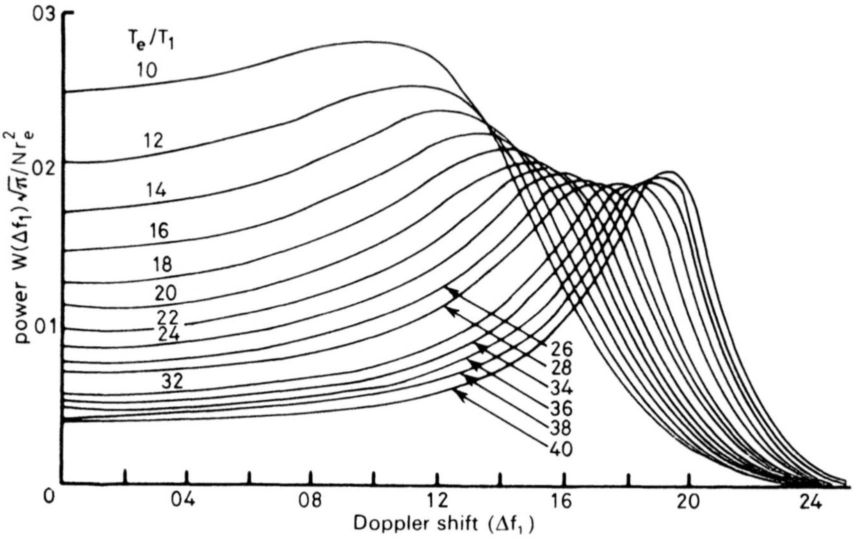

Fig. 4.10 *Dependence of the spectra of incoherent scatter on the temperature ratio T_e/T_i (after Evans 1969)*

Because incoherent scatter (IS) radars are expensive to build and operate, relatively few are, or have been, in operation. In 1986, there were some five IS sounders in operation (Table 4.3). The sounders at Jicamarca, Arecibo, Millstone Hill, and Sondrestrom are monostatic; that is, transmitter and receiver are colocated. The St. Santin Sounder is bistatic; the transmitter is at St. Santin and the receiver is at Nancay. The European Incoherent Scatter (EISCAT) facility's transmitter is near Tromsø, Norway; receivers are near Tromsø, Norway; Kiruna, Sweden; Sodankyla, Finland; and Murmansk, USSR.

The received power falls off as the square of the height, and the electron density in the topside decays exponentially with height; these two effects combine to cause echo power to decrease rapidly. This sets the upper height at which reliable profiles can be obtained. The lower height limit is set by the low electron densities in the D region and by equipment sensitivity; it can be as low as 80 km (e.g. Arecibo). Some sample $N(h)$ profiles are shown in Fig. 4.11. At Jicamarca, profiles have been obtained to heights of 5000 km or more.

Table 4.3 *Data on incoherent scatter sounders in operation in 1986*

Location	Geographic location		Geomagnetic	Modulation	Peak power (kW)	Frequency (MHz)	Antenna
	Latitude	Longitude					
Jicamarca	11·95°S	76·87°W	2°N	Vertical pulse	4000	49·92	290 × 290 m
Arecibo	18·3°N	77·75°W	30°N	Vertical pulse	2000	430·0	300 m spherical reflector
Millstone Hill	42·6°N	71·5°W	53°N	Vertical pulse	3000	440·0	parabola
				Oblique pulse	4000	1295·0	parabola
St. Santin	Transmitter						
	44·65°N	2·19°E	47°N	CW	75	935·0	
	Receiver			T_x vertical			20 × 100 m
	47·37°N	2·10°E		R_x oblique			40 × 200 m
EISCAT	Transmitter						
	69·7°N	19·0°E	67°N	Vertical and oblique pulse	2000	933·5	parabolic cylinder 120 m × 40 m
					5000	2240	32 m parabola
	Receivers						
	69·6°N	19·2°E		Vertical and oblique			
	67·8°N	20·4°E					
	67·4°N	26·6°E					
Sondrestrom	67·0°N	50·6°W	77°N	Vertical and oblique pulse	4000	1290	27 m parabola

Sample temperature contours (Fig. 4.12) show a predawn maximum around the F-layer peak.

In general, the maximum of the spectrum is displaced from the center frequency of the transmitted signal. This indicates a drift or movement of the scattering region. By steering the antenna beam it is possible to determine the

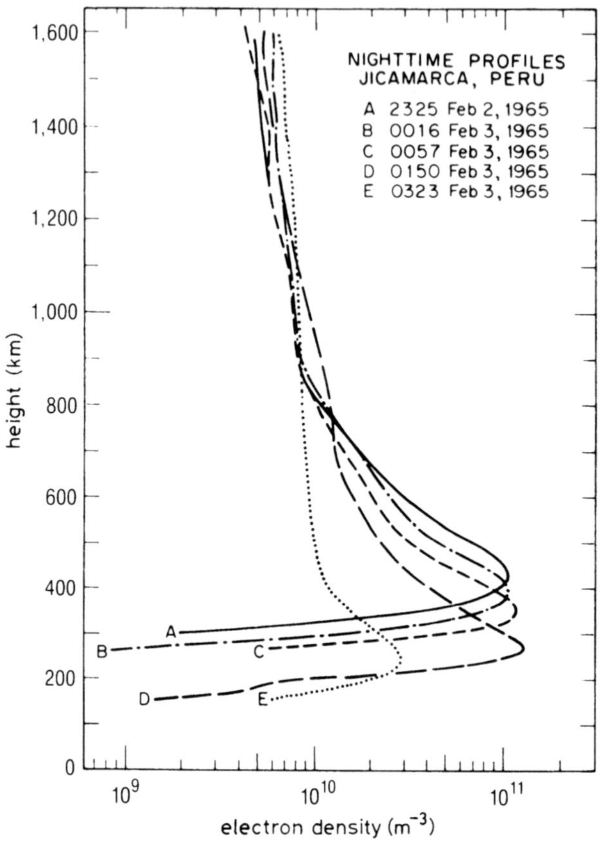

Fig. 4.11 *Sample electron density profiles at Jicamarca, Peru, obtained with the incoherent scatter technique (after McClure 1965)*

motion in three dimensions and, hence, to find the components of motion parallel to, and perpendicular to, the local geomagnetic field. On the assumption that the motion perpendicular to the magnetic field is caused by electromagnetic ($\boldsymbol{E} \times \boldsymbol{B}$) drift, the electric field can be determined. Similarly, the neutral wind can be determined from the component parallel to the geomagnetic field (see Sec. 2.9.3).

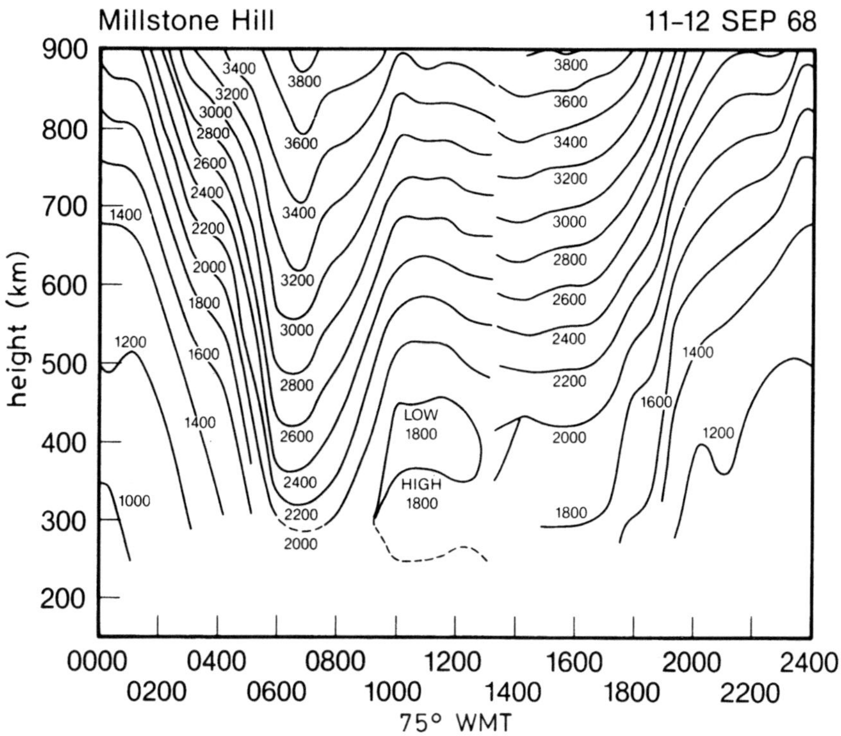

Fig. 4.12 *Sample electron temperature contours (kelvins) obtained with the incoherent scatter technique at Millstone Hill, Mass.* (*after Evans 1973*)
Note that, at a given altitude, the temperature peaks shortly after sunrise, owing to the presence of energetic photoelectrons

4.8 Other techniques

4.8.1 Virtual height–time recorder

This is a fixed frequency device, using a B-scan display that operates continuously. Provided that the signal does not penetrate the ionosphere, this system has the advantage of continuous monitoring of the ionosphere and is useful in detecting 'one-shot' events (e.g. an effect of a solar flare, a natural geophysical event such as an earthquake, or a man-made event such as an explosion). It has the advantage of simplicity and is inexpensive compared with an equivalent use of an ionosonde in the K-mode of operation.

4.8.2 Auroral radars

To study motions of auroral plasma structures in two dimensions, spaced radars have been set up such that the locations of the reflection regions can be accurately located by the intersections of beams. Such radars include, for example, the Scandinavian Twin Auroral Radar Experiment (STARE) (Nielsen 1982) and the Swedish and British Radar Experiment (SABRE) (Jones *et al.* 1981; Mattin and Jones 1986). From Doppler measurements these radars give the motion of auroral plasma structures that, with a model of the geomagnetic field, lead to the structure of ionospheric electric fields (Greenwald *et al.* 1985).

Radars in this category include also the Scandinavian and French Auroral Radar Investigation (SAFARI) (Villain *et al.* 1985), the Canadian Bistatic Auroral Radar System (BARS) (McNamara *et al.* 1983) and the Syowa Radar in Antarctica (Igarashi *et al.* 1982). Other fixed-frequency radars that are used to study the lower ionosphere and the neutral atmosphere include mesospheric-stratospheric-tropospheric (MST) radars (Balsley *et al.* 1979) and the SOUSY radar (Röttger *et al.* 1978).

4.8.3 Rockets and satellites

Rockets and satellites have been used extensively to study the electron-density structure of the ionosphere. In particular, these platforms make possible *in situ* measurements of vertical and horizontal irregularities (e.g. sporadic E) that are difficult to interpret from ionograms. Thus useful comparisons are possible. Rocket measurements have been of special value for electron-density profiles in the D region, which cannot be explored easily by remote sensing from the ground. Probably the most successful ionospheric satellites are the joint Canadian/US topside sounders Alouette and the International Satellites for Ionospheric Studies (ISIS, see Sec. 8.2). Satellites with radio beacons (such as the orbiting NNSS satellites) and several geosynchronous satellites (such as ATS 1, ATS 3, GOES, SIRIO, ATS 6, and ETS II) have been of considerable value in studies of total electron content and radio scintillation (see Sec. 8.3). *In situ* measurements of electron-density irregularities have made possible the construction of models of scintillation. In recent years the Dynamics Explorer 1 satellite has provided dramatic views of the expansion of the auroral zone(s) during disturbed periods. Satellite instrumentation has provided valuable data on the densities, compositions, temperatures, and motions of the neutral atmosphere.

4.8.4 D-region techniques

Unlike the E, F_1, and F_2 layers the D region cannot be easily studied by ionosondes because the electron densities are too low and because, on frequencies that are reflected below 90 km, the interference from other radio services, such as amplitude-modulated broadcasting, is prohibitive. Further, on frequencies below about 250 kHz, it is difficult to build efficient wideband transmitting and receiving systems. Amplitude and phase measurements on VLF navigation systems provide information on temporal, spatial, and disturbance variations in the D region (see Ch. 10). A considerable body of information on the D region has been obtained from the absorption of high-frequency radio waves reflected from the upper layers (see Ch. 7) and from a nonlinear interaction of high-power HF waves with the region (see Sec. 11.3.2). Because nondeviative absorption varies inversely as the square of the radio frequency, during periods of high absorption there is a low-frequency cutoff f_{min} on ionograms which, for a given sounding system, is roughly a function of ionospheric absorption. The parameter f_{min} (e.g. see Fig. 4.13) is scaled routinely at ionospheric sounding stations as an indication of ionospheric absorption, and its use is justified provided that the absorption changes are large compared with the other variable characteristics of the ionosphere. Partial reflections of medium-frequency signals from refractive index irregularities provide valuable information on electron-density structures and their motions (see Sec. 11.3.3).

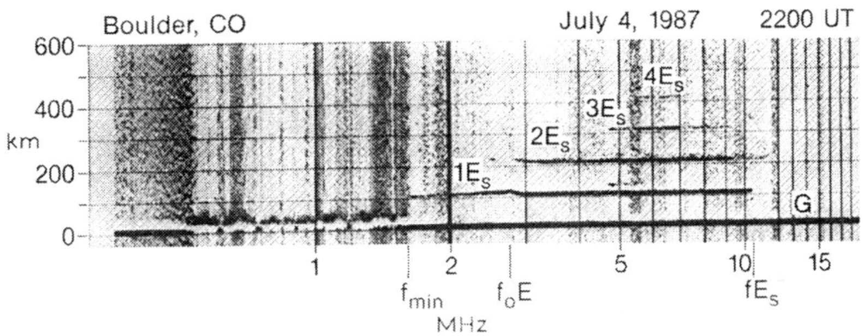

Fig. 4.13 *Ionogram with strong sporadic E (Courtesy of R. Conkright)*

4.9 Ionogram interpretation and scaling

One of the simplest types of ionogram (nighttime) is shown in Fig. 4.4*b*. It is seen that (i) the group height h' increases steadily with frequency until penetration occurs, and (ii) two distinct traces are present. The trace having the lower penetration frequency f_oF_2 is the ordinary wave trace, and the trace with the higher penetration frequency f_xF_2 is the extraordinary wave trace. The penetration frequencies are called critical frequencies. The two-hop echo has been reflected twice from the F_2 layer with an intermediate ground reflection. More complicated ionogram structures are shown in Figs. 4.4*a* and 4.7 in which there are reflections from the E, F_1, and F_2 layers. The ionogram structure can be further complicated by the presence of ordinary and extraordinary wave traces as can be seen for example in Figs. 4.4*b*, 4.6, 4.7, and 4.9. It is seen that, on frequencies just above f_oE and f_oF_1 and f_xF_1, the group height decreases with increase of frequency. This arises from retardation, below the level of reflection, which decreases with increase of frequency more rapidly than the retardation near reflection is increasing with frequency. The lowest heights of the ordinary (o) wave traces are scaled as the minimum virtual heights of the respective layers (e.g. $h'E$, $h'F_1$, $h'F_2$).

The ionogram in Fig. 4.13 shows a strong reflection near the 100 km level and penetration frequency above that of the normal E layer. In middle latitudes the critical frequency, fE_s, is highly variable and therefore the phenomenon is called sporadic *E*. In high latitudes this type of echo is common at night. Also, in high latitudes, three traces may be seen on an ionogram, such as that in Fig. 4.14.

The relationships between the z, o, and x critical frequencies are obtained from the reflection conditions $X = 1 + Y$, $X = 1$, and $X = 1 - Y$ (see Sec. 3.2). If $f_{N_{max}}$ is the value of the plasma frequency at the peak of the layer, we have

$$f_{N_{max}}^2 = f_z^2 + f_z f_H = f_o^2 = f_x^2 - f_x f_H \tag{4.16}$$

Hence,

$$f_x - f_z = f_H \tag{4.17}$$

Further, if f_o and f_x are much larger than f_H

$$f_x - f_o \approx \tfrac{1}{2} f_H \tag{4.18}$$

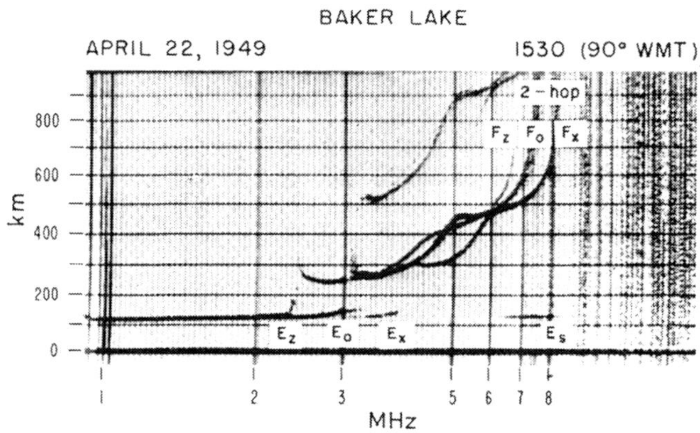

Fig. 4.14 *High-latitude ionogram showing x, o, and z reflections from the E, F_1, and F_2 layers*

We see that for a given f_H, which depends on geomagnetic location and altitude, the difference between the x and z critical frequencies is independent of the value of the critical frequencies, and the difference ($f_x - f_o$) between the extraordinary and ordinary wave critical frequencies is approximately so. Hence, in principle, the critical frequency separations give the magnetic field strength at the peak of the layer.

Ionograms vary greatly in appearance depending upon several factors such as geographic location, sunspot number, season, local time, and degree of disturbance. For examples of the wide variety of possible ionograms the reader is referred to Piggott (1975) and Shapley (1970).

4.10 Worldwide soundings

Although the earliest ionospheric sounding stations were established primarily for scientific purposes, a great expansion of the network during World War II resulted from the need to predict global radio propagation conditions. For the International Geophysical Year (1957–58) the number of sounding stations was increased to about 150 and, for effective interchange and use of data, uniform procedures for the interpretation of ionograms were established. Since the spatial and temporal variations of the ionosphere are so great, geographical coverage may be more important than accuracy of individual measurement. This point should be borne in mind in judging the value of ionospheric data. For example, the measurement of a critical frequency to say 1 kHz may be of value in a specific scientific study but this accuracy is of little value for long-term circuit planning because the critical frequencies are continuously changing in both space and time.

The global distribution of ionospheric stations has changed over the past 30 years but the total has remained about the same. The distribution of stations as of 1986 is shown in Fig. 4.15. Note that the distribution is far from uniform, there being large areas, mostly ocean, where there are no stations.

When measurements on a natural medium like the ionosphere are being considered, it is important to distinguish between controlled experiments, such as those conducted in a laboratory, and experiments in which man is essentially an observer. Ionospheric studies are usually observational, although some studies use modeling techniques involving laboratory plasmas and microwaves. Other exceptions include ionospheric modification by high-power radio waves, chemical modification, and perturbation by man-made explosions (see Ch. 14). It must be remembered that the natural uncontrolled changes are continuing during such modification experiments.

Comprehensive definitions and scaling rules have been developed (Piggott and Rawer 1978; Wakai *et al.* 1987). These include, for example, the selection of significant characteristics, scaling accuracy, and qualifying symbols for international data exchange. Observations are made hourly or at many stations every 15 min, and data are made available routinely on an hourly basis. The basic idea underlying ionogram scaling is to provide data sufficient for a user to visualize, to a greater or lesser extent, the original record. However, from a set of standard characteristics, the tabulated data cannot be an exhaustive description of the ionogram. Nevertheless they should represent the essential features of the once-reflected signals, for which multiple echoes and oblique reflections can be used as auxiliary guides to interpretation. With modern directional sounders, some of this information is readily available. With conventional sounders, skilled workers are required whose experience with ionogram sequences, multiple echoes, and peculiar echo shapes is vital to proper interpretation.

The following characteristics are usually scaled (Piggott and Rawer 1978): f_oF_2, f_oF_1, f_oE, f_oE_s, f_{min}, $h'F$, $h'F_2$, $h'E$, $h'E_s$, and $M(3000)F_2$. The MUF factor

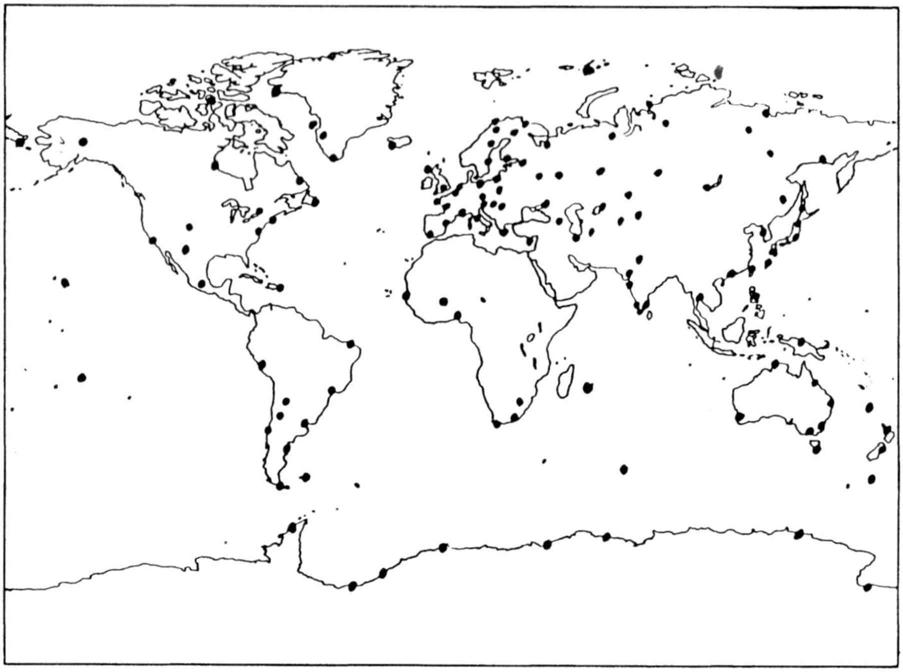

Fig. 4.15 *The global distribution of vertical sounding stations in 1986 (Courtesy of R. Conkright)*

for the F_2 layer $M(3000)F_2$ is the ratio of the maximum usable frequency for a distance of 3000 km to f_oF_2 (see Sec. 6.3.3.2). It is related to the height of the layer peak and the layer thickness (see eqn. 5.16). When a numerical scaling is questionable or impossible, qualifying letters are used to indicate the reason. For example: B indicates absorption, F shows spread echoes, C indicates equipment malfunction. A complete list of symbols is given in Piggott and Rawer (1978). Because the echo power varies greatly over a frequency sweep some parameters are gain sensitive, so the receiver gain setting that is best for scaling one parameter is not suitable for another. Thus ionogram interpretation and scaling can be facilitated by a gain run, that is, a sequence of three soundings at low, medium, and high receiver gains. These are particularly useful in the presence of absorption and spread F. When the ordinary wave trace is obscured (e.g. by interference) f_oF_2 may be obtained from f_xF_2 by using (4.18).

A particularly useful representation of ionogram data is the frequency- or *f*-plot, which was originally developed (by J. H. Meek) to enable the complicated and rapidly changing ionograms from high latitudes to be reduced with a minimum of interpretation. A sample *f*-plot is shown in Fig. 4.16; it provides an efficient means of recording and studying data that are too complicated or too numerous for conventional tabulation.

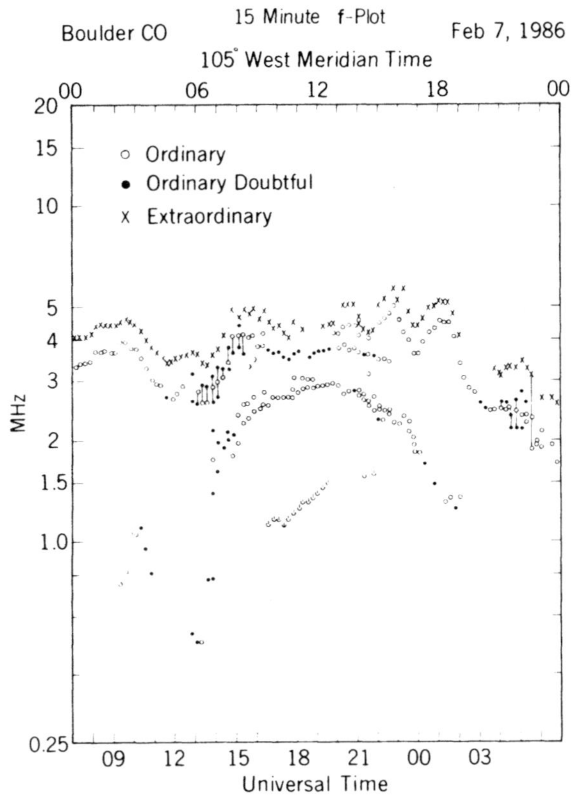

Fig. 4.16 *Sample f-plot of ionospheric data (Courtesy of R. Conkright)*

Ionospheric data are available from four World Data Centers set up under the auspices of the International Council of Scientific Unions: WDC-A in Boulder, Colorado, USA; WDC-B2 in Moscow, USSR; WDC-C1 in Chilton, England; and WDC-C2 in Tokyo, Japan. Catalogs of ionosphere vertical sounder data are available from each of the World Data Centers.

Ionospheric data are often presented in the form of monthly average values. The distributions of hourly values are nearly always skewed because some values are greatly displaced from the most probable average value, and sometimes it is possible to give only an upper or lower limit for an individual value. For these reasons the monthly median value has been adopted internationally. For the same reasons, the quartile values and the quartile range are adopted to show the variability of the hourly data.

4.11 Group height

The group (equivalent or virtual) height, as a function of frequency, is one of the most useful of ionospheric characteristics. It is measured relatively easily from ionograms, and a vast amount of data exists. The virtual height depends, in a complicated way, on the electron-density profile, the geomagnetic field, and the wave frequency. It is given by

$$h' = c\int_0^{h_r} \frac{dh}{u} = \int_0^{h_r} \mu' dh = h_0 + \int_{\phi_0}^{\phi_r} \frac{dh}{d\phi}\mu'(\phi)d\phi \qquad (4.19)$$

where h_r is the real height of reflection, u is group speed, and μ' is group refractive index. The parameter ϕ is some convenient function of the electron density (for example: $N, f_N, X, \log f_N$) and ϕ_0 and ϕ_r are the values of ϕ at the bottom of the ionosphere, h_0, and at the reflection height. For this change of variable it is essential that $dh/d\phi$ be finite everywhere along the path. This means that there can be no valleys in $\phi(h)$.

Equation (4.19) can be evaluated analytically only for certain special profiles: for transverse propagation of the ordinary wave and for longitudinal propagation of the x- and z-waves. Table 4.4 gives formulas for h' in some model layers. When there is no magnetic field the virtual heights are obtained by putting $Y = 0$. In the case of a realistic electron-density profile the virtual height is obtained numerically by summation over thin slabs:

$$h' = h_0 + \sum_i \overline{\mu_i'}\,\Delta h_i \qquad (4.20)$$

The group refractive index μ' is given by

$$\mu'\mu = \mu^2 + \frac{2}{D}\left[1 - \mu^2 - X^2 + \frac{(1-\mu^2)(1-X^2)\,Y_L^2}{A}\right]; \qquad (4.21a)$$

here

$$D = 2(1-X) - Y_T^2 - A \qquad (4.21b)$$

and

$$A^2 = Y_T^4 + 4Y_L^2(1-X)^2 \qquad (4.21c)$$

and μ^2 is given by (3.9). In calculating the virtual height, special precautions have to be taken near the level of reflection where μ' approaches infinity. Curves showing the variation of μ' with X for the ordinary wave have been given by Shinn and Whale (1952). The following approximations are useful near reflection:

Table 4.4 *Virtual heights for longitudinal propagation in model layers (x and z waves)*

$N(h)$	$h'(f)$	
$N = a(h - h_0)$	$h' = h_0 + \frac{2f^2}{ka}(1 \pm \frac{2}{3}Y)$	(1)
$f_N = \beta(h - h_0)$	$h' = h_0 + \frac{\pi}{2\beta}[1 \pm (f_H/f)^2]^{1/2}(f \pm \frac{1}{4}f_H)$	(2)
$f_N^2 = F^2 e^{h/H}$	$h' = 2H\left[\tanh^{-1}\mu_0 \mp \frac{Y}{2(1 \pm Y)}\mu_0\right]$ or	(3)
	$h' = H\left(\ln\frac{1+\mu_0}{1-\mu_0} \mp \frac{Y}{1 \pm Y}\mu_0\right)$, where $\mu_0^2 = 1 - \frac{(F/f)^2}{1 \pm Y}$	(4), (5)
$f_N^2 = f_c^2\left[\frac{2(h - h_0)}{y_m} - \frac{(h - h_0)^2}{y_m^2}\right]$	$h' = h_0 + \frac{(1 \pm Y)^{1/2}}{2}\left(\frac{f}{f_c}\right)y_m\left\{1 \pm \frac{Y}{4(1 \pm Y)}\left[1 + \frac{Y}{1 \pm Y}\left(\frac{f_c}{f}\right)^2\right]\right\}$ $\times \ln\frac{f_c + (1 \pm Y)^{1/2}f}{f_c - (1 \pm Y)^{1/2}f} \pm \frac{Yy_m}{4(1 \pm Y)}$	(6)

Note: When $\pm$ and $\mp$ are used, all top signs go together (denoting the z wave) and all bottom signs go together (denoting the x wave).

For the ordinary wave

$$\mu'_+ \approx \varepsilon^{1/2} \operatorname{cosec} \theta \tag{4.22a}$$

where

$$\varepsilon = 1 - X \tag{4.22b}$$

and for the extraordinary wave

$$\mu'_- \approx (2 - Y)[2(1 - Y)(1 + \cos^2\theta)\eta]^{-1/2} \tag{4.23a}$$

where

$$\eta = 1 - X - Y \tag{4.23b}$$

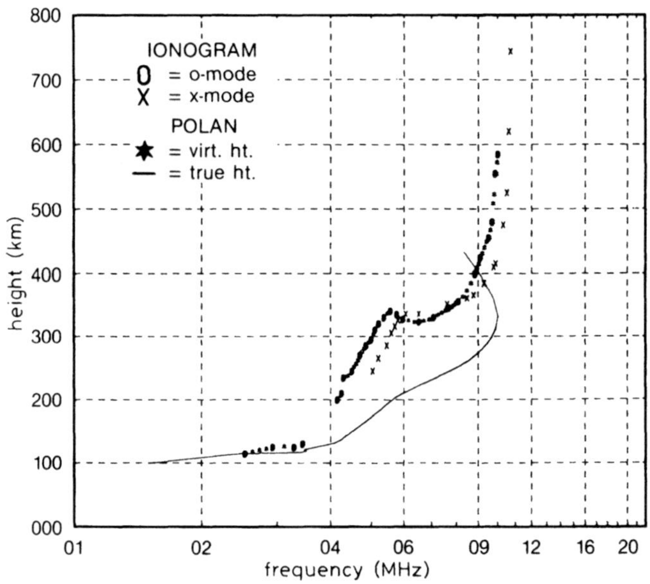

Fig. 4.17 *Sample scaled ionogram for middle latitudes and the real height profile obtained using POLAN (Courtesy of R. Conkright)*
The profile above the peak is extrapolated

4.12 Real height

When the influences of the Earth's magnetic field and collisions are neglected, the problem of inverting the group-height integral (4.19) is relatively straightforward (e.g. Davies 1969, Secs. 8.7.2 and 8.7.3). In the general case, including the Earth's magnetic field, a variety of numerical techniques are available for inversion of the ordinary wave curve to an equivalent monotonic electron-density profile (*Radio Science* 1967). One of the most comprehensive is the Polynomial Analysis program POLAN developed by Titheridge (1985), and designed for maximum accuracy and reliability. The program uses polynomial real height sections of any required degree, fitting any number of points. Use of data for the ordinary and extraordinary wave makes possible an estimate of the size of a valley between two layers. Physically unacceptable solutions are rejected by imposing limits on the profile parameters. The virtual height can be calculated from the profile and compared with the input data. Procedures are included to deal with the (unobserved) underlying ionization and valleys. The unobserved ionization results from the fact that the lower frequency limit of the ionogram is not zero. POLAN uses extrapolation and interpolation to fill in gaps in ionogram scaling (e.g. gaps that may exist near penetration, or inadequate density of data points). For routine work POLAN may be used as a 'black box' with the following input: virtual height, frequency, dip, and gyrofrequency. Optimized default procedures are then used. When the input data are not self-consistent, and thus imply some physically unacceptable feature in the profile, this is noted and corrected. Copies of the program are available on magnetic media from World Data Center–A. In POLAN the function ϕ in (4.19) is f_N. A sample electron-density profile obtained by using POLAN and the ionogram from which the profile was obtained are shown in Fig. 4.17.

4.13 References

BALSLEY, B. B., ECKLUND, W. L., CARTER, D. A., and JOHNSTON, P. E. (1979): 'The poker flat MST radar. First results', *Geophys. Res. Lett.*, **5**, p. 921

BARRY, G. H. (1971): 'A low-power vertical-incidence ionosonde', *IEEE Trans.*, **GE-9**, p. 86

BIBL, K., and REINISCH, B. (1978): 'The universal digital ionosonde', *Radio Sci.*, **13**, p. 519

BOWLES, K. L. (1961): 'Incoherent scattering by free electrons as a technique for studying the ionosphere and exosphere. Some observations and theoretical considerations', *J. Res. NBS*, **65D**, p. 1

DAVIES, K. (1969): 'Ionospheric radio waves' (Ginn-Blaisdell, Waltham, MA, USA)

DEVLIN, J. C., DYSON, P. L., and HAMMER, P. R. (1977): 'A pulse synthesis technique for improving ionosonde resolution', *Radio Sci.*, **12**, p. 767

EVANS, J. V. (1969): 'Theory and practice of ionosphere study by Thomson scatter radar', *Proc. IEEE*, **57**, p. 496

EVANS, J. V. (1973): 'Millstone Hill Thomson scatter results for 1968'. Technical Report 499, Lincoln Laboratory, MIT, Lexington, MA, USA

GETHING, P. J. D. (1978): 'Radio direction finding'. IEE Electromagnetic Wave Series 4 (Peter Peregrinus, Ltd., Stevenage, England)

GORDON, W. E. (1958): 'Incoherent scattering of radio waves by free electrons with applications to space exploration by radar', *Proc. IRE*, **46**, p. 1824.

GREENWALD, R. A., BAKER, K. B., HUTCHINS, R. A., and HANUISE, C. (1985): 'An HF phased-array radar for studying small-scale structure in the high-latitude ionosphere', *Radio Sci.*, **20**, p. 63

GRUBB, R. N. (1979): 'The NOAA SEL HF radar system (Ionospheric Sounder)'. NOAA Tech. Memo. ERL SEL-55

HUNSUCKER, R. D. (1990): 'Radio techniques for probing the terrestrial ionosphere' (Springer-Verlag, New York)

IGARASHI, K., OGAWA, T., OSE, R., FUJII, R., and HIRASAWA, T. (1982): 'A new VHF Doppler radar experiment at Syowa Station', *Mem. Natl. Inst. Polar Res.* Special Issue **22**, p. 258

JONES, T. B., SPRACKLEN, C., STEWART, C. P., and THOMAS, E. C. (1981): 'SABRE – a UK-German auroral radar'. Proc. IEE Conf. Ant. Prop., IEE Conf. Publ. 195, p. 269

KELLY, T. D. (1986): 'The need for real-time remote control of vertical incidence ionosondes' *in* SIMON, P. A., HECKMAN, G., and SHEA, M. A. (Eds): 'Solar-terrestrial predictions', NOAA/ERL, Boulder, CO, USA, p. 571

MATTIN, N., and JONES, T. B. (1986): 'SABRE observations of E-region vertical velocity structures', *J. Atmos. Terr. Phys.*, **48**, p. 625

McCLURE, J. P. (1965): 'Electron density studies at Jicamarca' in DE MENDONÇA, F. (Ed.): 'Equatorial aeronomy', Brazil Space Commission, Brazilia, Brazil, p. 170

McNAMARA, A. G., McDIARMID, D. R., SOFKO, G. J., KOEHLER, J. A., FORSYTH P. A., and MOORCROFT, D. R. (1983): 'BARS – a dual bistatic auroral radar system for the study of electric fields in the Canadian sector of the auroral zone'. *Adv. Space Res.*, **2**, p. 145

NIELSEN, E. (1982): 'The STARE system and some of its applications' in RUSSELL, C. T., and SOUTHWOOD, D. J. (Eds): 'The IMS source book'. American Geophysical Union, Washington, DC, p. 213

PATERSON, B., and BEVINS, C. (1989): 'The IPS fifth generation ionospheric monitoring system'. IPS Tech. Rep. TR-89-01, IPS Radio Space Services, Darlinghurst, NSW, Australia

PAUL, A. K. (1979), 'Radio pulse distortion in the ionosphere', *J. Geophys. Res.*, **46**, p. 15

PIGGOTT, W. R. (1975): 'High-latitude supplement to the URSI handbook of ionogram interpretation and reduction'. Report UAG-50, World Data Center-A, Boulder, CO, USA

PIGGOTT, W. R., and RAWER, K. (1978): 'URSI handbook of ionogram interpretation and reduction'. Report UAG-23A, World Data Center-A, Boulder, CO, USA

POOLE, A. W. V. (1985): 'Advanced sounding 1. The FMCW alternative', *Radio Sci.*, **20**, p. 1609

POOLE, A. W. V., and EVANS, G. P. (1985): 'Advanced sounding 2. First results from an advanced chirp ionosonde', *Radio Sci.*, **20**, p. 1617

Radio Science (1967): 'Special issue on analysis of ionograms for electron density profiles', *Radio Sci.*, **2**, (10), pp. 1119–1282

REINISCH, B. W. (1986): 'New techniques in ground-based ionospheric sounding and studies', *Radio Sci.*, **21**, p. 331

RÖTTGER, J., KLOSTERMEYER, J., CZECHOWSKY, P., RÜSTER, R., and SCHMIDT, G. (1978): 'Remote sensing of the atmosphere by VHF radar experiments', *Naturwissenschaften*, **65**, p. 285

SHAPLEY, A. H. (1970): 'Atlas of ionograms'. Report UAG-10, World Data Center-A, Boulder, CO, USA

SHINN, D. H., and WHALE, H. A. (1952): 'Group velocities and group heights from the magneto-ionic theory', *J. Atmos. Terr. Phys.*, **2**, p. 85

THOMSON, J. J. (1906): 'Conduction of electricity through gases' (Cambridge University Press)

TITHERIDGE, J. E. (1985): 'Ionogram analysis with the generalized program POLAN.' Report UAG-93, World Data Center-A, Boulder CO, USA

VILLAIN, J. P., CAUDAL, G., and HANUISE, C. (1985): 'A SAFARI-EISCAT comparison between the velocity of F-region small-scale irregularities and the ion drift', *J. Geophys. Res.*, **90**, p. 8433

VILLARD, O. G. (1976): 'The ionospheric sounder and its place in the history of radio science', *Radio Sci.*, **11**, p. 847

WAKAI, N., and ISHIZAWA, K. (1962): 'A method of simultaneous recording of the echo amplitude with h'-f curve in the ionosphere sounding and some application to the measurement of ionospheric parameters', *J. Radio Res. Labs.*, **9**, p. 421

WAKAI, N., OHYAMA, H., and KOIZUMI, T. (1987): 'Manual of ionogram scaling; Third version', Radio Res. Labs., Japan

WATTS, J. M., and DAVIES, K. (1960): 'Rapid frequency analysis of fading radio signals', *J. Geophys. Res.*, **65**, p. 2295.

WHALE, H. A. (1969): 'Effects of ionospheric scattering on very-long-distance radio communication' (Plenum Press, New York)

WRIGHT, J. W., and PITTEWAY, M. L. V. (1982): 'Data processing for the dynasonde', *J. Geophys. Res.*, **87**, p. 1589

Chapter 5

Morphology of the ionosphere

5.1 Synoptic variations of the ionosphere

5.1.1 Diurnal variations

Sample day and night $N(h)$ profiles for a middle latitude are shown in Fig. 5.1. Large diurnal changes occur particularly in the lower ionosphere. The daytime structure may be thought of as a single bank of electrons starting around 100 km, having a peak density near 300 km and decaying steadily at higher altitudes. The lower E and F_1 layers may be only inflections, or ledges, in the electron distribution, i.e. no valleys. At night, all vestige of the F_1 layer disappears, and the peak density of the E layer falls to about $1 \times 10^9\,m^{-3}$ ($f_oE \approx 0{\cdot}3$ MHz), thus producing a relatively simple structure. The electron density in the D region usually cannot be observed by an ionosonde but it falls from about $10^9\,m^{-3}$ at 80 km at noon to less than $10^8\,m^{-3}$ at night. Some sample electron-density profiles for the D region, shown in Fig. 5.2, illustrate the day-to-day variability.

Additional features of the diurnal variations in ionospheric structure are illustrated in Fig. 5.3. Note that the E layer appears promptly at sunrise and essentially disappears at sunset (except for the residual ionization at night). The electron densities are greater at all heights by day than by night except at Huancayo, where the density continues to increase for a few hours after sunset. The broken lines in Fig. 5.3 are extrapolations based on a Chapman model since the topside cannot be explored by ground-based ionosondes. Information in the topside can be obtained from topside sounders (Sec. 8.2), satellite radio beacons (Sec. 8.3) and incoherent scatter radars (Sec. 4.7).

There is a general tendency for the height of the F_2 peak, h_mF_2, to fall at dawn, and then to rise during the afternoon or evening. In low latitudes, h_mF_2 reaches a high level by about 1900 local time and then falls such that at midnight it is about 100 km lower than at noon. This evening uplift of the equatorial F_2 layer results from upward electromagnetic ($\boldsymbol{E} \times \boldsymbol{B}$) drift caused by a west-to-east electric field generated in the E region. This is called the fountain effect (Hanson and Moffett 1966). This height variation is especially noticeable at Huancayo where at 1900 LT to 2000 LT the F-layer peak can be above 500 km.

In the highest latitudes, the ionosphere may be in continuous sunlight or darkness for long periods, depending upon the season. In these circumstances the ionosphere will be typical of daylight or nighttime conditions, respectively.

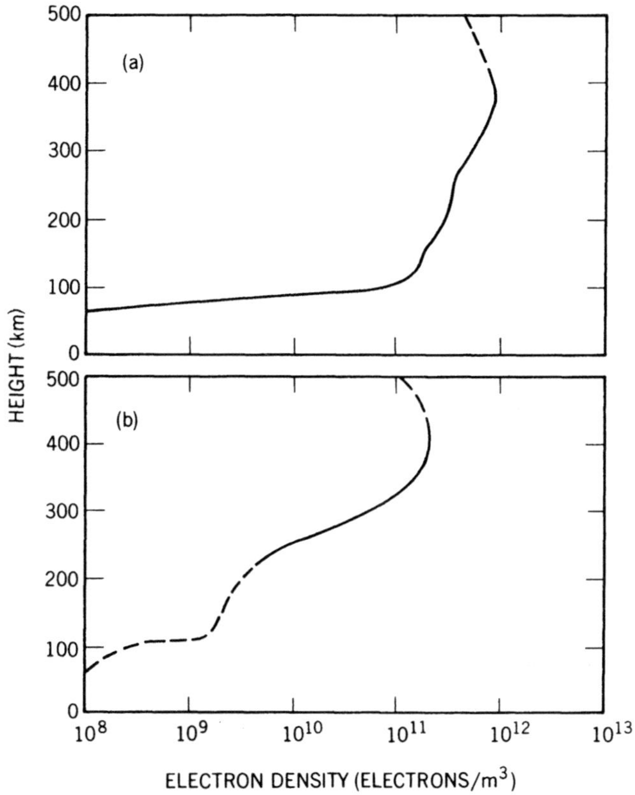

Fig. 5.1 *Sample electron-density profiles*
(*a*) Daytime
(*b*) Nighttime

A moderate diurnal variation may result from small variations of the solar zenith angle, but a diurnal variation is still detectable at the South Pole, where the solar zenith angle is essentially constant. This variation confirms that, in high latitudes at least, factors besides solar illumination have a role in determining the diurnal variations of the ionosphere. One important factor is the ($\boldsymbol{E} \times \boldsymbol{B}$) convection of ionization from lower daylit latitudes toward the nightside. This mechanism accounts also for the maintenance of the F_2 layer during the long winter night.

5.1.2 Seasonal variations

The seasonal variation in $N(h)$ profiles is illustrated in Figs. 5.4 and 5.5 for midnight and noon conditions, respectively. The nighttime F layer tends to be at higher heights in summer than in winter, the tendency being accentuated in lower latitudes. The nighttime F layer tends to be thicker when higher. In general, the maximum electron density and the total electron content of the nighttime F region are greater in summer than in winter. Turning to the noon profiles, the most distinctive feature is that, in middle latitudes, the peak density is considerably larger in winter than in summer; this is called the seasonal anomaly. It is more evident in high middle latitudes than in low and high latitudes. In the summer months, in middle latitudes, the F layer bifurcates into the F_1 and F_2 layers. Under these conditions the F_2 peak density is fairly small and is located at a relatively high altitude. The F_1 layer is not so much a distinct layer but rather a minor inflection in the profile at about 180 to 220 km.

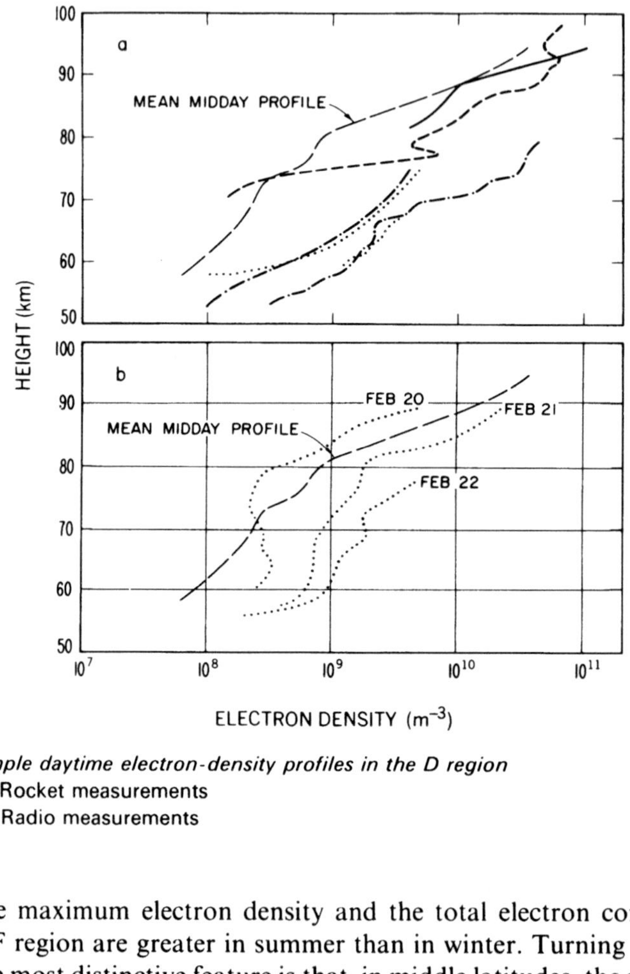

Fig. 5.2 *Sample daytime electron-density profiles in the D region*
(*a*) Rocket measurements
(*b*) Radio measurements

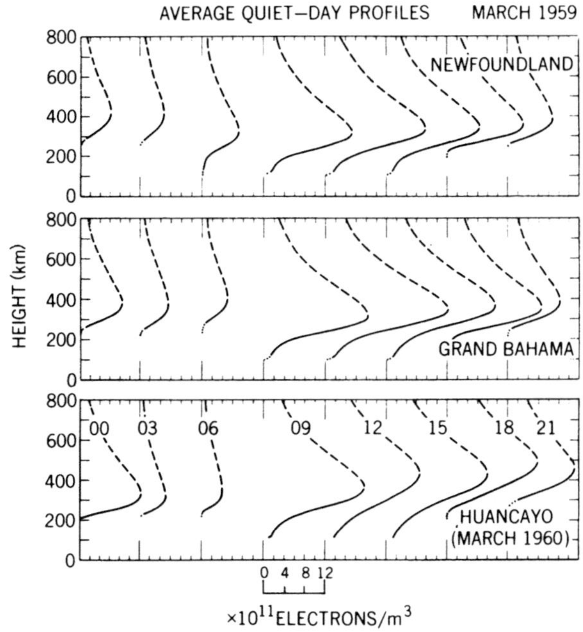

Fig. 5.3 *Diurnal changes in ionospheric profiles*

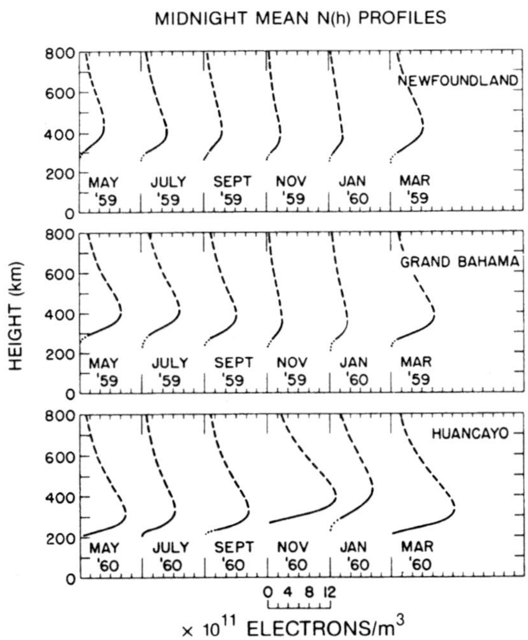

Fig. 5.4 *Seasonal changes in midnight N(h) profiles*

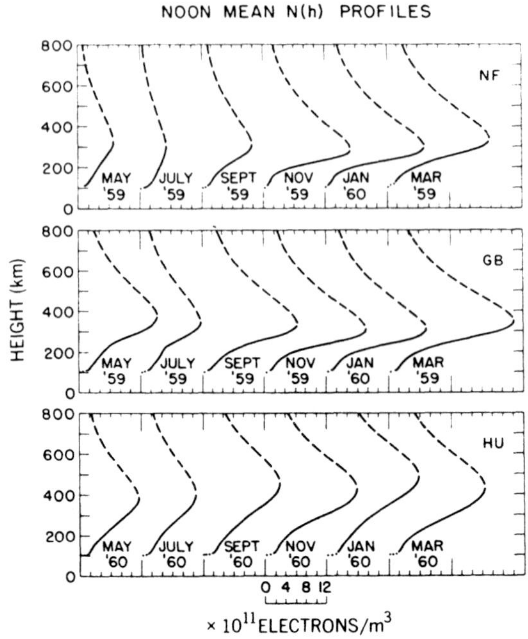

Fig. 5.5 *Seasonal changes in noon N(h) profiles*

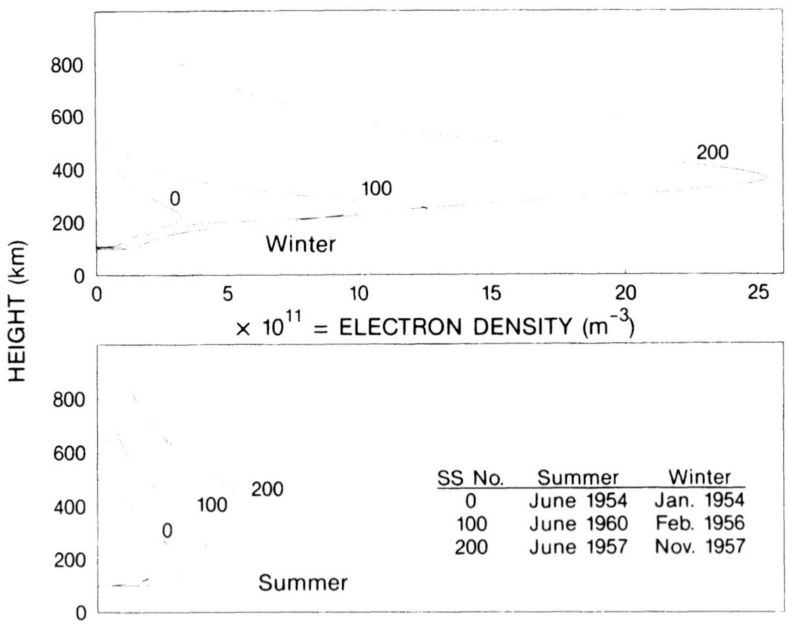

Fig. 5.6 *Typical N(h) profiles at sunspot numbers 0, 100, and 200 (after Wright 1962)*

5.1.3 Solar cycle variations

The long-term variation in sunspot occurrence has a profound effect on the noon electron distribution, as can be seen from Fig. 5.6. The graphs are arranged to show increments in ionization corresponding to increments of 100 in sunspot number. The electron densities above the peak are extrapolated. A striking feature of this variation is that relatively the largest increases in electron density occur at successively higher heights.

5.1.4 Geographical variations

The global structure of the ionosphere can be understood by examination of iso-ionic contours in a meridian plane together with contour maps of the worldwide distribution of critical frequencies. Figure 5.7 is an ionospheric

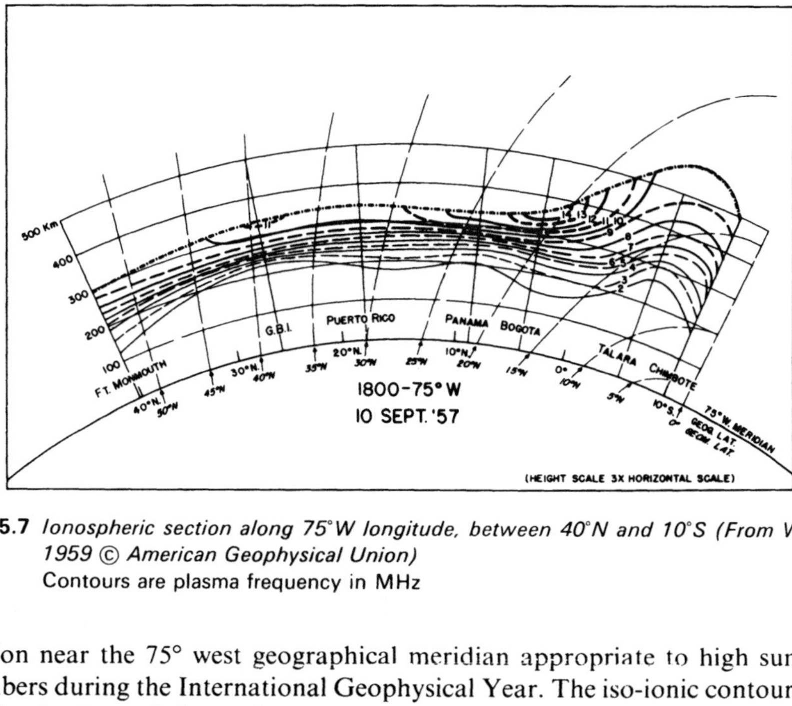

Fig. 5.7 *Ionospheric section along 75°W longitude, between 40°N and 10°S (From Wright 1959 © American Geophysical Union)*
Contours are plasma frequency in MHz

section near the 75° west geographical meridian appropriate to high sunspot numbers during the International Geophysical Year. The iso-ionic contours are for fixed values of plasma frequency up to the height of the F_2 maximum. The following features are important:

(i) The F_2 layer is much thicker near the Equator than elsewhere.
(ii) There are regions of high electron concentration at geomagnetic latitudes of about $\pm 20°$ during the early evening, which arise from the fountain effect mentioned in Sec. 5.1.1. This distortion of the ionosphere is important in transequatorial radio propagation, and is considered in Sec. 6.7.2.5. The maxima are referred to as the equatorial, or Appleton, anomaly (Appleton

1954). Many electron-density profiles are available from the various World Data Centers.

5.2 Temporal and spatial variations of critical frequency

5.2.1 General features

For many purposes, including radio communications, the most important characteristics of the ionospheric layers are their critical frequencies and virtual heights. In Fig. 5.8 are shown the critical frequencies of the E, F_1, and F_2 layers observed at a middle-latitude station (Washington, DC) over nearly 50 years. These graphs bring out some important features:

(i) The seasonal variation of f_oE and f_oF_1 are in phase with solar elevation whereas f_oF_2 is in antiphase (winter anomaly).
(ii) The F_1 layer disappears in winter.
(iii) There is a marked increase in the critical frequencies, in sympathy with the 11-year sunspot cycle.

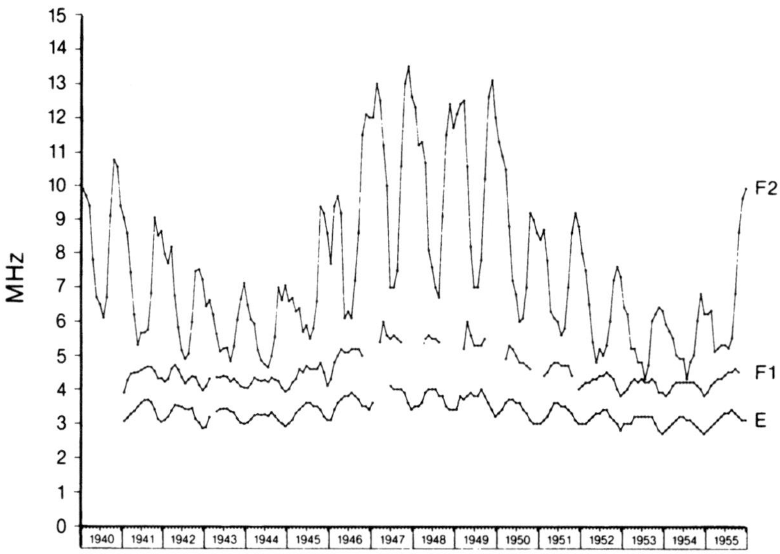

Fig. 5.8 *Monthly median noon critical frequencies at Washington, DC, from 1940 to 1987 (Courtesy C. M. Rush)*

5.2.2 The E layer

To a first approximation the daytime critical frequencies (in MHz) of the E layer are given by

$$f_oE = 0{\cdot}9[(180 + 1{\cdot}44R_{12}) \cos \chi]^{0{\cdot}25} \tag{5.1}$$

where R_{12} is the 12 month running-mean sunspot number and χ is the solar zenith angle. The exponent may vary between 0·1 and 0·4 for the diurnal variation but (5.1) is a reasonably good fit for seasonal and diurnal variations.

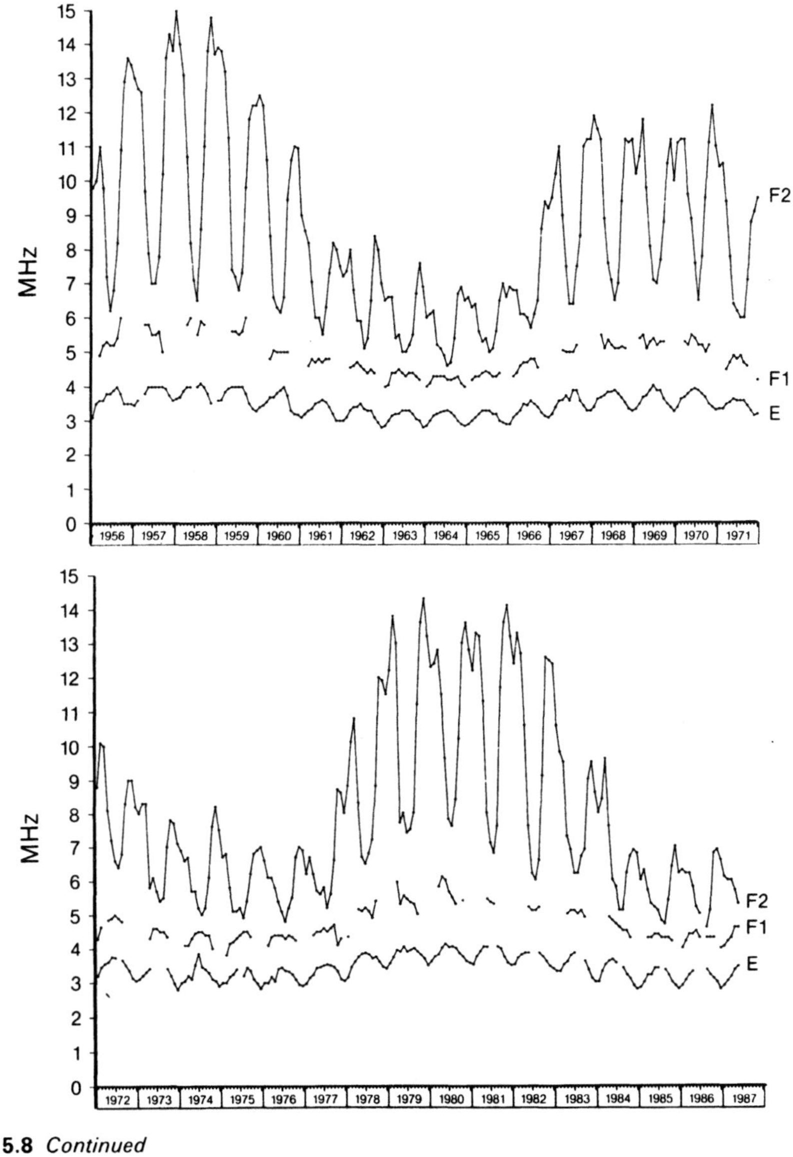

Fig. 5.8 *Continued*

Values of f_oE determined from (5.1) are usually within 0·2 MHz of the observed values. Muggleton (1975) developed an alternative equation. During the night, f_oE reaches a minimum of about 0·5 MHz (Wakai 1971). Leftin (1976) gave the following formulas for f_oE (in MHz) at midnight and at sunrise, sunset:

$$f_oE\text{ (midnight)} = 0{\cdot}36[1 + 0{\cdot}0098R_{12}]^{1/2} \tag{5.2}$$

$$f_oE\text{ (sunrise, sunset)} = 1{\cdot}05[1 + 0{\cdot}008R_{12}]^{1/2} \tag{5.3}$$

The nighttime f_oE varies by about 0·2 MHz from minimum to maximum of the sunspot cycle.

Formulas (5.1), (5.2), and (5.3) do not hold in high latitudes (i.e. above about 70° latitude). Further, in the auroral zone, nighttime ionization is produced by

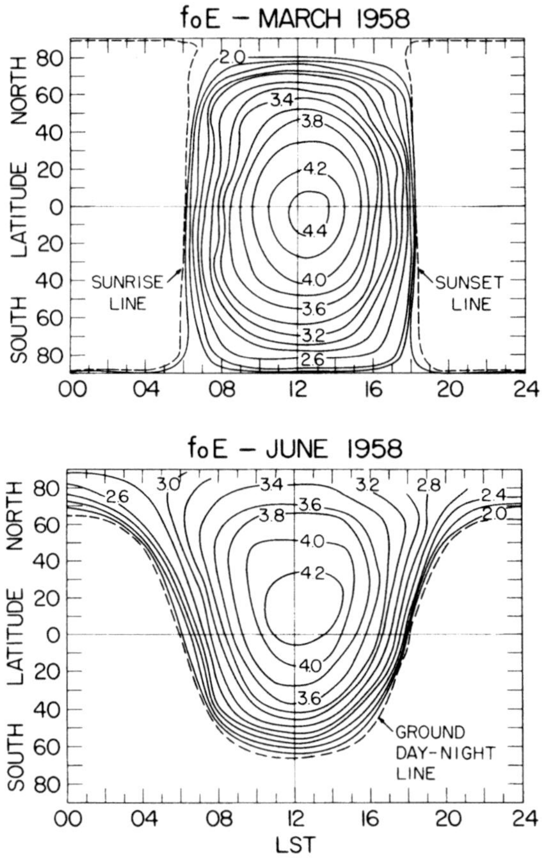

Fig. 5.9 *Variation of f_oE with local time and latitude according to (5.1)*

particles from the magnetosphere. Global maps of median f_oE, based on observations, for equinox, summer, and winter, for sunspot maximum (1958) have been given by Leftin (1976). Examples of the local time variations of f_oE with latitude are given in Fig. 5.9.

5.2.3 The F_1 layer

The F_1 layer is observed only during the day; it is more pronounced during the summer than during the winter, and at low sunspot numbers, and during ionosphere storms, that is, when f_oF_2 is low.

To a first approximation (in MHz),

$$f_oF_1 = (4{\cdot}3 + 0{\cdot}01R_{12}) \cos^{0{\cdot}2}\chi \tag{5.4}$$

A contour plot of f_oF_1 based on (5.4) with month and local solar time at Washington, DC, is shown in Fig. 5.10.

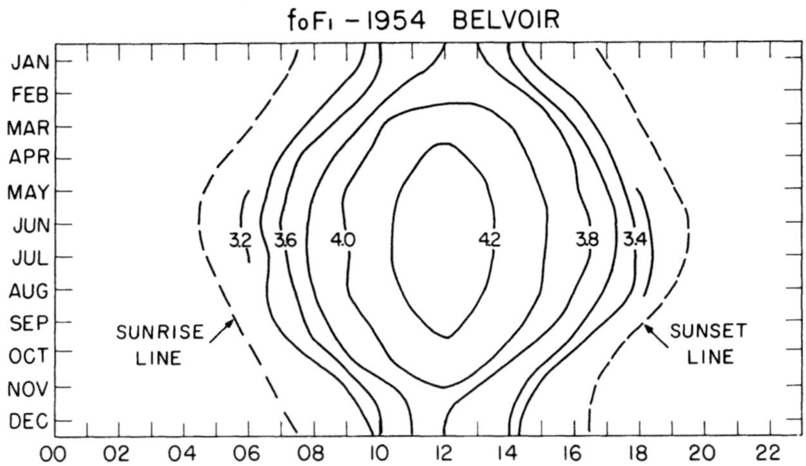

Fig. 5.10 *Variation of f_oF_1 with local time and season according to (5.4)*

A more comprehensive expression for f_oF_1 in terms of geomagnetic latitude Φ and an ionospheric index I_{F_2} (Minnis 1964) is given by Ducharme *et al.* (1971):

$$f_oF_1 = f_s \cos^n\chi \tag{5.5a}$$

where

$$f_s = f_o(f_{100} - f_o)\frac{I_{F_2}}{100} \tag{5.5b}$$

$$f_o = 4{\cdot}408 + 0{\cdot}0076\Phi - 0{\cdot}000152\Phi^2 \tag{5.5c}$$

$$f_{100} = 5{\cdot}365 + 0{\cdot}0129\Phi - 0{\cdot}000248\Phi^2 \tag{5.5d}$$

and

$$n = 0{\cdot}11 + 0{\cdot}0038\Phi - 0{\cdot}000045\Phi^2 + 0{\cdot}0003\, I_{F_2} \tag{5.5e}$$

These formulas apply for values of χ less than about 40° (with $I_{F_2} = 100$).

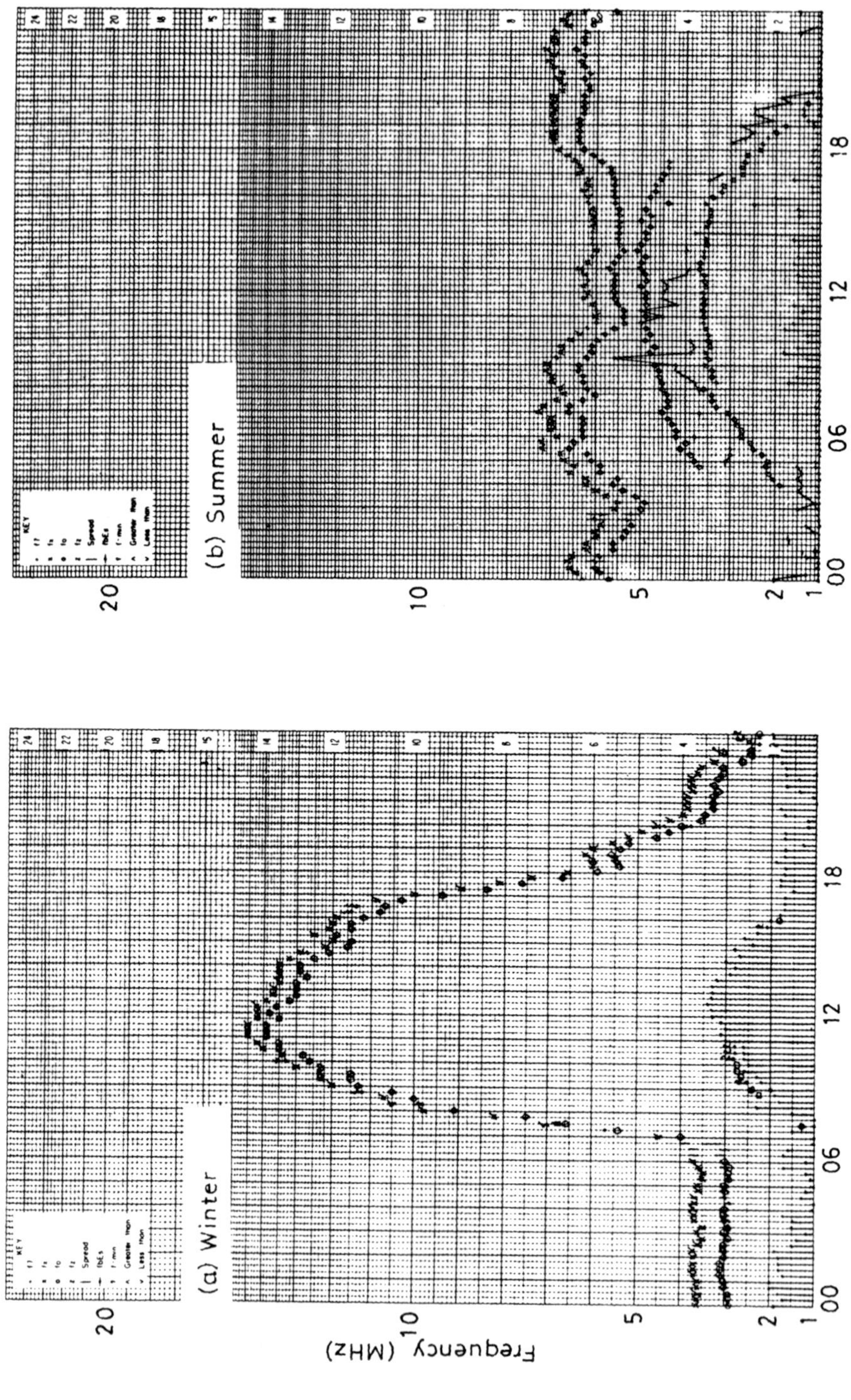

Fig. 5.11 *Sample f-plots showing diurnal and seasonal changes at Adak, Alaska*
(*a*) 14 January 1959
(*b*) 6 June 1959

Somewhat similar expressions have been given by Rosich and Jones (1973). Thus, a peak value of f_oF_1, for $I_{F_2} = 150$ and $\Phi \simeq 45°$, is about 6 MHz.

The I_{F_2} index is a solar-related index based on observations of f_oF_2; it was developed by Minnis (1964). It employs noontime monthly median values of f_oF_2 observed at 13 ionosonde stations located throughout the world. Linear regression equations representing the relationship between the monthly median noon f_oF_2 and the 3-month weighted sunspot number are used to convert to an equivalent sunspot number at each of the stations. The conversion process is carried out using the f_oF_2 data from the 13 stations, and the median of these calculated equivalent sunspot numbers is defined as I_{F_2}.

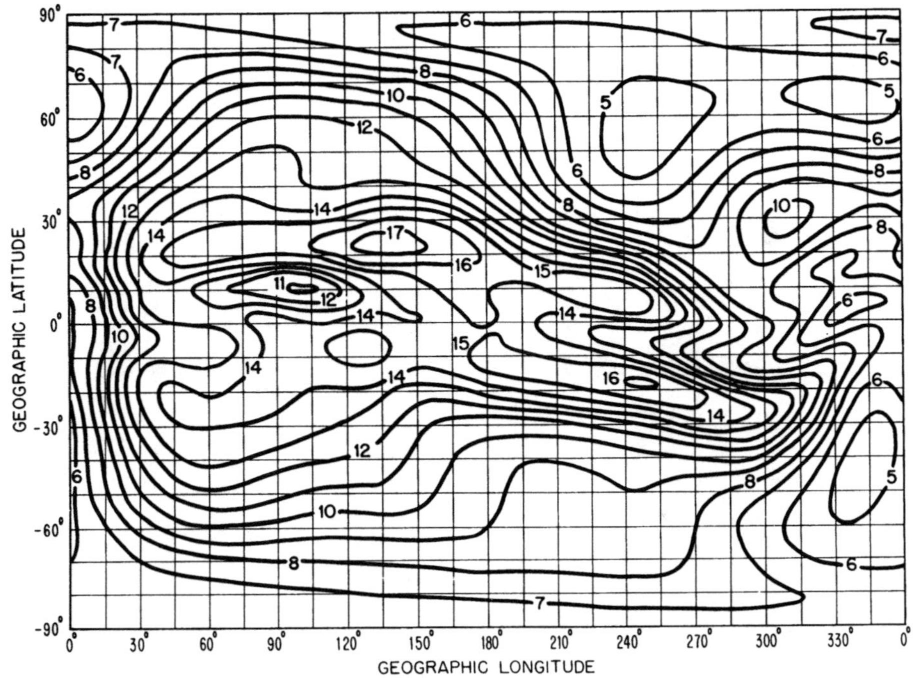

Fig. 5.12 *Contour map (in megahertz) of the global representation of the median value of f_oF_2 for March 1979 at 0600 UT*

5.2.4 The F_2 layer

From the point of view of HF radio communications, the F_2 layer is by far the most important layer and it is also the most variable. Unlike f_oE and f_oF_1, f_oF_2 does not follow a $\cos\chi$ law either diurnally or seasonally. This is illustrated in the f plots for Adak (Fig. 5.11). Although during January the Sun is at a low elevation ($\cos\chi = 0{\cdot}291$) compared with June ($\cos\chi = 0{\cdot}877$), the January noon f_oF_2 is more than twice the summer value. This implies a peak electron density ratio of more than 4 to 1. Notice that this 'winter anomaly' occurs in the daytime only; it is thought to be due to a large summer electron loss rate caused by an increase in the molecular-to-atomic composition of the neutral atmosphere. In summer the f_oF_2 shows remarkably little diurnal variation. Note also the presence of a consistent F_1 layer during the summer day and its complete absence in the winter day.

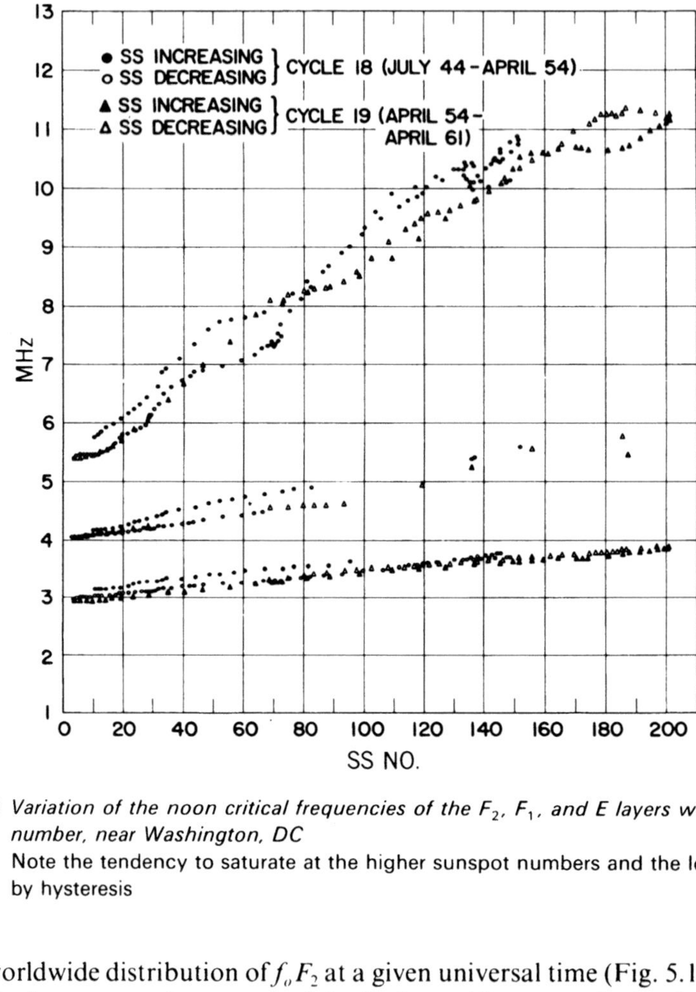

Fig. 5.13 *Variation of the noon critical frequencies of the F_2, F_1, and E layers with sunspot number, near Washington, DC*

Note the tendency to saturate at the higher sunspot numbers and the loop caused by hysteresis

The worldwide distribution of f_oF_2 at a given universal time (Fig. 5.12) reveals the marked geomagnetic control of the F_2 layer. The most distinctive features on the map are the two regions of high f_oF_2 lying at about $\pm 20°$ dip latitude, which is called the equatorial (or Appleton) anomaly. These 'highs' lag behind the subsolar point but move around with the Sun. Because of the geomagnetic control, the critical frequencies of the F_2 layer exhibit a marked longitudinal effect. The highs not only move from east to west but also vary in geographic latitude, tending to remain at a fixed magnetic dip.

Turning to the sunspot cycle dependence of f_oF_2 (Fig. 5.13), we see the variation of the 12-month running-average value of f_oF_2 at Washington, DC, plotted against the 12-month running-average Zurich relative sunspot number. Corresponding data for f_oF_1 and f_oE are included. The relationships between the F_2 critical frequencies and the sunspot numbers are roughly linear for sunspot numbers between 0 and about 150. There is evidence of hysteresis, indicating differences between the increasing and decreasing parts of the cycle, and there is evidence of 'flattening' at the highest sunspot numbers, which has to be taken into account when sunspot numbers are used to predict F_2 critical frequencies. These saturation and hysteresis effects appear to result from the definition of sunspot number in terms of sunspot groups and individual spots (see Sec. 2.2.1.2).

5.3 Model layers

5.3.1 Uses of models

For many purposes, particularly for computation, it is convenient to approximate an electron-density (or other) profile by analytical expressions that are amenable to mathematical manipulation. The advantages of a model are that it can be used to check computer programs (e.g. ray tracing) and it provides a common reference for different workers. Models can be categorized in several ways, for example, (*a*) numerical, physical, and semi-empirical, (*b*) simple or complicated. Examples of simple models include sharply bounded slab, linear in plasma frequency, linear in electron density, parabolic, quasi-parabolic layers, Chapman layer, sech-squared layer, and exponential layer. More complicated models include polynomial layers, layers made up of several simple layers (both separate and overlapping), and numerical (tabulated) profiles.

Besides electron-density profiles, there are numerous models of the global distribution of ionospheric characteristics. The most widely used are the numerical maps developed originally by the National Bureau of Standards and now maintained by the National Telecommunications and Information Administration (NTIA) in Boulder, Colorado (Jones and Gallet 1965). More extensive models such as the International Reference Ionosphere (Rawer 1981) include the NTIA maps.

5.3.2 Simple model layers

The following layers are extensively used in radio propagation work, and some expressions for virtual heights of reflection are given in Table 4.4.

(i) *Square-root layer*

$$f_N = \beta(h - h_0) \tag{5.6}$$

where β is the height gradient of plasma frequency and h_0 is the height of the bottom of the layer.

(ii) *Linear layer*

$$N = \alpha(h - h_0) \tag{5.7}$$

where α is the electron density gradient and h_0 the layer base.

(iii) *Exponential layer*

$$N = N_R \exp[(h - h_R)/H] \tag{5.8}$$

where N_R is the electron density at a reference height h_R, and H is a scale height that is negative in the topside of the ionosphere.

(iv) *Sech-squared layer*

$$N = N_m \operatorname{sech}^2[(h - h_m)/a] \tag{5.9}$$

where N_m is the maximum electron density at a height h_m and a is thickness.

(v) *Chapman layer*

$$N = N_0 \exp \tfrac{1}{2}[1 - z - \exp(-z)] \tag{5.10a}$$

$$z = (h - h_0)/H \tag{5.10b}$$

where N_0 is the peak density at a height h_0, and H is a scale height.

(vi) *Parabolic layer*

Near the peak of a Chapman layer z is small and (5.10a) reduces to

$$N = N_0\left[1 - \left(\frac{h - h_0}{y_m}\right)^2\right] \tag{5.11a}$$

where

$$y_m = 2H \tag{5.11b}$$

is the half-thickness of the parabola, i.e. the distance from peak to base h_0 of the layer. Typical values of y_m for the F_2 layers lie in the range 50–100 km.

(vii) *Quasi-parabolic layer*

$$N = N_m\left[1 - \left(\frac{r - r_m}{y_m}\frac{r_0}{r}\right)^2\right] \tag{5.12}$$

where r is the radial distance from the center of the Earth, r_m is the radial distance of the peak density N_m, $y_m = r_m - r_0$, and r_0 is the radial distance to the bottom of the layer.

5.3.3 Numerical maps

The term 'numerical map' is used to denote a function $\Omega(\phi, \lambda, t)$ of the variables latitude (ϕ), longitude (λ), and universal time (t), originally developed for ionospheric purposes by Jones and Gallet (1965). The hour angle T ranges from $-180°$ to $+180°$, $T = 0$ at noon, and $T = 180°$ at midnight. The function Ω has the form of a Fourier series in T:

$$\Omega(\phi, \lambda, t) = a_0(\phi, \lambda) + \sum_{j=1}^{H} [a_j(\phi, \lambda) \cos jT + b_j(\phi, \lambda) \sin jT] \tag{5.13}$$

where

$$a_j(\phi, \lambda) = \sum_{k=0}^{K} U_{2j,k} G_k(\phi, \lambda) \tag{5.14a}$$

$$b_j(\phi, \lambda) = \sum_{k=0}^{K} U_{2j-1,k} G_k(\phi, \lambda) \tag{5.14b}$$

In (5.13), H denotes the number of harmonics used to represent the diurnal variation. The geographical coordinate functions $G_k(\phi, \lambda)$ are of the form $\sin^q x \cos^m \phi \cos m\lambda$ in which an integer q depends on the harmonic order m of the longitude. These functions are tabulated in Jones *et al.* (1969). The variable x is a modified form of the magnetic dip I; it is defined by

$$\sin x = I/(I^2 + \cos \phi)^{1/2} \tag{5.15}$$

This parameter expresses the fact that the ionosphere, especially the F_2 layer, is controlled by the magnetic dip as well as the geographic latitude (see Sec. 5.2.4). Thus, a numerical map is defined by a set of coefficients, $U_{2j,k}$. The larger the number of coefficients used, the greater the degree of detail in the maps. Put another way, economizing on computer time by decreasing the number of coefficients gives a coarser map. Figure 5.12 is a graphical representation of a numerical map.

Important advantages of the numerical maps are that (i) as new data become available, the coefficients can be easily revised, and (ii) the maps can be reproduced exactly. Further, the coefficients can be updated by the incorporation of semi-theoretical results (Rush *et al.* 1984). The most common numerical maps are global contours of various monthly median values for fixed hours in UT.

Maps of the critical frequency of the E layer, f_oE (Leftin 1976), the critical frequency of the F_1 layer, f_oF_1 (Rosich and Jones 1973), and the critical frequency of the F_2 layer, f_oF_2 (Jones *et al.* 1969; Jones and Obitts 1970) have been used to represent ionospheric structure (Nisbet 1971), HF propagation conditions (Barghausen *et al.* 1969; CCIR 1986), and transionospheric propagation factors (Bent *et al.* 1978). The maps have been modified with daily ionospheric observations to represent daily maps (Rush 1978) and to represent storm departures from monthly median values (Jones *et al.* 1973).

5.3.4 *Composite models*

One of the simplest composite models is a two-parabola model; one parabola representing the E layer, and one the F_2 layer is shown in Fig. 5.14*a*, e.g. Appleton and Beynon (1947). The parabolas may be distinct or overlapping. An ionospheric profile developed by Bradley and Dudeney (1973) and shown in Fig. 5.14*b* consists of the following:

(i) A parabolic E layer below its height of maximum density h_mE, with semi-thickness y_mE. The height of the peak h_mE is fixed at 110 km and $y_mE = 20$ km

(ii) A parabolic F_2 layer with peak density at a height $h_m F_2$, and semi-thickness $y_m F_2$

(iii) A linear increase of electron density with height between $h_m E$ and the point on the parabolic F_2 layer where the plasma frequency f_j is $1{\cdot}7\, f_0 E$.

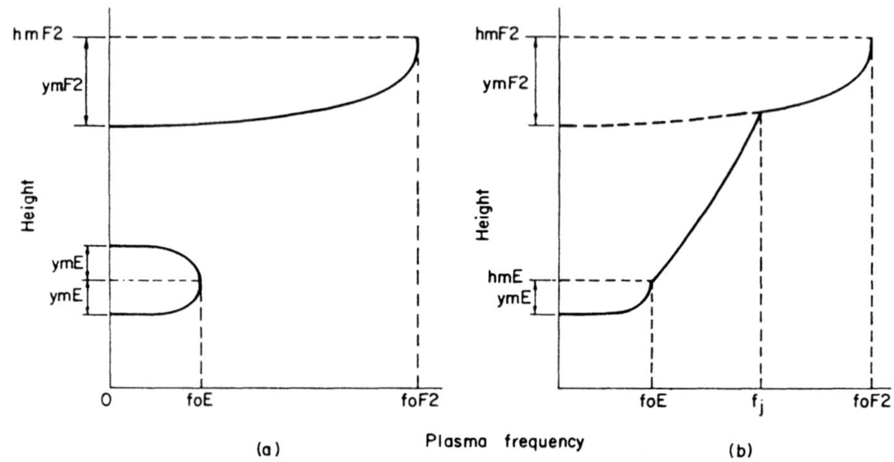

Fig. 5.14 *(a) Electron density profile composed of a parabolic E layer and a semi-parabolic F layer*
(b) Electron density profile adopted by Dudeney and Bradley consisting of semi-parabolic E and F layers with linear height gradient between the peak of the E layer and the height, in the F layer, at which the plasma frequency equals $1{\cdot}7\ f_o E$
(From Bradley 1979 with permission)

The F_1 layer is omitted on the basis that the cusps on ionograms, which are scaled as F_1-layer characteristics, result from only minor fluctuations in electron density. This model has the advantage that there is no abrupt change when $f_o F_1$ is not given.

The height of the peak density of the F_2 layer is derived from M, the $M(3000)F_2$ (see Sec. 6.6):

$$h_m F_2 = \frac{1490}{M + D} - 176 \qquad (5.16a)$$

with

$$D = \frac{0{\cdot}18}{x - 1{\cdot}4} \qquad (5.16b)$$

and

$$x = f_o F_2 / f_o E \qquad (5.16c)$$

The semi-thickness $y_m F_2$ is obtained from the measured minimum height $h'F$

of the F_2 layer:

$$y_m F_2 = h_m F_2 - h'F + C \tag{5.17a}$$

where

$$C = \left[\frac{0{\cdot}618}{x - 1{\cdot}33} \right]^{0{\cdot}86} (h_m F_2 - 104) \tag{5.17b}$$

The parameters $f_o E$, $f_o F_2$, and $h_m F_2$ can be obtained from ionograms or from numerical maps.

Bent *et al.* (1978) developed a model of the ionosphere, designed for trans-ionospheric propagation calculations, with particular emphasis on the topside. The bottomside is described by a bi-parabola:

$$N(h) = N_m F_2 \left[1 - \left(\frac{h - h_m F_2}{y_m} \right)^2 \right]^2 \tag{5.18}$$

The topside profile below 1000 km is subdivided into four intervals. The upper three cover equal height intervals and assume in each a constant logarithmic decrement which depends on the 10·7 cm solar noise flux. In the fourth interval, just above the peak, a parabolic shape is used; it meets the lowest of the exponential intervals such that the gradient is continuous at the junction height h_0, where

$$h_0 = h_m F_2 + y_m/4 \tag{5.19a}$$

and

$$N_0 = 0{\cdot}8640\ N_m F_2 \tag{5.19b}$$

Here N_0 is the electron density at height h_0 and $h_m F_2$ is given by (5.16*a*) (Bilitza *et al.* 1979).

5.3.5 The International Reference Ionosphere

The International Reference Ionosphere (IRI) is a compendium of height profiles for four plasma parameters: plasma density, plasma temperatures of electrons and ions, and ion composition (Rawer *et al.* 1978; Rawer 1981; Rawer and Bradley 1987). In essence the IRI is a computer program that produces profiles. The program is available from a World Data Center. The worldwide values of the main F_2 plasma parameters, peak density and height, are obtained from numerical maps (CCIR 1967). The peak heights $h_m F_2$ are obtained from maps of $M(3000)F_2$ and a modified form of (5.16).

The electron-density profile is shown in Fig. 5.15. The bottomside is composed of five height intervals, numbered 2 through 6; a topside profile is numbered 1. Above the E layer (6) it contains a valley (5), which is small near noon but deeper at low solar elevations and at night. Above the valley is an 'intermediate region'; it bridges the gap between the valley and the F layer (2), which may or may not contain an F_1 layer (3). The IRI contains no information on sporadic layers but includes a D layer together with profiles of electron temperature (T_e), ion temperature (T_i), and ion composition.

5.3.6 Four-dimensional model

The values of f_0F_2 and h_mF_2 obtained from numerical maps can be combined with suitable layer shapes to provide electron-density profiles (e.g. Tascione *et al.* 1988) at any latitude, longitude, and time. This program includes models of the ionosphere in the auroral zones and in the trough.

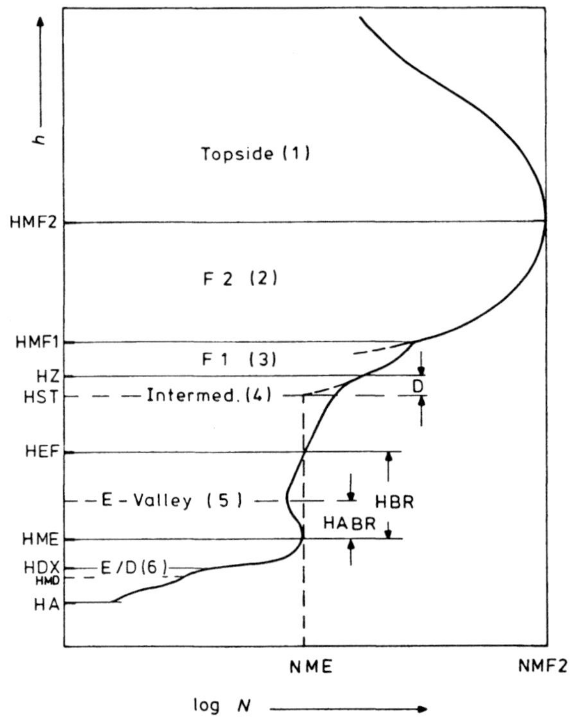

Fig. 5.15 *The electron-density profile model used in the International Reference Ionosphere (From Rawer 1981)*

5.4 The D region

The D region is important in radio propagation because (*a*) it absorbs energy from waves at MF, HF, and VHF, and (*b*) it reflects LF and VLF signals. Because of the small electron densities in the D region ($\approx 10^9$ el.m^{-3}) and high electron-neutral collision frequencies ($\approx 10^6$ s^{-1}) the D region cannot be studied using conventional ionosondes. Profile information can be obtained by high-power radio technqiues, for example, wave interaction (Lee and Ferraro 1969;

Sec. 11.3.2), partial reflection (Belrose 1970), and incoherent scatter (Farley 1970), and by rocket measurements (see Sec. 11.3.4). Some samples of daytime D-region profiles are given in Fig. 5.2, from which we see the high variability of the D region. George (1971) showed that the D region is partly under geomagnetic control; there are zones of high absorption of high-frequency waves near 20° north and 20° south of the magnetic equator.

5.5 Sporadic E

In addition to the regular layers of the ionosphere there are several that are transient or irregular. The most important of these is the sporadic-E layer, often designated E_s. Sporadic E is important because it can reflect radio waves up to frequencies of about 100 MHz. Some sample $N(h)$ profiles for the E region are shown in Fig. 5.16. Notice the presence of detailed structure. Unlike the normal E-layer critical frequency, the sporadic-E critical frequency ($f_o E_s$) is highly variable in time and space. It can vary from less than $f_o E$ (≈ 2 to 3 MHz) to more than 30 MHz at a given location (Matsushita and Smith 1975). Leftin *et al.* (1968) constructed monthly median maps of $f_o E_s$ for various universal times. Because of the high variability, these authors also gave maps of the lower and upper deciles of the monthly distributions. Seasonal maps of the percentage of time $f_o E_s$ exceeds 7 MHz were constructed by Smith (1978) and appear in Recommendation 534–2 of CCIR (1986). Worldwide patterns of occurrence of E_s indicate that the variations of the different types of E_s depend on geographical location (Fig. 5.17). Close to the dip equator, E_s is essentially a daytime phenomenon with little seasonal variation. On the other hand, in the auroral zone, E_s is most prevalent during night but again there is little seasonal variation. Auroral sporadic E is produced by energetic electrons from the magnetosphere. In middle latitudes $f_o E_s$ is less than it is elsewhere, and the occurrence is subject to diurnal and seasonal variations. In general, E_s is more prevalent in local summer than in winter and by day rather than by night.

Some examples of sporadic-E reflections on standard ionograms are shown in Fig. 5.18. Note that the maximum frequency returned from the layer can be much greater than that from the normal E layer and, indeed, from any of the normal layers. Sometimes the E_s traces show retardation near the maximum frequency, indicating a relatively thick layer with a well-defined maximum of electron density. Under other conditions the echoes exhibit no retardation, suggesting that the layer is extremely thin or else that the reflection is due to a sharp change in electron density gradient. On some occasions the E_s layer is opaque and blankets the upper layers; on other occasions the upper layers can be seen through the E_s, which suggests that the E_s is patchy and that the radio waves are penetrating through the gaps. Sometimes magnetoionic splitting is evident; at other times there is none. Sporadic E can occur in patches about 100 km wide.

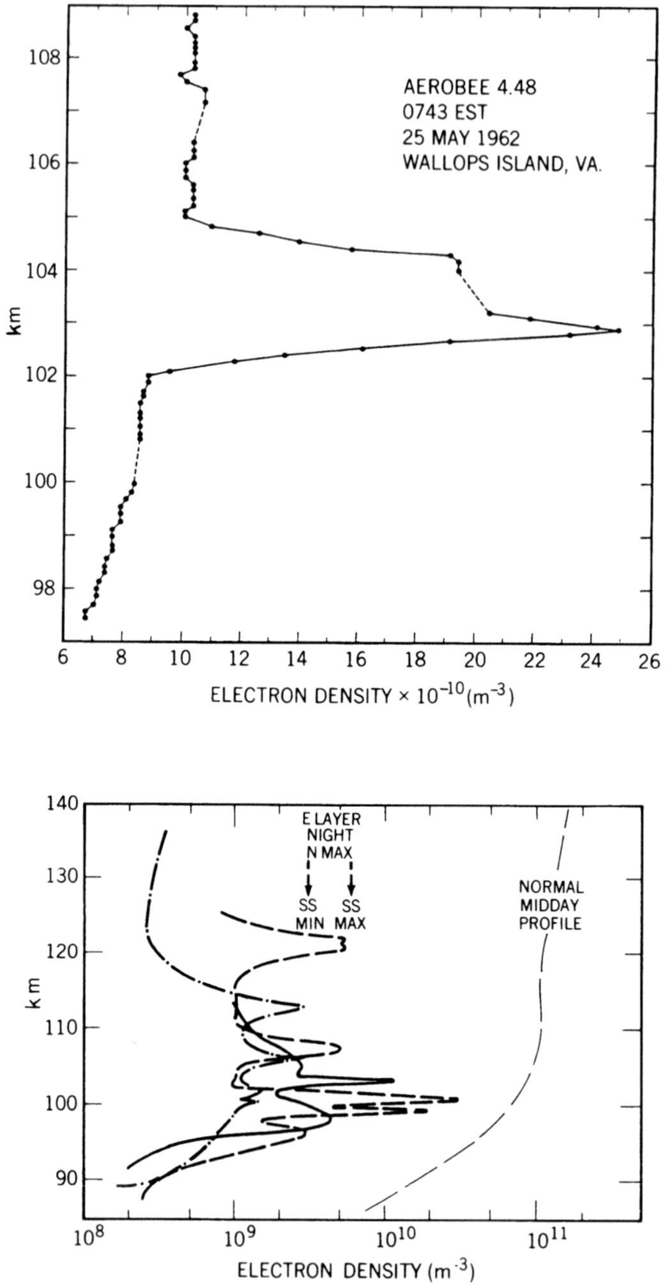

Fig. 5.16 *(a) Electron-density profile of daytime sporadic E, obtained with a rocket sounding (From L. G. Smith 1966)*

(b) Electron-density profile of the nighttime E layer, showing structure of sporadic E

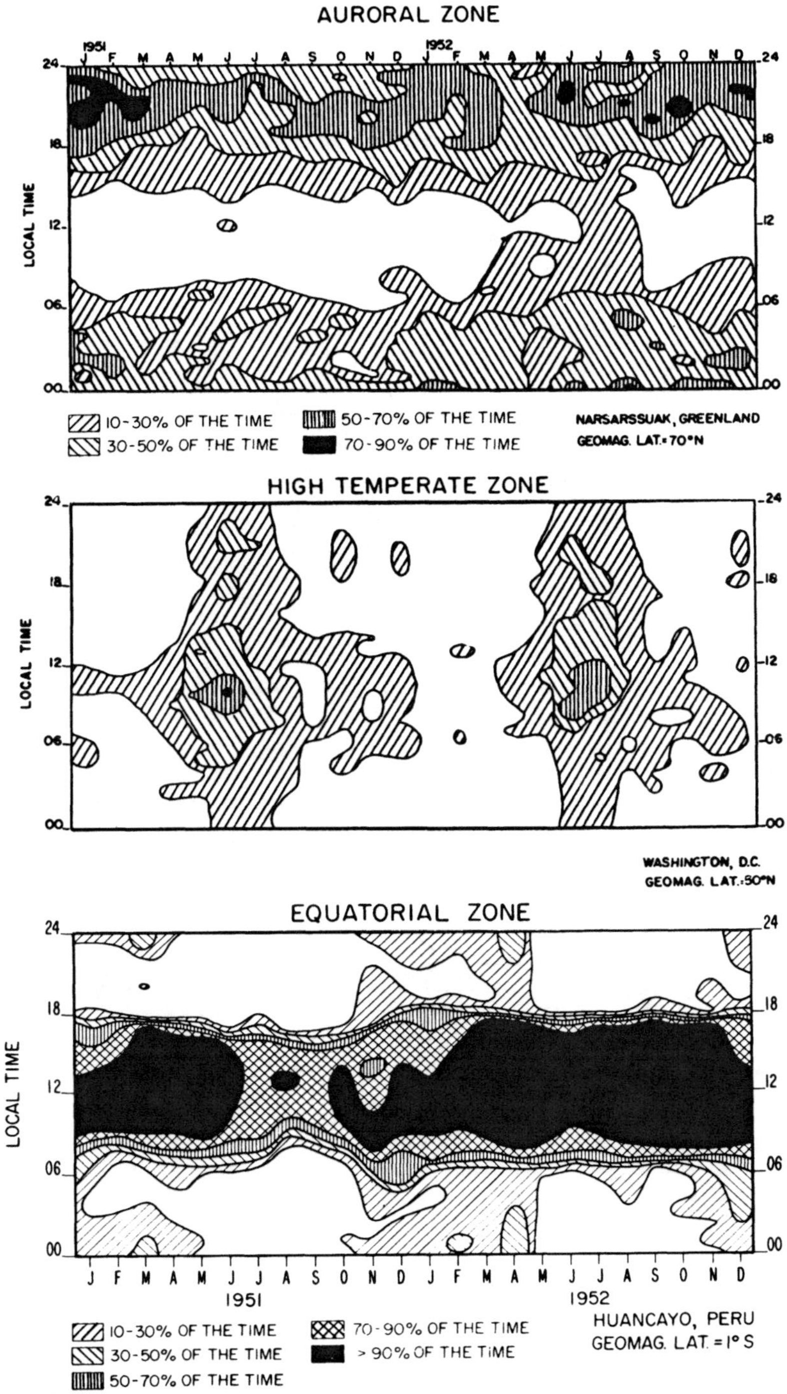

Fig. 5.17 *Seasonal and diurnal variations in percentage occurrence of $fE_s > 5$ MHz in three latitude zones (auroral, high temperate, and equatorial) (From E. K. Smith 1957)*

Sporadic E can occur over a range of heights from about 90 to 120 km or more, and evidence has been advanced for preferred heights (Smith and Matsushita 1962). There is some evidence also for believing that one type of E_s is due to ionization by meteors. It is almost certain that several different physical phenomena are lumped together under the general heading of sporadic E. For the scaling of E_s characteristics see Piggott and Rawer (1978). Some types of E_s, such as that encountered near the magnetic equator, are well behaved and are regular features of the ionosphere. One characteristic of sporadic E, its partial transparency, gives rise to a 'blanketing' frequency $f_b E_s$ below which E_s is opaque and above which it is transparent. Leftin and Ostrow (1969) constructed maps of $f_b E_s$ (medians and deciles).

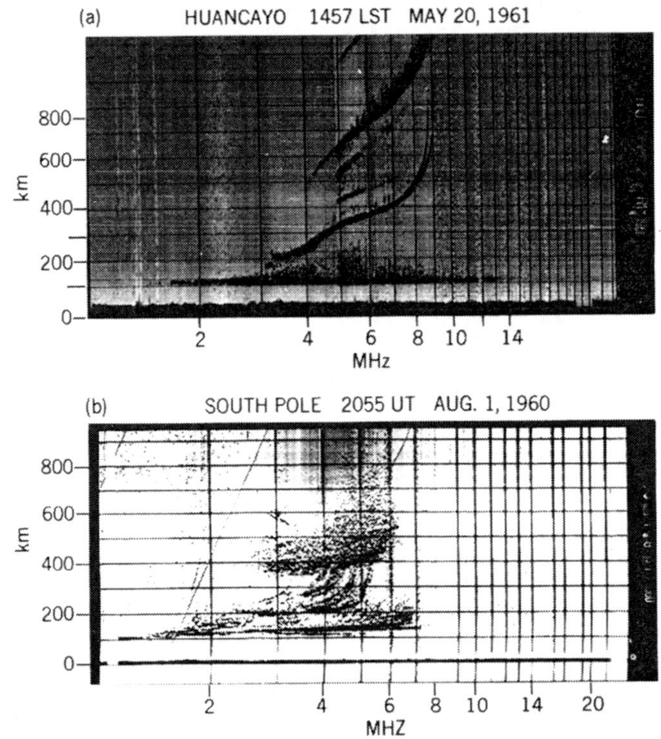

Fig. 5.18 *Examples of sporadic-E echoes*
(*a*) Near the magnetic equator
(*b*) at the South Pole

5.6 Spread F

5.6.1 General features of spread F

It is often observed that the echo pulse reflected from the F_2 layer has a much longer (10 times) duration than the transmitted pulse; this phenomenon is called spread F. Undoubtedly, spread F is caused by the scattering of the signal from irregularities embedded in the ionosphere both in depth and away from the zenith. Irregularities are important in HF and VHF radio propagation because the time spreading limits the rate of information transfer and because it produces high fading rates. Some examples of ionograms with spread-F echoes are presented in Figs. 5.19 and 5.20 from which it is seen that the phenomenon occurs in high and low latitudes. Spread F is generally defined in terms of the appearance of an ionogram rather than in terms of the physical mechanism involved. This has led to a division into 'range spread' and 'frequency spread'.

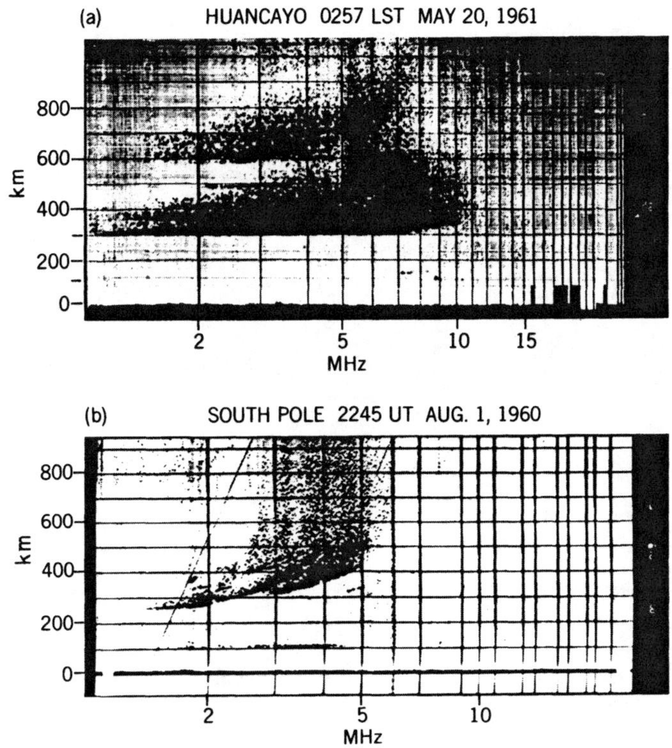

Fig. 5.19 *Two examples of spread F*
(*a*) Range spread commonly observed near the Equator
(*b*) Frequency spread usually observed in high latitudes

Frequency spread refers to diffuseness near the critical frequency of the F_2 layer whereas range spread refers to diffuseness near the horizontal part of the trace. The appearance of spread F on an ionogram is denoted, in scaling, by the letter F either as a qualifying symbol or used alone when numerical scaling is not possible. The presence of spread F on ionograms is an indicator of the occurrence of plasma irregularities in the ionosphere. An extensive study of spread F and ionospheric F-region irregularities has been made by Herman (1966). Although spread F is an impediment to accurate scaling of ionograms it gives valuable information on the occurrences, both geographical and temporal, of F_2-layer irregularities. In middle and high latitudes much spread F appears to occur in patches, as wide as several hundreds of kilometers, of magnetic field-aligned irregularities whose sizes transverse to the magnetic field are about 1 km. These patches of ionization drift horizontally with speeds of about $100\,\mathrm{m\,s^{-1}}$. Near the Equator, the irregularities drift eastward. They are associated with rising plumes in the topside observed with the incoherent scatter radar in Jicamarca, Peru (Fig. 5.21). These plumes are seen up to heights of about 800 km.

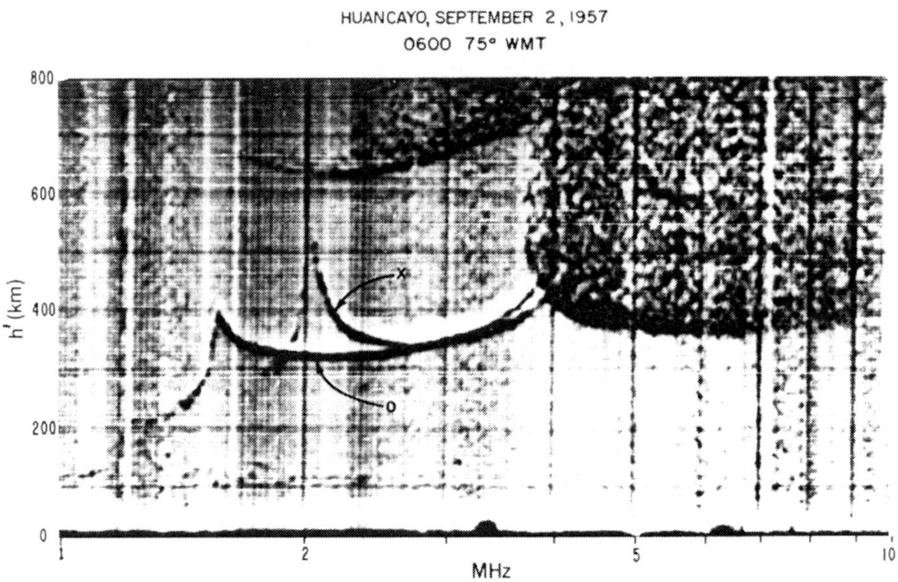

Fig. 5.20 *An example of spread F observed near the magnetic equator*

Some ionograms (e.g. Fig. 5.22) exhibit spread on multiple echoes only. This is believed to be due to ground backscatter and two F reflections. Spread F on ionograms has been synthesized (e.g. by Calvert and Cohen 1961).

5.6.2 Geographical variations of spread F

The dominant feature in spread-F occurrence is the existence of two areas: (i) an equatorial region, invariably on the night side, and (ii) a high-latitude region mostly on the dark side but sometimes on the dayside. Equatorial spread F occurs in latitudes bounded by $\pm 20°$, and high-latitude spread F begins at about 40° geomagnetic latitude; its probability of occurrence increases with latitude. There is little spread-F occurrence in geomagnetic latitudes between 20° and 40°.

5.6.3 Diurnal variations of spread F

Near the magnetic dip pole there is a region of permanent spread F where, during summer, spread F occurs 90–100% of the time (Herman 1966) at night

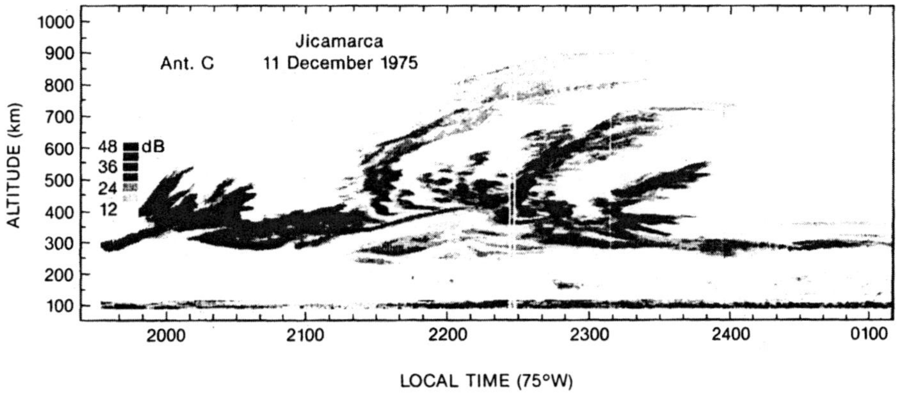

Fig. 5.21 *Incoherent scatter record at Jicamarca, Peru, showing rising plumes of ionospheric irregularities after sunset near the magnetic equator*

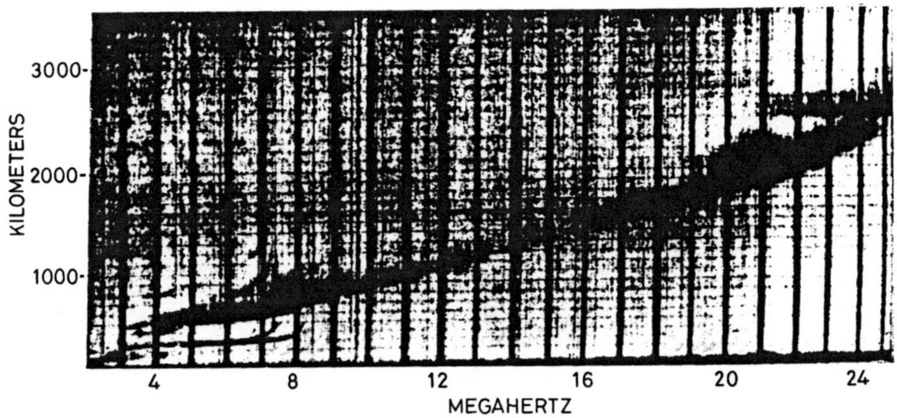

Fig. 5.22 *Ionogram obtained at Sterling, Virginia, on December 15, 1952, at 1257 75° West Meridian Time (WMT), using horizontal rhombic antennas pointing west, showing ground backscatter (From Silberstein 1954 © IEEE; reprinted with permission)*

but still occurs 50–60% around noon. During winter in this region, no diurnal variation is observed, and spread F occurs almost 100% of the time both day and night. Near the geomagnetic equator, spread F is an evening and nighttime phenomenon beginning near sunset and having a most likely time of occurrence between about 2100 and 0100 local time, and being earlier at sunspot maximum. It appears on all quiet nights from November through January and is correlated with the evening rise in the height of the F layer (see Fig. 5.21).

5.6.4 Seasonal variations of spread F

In polar latitudes the seasonal variations change widely from year to year. In general, during nighttime the maximum occurrences are near the equinoxes, the next highest in winter, and the lowest in summer. In equatorial latitudes spread F occurs mostly near the equinoxes, or during local summer, in longitudes where the geographical and magnetic-dip equators are far apart (e.g. in South America).

5.6.5 Sunspot dependence

In geomagnetic latitudes higher than 60°N there is an increase of occurrence of spread F with sunspot number (Shimazaki 1962). An inverse correlation has been found for Slough, England (Briggs 1964) where the sunspot dependence is greater in summer than in winter. Near the magnetic equator the sunspot influence appears through the effect of magnetic disturbance on spread F. The strong negative correlation at sunspot maximum between the appearance of spread F and magnetic disturbance disappears in years of low sunspot number. The disturbed equatorial region expands in size with increasing sunspot number and also in intensity (Singleton 1980).

5.6.6 Dependence on magnetic acitivity

In equatorial regions spread F occurs on magnetically quiet days and disappears with the onset of a magnetic storm, whereas in middle latitudes spread F is essentially a storm phenomeon. Near the magnetic equator the occurrence of spread F is negatively correlated with magnetic activity (Herman 1966), the correlation being strongest at the autumnal and vernal equinoxes. The correlation is moderate near the north solstice (summer) and poor in the winter. Spread-F statistics derived from ground-based ionogram data, in high latitudes, are affected by blackout (complete absorption of signals) during disturbances (Shimazaki 1962). Observations of spread F by topside sounders (Calvert and Schmid 1964) are not affected by D-region absorption.

5.6.7 Occurrence of spread F

Spread-F occurrence increases with decrease of f_oF_2, and there appears to be a threshold of f_oF_2 above which spread F is not observed. In equatorial regions spread-F occurrence decreases with increase of layer semi-thickness up to about 100–150 km and to increase again as the semi-thickness exceeds 200 km (Herman 1966). Spread F is closely correlated with the scintillation of radio signals from satellites and from radio stars (see Sec. 8.5).

5.6.8 Refractive scattering

Scintillation theory has been used (Booker *et al.* 1986) to explain frequency spread on ionogram F-layer traces on the basis of large-scale irregularities (between Fresnel radius ≈ 2 km and outer scale of tens of kilometers). Ionospheric scattering on high frequencies involves major ray refraction. For fractional mean-square fluctuations of electron density $\overline{(\Delta N/N)^2}$ of 0·01, Booker *et al.* (1985) showed that the frequency spread around f_oF_2 extends from 7% in the downward (frequency) direction to more than 20% in the upward direction. For $\overline{(\Delta N/N)^2} = 0{\cdot}1$, the spread runs from about three-quarters of the undisturbed value to almost twice the undisturbed f_oF_2.

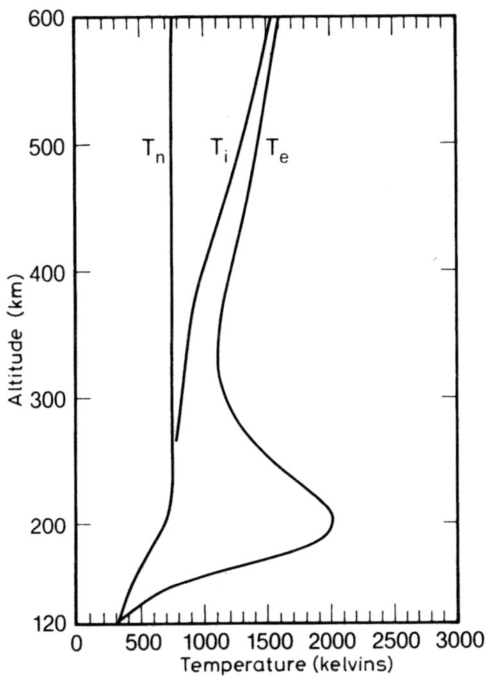

Fig. 5.23 *Height variations of electron temperature T_e, ion temperature T_i, and neutral gas temperature T_n*
Note that at low heights $T_e \approx T_i \approx T_n$ whereas at high altitudes $T_i \approx T_e > T_n$. The peak in T_e near 200 km is due to energetic photoelectrons

5.7 Electron and ion temperatures

Electron and ion temperatures can affect some types of radio propagation (see Sec. 4.7). Electron temperature T_e and ion temperature T_i can differ from each other and from the neutral temperature T_n. This results from the fact that, in the process of photoionization, electrons are ejected from the parent molecules with energies greater than the thermal energy of the ambient gas. The height distributions of T_e, T_i, and T_n are illustrated in Fig. 5.23. The neutral temperature increases rapidly between 120 km and about 200 km and then asymptotically approaches an exospheric limit T_∞. The electron temperature is a maximim at around 200 km near the level of peak production. Because of the large ion–neutral collision frequency the ions are essentially in thermal equilibrium with the neutrals at heights below 200 km. On the other hand, above 600 km electron-ion collisions dominate and the ion temperature approaches the electron temperature. Electron temperatures, as measured with incoherent scatter radars, can be several thousand kelvins and vary with sunspot number, local time, etc. (Fig. 4.12).

5.8 References

APPLETON, E. V. (1954): 'The anomalous equatorial belt in the F_2-layer', *J. Atmos. Terr. Phys.*, **5**, p. 349

APPLETON, E. V., and BEYNON, W. J. G. (1947): 'The application of ionospheric data to radio communication problems, II', *Proc. Phys. Soc.*, **59**, p. 58

BARGHAUSEN, A. F., FINNEY, J. W., PROCTOR, L. L., and SCHULTZ, L. D., (1969): 'Predicting long-term operational parameters of high-frequency sky-wave telecommunications systems'. ESSA Tech. Rep. ERL 110–ITS 78 (US Govt. Printing Office, Washington, DC 20402)

BELROSE, J. S. (1970): 'Radio wave probing of the ionosphere by the partial reflection of radio waves (from heights below 100 km)', *J. Atmos. Terr. Phys.*, **32**, p. 567

BENT, R. B., LEPOFSKY, J. R., LLEWELLYN, S. K., and SCHMID, P. G. (1978): 'Ionospheric range-rate effects in satellite-to-satellite tracking' *in* SOICHER, H. (Ed.): 'Operational modelling of the aerospace propagation environment'. Vol. I, AGARD Conf. Proc. No. 238, p. 9–1

BILITZA, D., SHEIKH, M. N., and EYFRIG, R. (1979): 'A global model for the height of the F_2-peak using M3000 values from the CCIR numerical map', *Telecom. J.*, **46**, p. 549

BOOKER, H. G., FERGUSON, J. A., and VATS, H. O. (1985): 'Comparison between the extended-medium and the phase screen scintillation theories', *J. Atmos. Terr. Phys.*, **47**, p. 381

BOOKER, H. G., PASRICKA, P. K., and POWERS, W. J. (1986): 'Use of scintillation theory to explain frequency-spread on F-region ionograms', *J. Atmos. Terr. Phys.*, **48**, p. 327

BRADLEY, P. A. (1979): 'Propagation at medium and high frequencies — 2: Long- and short-term models' *in* 'Aerospace propagation media modelling and prediction schemes for modern communications, navigation, and surveillance systems'. AGARD Lecture Series No. 99, p. 3–1

BRADLEY, P. A., and DUDENEY, J. R. (1973): 'A simple model for the vertical distribution of electron concentration in the ionosphere', *J. Atmos. Terr. Phys.*, **35**, p. 2131

BRIGGS, B. H. (1964): 'Observations of radio star scintillations and spread-F echoes over a solar cycle', *J. Atmos. Terr. Phys.*, **26**, p. 1

CALVERT, W., and COHEN, R. (1961): 'The interpretation and synthesis of certain spread-F configurations appearing on equatorial ionograms', *J. Geophys. Res.*, **66**, p. 3125

CALVERT, W., and SCHMID, C. W. (1964): 'Spread-F observations by the Alouette topside sounder satellite', *J. Geophys. Res.*, **69**, p. 1839

CCIR (1967): 'Atlas of ionospheric characteristics', CCIR Report 340, Int. Telecom. Union, Geneva

CCIR (1986): 'Propagation in ionized media, Recommendations and Reports of the CCIR, 1986, VI'. Int. Telecom. Union, Geneva

DUCHARME, E. D., PETRIE, L. E., and EYFRIG, R. (1971): 'A method of predicting the F_1 layer critical frequency', *Radio Sci.*, **6**, p. 369

FARLEY, D. T. (1970): 'Incoherent scattering at radio frequencies', *J. Atmos. Terr. Phys.*, **32**, p. 693

GEORGE, P. L. (1971): 'The global morphology of the quantity $\int Nvdh$ in the D- and E-regions of the ionosphere', *J. Atmos. Terr. Phys.*, **33**, p. 1893

HANSON, W. B., and MOFFETT, R. J. (1966): 'Ionization transport effects in the equatorial F region', *J. Geophys. Res.*, **71**, p. 5559

HERMAN, J. R. (1966): 'Spread F and ionospheric F-region irregularities', *Rev. Geophys.*, **4**, p. 255

JONES, W. B., and GALLET, R. M. (1965): 'Representation of diurnal and geographic variations of ionospheric data by numerical methods', *Telecom. J.*, **32**, p. 18

JONES, W. B., and OBITTS, D. L. (1970): 'Global representation of the annual and solar cycle variation of f_oF_2 monthly median 1954–1958'. OT Report 3, Inst. Telecom. Sci., Boulder, CO 80303

JONES, W. B., GRAHAM, R. P., and LEFTIN, M. (1969): 'Advances in ionspheric mapping by numerical methods'. ESSA Tech. Rep. ERL 107–ITS 75, US Govt. Printing Office, Washington, DC 20402

JONES, W. B., GALLET, R. M., LEFTIN, M., and STEWART, F. G. (1973): 'Analysis and representation of the daily departures of the f_oF_2 from the monthly median'. OT Report 73–12. US Dept. of Commerce, Boulder, CO 80303

LEE, H. S., and FERRARO, A. J. (1969): 'Winter D-region electron concentration and collision frequency features obtained with high power interaction measurements', *J. Geophys. Res.*, **74**, p. 1184

LEFTIN, M. (1976): 'Numerical representation of monthly median critical frequencies of the regular E region (f_oE)'. OT Report 76–88, US Govt. Printing Office, Washington, DC 20402

LEFTIN, M., and OSTROW, S. M. (1969): 'Numerical maps of f_oE_s for solar cycle minimum'. ESSA Tech. Rep. ERL 124–ITS 87, Dept. of Commerce, Boulder, CO 80303

LEFTIN, M., OSTROW, S. M., and PRESTON, C. (1968): 'Numerical maps of f_oE_s for sunspot cycle minimum and maximum'. ESSA Tech. Rep. ERL 73–ITS 63, US Dept. of Commerce, Boulder CO 80303

MATSUSHITA, S., and SMITH, L. G. (Eds.) (1975): 'Special Issue on Recent Advances on the Physics and Chemistry of the E region', *Radio Sci.*, **10**, p. 229

MINNIS, C. M. (1964): 'Ionospheric indices' *in* SAXTON, J. A. (Ed.): 'Advances in radio research, Vol. II' (Academic Press, London and New York) p. 1

MUGGLETON, M. L. (1975): 'A method of predicting f_oE_s at any time and place', *Telecom. J.*, **42**, p. 413

NISBET, J. S. (1971): 'On the construction and use of a simple ionospheric model', *Radio Sci.*, **6**, p. 437

PIGGOTT, W. R., and RAWER, K. (1978): 'URSI Handbook of ionogram interpretation and reduction'. Report UAG-23A, World Data Center–A, US Dept. of Commerce, Boulder, CO 80303

RAWER, K. (1981): 'International reference ionosphere, IRI-79'. Report UAG–82, World Data Center–A, Boulder, CO 80303

RAWER, K., and BRADLEY, P. A. (Eds.) (1987): 'International reference ionosphere – Status 1986/87', *Advances in Space Research* Vol. 7, pp. (6)1–(6)129

RAWER, K., BILITZA, D., and RAMAKRISHNAN, S. (1978): 'Goals and status of the international reference ionosphere', *Rev. Geophys. Space Phys.*, **16**, p. 177

ROSICH, R. K., and JONES, W. B. (1973): 'The numerical representation of the critical frequency of the F_1 region of the ionosphere'. OT Report 73–22, US Govt. Printing Office, Washington, DC 20402

RUSH, C. M. (1978): 'An ionospheric observation network for use in short-term propagation predictions', *Telecom. J.*, **43**, p. 544

RUSH, C. M., POKEMPNER, M., ANDERSON, D. N., PERRY, J., STEWART, F. G., and REASONER, R. (1984): 'Maps of f_oF_2 derived from observations and theoretical data', *Radio Sci.*, **19**, p. 1083

SHIMAZAKI, T. (1962): 'A statistical study of occurrence probability of spread F at high latitudes', *J. Geophys. Res.*, **67**, p. 4617

SILBERSTEIN, R. (1954): 'Sweep frequency backscatter – some observations and deductions', *IRE Trans. Antennas Propag.*, **AP-2**, p. 56

SINGLETON, D. G. (1980): 'Predicting transionospheric propagation conditions' *in* DONNELLY, R. F. (Ed.): 'Solar–terrestrial predictions proceedings Vol. 4', US Govt. Printing Office, Washington, DC 20402, D1-D16

SMITH, E. K. (1957): 'Worldwide occurrence of sporadic E'. Circular 582, National Bureau of Standards

SMITH, E. K. (1978): 'Temperate zone sporadic E maps (f_oE_s > 7 MHz)', *Radio Sci.*, **13**, p. 571

SMITH, E. K., and MATSUSHITA, S. (Eds.) (1962): 'Ionospheric sporadic E', (Macmillan, New York)

SMITH, L. G. (1966): 'Rocket observations of sporadic E and related features of the E region', *Radio Sci.*, **1**, p. 178

TASCIONE, T. F., KROEHL, H. W., CREIGER, R., FREEMAN, J. W., WOLF, R. A., SPIRO, R. W., HILMER, R. V., SHADE, J. W., and HAUSMAN, B. A. (1988): 'New ionospheric and magnetospheric specification models', *Radio Sci.*, **23**, p. 211

WAKAI, N. (1971): 'Study of the nighttime E region and its effect on the radio wave propagation', *J. Radio. Res. Labs.*, **18**, p. 245

WRIGHT, J. W. (1959): 'Note on quiet-day vertical cross sections of the ionosphere along 75° west geographic meridian', *J. Geophys. Res.*, **64**, p. 1631

WRIGHT, J. W. (1962): 'Dependence of the ionospheric F region on the solar cycle', *Nature*, **194**, p. 461

Chapter 6

Oblique propagation

6.1 Characteristics of HF propagation

Although vertical soundings are of great value in the study of ionospheric structure and dynamics, radio users are concerned with oblique propagation, e.g. point-to-point, broadcasting, and surveillance. In this Chapter we concern ourselves with the applications of vertical data to problems encountered in oblique propagation, e.g. with methods of calculating maximum usable frequencies and with identification of echo structures.

High-frequency (3 to 30 MHz) radio waves have some advantages:

(i) Low cost of terminal equipment
(ii) Low power requirements
(iii) Adequate bandwidths
(iv) Adequate signal strengths.

In contrast, it may be noted that medium-wave frequencies (300 kHz to 3 MHz) suffer very heavy absorption during the day, whereas on lower frequencies there is insufficient bandwidth for more than a few voice channels. Frequencies much higher than 30 MHz are not normally reflected from the ionosphere (see Ch. 13).

High-frequency radio waves also have some disadvantages:

(i) The variability of propagation conditions. For optimum results, this requires frequent changes in the operating frequency. Even on the optimum frequency, communications are often subject to interruption by ionospheric storms
(ii) The large number of possible propagation paths and the resulting time dispersion of a single signal
(iii) The large and rapid amplitude and phase fluctuations
(iv) The high interference
(v) Frequency distortion of wide-band signals

In the lower part of the high-frequency spectrum, reflections from the E region (particularly E_s) contribute to the complexity of the echo structure. The range of high frequencies useful for medium- and long-distance radio communications is limited, on the upper end, by the height and maximum electron density in the controlling layer (usually the F_2 layer) and, on the lower end, by absorption in the D region. For propagation over long distances the F_2 layer is of fundamental importance (Boithias 1987).

6.2 Theoretical considerations

6.2.1 The wave equation with oblique propagation

Consider a plane wave propagating in a horizontally stratified layer with its $\boldsymbol{k}$ vector at an angle ϕ to the vertical. The wave is of the form exp $i(\omega t - k_x x - k_z z)$ where k_x and k_z are the horizontal and vertical components of $\boldsymbol{k}$ respectively. Let E represent the wave amplitude; then

$$\frac{d^2E}{dx^2} + \frac{d^2E}{dz^2} = \frac{1}{v^2}\frac{d^2E}{dt^2} \qquad (6.1)$$

Substituting $d^2/dx^2 = -k_x^2 = -k^2 \sin^2\phi$ and $d^2/dt^2 = -\omega^2$ we obtain

$$\frac{d^2E}{dz^2} + k_0^2\,[\varepsilon - U(z)]E = 0 \qquad (6.2a)$$

where

$$\varepsilon = -\sin^2\phi_o = S^2 \qquad (6.2b)$$

and

$$U(z) = \mu^2(z) \qquad (6.2c)$$

Here ϕ_o is the angle of incidence on the layer (initial condition) and $U(z)$ is a potential function describing the properties of the medium. Equation (6.2a) is a form of Schrödinger's equation and has been used in modified form (e.g. by Gurevich and Tsedilina 1985) to study the propagation of radio waves in ionospheric ducts.

6.2.2 The Booker quartic

Substitution of the quantity

$$q = \mu \cos\phi = (\mu^2 - S^2)^{1/2} \qquad (6.3)$$

into

$$E = E_m \exp\,[-i(k_x x + k_z z)]$$

gives a wave function of the form

$$E = E_m \exp\,[-ik_0(Sx + qz)] \qquad (6.4)$$

where $k_0 = \omega/c$. When q varies with height this becomes

$$E = E_m \exp\left[-ik_0\left(Sx + \int q dz\right)\right] \tag{6.5}$$

For a cold magnetoionic plasma q is given by the roots of the Booker quartic, which has been discussed extensively (e.g. Budden 1985). The four roots of the quartic apply to the upgoing and to the downgoing ordinary and extraordinary waves. The Booker quartic has been used to study propagation paths (Booker 1949). Note that q varies from $\cos\phi_o$ at entry to zero at reflection.

6.3 Plane earth and plane ionosphere

6.3.1 Equivalence relationships

6.3.1.1 The 'secant law'

In this Section we consider the relationships that exist between the frequencies, virtual or group, or equivalent paths, and absorptions, of two waves, one reflected with oblique incidence and the other reflected with normal incidence from the same true height. For this purpose consider a ray incident on a plane ionosphere at an angle ϕ_o, as shown in Fig. 6.1, in which the electron density is increasing with height so that total internal reflection takes place.

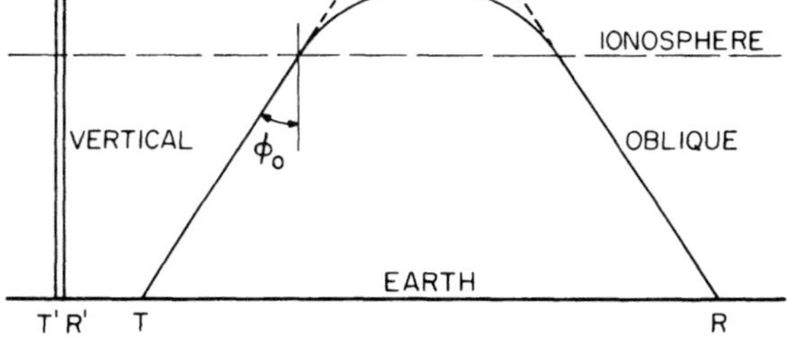

Fig. 6.1 *Equivalent waves reflected obliquely and vertically at the same real height (B and B') and the same virtual heights (A and A'), plane ionosphere and plane earth*

In the absence of both collisions and an imposed magnetic field, we have from (3.11) and Snell's law that the relationship between the frequency f_o of a wave reflected obliquely at given real height and the frequency f_v of a wave incident vertically and reflected at the same real height is

$$f_o = f_v \sec \phi_o \tag{6.6}$$

The frequency f_v is called the 'equivalent vertical frequency' corresponding to f_o. Equation (6.6) is called the secant law. It shows that the ionosphere can reflect much higher frequencies with oblique propagation than with vertical propagation.

6.3.1.2 Breit and Tuve's theorem

This theorem states that group (also called equivalent and virtual) path P', for propagation between a transmitter T and a receiver R (Fig. 6.1) is given by the length of the equivalent triangle TAR; that is,

$$\begin{aligned} P' &= \int_{TBR} \frac{ds}{\mu} = \int_{TBR} \frac{dx}{\mu \sin \phi} = \frac{1}{\sin \phi_o} \int_{TBR} dx \\ &= \frac{TR}{\sin \phi_0} = TA + AR \end{aligned} \tag{6.7}$$

Note that the true height of reflection (at B) is always less than the equivalent height at A and that both T and R must lie outside the ionosphere.

6.3.1.3 Martyn's (equivalent path) theorem

If f_o and f_v are equivalent frequencies of waves reflected obliquely and vertically from the same real height in a flat ionosphere, then the virtual height of reflection of f_v is equal to the height of the equivalent triangular path for the oblique signal.

$$P'(f_o) = 2h'(f_v) \sec \phi_o \tag{6.8}$$

This theorem expresses the important relationship that the virtual height of reflection of an obliquely incident wave is the same as that of the equivalent vertical wave. This theorem follows from the relationship between the refractive indices μ_o and μ_v for the oblique angle (ϕ) and vertical waves at the same real height, viz.

$$\mu_o \cos \phi = \mu_v \cos \phi_o \tag{6.9}$$

6.3.2 Maximum frequencies

The theorems of Breit and Tuve and of Martyn stated above show that, for a plane ionosphere, the reflection process is equivalent to mirror-type reflection at a height equal to the virtual height h' of reflection of the equivalent vertical frequency. The variation of the virtual height with frequency can be determined experimentally by an ionosonde.

The relationship between oblique and equivalent-vertical frequencies, in the case of a plane reflector at an equivalent height h', for transmission over a distance D, is given by

$$f_o = f_v \sec \phi_o = f_v \left[1 + \left(\frac{D}{2h'} \right)^2 \right]^{1/2} \tag{6.10}$$

To obtain f_o for a given value of f_v, we need to know h' as a function of f_v (from an ionogram) and the relationship between sec ϕ_o and h' which, when presented graphically, is called a transmission curve (Smith 1939). The application of a transmission curve to an $h'f$ curve is, essentially, a graphical solution of these two equations. The family of such transmission curves shown in Fig. 6.2 is

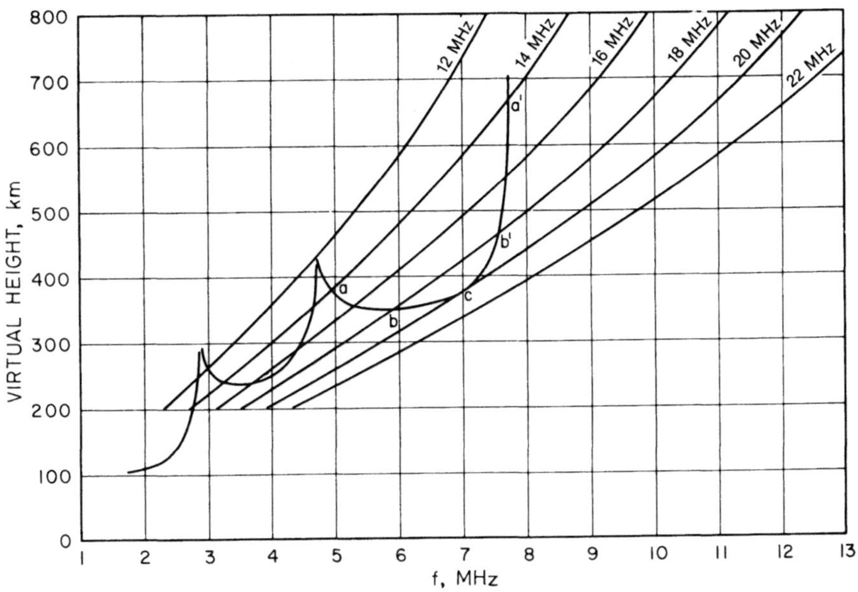

Fig. 6.2 *Family of transmission curves parametric in frequency for a fixed distance of 2000 km superimposed on an h'f curve, for a flat earth and a flat ionosphere*

plotted on the same h'–f grid as a conventional $h'f$ curve. The intersections of a transmission curve, for a given (oblique) frequency f_o, with the vertical $h'(f_v)$ curve give the virtual heights of reflection of the transmitted waves. In Fig. 6.2 the three sections of the $h'(f_v)$ curve would correspond to reflections from the E, F_1, and F_2 layers. A frequency of 14 MHz, for example, propagating over a distance of 2000 km could travel by several different paths, two of which would be reflected from the F_2 layer at heights corresponding to the intersections marked a and a' on the $h'(f_v)$ curve with the transmission curve marked 14 MHz. The same transmission curve also intersects the F_1 section in two places and the E section in two places when the transmission curve is extrapolated down to 100 km. The ray corresponding to the lower virtual height of each pair is called the low-angle ray, whereas the path corresponding to the higher virtual height is known as the high-angle or Pedersen ray. There would be six different ray paths for a 14 MHz wave. The 18 MHz transmission curve misses the F_1 section; thus no reflection from the F_1 layer is possible on this frequency.

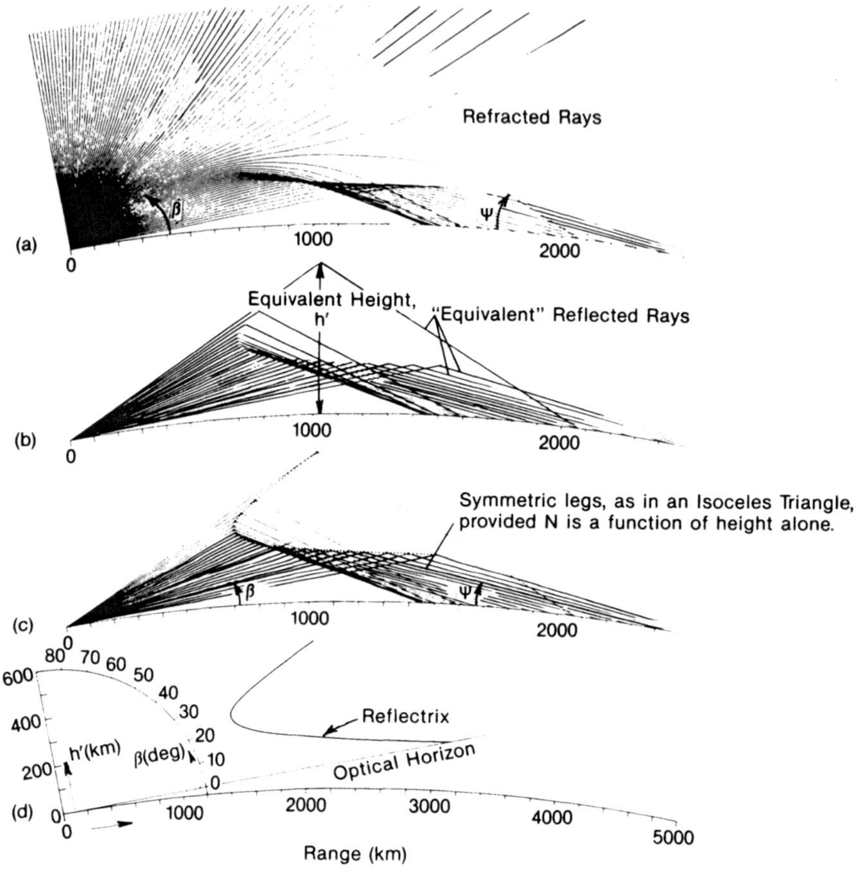

Fig. 6.3 *(a) Real rays in a concentric Earth and Chapman ionosphere, f = 8 MHz, f_c = 4 MHz, h_{max} = 300 km, H = 100 km*
(b) Equivalent or virtual paths reflected at the apex of the fictitious triangular paths
(c) The reflectrix as the locus of the virtual heights
(d) The reflectrix as a function of virtual height, range, and elevation angle
(From Croft 1967 © American Geophysical Union)

As the frequency f of the parametric curves increases, the points of intersection (bb') in Fig. 6.2 come closer together, merging in a tangent contact at one point c. At this point only one ray path would be possible, and 20 MHz is the highest frequency that could propagate over 2000 km under these ionospheric conditions. The frequency of the tangent curve is called the maximum usable frequency (MUF). At this point, where the high and low waves merge, focusing takes place and the received signal is enhanced.

For frequencies higher than 20 MHz there is no value of $f_v \sec\phi_o$ that will satisfy (6.10) and hence no level of reflection. Therefore, the signal is said to skip at the distance concerned (2000 km). The distance for which a given frequency is the maximum frequency is called the 'skip' distance for that frequency. Within this distance no signals are received from the ionosphere by the regular process of ionospheric refraction. For a fixed ground range, as the operating frequency is increased, the real height of reflection of the low-angle ray increases, whereas the height of the high-angle ray decreases until the low-angle and high-angle rays coalesce at the reflection height of the maximum frequency.

In Fig. 6.3*a* we see what happens to the ray paths (on a given frequency) as the angle of elevation (takeoff) slowly increases. For low angles the ground range is long. As the elevation increases, the ground range decreases until the skip is reached, after which the range increases rapidly. Eventually penetration occurs. Figure 6.3*b* shows the equivalent triangular paths for different angles of elevation. The apexes of the equivalent triangles lie on a smooth curve, called a reflectrix, as shown in Fig. 6.3*c*. The ray with its apex at the 'nose' of the reflectrix is the skip-ray and its frequency is the MUF. Rays reflected at lower virtual heights are low-angle rays, and rays reflected on the upper side of the reflectrix are high-angle rays. For a given reflectrix, the relationship between virtual height and angle of elevation (β) is given by Fig. 6.3*d*. The variation of ground range D with angle of incidence ϕ_o is sketched in Fig. 6.4, for the case of a single layer, for various ratios of the operating frequency to the penetration frequency. For a given distance there are two values of ϕ_o, one corresponding to the low-angle path and the other to the high-angle path. The minimum distance is the skip distance. The high-angle ray is very sensitive to changes in ϕ_o and, therefore, power contained in a small range of angles is spread over a large area of ground and the rays are defocused.

6.3.3 Parabolic layer theory

An alternative to the transmission-curve method for calculating characteristics of oblique propagation is to use model layers in which the reflection height, equivalent path, ground range, etc., can be determined analytically. The parameters specifying the model can be obtained from models such as the International Reference Ionosphere (see Sec. 5.3) or directly from ionograms. The most widely used model is the parabola (see Sec. 5.3.2), which has the advantage of simplicity and the fact that it can be fitted near the peak of an ionospheric layer reasonably well at those heights at which the MUF is determined.

6.3.3.1 Virtual height

With vertical propagation, no imposed magnetic field, and negligible electron collision frequency, the virtual height of reflection of a wave of frequency f in a parabolic layer with critical frequency f_c and semi-thickness y_m is

$$h' = h_0 + F\,y_m \tag{6.11a}$$

where

$$F = \tfrac{1}{2} \times \ln \frac{1 + x}{1 - x} \tag{6.11b}$$

and

$$x = f/f_c \tag{6.11c}$$

and h_0 is the height of the bottom of the layer. From an ionogram we can determine f_c and $h'(x)$. A plot of $h'(x)$ versus $F(x)$ yields a straight line whose slope gives y_m; its intercept on the h' axis gives h_0 (Davies 1969, Sec. 12.4).

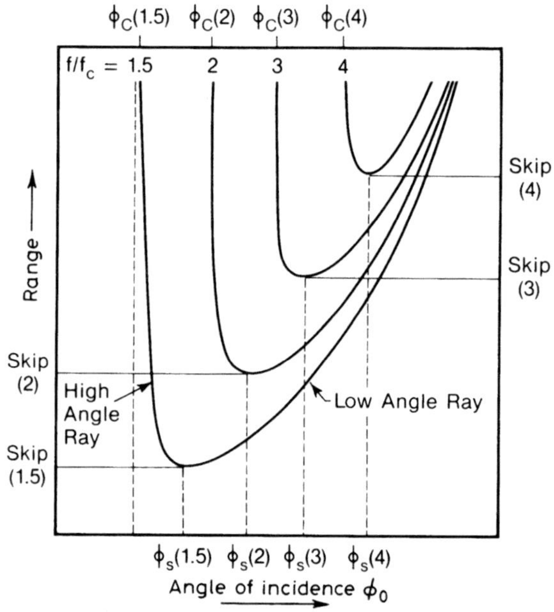

Fig. 6.4 *Variation of ground range D with angle of incidence ϕ_o for a wave of frequency f in a plane ionospheric layer with penetration frequency f_c*
Note the minimum range, or skip distance, and the minimum angle ϕ_c for reflection

6.3.3.2 Ground range

For a wave of frequency f incident at an angle ϕ_o on a plane layer, the ground range D is given by

$$D = 2h_0 \tan \phi_o + 2 \sin \phi_o \int_{h_0}^{h_r} \frac{dh}{(\mu^2 - \sin^2 \phi_o)^{1/2}} \tag{6.12}$$

where h_r is the height of reflection. Put

$$\mu^2 = 1 - a(h - h_0) + b(h - h_0)^2 \tag{6.13a}$$

$$a = 2f_c^2/y_m f^2 \tag{6.13b}$$

and

$$b = f_c^2/y_m^2 f^2 \tag{6.13c}$$

Integration gives

$$D = 2h_0 \tan \phi_o + Qy_m \tag{6.14}$$

where

$$Q = x \sin \phi_o \ln[(1 + x \cos \phi_o)/(1 - x \cos \phi_o)]. \tag{6.15}$$

From (6.14) we see that there is a lower limit, ϕ_c, to ϕ_o, given by

$$\phi_c = \cos^{-1}(f_c/f), \tag{6.16}$$

below which the rays penetrate the layer. This minimum angle is different from the angle ϕ_s which refers to the skip path for which D is a minimum (see Fig. 6.4). The skip condition is that $dD/d\phi_o = 0$. The ratio f_{max}/f_c for the distance D_s is called the maximum-usable-frequency (MUF) factor or the $M(D_s)$ factor. The relationship between the M factor, for a given D, and ϕ_o depends only on the ratio y_m/h_0 and not on the absolute magnitudes of the height and thickness of the layer.

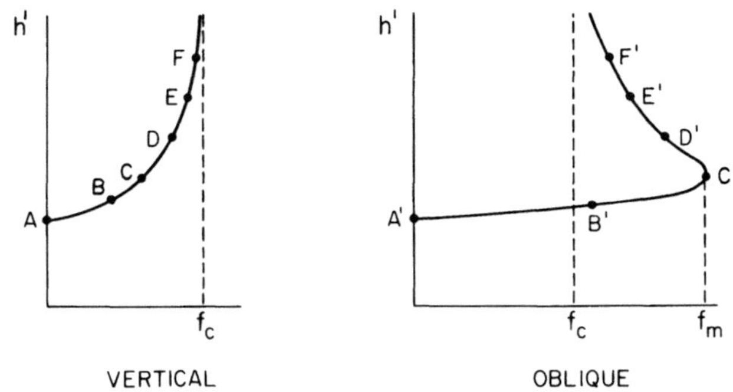

Fig. 6.5 *Corresponding vertical and oblique ionograms for a single ionospheric layer with a critical frequency f_c*

The virtual heights *A*, *B*, etc., on the vertical ionogram are the same as *A'*, *B'*, etc., on the oblique ionogram. Note that the MUF (f_m) on the oblique ionogram is not reflected at the level of maximum electron density

6.3.4 Relationship of a vertical h'(f) curve to an oblique P'(f) curve

The ionosonde technique, discussed in Sec. 4.3, has been used, with transmitter and receiver separated, to study the structure of oblique echo signals. The variation of group path P' with frequency is often called an oblique ionogram. The relationship between a vertical and an oblique ionogram is sketched in Fig. 6.5. For any point A on the vertical ionogram there will be a corresponding point A' on the oblique ionogram for which $P' = 2h' \sec\phi_o$ where ϕ_o is the angle of incidence of the oblique signal. As the equivalent height increases, the frequency of the vertical ionogram increases but $\sec\phi_o$ decreases. The net result is that $f_o = f_v \sec\phi_o$ passes through a maximum value such that at lower equivalent heights f_v decreases more rapidly than $\sec\phi_o$ increases. Note that $C'F'$ (Fig. 6.5) refers to the high-angle or Pedersen path and $A'C'$ refers to the low-angle path. The maximum frequency is not reflected at the height of maximum electron density but where $f_v \sec\phi_o$ is a maximum. As the ground range increases, the maximum frequency f_{max} increases and the low-angle trace becomes sharper as shown in Fig. 6.6. The 'nose' of the oblique ionogram is sharper for a long distance than for a short distance.

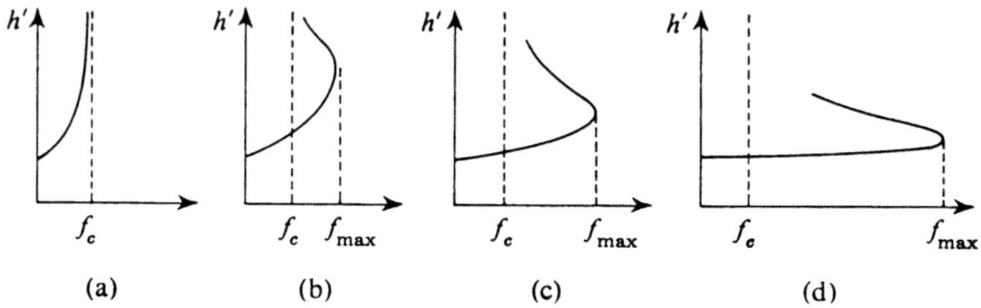

Fig. 6.6 *Oblique ionograms for different distances between transmitter and receiver*
(*a*) *Vertical*
(*b*) *Short distance*
(*c*) *Medium distance*
(*d*) *Long distance*
Note that the trace is asymptotic to the critical frequency (broken) line and that the maximum frequency increases with increasing distance

6.4 Effect of the Earth's magnetic field

The main effect of the Earth's magnetic field is to split the incident wave into a left-handed (with respect to the magnetic field) ordinary wave and a right-handed extraordinary wave (see Sec. 3.4.2). These waves travel different paths in the ionosphere. As a result all the traces on an oblique ionogram are split in two, complicating the echo structure. When the magnetic field is included the equivalence relations are not strictly valid. It is not then possible to integrate analytically to obtain the ray path but it is practical to do so by means of point-by-point ray tracing.

Some idea of the relationships between the frequencies of the ordinary (f_o) and extraordinary (f_x) waves reflected at the same equivalent height can be obtained for the cases of quasi-longitudinal (QL) and quasi-transverse (QT) propagation. In the case of north–south or south–north propagation, especially near the dip equator, the angle θ between the wave normal and the magnetic field is small, so the propagation is quasi-longitudinal, and

$$f_x - f_o \simeq f_H \cos\theta \tag{6.17}$$

where $\theta = I$ the magnetic dip. Thus the shape of the extraordinary trace is similar to the shape of the ordinary trace but is shifted, by $f_H \cos\theta$, to higher frequencies. This QL approximation is fairly good near the maximum frequency because there the wave frequency is usually much higher than the gyrofrequency f_H, and this ensures the validity of the QL approximation. Equation (6.17) does not hold for very short distances and breaks down completely with vertical propagation because, near reflection, the propagation of the ordinary wave is quasi-transverse.

In the case of magnetic east–west propagation, the conditions at reflection are transverse (i.e. $\theta \approx 90°$) and

$$f_x - f_o \approx f_H^2/2f_o \tag{6.18}$$

In this case, the frequency separation decreases with increase of wave frequency and is, therefore, smallest at the maximum frequency. For a 2000 km E–W path in middle latitudes, where $f_H \approx 1{\cdot}5$ MHz and f_o lies between 5 MHz and 10 MHz, (6.18) gives frequency separations between 0·23 MHz and 0·12 MHz; that result is in general agreement with observations by Agy and Davies (1959).

6.5 Some effects of horizontal gradients

Worldwide maps of ionospheric characteristics, especially those of the F_2 layer, show the presence of horizontal variations. Horizontal gradients give rise to asymmetric ray paths and to reflections out of the great-circle plane through transmitter and receiver. In realistic profiles, paths are determined by ray tracing. In some cases models can be used (Kelso 1964; Nielson 1968) for which analytic expressions can be obtained. Davies and Rush (1985) used a model in which a slab, uniform in height and in which the electron density increases linearly with distance, is superimposed on a plane parabolic layer such that the refractive index μ is given by

$$\mu^2 = Ax + B + Cz^2 \tag{6.19a}$$

where

$$A = -A_0/f^2 \tag{6.19b}$$

$$B = 1 - (f_c/f)^2 \tag{6.19c}$$

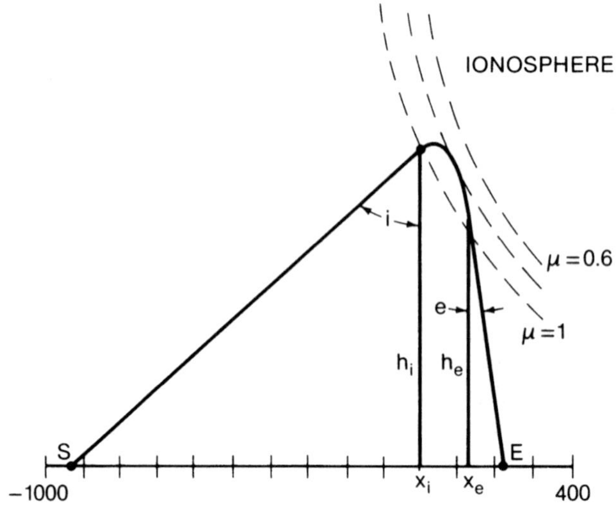

Fig. 6.7 *Asymmetric ray path in a linear-parabolic layer [defined by eqn. (6.19) in text] with a horizontal tilt*
The angle of incidence i is different from the angle of exit e, and the heights of entry h_i and exit h_e are different

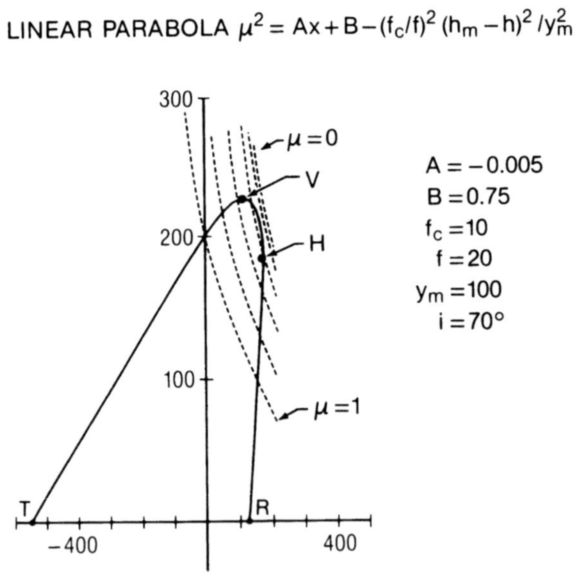

Fig. 6.8 *Ray path in a linear-parabolic layer showing vertical (V) and horizontal (H) reflections*
A = horizontal gradient of the square of the refractive index per kilometer, $B = 1 - (f_c/f)^2$, f_c = critical frequency in megahertz, f = wave frequency in megahertz, and i = angle between the ray and the vertical at entry

$$C = (f_c/fy_m)^2 \tag{6.19d}$$

$$z = h_m - h \tag{6.19e}$$

Here A_0 is the gradient of $(f_oF_2)^2$ in the x (horizontal) direction, and other quantities are as defined above.

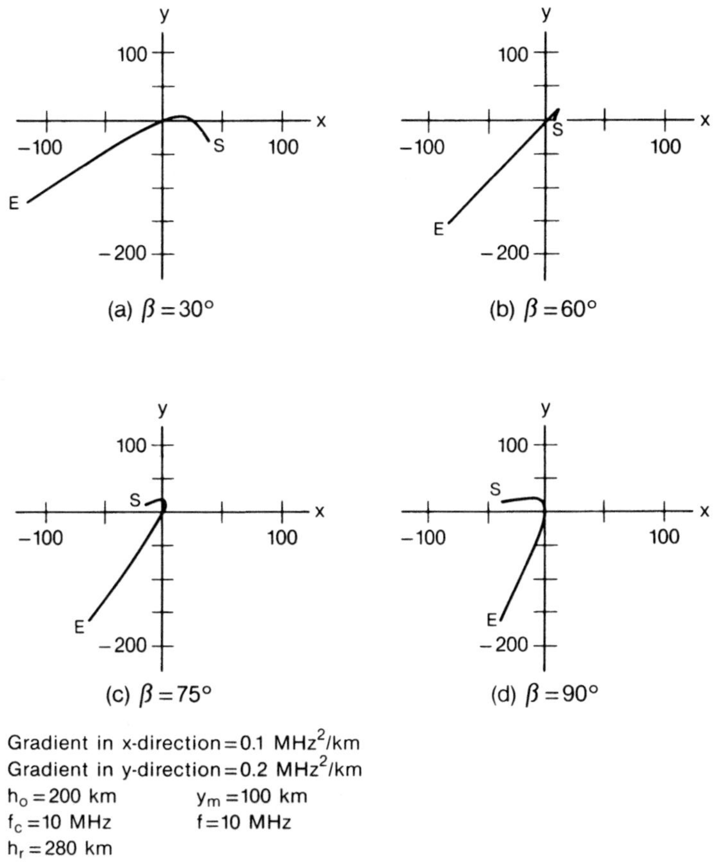

Fig. 6.9 *Projections onto the horizontal plane of ray paths in a linear-parabolic layer with gradients in both horizontal (x and y) directions*

For a ray path that makes an angle β to the x–z plane at the apex of the path (x_r, y_r, z_r) the coordinates of any other point (x, y, z) on the path in the ionosphere are given by

$$y = y_r \pm \frac{(B + \alpha_2)^{1/2}}{C^{1/2}} \ln [Z + (Z^2 - 1)^{1/2}] \tag{6.20a}$$

$$y = y_r \pm \frac{2(B + \alpha_2)^{1/2}}{A} [(Ax + \alpha_1)^{1/2} - (Ax_r + \alpha_1)^{1/2}] \quad (6.20b)$$

where

$$Z = z/z_r. \quad (6.20c)$$

$$\alpha_1 = (Cz_r^2 + B)\cos^2\beta - Ax_r \sin^2\beta \quad (6.20d)$$

$$\alpha_2 = (Ax_r + Cz_r^2)\sin^2\beta - B\cos^2\beta \quad (6.20e)$$

The ray path coordinates are determined, for given values of x_r, y_r, z_r, and β, by insertion of values of Z into (6.20*a*) and solving for $(y - y_r)$; the result is substituted into (6.20*b*) to give x. The plus and minus signs in (6.20*a*) and (6.20*b*) give the two legs of the path. Note that the ionosphere is bounded by the surface at which $\mu = 1$.

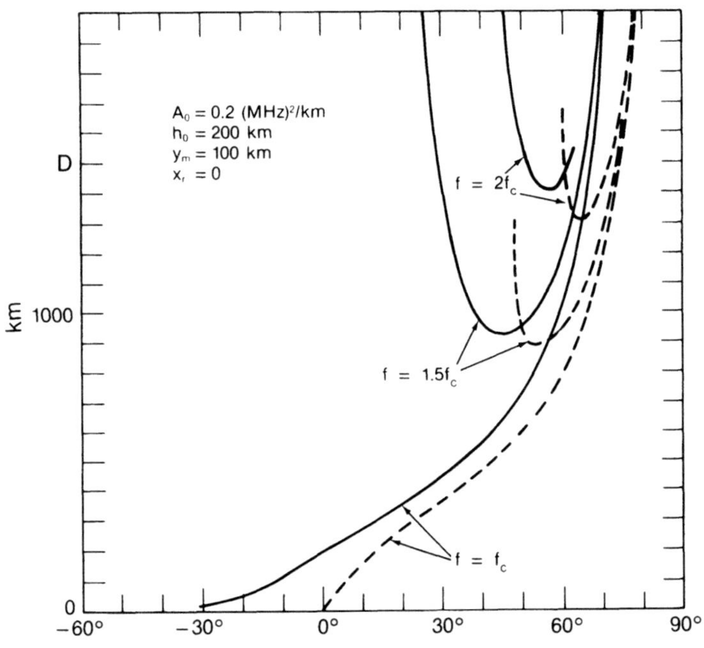

Fig. 6.10 *Range versus angle of incidence in a linear-parabolic layer (full line) and in a flat parabola (broken line), with a flat Earth for various frequencies in terms of the critical frequency*

These ray paths have several interesting features (see Fig. 6.7):

(i) The angle of incidence i and the angle of exit e are, in general, unequal. The incident and emergent angles are related by

$$\sin^2 e - \sin^2 i = A(x_e - x_i) \quad (6.21a)$$

and

$$\cos^2 e - \cos^2 i = -C(z_e^2 - z_i^2) \tag{6.21b}$$

The ground ranges outside the ionosphere are $h_i \tan i$ and $h_e \tan e$.

(ii) When the gradient A_0 is sufficiently large, a horizontal reflection is possible as shown in Fig. 6.8.

(iii) Path projections onto a horizontal plane (Fig. 6.9) show that, with a propagation component in the y-direction, the ray path is deviated out of the plane of incidence. In Fig. 6.9 vertical reflection is at the origin and there are gradients in the x and y directions leading to reflections in these directions. Radio location of E from measurements at S, and vice versa, will be in error (Davies and Rush 1985).

The variations of ground range D with angles of incidence for a tilted layer are shown in Fig. 6.10. For comparison the variations are given for the flat ionosphere. The following features are of interest:

(i) With vertical incidence ($i = 90°$) and $f = f_c$ the ray is not returned to the transmitter but reaches the ground about 200 km to the left.

(ii) The echo reflected back to the transmitter is incident at an angle of about 30°.

(iii) There is relatively little change in the skip distance, compared with the flat-Earth and flat-ionosphere case. This results from the fact that the increase in ground range on one leg roughly compensates for the decrease on the other leg (Davies and Rush 1985).

6.6 Inclusion of Earth–ionosphere curvature

6.6.1 Propagation in a thin layer

The effect of Earth curvature, which is important for ground ranges greater than 500 km, is illustrated in Fig. 6.11. The ionosphere curvature can be neglected but the Earth curvature has the effect of shortening the distance between transmitter and receiver (chord TCR), and the 'effective' height of the ionosphere is increased by the distance BC. From Fig. 6.11 we have

$$BC = a\left(1 - \cos\frac{\varphi}{2}\right) \tag{6.22}$$

$$\tan\phi_o = \frac{\sin\dfrac{\varphi}{2}}{1 + h_0/a - \cos\varphi/2} \tag{6.23}$$

where φ is the angle at the center of the Earth subtended by the path.

$$f_{\max} = f_c \sec\phi_o = f_c M(D) \tag{6.24}$$

where $M(D)$ is the MUF factor for the distance D. It is seen that, for a given h_0 and D, $M(D)$ is smaller for a curved Earth than for a flat Earth, so that the net effect of curvature is to decrease the MUF.

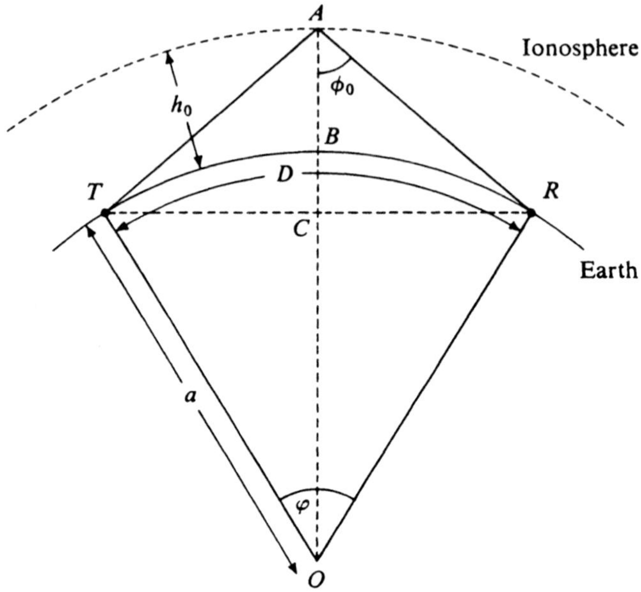

Fig. 6.11 *Geometry of reflection from a thin curved ionosphere*

There is a maximum value ϕ_m of ϕ_o and, hence, a maximum one-hop range D_m when the angle of elevation at the ground is zero, where

$$\sin\phi_m = \frac{a}{a + h_0} \tag{6.25}$$

and

$$D_m^2 = 8\,ah_0 \tag{6.26}$$

Hence

$$f_{max} = f_c\left(1 + \frac{a}{2h_0}\right)^{1/2} \approx f_c\left(\frac{a}{2h_0}\right)^{1/2} \tag{6.27}$$

This thin-layer theory is useful in the case of propagation by sporadic E, for which $h_0 = 100\,\text{km}$ and

$$f_{max} \approx 5{\cdot}4\,f_c \tag{6.28}$$

For a height of 300 km and $D_m \approx 3900\,\text{km}$, $f_{max} \approx 3{\cdot}4\,f_c$; that is, for F-layer propagation $M(3900) \approx 3{\cdot}4$. Thus a useful practical limit for one-hop propagation is 4000 km, which corresponds to a layer height of 413 km.

6.6.2 Modified secant law

When ionosphere curvature is present the equivalence relations of Sec. 6.3.1 are no longer valid; however, for many practical purposes a modified secant law is used:

$$f_o = kf_v \sec \phi_o \qquad (6.29)$$

where the value of the correction factor k varies between 1·0 and 1·2. The quantity $k \sec \phi_o$ is often called 'sec ϕ_o corrected' and is used for the construction of transmission curves (Smith 1939).

The primary purpose of transmission curves is the determination of maximum frequencies for a given radio circuit. For this purpose it is possible to construct the type of transmission curve shown in Fig. 6.12 in which a single curve suffices for each distance and a family of curves parametric in distance can be drawn on a single transparent overlay for use with standard ionograms. This is done by making the frequency scale logarithmic, and requires that the frequency scale of the vertical ionogram be logarithmic also.

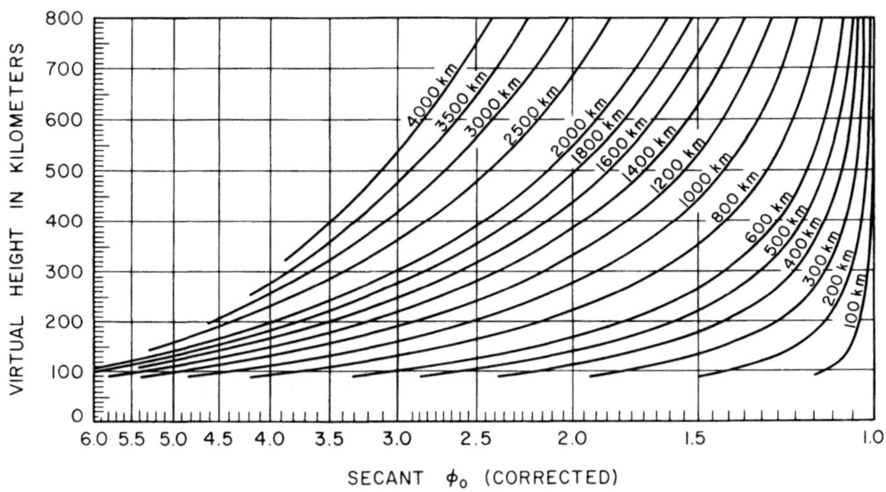

Fig. 6.12 *Logarithmic transmission curve for a curved ionosphere*

In Fig. 6.12, the factor $k \sec \phi_o$ is plotted on the same logarithmic scale as the ionogram, but the numbers increase in the direction opposite to that of the frequency of the ionogram. The virtual height h' is the same as that on the ionogram. These curves are plotted on a transparent overlay and placed over the ionogram so that the height scales coincide and, in general, the curves intersect in two places. The values of the product $kf_v \sec\phi_o$ or f_o appear opposite the abscissa 1·0. As the overlay slides along the horizontal axis, different values of f_o are obtained. Note that because of the logarithmic frequency scale, the same number, except for the decimal point, appears on the transparency opposite the 1·0 index on the logarithmic ionogram scale. The transmission curve for a given distance can be slid along the abscisssa until it is just tangent to the $h'f$ curve. The product, as then read at the 1·0 index, is the value of the maximum frequency.

For many purposes, the maximum-usable-frequency factor (M factor) (see Sec. 6.3) is desired, that is, the ratio of the maximum frequency to the critical frequency of the layer. This may be found directly from the logarithmic scale on the transparency opposite the critical frequency cusp when the transparency is set to read the maximum frequency (i.e. tangent condition). The transmission curve for a distance of 3000 km is often used as a reference. It is given by

$$M(3000) \quad = \quad \mathrm{MUF}(3000)/f_c \quad = \quad (67{\cdot}6542 \; - \; 0{\cdot}014938h')/\sqrt{h'} \qquad (6.30)$$

The transmission curves given in Fig. 6.12 are most accurate for F_2-layer heights because the factor k is based on typical F_2-layer profiles. Over short distances they are accurate for all layers because, as the distance approaches zero, k approaches the value 1·0. The effects of the Earth's magnetic field have been neglected in constructing these transmission curves.

6.6.3 *Curved parabolic layer*

In a curved ionosphere Snell's law takes the form

$$\mu r \sin i \quad = \quad \mu_0 r_0 \sin\phi_o \quad = \quad a \cos\Delta \qquad (6.31)$$

where r is the length of the radius vector from the center of the Earth to the level at which the refractive index is μ; the angle between the radius vector and the ray is i, as shown in Fig. 6.13; μ_0, r_0, and ϕ_o are reference values at the bottom of the ionosphere; and Δ is the angle of elevation at the ground.

The total ground range D (Fig. 6.14) is made up of two parts: (i) D_1, corresponding to the path in the ionosphere and (ii) D_2, corresponding to the path outside the ionosphere. Let an element of ray path at a height h, making an angle i with the radius vector, subtend and angle $d\varphi$ at the Earth's center. From geometry and (6.31) we have

$$(a + h)d\varphi \quad = \quad \tan i \, dh \quad = \quad \frac{(a + h_0)\sin\phi_o \, dh}{[(a + h)^2 \mu^2 - (a + h_0)^2 \sin^2\phi_o]^{1/2}} \qquad (6.32a)$$

and, hence

$$\varphi \quad = \quad 2\int_{h_0}^{h_r} \frac{(a + h)\sin\phi_o \, dh}{(a + h)\,[(a + h)^2 \mu^2 - (a + h_0)^2 \sin^2\phi_o]^{1/2}} \qquad (6.32b)$$

where h_r is the height of reflection.

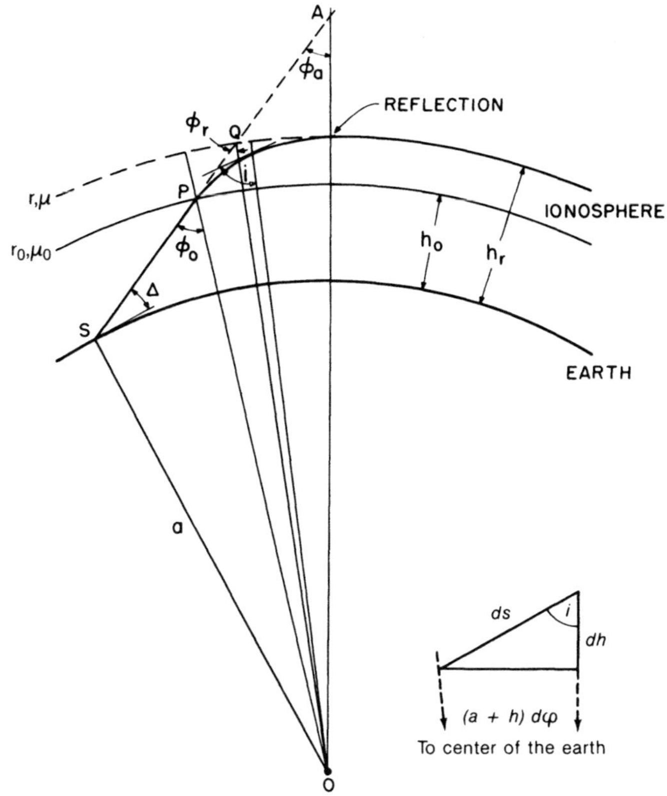

Fig. 6.13 *Ray refraction in a thick curved ionosphere*

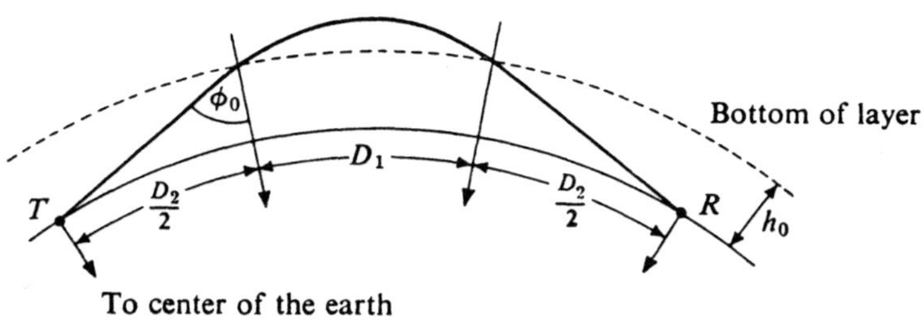

Fig. 6.14 *Segments of the total ground range for a curved Earth and curved ionosphere*

Equation 6.32*b* is the basic expression for the determination of the ground range $D(=a\varphi)$. It can be integrated analytically in one of several ways:

(i) Select an electron density model such that (6.32*b*) has an analytic solution
(ii) Fit the $N(h)$ profile by a few segments with models that have analytical solutions

(iii) Use an arbitrary (tabulated) electron density profile and ray trace point by point through the ionosphere between entry and exit

Appleton and Beynon (1940, 1947) substituted (6.13) into (6.32) and eliminated terms in h_0/a to obtain

$$\varphi_1 = \frac{2 \sin \phi_o}{a + h_0} \int_{h_0}^{h_r} \frac{dh}{(\cos^2 \phi_o - \alpha y + \beta y^2)^{1/2}} \tag{6.33a}$$

where

$$\alpha = 2f_c^2/f^2 y_m - 2 \sin^2 \phi_o/(a + h_0) \tag{6.33b}$$

and

$$\beta = f_c^2/y_m^2 f^2 \tag{6.33c}$$

On integration we get

$$D_1 = \frac{a}{a + h_0} \times y_m \sin \phi_o \ln \frac{u + x \cos \phi_o}{u - x \cos \phi_o} \tag{6.34a}$$

where

$$u = 1 - x^2 [y_m/(a + h_0)] \sin^2 \phi_o \tag{6.34b}$$

and $x = f/f_c$. For the path outside the ionosphere

$$D_2 \approx 2a \cot \phi_o - 2a(\cot^2 \phi_o - 2h_0/a)^{1/2} \tag{6.35}$$

Comparison of (6.34*a*) and (6.14) shows that a curved ionosphere is like a flat ionosphere in which the effective critical frequency is $f_c - [f^2 y_m/f_c(a + h_0)] \sin^2 \phi_o$. For a description of how to determine the MUF, skip distance, M factor, etc., see papers by Appleton and Beynon (1940, 1947).

6.6.4 *Quasi-parabolic layer*

By dropping terms of the order h_0/a we are, in a sense, altering the parabolic layer in such a way as to make (6.32*b*) integrable. An equivalent approach is used with the quasi-parabolic layer (see 5.12) which when substituted into (6.32*b*) leads to

$$D_1 = \frac{ar_0 \sin \phi_o}{\sqrt{C}} \ln \left[\frac{(B + 2\, C/r_0 + 2\sqrt{C} \cos \phi_o)^2}{B^2 - 4\, AC} \right] \tag{6.36a}$$

in which

$$A = 1 + x_m \left[\left(\frac{r_0}{y_m} \right)^2 - 1 \right] \tag{6.36b}$$

$$B = -x_m r_m \left(\frac{r_0}{y_m} \right)^2 \tag{6.36c}$$

$$C = x_m \left(\frac{r_0}{y_m}\right)^2 r_m^2 - a^2 \sin^2 \phi_o \tag{6.36d}$$

$$x_m = (f_c/f)^2 \tag{6.36e}$$

and D_2 is given by (6.35). Since (6.36*a*) is an exact solution for the ionospheric part of the path, it is useful in checking the accuracy of ray-tracing programs (Croft and Hoogasian 1968).

The skip distance and the maximum frequency can be evaluated from the condition that $dD/d\phi_o = 0$. Figure 6.15 shows the variation of the M factor with ground range for different heights of maximum density h_m and for different values of the ratio y_m/h_0. When these parameters are known, together with f_c, it is possible to determine the maximum frequency by interpolation between curves. To get a reliable estimate of the M-factor at any time, it is desirable that h_0 and y_m should be determined directly from ionograms at the midpoint of the path. Such a course is, however, rarely practical, and it is necessary to predict the most probable values of y_m and h_0 from history. From the network of ionospheric sounding stations the diurnal, seasonal, and sunspot cycle variations of these parameters are known over most of the world. Comparisons have shown that the results obtained by the parabolic method and by Smith's transmission curves are essentially the same. For two-hop propagation it is necessary to calculate the factor for half the distance.

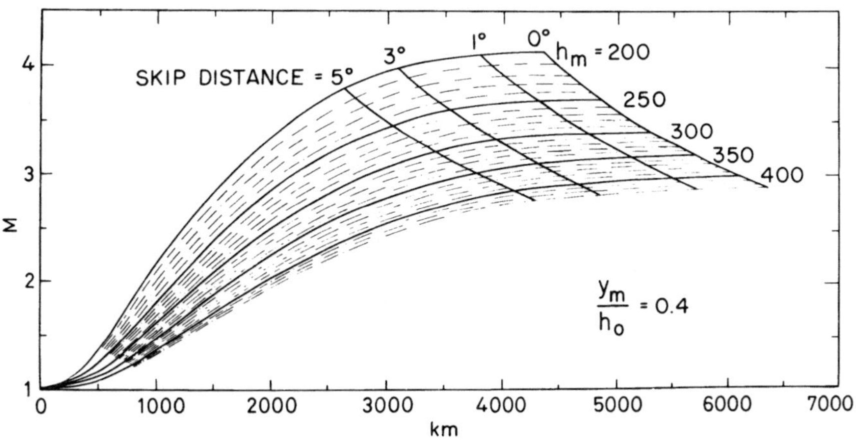

Fig. 6.15 *Variation of the M factor with distance, for a parabolic layer (From Appleton and Beynon 1940, © The Physical Society, reprinted with permission)*

The transmission theory presented above applies to the case of a spherically stratified ionosphere only. Now, over long transmission paths, the ionospheric parameters may vary appreciably and so the various hops will be of unequal length. Under these conditions it is usually assumed that the parameters vary from one hop to the next but that the layer is uniform over the part of the hop in the ionosphere. However, ray-tracing programs are in existence that take into account the hop tilts (Hatfield 1980).

6.7 Oblique ionograms

6.7.1 Technique

The technique for producing oblique ionograms is similar to that for vertical ionograms except that the transmitter and receiver are separated. With modern digital sounders, which have highly stable and accurate crystal oscillators, it is easy to ensure that:

(i) The transmitter and receiver are closely tuned
(ii) The timing of the pulsing is stable so that the received pulses do not drift
(iii) Satellite time signals can be used for synchronization.

6.7.2 Experimental records

6.7.2.1 Nomenclature

In the description of oblique ionograms we are not concerned with whether a frequency is 'usable' but rather with what is observed. The following terms are used:

(i) Maximum observed frequency (MOF) means the highest frequency on which the sounder transmitter signals are observed on the ionogram, regardless of the propagation path involved.
(ii) Lowest observed frequency (LOF) means the lowest frequency on which the sounder transmitter signals are observed on the ionogram, regardless of the propagation path involved.
(iii) MOF and LOF may be used also to describe identifiable modes. For example, $2F_2$LOF means the lowest frequency, observed on the ionograms, that is propagated by two reflections at the F_2 layer with an intermediate ground reflection. Similarly, $2F_2$ MOF means the highest observed frequency of the two-hop F_2 trace.
(iv) The lowest observed frequency of the high-angle ray and that of the low-angle ray may be denoted by the letters H and L, respectively. Thus F_2HLOF is the lowest frequency of the high-angle signal that propagates along the one-hop F_2 path.
(v) When it is required to distinguish between the ordinary and extraordinary raypaths, an o or x may follow in parentheses. Thus the F_2MOF(x) is the maximum observed frequency of the extraordinary wave that is reflected once at the F_2 layer.

(vi) The frequency at which the high- and low-angle rays join is called the junction frequency and is denoted by JF.

Note that capital letters are used on oblique work in contrast to the lower-case letters agreed upon in vertical soundings.

6.7.2.2 Short distances

Figure 6.16 shows an oblique ionogram for a relatively short distance. There is a close similarity to the expected ionogram (i.e. Fig. 6.6) in that, on a given frequency, there are two echoes for each layer (high-angle and low-angle). Further, there are distinct ordinary and extraordinary wave traces with well-defined maximum frequencies. In this ionogram, the highest observed frequency and the lowest observed frequency are associated with the E-layer echo.

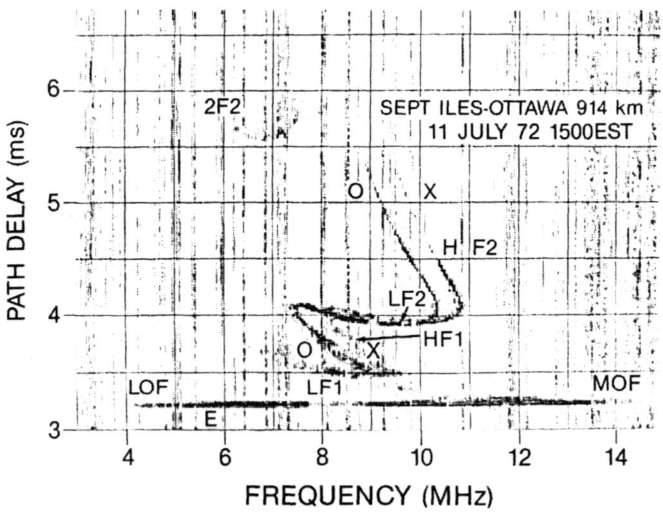

Fig. 6.16 *Oblique ionogram for a distance of 914 km (From Rice 1976)*
Note the presence of ordinary and extraordinary traces and high- and low-angle traces. For this particular ionogram the maximum frequency is determined by the E layer

6.7.2.3 Medium distances

Figure 6.17 shows a sample oblique ionogram taken, during magnetically quiet conditions, over a geographically east–west path. The traces are sharp and the maximum (or junction) frequency of the one-hop F_2 trace is well defined. The high-angle, or Pedersen, trace can be seen for the one-hop trace. Splitting of the high-angle two-hop trace into ordinary and extraordinary traces can be seen.

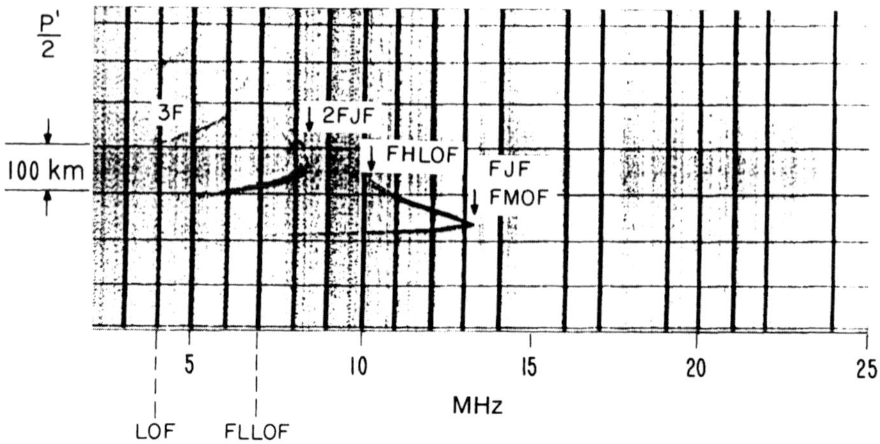

Fig. 6.17 *Oblique ionogram for the 2370 km, geographic approximately east–west path from Boulder, Colo., to Sterling, Va., September 1, 1954, 2112 (90° WMT)*
Here the junction frequency (FJF) or basic MUF is the same as the maximum observed frequency (MOF)

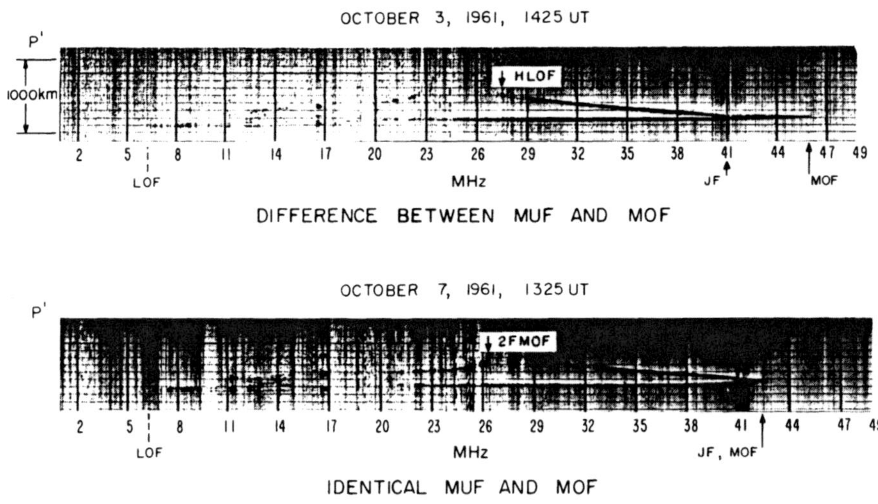

Fig. 6.18 *Oblique ionogram for the 3300 km path from Tripoli, Libya, to Accra, Ghana, illustrating a case in which the maximum observed frequency is the same as the junction frequency, or basic MUF, and one in which they differ by several megahertz*

Figure 6.18 shows an example in which the maximum observed frequency in essentially the same as the junction frequency and an example in which the MOF is greater than the junction frequency. Differences between the F_2JF and F_2MOF depend on path length, geographical location, season, etc., and are caused by scattering by ionospheric irregularities. For east–west propagation over 2400 km in the USA the ratio MOF/JF is about 1·04 during the day and about 1·00 at night.

In low magnetic latitudes during evening hours near the equinoxes the nose of the oblique ionogram is drawn out to higher frequencies and the ratio MOF/JF can be as high as about 2 (see Fig. 6.19). This is the oblique equivalent of spread F and, under these conditions, the maximum observed frequency depends, to some extent, on equipment sensitivity.

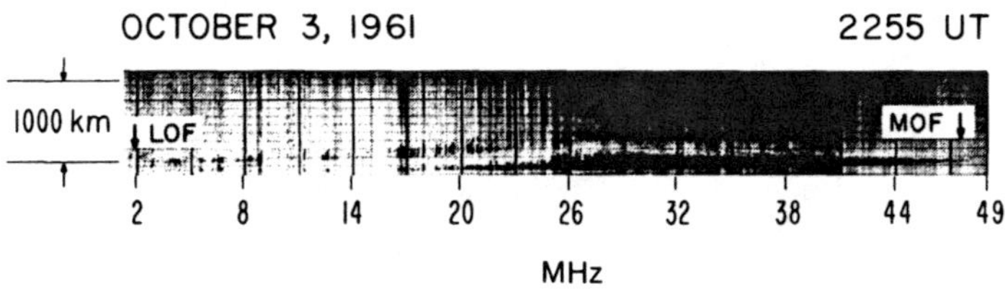

Fig. 6.19 *Oblique ionogram for the 3300 km path from Tripoli, Libya, to Accra, Ghana, with spread echoes*
In this case the maximum observed frequency is much larger than the basic MUF

6.7.2.4 Definitions of maximum transmission frequencies

Because there are different maximum frequencies, the International Radio Consultative Committee (CCIR 1986, Recommendation 373-5) recommended the following definitions:

(i) Operational MUF, or simply MUF, is the highest frequency that would permit acceptable operation of a radio service between given terminals at a given time under specified working conditions (such as antennas, transmitter power, class of emission, information rate, and required signal-to-noise ratio).

(ii) Basic MUF is the highest frequency by which a radio wave can propagate between given terminals, on a specified occasion, by ionospheric refraction alone (junction frequency).

It should be noted that the operational MUF refers to propagation in the actual ionosphere and includes effects of scattering, partial reflections, etc. It therefore depends on the characteristics of the radio system used. On the other hand, the basic MUF is essentially independent of power.

Another definition is the monthly median MUF, which is the highest frequency received on 50% of the days at a given hour. That is, higher frequencies are received less often, and lower frequencies are received more often. The optimum working frequency, designated FOT from the initials of the French version, is the lower decile of the daily values of the operational MUF at a given time over a specified period, usually a month. That is, it is the frequency exceeded by the basic MUF during 90% of the specified period. Generally,

$$\text{FOT} = 0{\cdot}85\ \text{MUF} \tag{6.37}$$

The basic MUF may be modified by the number of hops, magnetoionic component, and distance involved. For example, $1F_2$ (4000) MUF (x) path via the F_2 layer by the extraordinary wave.

6.7.2.5 Long distances

As the transmission distance increases, the complexity of the echo structure tends to increase (Fig. 6.20). For distances of about 4000 km and more, many-

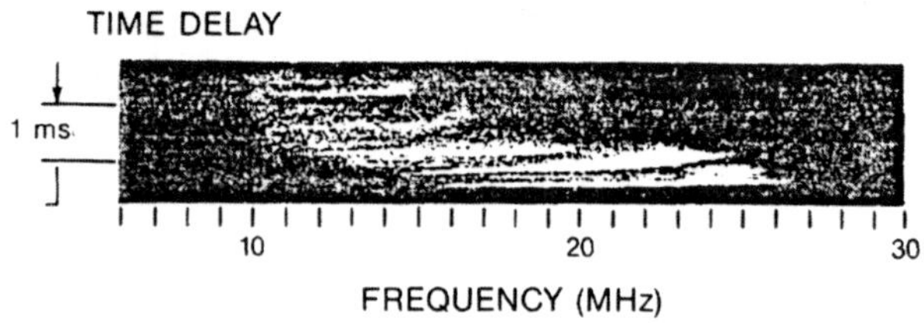

Fig. 6.20 *Oblique chirp ionogram for the 8000 km path from Pahoa, Hawaii, to South Dartmouth, Mass. (From Barry and Fenwick 1969, with permission)*

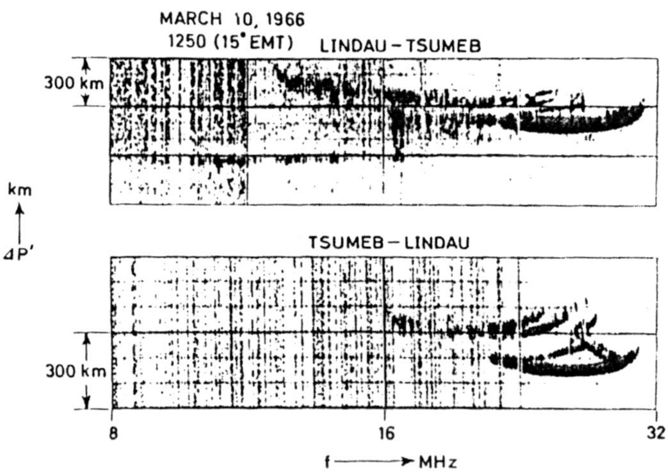

Fig. 6.21 *Oblique pulse ionograms for the 7915 km circuits from Lindau, Germany, to Tsumeb, South-West Africa (100 kW) and Tsumeb to Lindau (10 kW), illustrating the effect of different transmitter powers (From Möller 1969, with permission)*

hop echoes are observed as well as echoes reflected from different layers (e.g. E_s and F_2) with several hops. Thus, the composite signal is quite complicated. Furthermore, with long pulses the echoes can overlap (see Fig. 6.20) because the lengths of the different paths are nearly the same. As the frequency increases the time dispersion of the echo decreases and hence, for minimum signal distortion, a frequency close to the MOF should be used. That the apparent structure of an oblique ionogram depends, to some extent, on the sensitivity of the terminal equipment has been shown by Möller (1969) over the 7915 km paths Lindau–Tsumeb (100 kW power) and Tsumeb–Lindau (10 kW) as seen in Fig. 6.21.

For distances greater than the one-hop limit for F_2-layer propagation ($\approx$ 4000 km) the high-angle, or Pedersen, ray can determine the maximum frequency (Fig. 6.22). For propagation across the north Atlantic, the occurrence of the Pedersen ray tends to peak near noon at the midpoint when horizontal gradients of electron density along the path are at a minimum. This is to be expected because a gradient along the path will give rise to an asymmetric path,

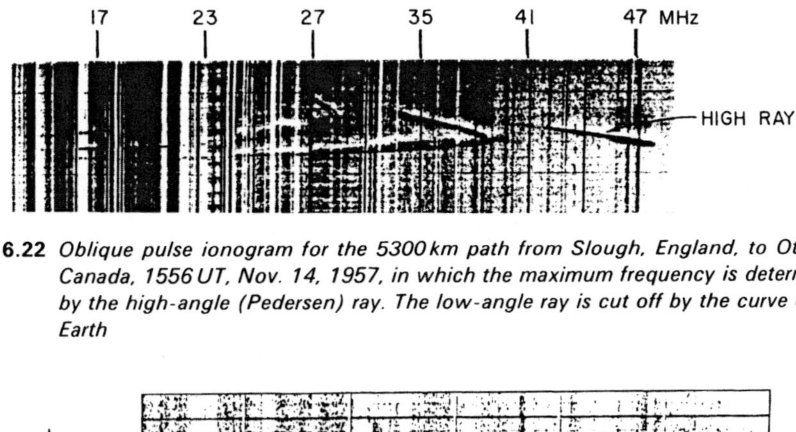

Fig. 6.22 *Oblique pulse ionogram for the 5300 km path from Slough, England, to Ottawa, Canada, 1556 UT, Nov. 14, 1957, in which the maximum frequency is determined by the high-angle (Pedersen) ray. The low-angle ray is cut off by the curve of the Earth*

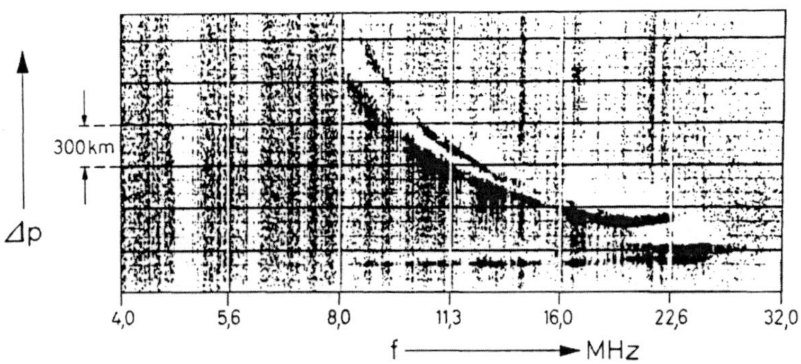

Fig. 6.23 *Oblique pulse ionogram for the 7915 km path from Lindau, Germany, to Tsumeb, South-West Africa, August 13, 1964; 1240, 15° EMT (From Dieminger 1968, with permission)*

thus decreasing the (already small) angle of elevation at one end of the path. An oblique ionogram for the 7915 km, north–south, transequatorial path Lindau–Tsumeb (Fig. 6.23) suggests the presence of a high-angle trace having a group

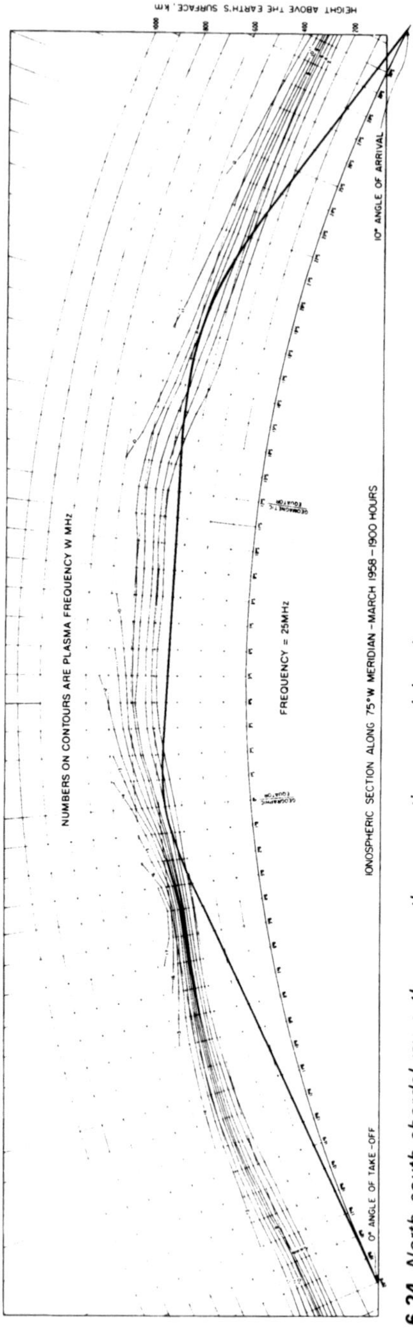

Fig. 6.24 *North–south chordal ray path across the magnetic equatorial region*
The path shows two ionospheric reflections without an intervening ground reflection

path that decreases with increasing frequency. At these long distances the exact path structure is very difficult to identify. Numerous examples of long- and medium-distance oblique ionograms are given in Jones (1969).

From Fig. 5.7 we see that spherical stratification of the F_2 layer does not necessarily hold. Distortion of the ionosphere occurs near sunrise and sunset, in troughs and in the equatorial anomaly (see Sec. 5.1.1). Under these circumstances radio waves can travel to great distances in or below the ionosphere, with low absorption, without intermediate ground reflections. Such a transequatorial path is illustrated in Fig. 6.24; it is sometimes called a chordal path. Chordal propagation is probably important in transmission to the antipodal point (Bold 1969). Fenwick and Villard (1963) showed that this type of propagation could account for round-the-world (RTW) echoes, which may have low dispersion and attenuation (see Rumi 1975 for review). This has been confirmed as seen in the ionogram in Fig. 6.25.

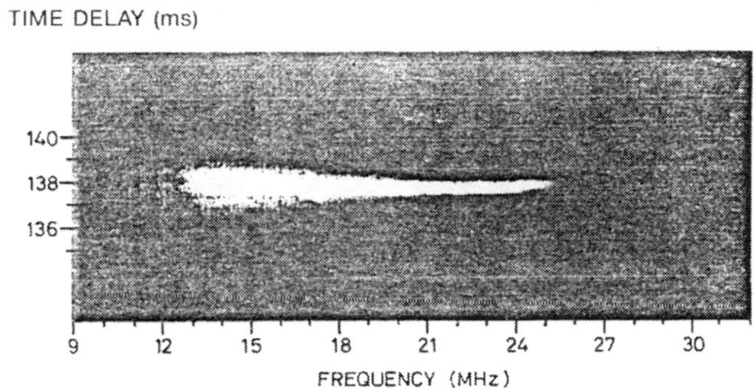

Fig. 6.25 *Oblique chirp ionogram with round-the-world echoes ($\approx$ 40 000 km) from Stanford, Calif., to Stanford (From Barry and Fenwick 1969, with permission)*
Power $\approx$ 100 W, antenna is 13–30 MHz log periodic

Guiding of radio waves in the ionosphere gives rise to 'whispering gallery' paths similar to those for sound waves. This type of propagation is possible with a spherically stratified layer, provided that the incident power can be refracted or scattered into the path as might occur at the day-to-night transitions, irregularities, and discontinuities (e.g. E_s, auroral ionized columns, meteor trails, equatorial tilts). The propagation of guided modes has been studied by Grossi and Langworthy (1966), Chvojkova (1965), and Gurevich and Tsedilina (1985). The MUFs in a whispering gallery are roughly twice those for multi-hop paths between ground-based terminals, and whispering gallery modes occur during both day and night. Ducting in the E–F valley occurs when the power is successively reflected at the lower surface of the F layer and the upper surface of the E layer (see Fig. 6.26). This ducting has been analyzed by waveguide techniques (Chang 1971).

6.7.2.6 Path structure nomenclature

With propagation over long distances the net path can be a complicated summation of various hops including several layers, particularly involving the F_2 layer and sporadic E. Some of these paths are sketched in Fig. 6.27 which includes a descriptive symbolism in which the reflecting layers are indicated in sequence. A hyphen between two letters indicates a ground reflection; the absence of a hyphen between two letters indicates two successive ionospheric reflections (e.g. a chordal hop). Only two layers are indicated in Fig. 6.27 but

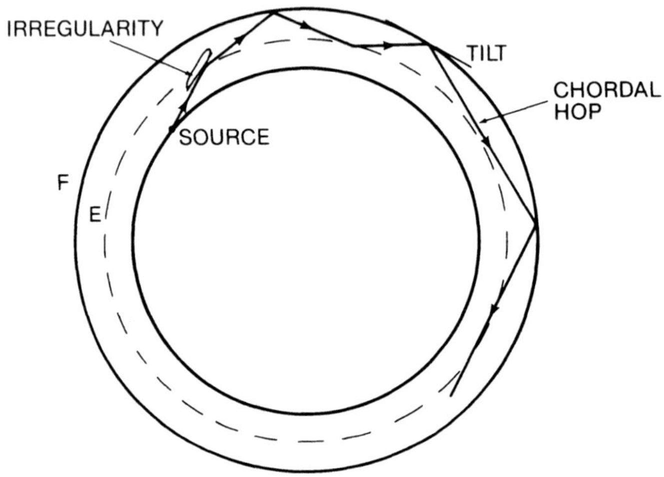

Fig. 6.26 *Long-distance paths involving scattering, reflections from the top of the E (or E_s) layer, and chordal FF propagation (whispering-gallery modes)*

E might be replaced by E_s and F by F_1 or F_2. In the case of *n* successive reflections from the same layer, with intermediate ground reflections, the use of the symbol *nF* (or *n*E) is recommended.

6.7.3 Sporadic E echoes

Figure 6.28 shows that sporadic E(E_s) reflections can have a dominant role in the determination of the MOF and, indeed, the entire structure of the ionogram. Because intense sporadic E can be highly reflecting, the net signal can be strong. Because the time dispersion is small, sporadic E reflection can be efficient. Over long distances, however, the role of sporadic E is not known in detail, partly because it is rare for a patch of E_s to extend over a sufficiently large area of the world to be the sole reflecting layer and partly because the propagation mechanism depends on the type of sporadic E. It appears to have an important part in modifying propagation paths by reflecting waves off the topside of the E_s layer to produce FEF paths. The paths shown in Fig. 6.27*e* avoid a double traverse of the absorbing D region which occurs in two-hop F_2 paths. Figure 6.27 also shows other mixed E_s–F_2 paths caused by reflection (Fig. 6.27*d* and Fig. 6.27*f*) and by scatter (Fig. 6.27*c*).

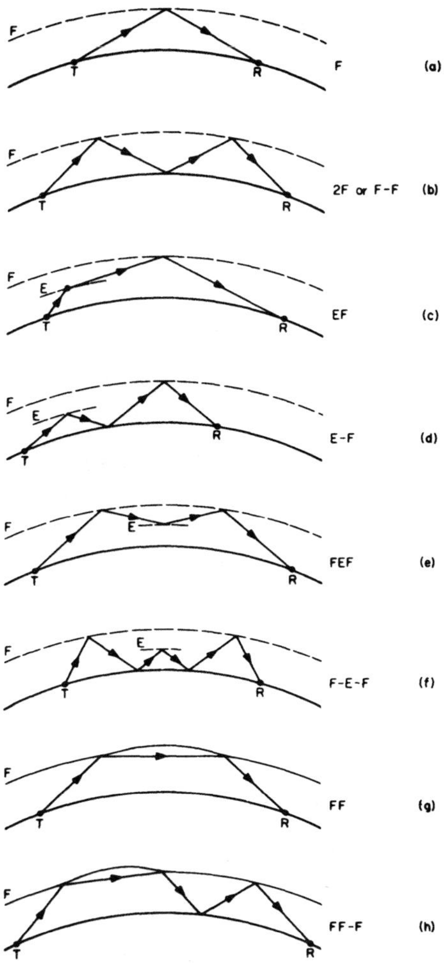

Fig. 6.27 *Descriptive nomenclature for oblique ray paths*

Sporadic E is subdivided into auroral E or night E (CCIR 1986), and it can provide improved communication in the Arctic (Lied 1967). The E_s layer in the auroral zone is mainly observed at night, and it is believed to be caused by electron precipitation from the magnetosphere. Ionospheric records indicate two types of E_s: (i) retardation and (ii) nonretardation. The former is produced by a fairly thick layer whereas the latter is probably produced by a thin layer, by blobs, or by partial reflections. The retardation E_s may be fairly uniform over horizontal distances of the order of 100 km whereas nonretardation E_s is patchy. The average maximum nighttime fE_s is about 5 MHz. For practical purposes it has been useful to provide statistical information on the E_s layer such as the percentage of the number of hours in a month when fE_s exceeds a given value (Smith 1957). The occurrence frequency of fE_s greater than 5 MHz can range from more than 30% near the auroral zone to less than 10% near the pole. Also it appears to be more frequent over North America than over Siberia (Smith 1957). The zone of maximum fE_s does not coincide with the maximum of the visible auroral zone. In the polar cap, sporadic E extends in bands or ribbons across the polar cap, roughly in the direction of the Sun.

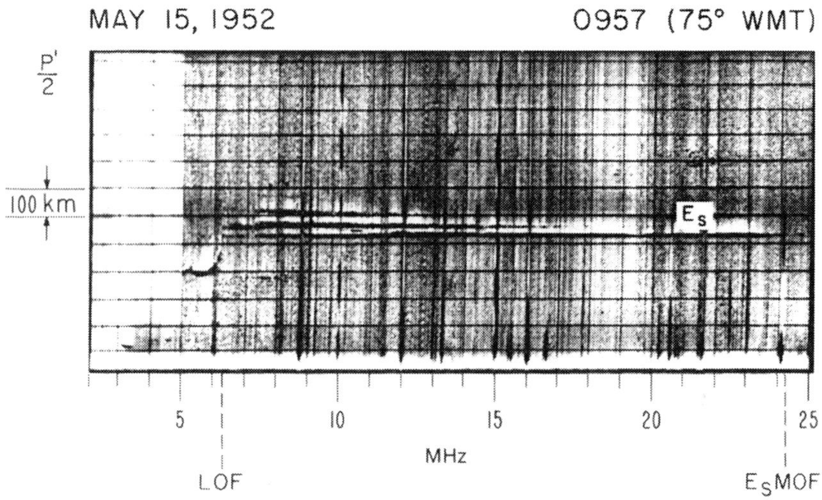

Fig. 6.28 *Oblique pulse ionogram for a 1150 km path from Sterling, Va., to St. Louis, Mo., in which the maximum frequency is determined by sporadic E*

Sporadic E propagation may enhance the quality of a signal and/or increase the usable spectrum. On the other hand, it can screen reflections from the F_2 layer (see Fig. 6.27*d*), which can have adverse consequences.

The E_s layer does not necessarily provide a good medium for communication just because it reflects radio signals. The presence of irregularities can produce rapid and deep fading, leading to serious degradation in communications. Irregularity sizes smaller than 300 m have been observed at Arecibo (Miller and Smith 1975).

In middle latitudes one type of sporadic E appears to be produced by plasma compression due to gradients in the vertical structure of neutral winds (Whitehead 1989). For further information on the physics and chemistry of sporadic E the reader is referred to special issues of *Radio Science* (1972; 1975).

6.8 Scatter propagation

6.8.1 Ionospheric scatter

Scattering of radio wave power by ionospheric irregularities has an important effect on ionogram interpretation. For example, in the presence of spread F the following questions arise: (i) What is the definition of a critical frequency? and (ii) Where does one place a transmission curve to obtain an M-factor? In addition, scatter in a lower layer can increase the angle of incidence of a beam (e.g. see Fig. 6.27*d*), which leads to a higher maximum frequency. This phenomenon is particularly important in (*a*) the equatorial irregularity zone and (*b*) the auroral irregularity zone.

Spreading is due to a multiplicity of paths for a given signal, which can give rise to rapid fading and, hence, signal distortion. As far as scattering is concerned the most important layers are the F (spread F) and sporadic E layers. Irregularities in the F layer tend to be magnetic-field aligned in the form of either rods with axial ratios of about 20 : 1 or sheets. Near the magnetic equator, patches of irregularities cover large areas, 100 to 2000 km in the north–south direction and 500 to 3000 km in the east–west direction. These patches develop shortly after sunset and generally move from west to east. Equatorial irregularities have been seen up to 1000 km on Alouette–1 ionograms (see Sec. 8.2.3).

In the polar ionosphere, F-region irregularities are convected over the polar cap with speeds up to 1·0 km/s; the convection gives rise to severe fading (often called 'flutter') of up to 10 Hz. Time-delay spreads between 0·1 and 0·2 ms have been observed for great-circle paths, i.e. those in the plane containing transmitter, receiver, and the Earth's center, and 1 to 2 ms for non-great-circle paths (Lomax 1972). Hunsucker and Bates (1969) reported a fading correlation bandwidth greater than 1 kHz for 90% of the time, in which amplitude-modulated signals suffer severe distortion because of destructive interference between the two sidebands.

Scattering from field-aligned irregularities is generally aspect sensitive in that it is strongest when the incident beam is perpendicular to the magnetic field and decreases rapidly as the ray departs from perpendicularity. This condition of perpendicularity is often met because of refraction of the radio beam in the F layer. It appears that (i) for VHF and UHF the aspect sensitivity is nearly independent of frequency; (ii) the scatter decreases about 10 dB/degree of aspect angle; (iii) the scattering cross-section varies exponentially with frequency with a decay of 33 dB/1000 MHz; and (iv) there is a moderate correlation between visible aurora and radio aurora (Chesnut 1972). Ionospheric scattering can lead to non-great-circle propagation, and, on transmissions between Europe and southern Africa, azimuthal deviations up to $\pm 50°$ have been observed (CCIR 1986, Report 890-1). Comprehensive discussions of various types of ionospheric scatter are contained in AGARD Conference Proceedings No. 37 (Thrane 1968), and an extensive bibliography is given by Hayden (1968).

6.8.2 Ground sidescatter

Non-great-circle signals result generally from multi-hop paths in which side-scatter occurs from irregularities at the Earth's surface as illustrated in Fig. 6.29

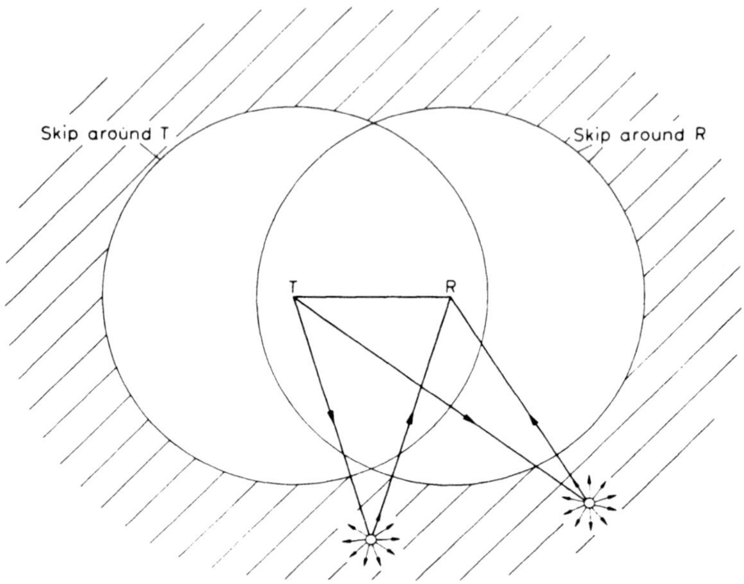

Fig. 6.29 *Propagation paths showing reception inside the skip zone by sidescatter (From Dieminger 1968, with permission)*

so that reception is possible within a skip zone. The azimuth of the strongest signal depends on ionospheric focusing (Sec. 7.3) and on the transmitter antenna pattern. Large bearing deviations have been observed between North America and England (Hagg and Rolf 1963) and between Europe, North America, and Japan (Kanaya and Yokoi 1960).

6.8.3 Practical aspects

Ground and ionospheric sidescatter can lead to an enhancement in MUF above the basic MUF. Further, if the sidescatter path is away from the auroral zone, it can provide communications during disturbed periods. However, communications can be degraded by the increased complexity of the sidescattered signals.

6.8.4 Ground backscatter

Radio signals can be backscattered from a rough ground or sea as illustrated in Fig. 6.30. The backscattered signals can return to the sender along either the same (reciprocal) path or along different paths. The ionogram in Fig. 5.22 shows

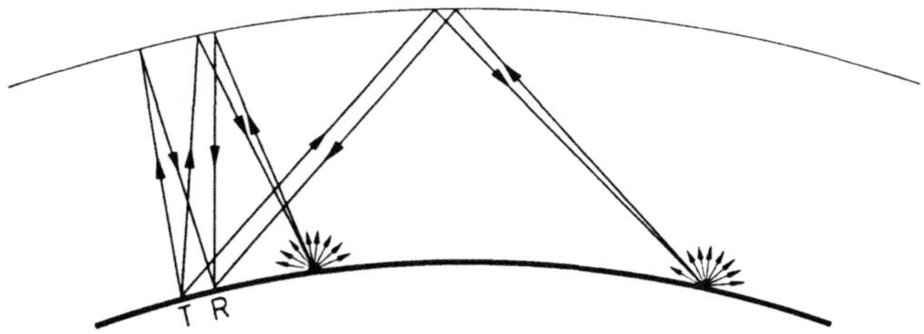

Fig. 6.30 *Geometry of backscatter for various ground ranges (From Dieminger 1968, with permission)*

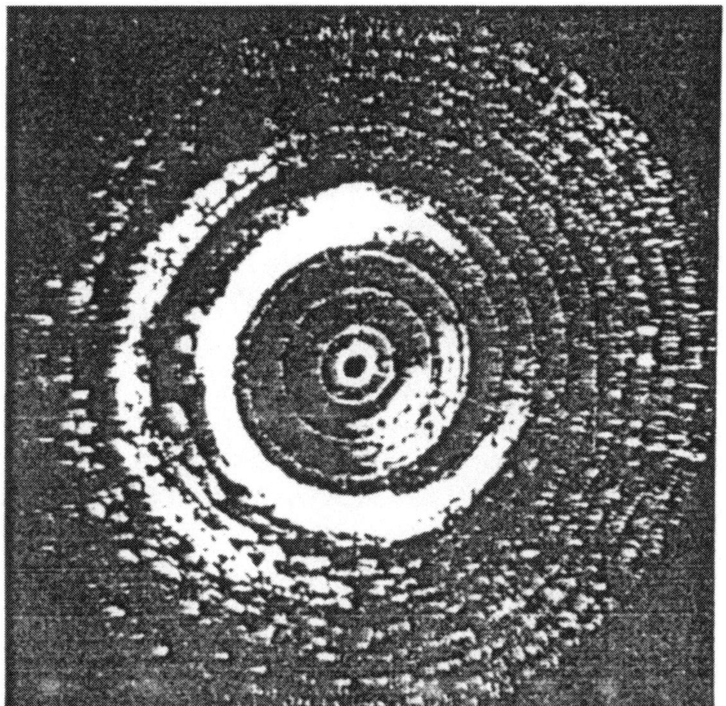

Fig. 6.31 *Plan position backscatter display taken at Stanford, Calif., on 14 MHz (From Villard and Peterson 1952 © IEEE; reprinted with permission)*
Range circles are at 500 km intervals

a backscatter trace taking off from the two-hop F_2 trace and extending to high frequencies and ranges. In this type of reflection the ionosphere is essentially smooth and the scatter occurs at the ground. The sharp lower edge of the backscatter trace results from skip-distance focusing (see Sec. 7.3). This type of ionogram provides a convenient means of determining the maximum frequency,

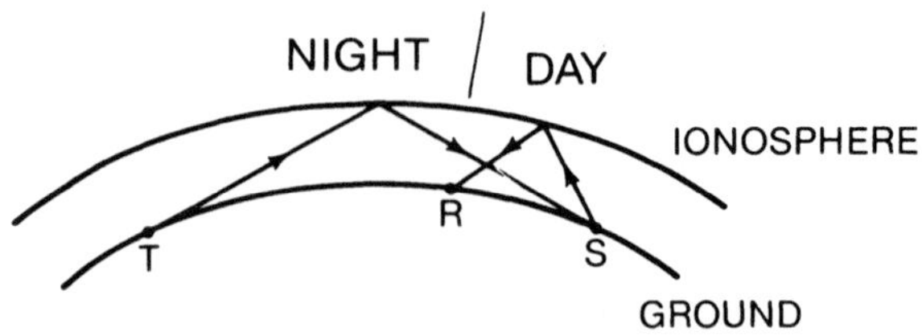

Fig. 6.32 *Signals received on a frequency higher than the MUF at the path midpoint by ground scatter behind the receiver, where the ionosphere is in daylight*

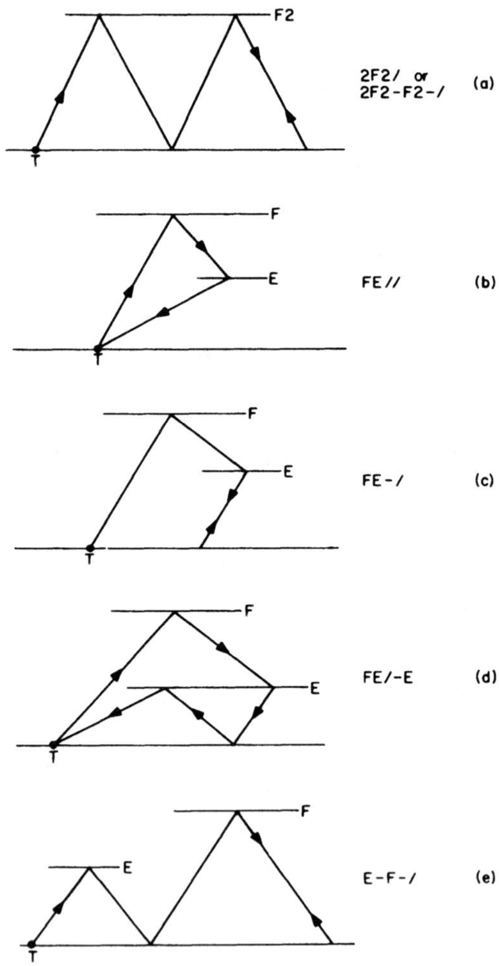

Fig. 6.33 *Terminology for describing ray paths that involve ground and/or ionosphere scatter*

over a geographical area, from a single ionosonde. The analyses of backscatter ionograms have been discussed by Croft (1972) and Maliphant (1968). Another type of backscatter record (Fig. 6.31) is obtained with a fixed-frequency (usually above the critical frequency) radar having an antenna rotating in azimuth to provide direction information. This is called a plan position display.

Ionospheric echoes backscattered from the surface of the sea, sometimes called seascatter, have been used to identify sea surface conditions (e.g. wave spectrum, speed) at distances up to several thousand kilometers (Jones *et al.* 1986). However it is desirable, if not essential, to remove the adverse effects of the ionosphere from the observations. Backscatter can return echoes from behind the receiver (Fig. 6.32) when the receiver is within the skip zone.

A proposed terminology for the description of backscatter paths has been given by Bartholomew (1969) and is illustrated in Fig. 6.33. The terminology is similar to that described in Sec. 6.7.2; a dash (–) indicates the ground reflection, and a slash (/) follows the letter designating ionospheric scatter; e.g. an E–F/ describes an E reflection followed by an F reflection followed by backscatter at the ground and returned along a reciprocal path.

6.9 Multipath propagation times

Oblique ionograms show that, as the maximum frequency is approached, the time spread or multipath of the signal decreases. This time spread can be of considerable importance in the transmission of information at high speeds because the maximum rate (in binary units per second) is roughly equal to the reciprocal of the range of multipath propagation time ΔT_m, where ΔT_m is a function of frequency, path length and location, local time, and season.

The multipath reduction factor (MRF) is defined as the lowest percentage of the MUF for which the range of multipath propagation times is less than a specified value (Bailey 1959). The MRF thus defines a frequency above which a specified minimum protection against multipath is provided. From experimental data (oblique ionograms) for a number of paths in different geographic locations, MRF curves have been deduced statistically (Fig. 6.34). In the oblique ionograms shown above (Figs. 6.16–6.23), the multipath spread approaches zero at the maximum frequency. The spread may be significantly reduced when operation is on lower frequencies, and signals such as the high-angle ray in Fig. 6.22 or on the E ray in Fig. 6.28 are used. It is conceivable that the rate of transmission of information could be increased by a factor of 100 over normal values by a judicious choice of operating frequency.

To illustrate the use of the MRF curve, consider a path length of about 2500 km. The multipath-time spread will range between about 0·5 ms for an f/MUF ratio of about 0·85, to a maximum value of about 4 ms for f/MUF $\approx$ 0·43. Thus, for a given MUF, there is a frequency on which the spread is a maximum. The value of the maximum multipath spread is shown in Fig. 6.35.

For a 2500 km path, the maximum time dispersion is about 3 ms. For a 1000 km path, the maximum dispersion is about 5 ms, and occurs at a value of MRF $\approx$ 56%.

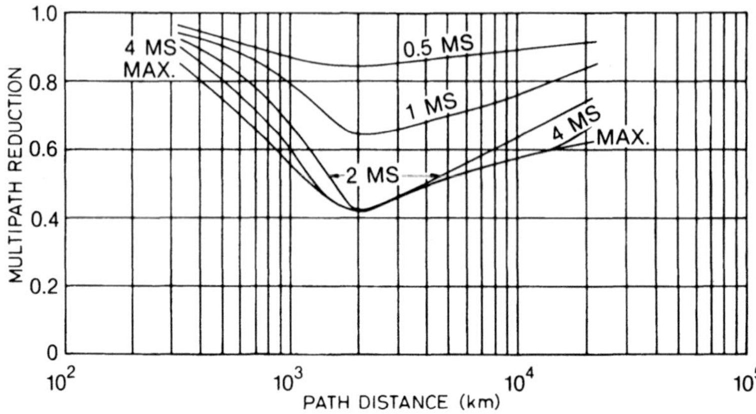

Fig. 6.34 *Multipath reduction factor (f/MUF)*

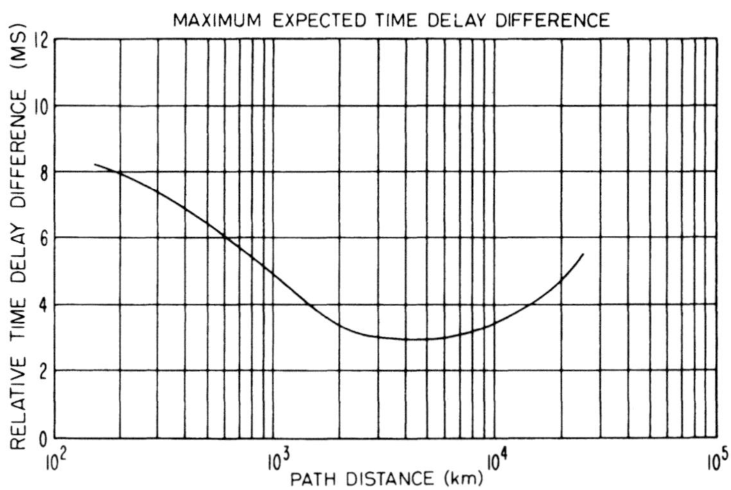

Fig. 6.35 *Maximum expected time delay difference*

For a single hop of distance D (km) the excess group path $\Delta P'$ (km) of the high-angle ray over the low-angle ray is given by

$$\Delta P' = \frac{850\,000}{D}(1 - f/f_m)^{1/2} \tag{6.37}$$

for $D < 1500$ km, where f is wave frequency and f_m is the MUF. The excess attenuation of the high-angle ray relative to that of the low-angle ray is given by

$$\Delta L(\text{dB}) \approx 0{\cdot}1\ \Delta P' \tag{6.38}$$

(Bradley 1979).

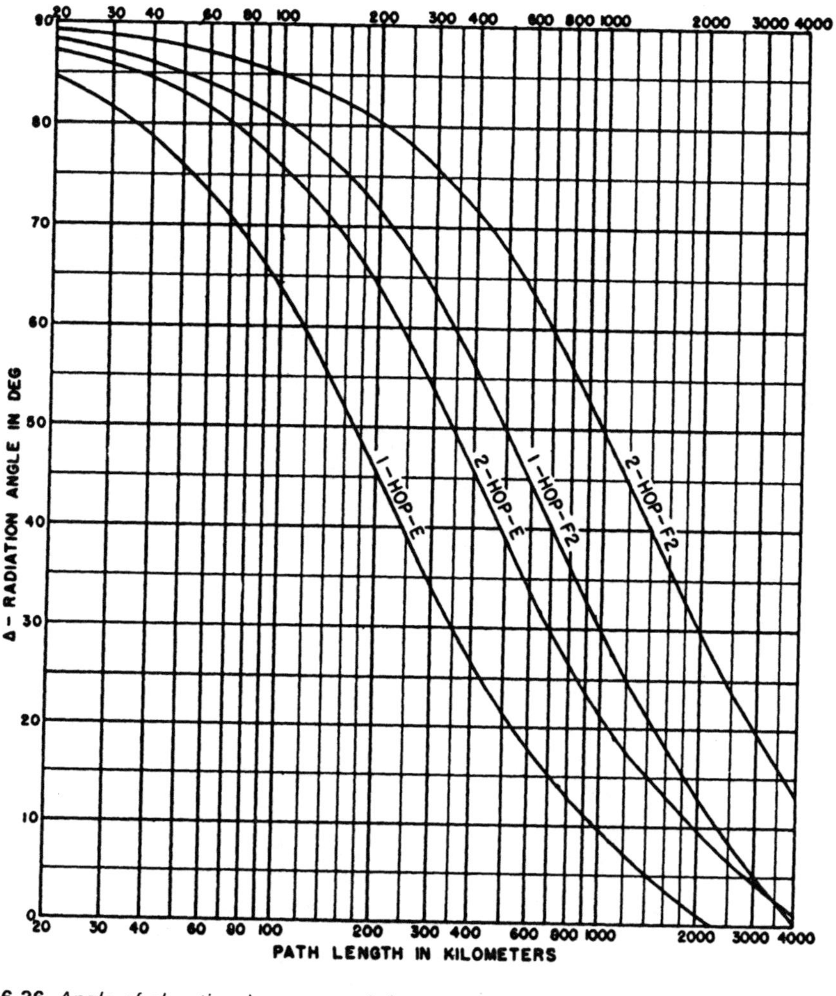

Fig. 6.36 *Angle of elevation Δ versus path length, based on virtual or equivalent reflection heights*
E layer at 100 km, and F_2 layer at 320 km

6.10 Angles of elevation

Angles of elevation, above the horizon, are of importance in the design of antennas. In general, for long distances (e.g. > 3000 km) of transmission, angles of elevation are on the order of 0° to 15° (Fig. 6.36). Low-angle beams are subject to reflection, or cutoff, by the E and/or E_s layer which can have a marked effect on the nature of the received signal. Figure 6.37 shows propagation conditions between Ascension Island and Slough (6750 km) on 19·87 MHz with angles of elevation up to 30° (Kift 1960). Angles of elevation are included in some computed oblique ionograms (Fig. 6.38).

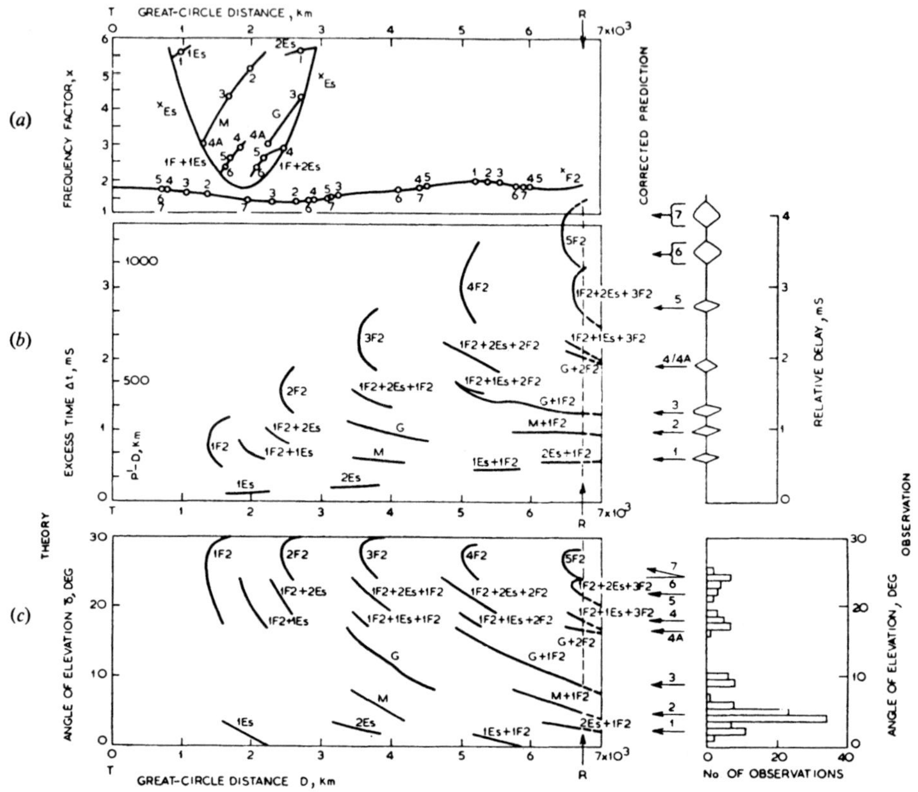

Fig. 6.37 *Angles of elevation, excess time delays, and frequency factors for a 19·87 MHz signal over the 6750 km path from Ascension Island to Slough, England, April 11, 1956, 0930–1030 UT (From Kift 1960) © IEE; reprinted with permission)*
The received signal may have traveled along a mixture of guided and reflected-hop paths

6.11 Inversion of $P'(f)$ curves to $N(h)$ profiles

In principle, an oblique ionogram can be used to determine the electron-density profile of the ionosphere, provided the profile is uniform along the path. Hence, we have a means of exploring ionospheric structures over 'inaccessible' areas of the world such as the oceans or the Arctic and Antarctic. Several approaches have been tried. For example:

(i) An oblique ionogram is converted to an equivalent vertical ionogram (using the relationships in Sec. 6.3.1), which is then reduced to an electron-density profile. Since the effect of the Earth's magnetic field is smaller for oblique propagation than for vertical propagation it is somewhat more accurate to neglect the Earth's magnetic field effect (Davies 1959) in deriving the vertical curve.

(ii) The ionosphere can be represented by parabolic or quasi-parabolic layers, and by iteration a best fit to the observed ionogram can be calculated.

Dubroff *et al.* (1979) and Rao (1974) applied approach (ii) to the analysis of the leading edges of backscatter records to provide three-dimensional distributions (Basler and Scott 1973). Although the electron-density profiles obtained by these means are not as reliable as those from vertical soundings, the data provide information over regions where vertical sounders are absent.

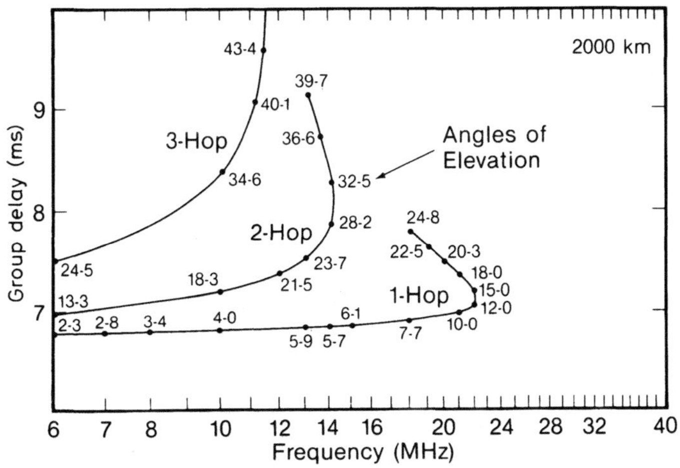

Fig. 6.38 *Synthetic oblique ionogram with angles of elevation (From Croft 1969, with permission)*

6.12 Ray paths

6.12.1 Need for ray path information

To determine certain features of the received signal (e.g. amplitude, polarization, relative phase, time of flight, Doppler) it is necessary to know the ray path between transmitter and receiver. Ray tracing is essential in electron density profiles derived from vertical soundings (for the calculation of focusing, angles of elevation, etc.) and, in general, is done by a step-by-step integration. Ray path information is also important in the determination of the limiting polarizations of the incident and received signal since they depend on the angles between the ray and the magnetic field near the base of the ionosphere; polarization is important in the design of transmitting and receiving antennas. Knowledge of ray paths is essential to the determination of Faraday rotation (i.e. polarization fading) and signal strength. Ray-tracing techniques used in the ionosphere are similar to those used in other branches of wave propagation (such as optics, and atmospheric waves) in which the medium is anisotropic and dispersive.

6.12.2 Ray paths in simple cases

From (6.3) we have, for a flat layer,

$$x(h) = \int \tan\phi \, dh = \sin\phi_o \int \frac{dh}{q} \tag{6.39a}$$

where

$$q^2 = \mu^2 - \sin^2\phi_o \tag{6.39b}$$

At reflection, $q_r = 0$.

It is sometimes convenient to solve (6.39*a*) by using

$$x(h) = h_0 \tan\phi_o + \sin\phi_o \int \left(\frac{dh}{dq}\right) \frac{dq}{q} \tag{6.40a}$$

and

$$P' = x/\sin\phi_o = h_0 \sec\phi_o + \int \left(\frac{dh}{dq}\right) \frac{dq}{q} \tag{6.40b}$$

For a few simple models (e.g. linear, parabola) in which curvature, electron collisions, and the Earth's magnetic field are neglected, the ray path can be determined analytically (see Table 6.1). It is left as an exercise for the reader to determine the total ranges and group paths in a square-root layer, linear layer, exponential layer, and parabolic layer given in Sec. 5.3.2 by putting $\mu_r = \sin\phi_o$ (Budden 1985, Ch. 12).

The phase path is given by

$$P = \int \mu ds = 2h_0/C + 2\int_C^0 \frac{(q^2 + S^2)}{q} \left(\frac{dh}{dq}\right) dq \tag{6.41}$$

where $S = \sin\phi_o$ and $C = \cos\phi_o$.

6.12.3 Inclusion of curvature

When curvature of the ionosphere is introduced, the problem of finding an analytical expression for a ray path becomes very difficult. Expressions have been derived (Chvojkova 1965; Woyk 1978) in which a realistic profile has been approximated by quasi-parabolic segments. The total angle of refraction τ

Table 6.1 *Oblique ray paths in plane layers*

Layer	Ray path $x(h)$	Group path $P'(h)$
$f_N = \alpha(h - h_0)$	$h_0 \tan\phi_o + \dfrac{\sin\phi_o}{\alpha}[\cos\phi_o - \sqrt{\cos^2\phi_o - \alpha(h - h_0)}]$	$h_0 \sec\phi_o + \dfrac{2}{\alpha}[\cos\phi_o - \sqrt{\cos^2\phi_o - \alpha(h - h_0)}]$
Linear $f_N^2 = \beta(h - h_0)$	$h_0 \tan\phi_o + \sin\phi_o \sin^{-1}[\beta \sec\phi_o(h - h_0)]$	$h_0 \sec\phi_o + \sin^{-1}[\beta \sec\phi_o(h - h_0)^2]$
Parabola $(f_N/f)^2 =$ $\gamma(h - h_0) - \delta^2(h - h_0)^2$	$h_0 \tan\phi_o + \dfrac{\sin\phi_o}{\delta}\ln\left[\dfrac{A + \delta(h - h_0) - u}{\cos\phi_o - u}\right]$ where $\gamma = 2f_c^2/y_m f^2, \quad \delta = f_c/y_m f$ $A^2 = \cos^2\phi_o - \gamma(h - h_0) + \delta^2(h - h_0)^2, \quad u = f_c/f$	$P'(h) = x(h) \operatorname{cosec}\phi_o$

between two points r_1 and r_2 is given by

$$\tau = -\int_{r_1}^{r_2} \frac{\tan i}{\mu}\, d\mu = -\int_{r_1}^{r_2} \frac{\tan i}{\mu} \frac{d\mu}{dr}\, dr \tag{6.42}$$

where i is the angle between the ray and the vertical. The ray path segements are of the form

$$r = \frac{p}{1 - \varepsilon \cos k\phi} \tag{6.43}$$

where ϕ is the angle subtended at the Earth's center by the ray path, and p, k, and ε depend on the profile and geometry. The condition for reflection is

$$\frac{d\mu}{dr} = -\frac{\mu}{r} \tag{6.44}$$

This treatment has been applied to waves incident on the ionosphere from space and to the ducting of waves.

Croft and Hoogasian (1968) gave formulas for the ground range, group path, and phase path in a quasi-parabola. Reflection contours (reflectrices) (see Fig. 6.39) show takeoff angles, ground distances, and group heights parametric in wave frequency. The group delay is given by the distance from the origin to the point on the reflectrix. Note that, for a given distance, there are in general two reflection points, corresponding to the low- and high-angle rays. Ray paths in the presence of tilted layers have been discussed by several authors (e.g. Nielson 1968). For further discussion of ray path calculations the reader is referred to a special issue of *Radio Science* (1968).

6.12.4 Homing

One reason for ray tracing is the determination of ray paths between a given transmitter and a receiver on a particular frequency. In ray tracing one normally starts from a given transmitter, with specified angles of elevation and azimuth, so, in general, the ray misses a specific target. Unfortunately, there is no general solution to this problem and the values of the appropriate angles, for which the ray will reach the specified receiver, have to be determined by trial and error. One approach is to make a rough range-versus-elevation curve in the vicinity of the receiver and inspect it for likely 'homing' angles. This procedure can be time consuming because the search is, in general, in two dimensions and, in the case of a satellite-borne receiver, in three dimensions (Jones and Stephenson 1975).

6.12.5 Backscatter rays

In the case of waves scattered back from the ground, or sea, to the transmitter, homing is not required. An extensive discussion of ray tracing for backscatter interpretation has been given by Croft (1972) and Georges and Stephenson (1969), who have simulated backscatter records such as those associated with the passage of traveling ionospheric disturbances.

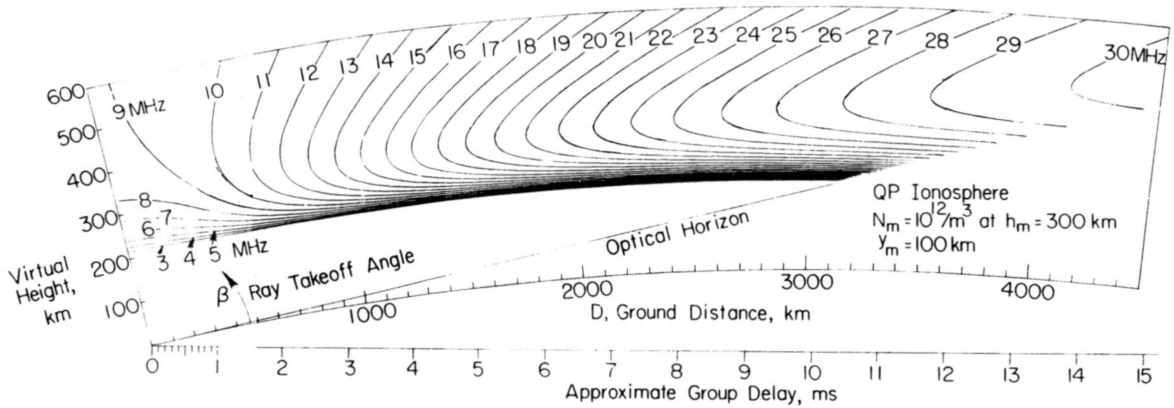

Fig. 6.39 *Reflectrixes showing propagation in a quasi-parabolic layer (From Croft and Hoogasian 1968)*
From any point on a reflectrix for a given frequency, drop a radial onto the curved baseline to find the ground distance of the ray. The length of the radial above the baseline gives the virtual height on the scale at the left. From the point on the reflectrix draw a line to the distance–height origin; the angle between this line and the optical horizon is the elevation angle. The length of this line gives the group delay on the lower scale

6.13 Inclusion of the Earth's magnetic field

6.13.1 Ray direction

The Earth's magnetic field renders the ionosphere anisotropic which, besides producing ordinary and extraordinary waves, makes the direction of energy flow (i.e. the ray direction) different from that of the wave normal (see Sec. 3.6). The problems associated with the tracing of rays in anisotropic media have been discussed extensively (e.g. Budden 1985), and several ray-tracing methods are available (e.g. Booker 1949; Hazelgrove 1955). Essentially, these methods require the step-by-step integration of a set of simultaneous differential equations. One such computer program extensively used by ionospheric workers is that of Jones (1968), described by Jones and Stephenson (1975). This program involves the integration of six differential equations, in polar coordinates, relating the spatial coordinates, time, and the refractive indices and their spatial gradients. Standard subroutines are included for models of electron density, electron collision frequency, and geomagnetic field. The program provides, up to a given point on the path, the spatial coordinates, group path, phase path, polarization, absorption, and (where appropriate) Doppler shift. Ray paths calculated in the above manner, or by other means, have interesting properties that are described below.

First, note that in the ionosphere it is not possible to apply Snell's law directly to the refraction of the ray. In the absence of collisions the rules given in Sec. 3.6 are useful for visualizing ray behavior. Before some actual ray paths are considered, it is of interest to consider the implications of these rules. First, they imply that, as a wave packet penetrates into the ionosphere, it 'slips' sideways in a direction parallel to the magnetic meridian. For upgoing waves the extraordinary ray is (for $\mu < 1$) deflected toward the magnetic equator, and the ordinary ray is deflected toward the nearer pole. Secondly, let us consider (3.37) in the case of waves propagating vertically. We see that $\tan\alpha$ is zero under the following conditions:

$Y = 0$	no magnetic field
$\theta = 0°, 180°$	propagation at the magnetic poles
$\theta = 90°$	propagation at the magnetic equator
$\mu^2 = 1.$	

At the levels of reflection, $\mu = 0$ and $X = 1$ for the ordinary wave, and $X = 1 \pm Y$ for the extraordinary wave. This gives

$$\tan\alpha_o = -\cot\theta \tag{6.45}$$

and

$$\tan\alpha_x = \frac{\sin\theta\cos\theta}{1 + \cos^2\theta}. \tag{6.46}$$

Hence, at reflection, the ordinary ray is perpendicular to the Earth's magnetic field. Note that, at reflection, α_x depends on the direction, but not on the magnitude, of the Earth's magnetic field.

6.13.2 Vertical propagation

Applying Snell's law in a plane stratified ionosphere, we find that the wave normal is always vertical. The lateral deviation D at a height h is given by

$$D = \int_0^h \tan \alpha \, dh \qquad (6.47)$$

which can be integrated numerically.

Ray paths over Boulder, Colo. ($f_H = 1{\cdot}48$ MHz, $I \approx 68°$) are shown in Fig. 6.40 for 'vertical incidence' in a monotonic profile (no valley) on frequencies above the gyrofrequency. The following features should be noted:

(i) The ray paths lie entirely in the magnetic meridian.
(ii) On a given frequency lower than the penetration frequency the poleward deviation of the ordinary wave at reflection is greater than the equatorward deviation of the extraordinary wave at reflection.
(iii) On a given frequency and for the same real height, the deviation of the extraordinary ray is greater than the deviation of the ordinary ray.
(iv) Most of the deviation of the ordinary ray occurs near the level of reflection where the propagation is quasi-transverse.

Since the deviations of the rays are zero at the poles and at the equator, there is a maximum deviation. Actually the deflection of the ordinary wave increases up to a magnetic pole at which there is a discontinuity, whereas the deflection of the extraordinary waves maximizes around a dip of 35°.

On frequencies close to the penetration frequencies the separation between the ordinary-wave (o-wave) reflection point and the extraordinary-wave (x-wave) reflection point can be 50 to 60 km. This condition has an interesting effect on a sequence of ionograms taken during the passage of a traveling ionospheric disturbance: a 'kink' is usually seen on the o-trace before it appears on the x-trace. From the time intervals between the appearances of similar kinks on the traces and the calculated separations of the reflection points, estimates have been made of the speeds of TIDs in the meridian plane. In the presence of a horizontal gradient of electron density the o-waves and x-waves will 'see' different profiles. The separation depends on the height and thickness of the layer and varies with time of day, and so will the frequency separation between f_oF_2 and f_xF_2.

On frequencies well above the penetration frequency, and when quasi-longitudinal propagation is valid, (3.37) gives

$$D_{o,x} \approx \mp 40{\cdot}3 \; N_T f_H f^{-3} \sin \theta \qquad (6.48)$$

so the total deviation is proportional to the total electron content N_T traversed.

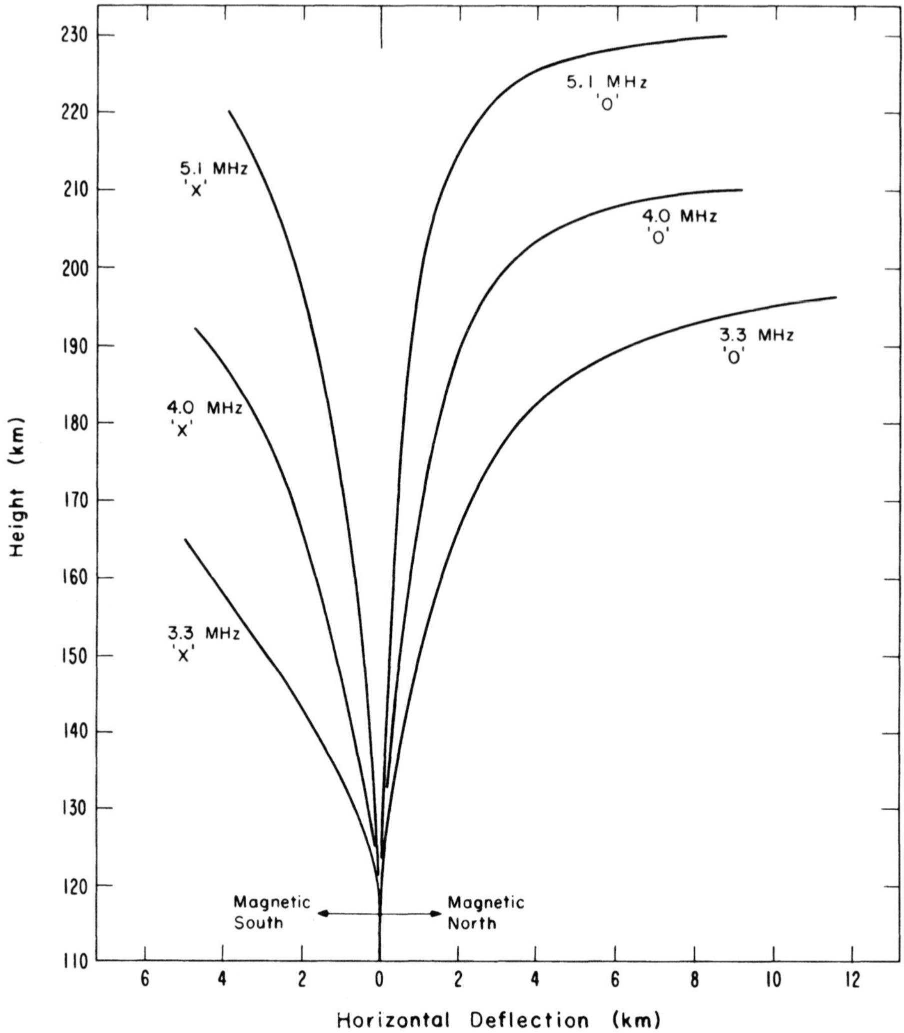

Fig. 6.40 *Ordinary- and extraordinary-ray paths for vertical propagation at Boulder, Colo., July 1967, 0200 UT, f_H = 1·48 MHz, I = 68°*
Note that the ordinary ray is deflected toward the magnetic pole whereas the extraordinary ray is deflected toward the magnetic equator

It is of interest to note, from (3.37), some examples of conditions under which $\tan\alpha$, and hence the deviation, can be large:

(i) For an *o*-wave, near penetration when $X \to 1$ and $\mu \to 0$, $\alpha = 90° - \theta$
(ii) For an *x*-wave, on frequencies below the gyrofrequency and for whistler propagation when $\mu > 1$ (see Chs. 10 and 11)

6.13.3 Oblique paths

6.13.3.1 Propagation in a magnetic meridian

With magnetic north–south propagation the ray remains in the meridian plane. However, as shown in Fig. 6.41, in general, the level of reflection (P) of the wave normal ($\boldsymbol{k}_r$) does not coincide with height of reflection (A) of the ray. One can

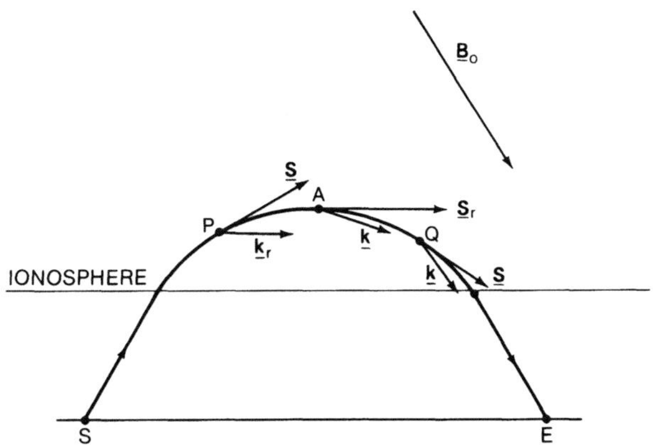

Fig. 6.41 *Oblique ray path in a magntic meridian showing different reflection levels of the ray ($\boldsymbol{S}$) and the wave normal ($\boldsymbol{k}$)*

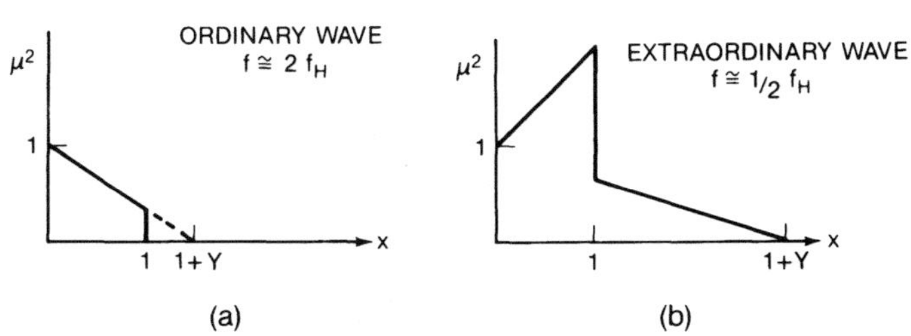

Fig. 6.42 *Dispersion curves $\mu^2(X)$ for propagation parallel to the magnetic field, $\theta = 0$*
(*a*) Frequency higher than the gyrofrequency
(*b*) Frequency lower than the gyrofrequency

determine the angle between the ray and the wave normal at the reflection level of the wave normal by applying Snell's law, since the angle with the magnetic field is equal to the dip and μ^2 can be found from the Appleton equation. The elevation angle of the ray is given by substitution in (3.37). This does not apply when there is a discontinuity in the refractive index as occurs at $X = 1$ and $\theta = 0$ for the ordinary wave (with $f > f_H$) (Fig. 6.42). This discontinuity occurs when the wave normal is refracted such that the $X = 1, \theta = 0$ condition occurs below the level at which the wave normal is horizontal (Fig. 6.43), and the refractive index drops suddenly to the reflection condition. This produces a discontinuity or cusp in the ray path, called the spitze. The critical angle of incidence for the spitze for the *o*-wave is given by

$$\phi_c = \sin^{-1}\{[Y/(1 + Y)]^{1/2} \cos I\} \tag{6.49}$$

When $\phi_o > \phi_c$ the spitze is absent. The largest value of ϕ_c occurs with trans-equatorial propagation for which $I = 0$ and $\phi_c = \sin^{-1}\{[Y/(1 + Y)]^{1/2}\}$ (e.g. for $Y = \frac{1}{2}$, $\phi_c = 35°$). That is, $\alpha = 90°$, so the ray is vertical. Ray paths of *x*-waves on frequencies lower than the gyrofrequency are discussed in Chapter 11.

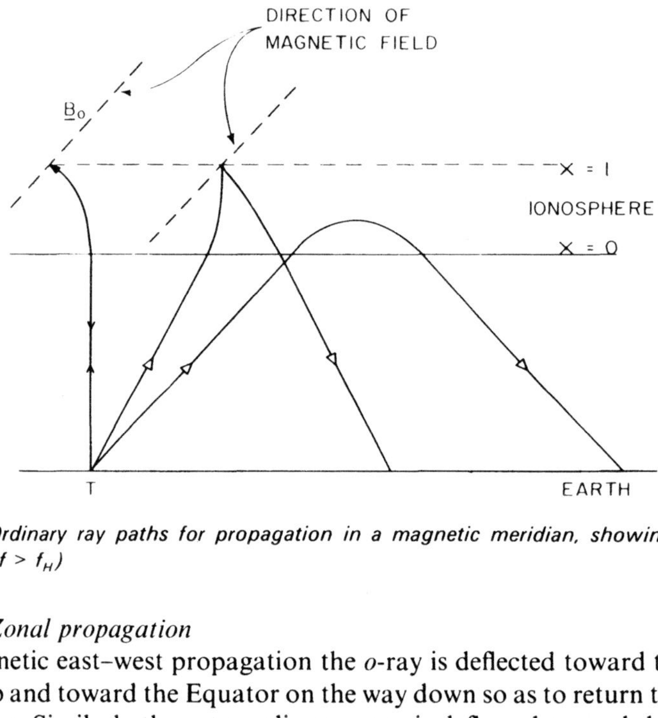

Fig. 6.43 *Ordinary ray paths for propagation in a magnetic meridian, showing the spitze ($f > f_H$)*

6.13.3.2 Zonal propagation

With magnetic east–west propagation the *o*-ray is deflected toward the pole on the way up and toward the Equator on the way down so as to return to the plane of incidence. Similarly the extraordinary wave is deflected toward the Equator. The magnitudes of the deviations depend, in a complicated way, on the wave frequency, the magnetic field, and the electron-density profile, that is, on the depth of penetration into the layer. Over a 2000 km path the maximum deviation is about 1 km for the low-angle ray and larger for the high-angle ray. For most practical purposes magnetoionic deviations of the rays are unimportant on frequencies in the HF band and on higher frequencies.

6.13.3.3 Propagation at an angle with the magnetic meridian

In general, the planes of incidence and emergence are approximately parallel but not coincident because of ray deviation. For propagation at an angle of about 45° at the dip equator the displacements of the emergent planes can be estimated by means of (6.48). Taking $f_H = 1 \times 10^6$ MHz, $f = 5 \times 10^6$ MHz, and $N_{\text{total}} = 2 \times 10^{16}$ electrons m^{-2} gives $D_{o,x} \approx \pm 5$ km. For examples of magnetoionic ray paths and the methods of calculation the reader is referred to Budden (1985).

6.14 References

AGY, V., and DAVIES, K. (1959): 'Ionospheric investigations using the sweep-frequency pulse technique at oblique incidence', *J. Res. NBS*, **63D**, p. 151

APPLETON, E. V., and BEYNON, W. J. G. (1940): 'The application of ionospheric data to radio communications problems; Pt I', *Proc. Phys. Soc.*, **52**, p. 518

APPLETON, E. V., and BEYNON, W. J. G. (1947): 'The application of ionospheric data to radio communications problems; Pt II', *Proc. Phys. Soc.*, **59**, p. 58

BAILEY, D. K. (1959): 'The effect of multipath distortion on the choice of operating frequencies for high-frequency communications circuits', *IRE Trans.*, **AP-7**, p. 398

BARRY, G. H., and FENWICK, R. B. (1969): 'Oblique chirp sounding' *in* JONES, T. B. (Ed.), 'Oblique ionospheric radiowave propagation'. AGARD Conf. Proc. No. 13, Advisory Group for Aerospace Research and Development, North Atlantic Treaty Organisation, p. 487

BARTHOLOMEW, R. (1969): 'A proposed terminology for ionospherically propagated backscatter modes' *in* JONES, T. B. (Ed.), 'Oblique ionospheric radiowave propagation'. AGARD Conf. Proc. No. 13. Advisory Group for Aerospace Research and Development, North Atlantic Treaty Organisation, p. 467

BASLER, R. P., and SCOTT, T. D. (1973): 'Ionospheric structure from oblique-backscatter soundings', *Radio Sci.*, **8**, p. 425

BOITHIAS, L. (1987): 'Radiowave propagation'. Translated by D. Beeson (McGraw-Hill Book Company, New York)

BOLD, G. E. J. (1969): 'Power distribution near the antipode of a short wave transmitter', *J. Atmos. Terr. Phys.*, **31**, p. 1391

BOOKER, H. G. (1949): 'Application of the magneto-ionic theory to radio waves incident obliquely upon a horizontally-stratified ionosphere', *J. Geophys. Res.*, **54**, p. 243

BRADLEY, P. A. (1979): 'Propagation at medium and high frequencies, long- and short-term models' *in* 'Aerospace propagation media modeling and prediction schemes for modern communications, navigation, and surveillance systems'. AGARD Lecture Series Vol. 99, Advisory Group for Aerospace Research and Development, North Atlantic Treaty Organisation, p. 9–1

BUDDEN, K. G. (1985): 'The propagation of radio waves' (Cambridge University Press, Cambridge)

CCIR (1986): 'Recommendations and reports of the CCIR, 1986: VI. Propagation in ionized media'. Int. Telecom. Union, Geneva

CHANG, H. (1971): 'Whispering gallery propagation in the E-region of the ionosphere at HF and VHF', *Radio Sci.*, **6**, p. 465

CHESNUT, W. G. (1972): Polar propagation effects on VHF–UHF radars' *in* FRIHAGEN, J. (Ed.): 'Radar propagation in the Arctic'. AGARD Conf. Proc. No. 97, Advisory Group for Aerospace Research and Development, North Atlantic Treaty Organisation, p. 5–1

CHVOJKOVA, E. (1965): 'Analytical formulas for radio paths in spherically stratified ionospheres', *J. Res. NBS*, **69D**, p. 453

CROFT, T. A. (1967): 'HF radio focussing caused by the electron distribution between ionospheric layers', *J. Geophys. Res.*, **72**, p. 2343

CROFT, T. A. (1969): 'A review of oblique ray tracing and its application to the calculation of signal strength' *in* JONES, T. B. (Ed.): 'Oblique ionospheric radiowave propagation'. AGARD

Conf. Proc. No. 13. Advisory Group for Aerospace Research and Development, North Atlantic Treaty Organisation, p. 137

CROFT, T. A. (1972): 'Sky-wave backscatter: A means for observing our environment at great distances', *Rev. Geophys. Space Phys.*, **10**, p. 73

CROFT, T. A., and HOOGASIAN, H. (1968): 'Exact ray calculations in a quasi-parabolic ionosphere with no magnetic field', *Radio Sci.*, **3**, p. 69

DAVIES, K. (1959): 'The effect of the earth's magnetic field on muf calculations', *J. Atmos. Terr. Phys.*, **16**, p. 187

DAVIES, K. (1969): 'Ionospheric radio waves' (Blaisdell Publishing Co., Waltham, MA)

DAVIES, K., and RUSH, C. M. (1985): 'Reflection of high-frequency radio waves in inhomogeneous ionospheric layers', *Radio Sci.*, **20**, p. 303

DIEMINGER, W. (1968): 'Ionospheric propagation of high frequency radio waves' *in* DAVIES, K. (Ed.): 'Radio wave propagation'. AGARD Lecture Series XXIX, Advisory Group for Aerospace Research and Development, North Atlantic Treaty Organisation, p. III–1

DUBROFF, R. E., RAO, N. N., and YEH, K. C. (1979): 'Methods of determining ionospheric structure from oblique sounding data' *in* COYNE, V. J. (Ed.): 'Special topics in HF propagation'. AGARD Conf. Proc. No. 263, Advisory Group for Aerospace Research and Development, North Atlantic Treaty Organisation, p. 13–1

FENWICK, R. B., and VILLARD, O. G., Jr. (1963): 'A test of the importance of ionosphere–ionosphere reflections in long distance and around-the-world high-frequency propagation', *J. Geophys. Res.*, **68**, p. 5659

GEORGES, T. M., and STEPHENSON, J. J. (1969): 'HF radar signatures of traveling ionospheric irregularities, 3D ray-tracing simulation', *Radio Sci.*, **4**, p. 679

GROSSI, M. D., and LANGWORTHY, B. M. (1966): 'Geometric optics investigation of HF and VHF guided propagation in the ionospheric whispering gallery', *Radio Sci.*, **1**, p. 877

GUREVICH, A. V., and TSEDILINA, E. E. (1985): 'Long distance propagation of HF radio waves' (Springer-Verlag, Berlin)

HAGG, E. L., and ROLF, W. (1963): 'A study of transatlantic radio propagation modes at 41·5 mc/s', *Canad. J. Phys.*, **41**, p. 220

HATFIELD, V. E. (1980): 'HF communications predictions 1978' *in* DONNELLY, R. F. (Ed.): 'Solar–terrestrial predictions proceedings, Vol. IV', D2–1, US Govt. Printing Office, Washington, DC 20402

HAYDEN, E. C. (1968): 'Ground scatter in review' *in* THRANE, E. (Ed.): 'Scatter propagation of radio waves', AGARD Conf. Proc. No. 37, Advisory Group for Aerospace Research and Development, North Atlantic Treaty Organisation, p. 1–1

HAZELGROVE, J. (1955): 'Ray theory and a new method for ray tracing'. *Physics of the ionosphere*, Phys. Soc., London, p. 355

HUNSUCKER, R. D., and BATES H. F. (1969): 'Survey of polar and auroral region effects on HF propagation', *Radio Sci.*, **4**, p. 347

JONES, R. M. (1968): 'A three-dimensional ray-tracing computer program'. Digest of ESSA Tech. Rep. ITSA No. 17. *Radio Sci.*, **3**, p. 93

JONES, R. M., and STEPHENSON, J. J. (1975): 'A versatile three-dimensional ray tracing computer program for radio waves in the ionosphere'. OT Report 75–76. US Govt. Printing Office, Washington, DC 20402

JONES, R. M., GEORGES, T. M., and RILEY, J. P. (1986): 'Measured ionospheric distortion of HF ground-backscatter spectra, *IEEE Trans.*, **AP–34**, p. 563

JONES, T. B. (Ed.) (1969): 'Oblique ionospheric radiowave propagation'. AGARD Conf. Proc. No. 13, Advisory Group for Aerospace Research and Development, North Atlantic Treaty Organisation

KANAYA, S., and YOKOI, H. (1960): 'The lateral deviation of radio waves propagated along the great-circle path from Europe', *Rep. Iono. Res. Japan*, **14**, p. 483

KELSO, J. M. (1964): 'Radio ray propagation in the ionosphere' (McGraw-Hill Co., Inc., New York)

KIFT, F. (1960): 'The propagation of high frequency radio waves to long distances', *Proc. IEE*, **107B**, p. 127

LIED, F. (Ed.) (1967): 'High frequency radio communications'. AGARDograph **104**, Technivision, Maidenhead, England

LOMAX, J. R. (1972): 'Frequency distortion in auroral HF propagation' *in* FRIHAGEN, J. (Ed.): 'Radar propagation in the Arctic'. AGARD Conf. Proc. No. 97, Advisory Group for Aerospace Research and Development, North Atlantic Treaty Organisation, p. 22–1

MALIPHANT, R. G. (1968): 'HF ground backscatter analysis with the aid of computer ray tracing' *in* THRANE, E. (Ed.): 'Scatter propagation of radio waves'. AGARD Conf. Proc. No. 37, Advisory Group for Aerospace Research and Development, North Atlantic Treaty Organisation, p. 7–1

MILLER, K. L., and SMITH, L. G. (1975): 'Horizontal structure of midlatitude sporadic-E layers observed by incoherent scatter radar', *Radio Sci.*, **10**, p. 271

MÖLLER, H. G. (1969): 'Comparison of predicted and observed luf' *in* JONES, T. B. (Ed.): 'Oblique ionospheric radiowave propagation'. AGARD Conf. Proc. No. 13, Advisory Group for Aerospace Research and Development, North Atlantic Treaty Organisation, p. 335

NIELSON, D. L. (1968): 'Ray-path equations for an ionized layer with a horizontal gradient', *Radio Sci.*, **3**, p. 101

Radio Science (1968): 'Special Issue on ray tracing', **3**, (1), p. 1

Radio Science (1972): 'Special Issue on sporadic E', **7**, (3), p. 345

Radio Science (1975): 'Special Issue on recent advances in the physics and chemistry of the E region', **10**, (3), p. 229

RAO, N. N. (1974): 'Inversion of sweep-frequency sky-wave backscatter leading edge for quasi-parabolic ionospheric layer parameters', *Radio Sci.*, **9**, p. 845

RICE, D. W. (1976): 'High resolution measurement of time delay and angle of arrival over a 911 km HF path' *in* BLACKBAND, W. T. (Ed.): 'Radio systems and the ionosphere'. AGARD Conf. Proc. No. 173, Advisory Group for Aerospace Research and Development, North Atlantic Treaty Organisation, p. 33–1

RUMI, G. C. (1975): 'Around-the-world propagation', *Radio Sci.*, **10**, p. 711

SMITH, E. K. (1957): 'Worldwide occurrence of sporadic E', NBS Circ. **582**

SMITH, N. (1939): 'The relation of radio sky-wave transmission to ionosphere measurements', *Proc. IRE*, **27**, p. 332

THRANE, E. (Ed.) (1968): 'Scatter propagation of radio waves, Parts I and II'. AGARD Conf. Proc. No. 37, Advisory Group for Aerospace Research and Development, North Atlantic Treaty Organisation

VILLARD, O. G., and PETERSON, A. M., (1952): 'Scatter-sounding: A technique for study of the ionosphere at a distance', *IRE Trans.*, **AP1–4**, p. 186

WHITEHEAD, J. D. (1989): 'Recent work on mid-latitude and equatorial sporadic E', *J. Atmos. Terr. Phys.*, **51**, p. 401

WOYK, E. (1978): 'Ray tracing theory and mirage occurrence conditions', *Appl. Optics*, **17**, p. 2108

Chapter 7

Amplitude and phase

7.1 Importance of amplitude and phase

A radio operator is concerned with characteristics of the received signal such as its strength, depth and rate of fading, time dispersion, phase and/or frequency dispersion, and polarization. The amplitude of a radio signal depends on several factors, among the more important being (i) ionospheric absorption, (ii) distance attenuation, (iii) focusing or defocusing, and (iv) the coupling between the wave and the antenna. Furthermore, because the ionospheric echoes fluctuate, or fade, with time we need to distinguish between average signal strength and instantaneous signal strength.

Amplitude is important because of the necessity to establish a sufficient signal-to-noise, or signal-to-interference, level. It is the variability of the amplitude that gives ionospheric radio propagation its peculiar characteristics and, hence, problems. The amplitude varies with rates that range from less than a second to the 11-year sunspot cycle. The short-term (less than 1 h) fluctuations are usually caused by wave interference effects coupled with ionospheric motion whereas the long-term changes are caused by variations in absorption in the D region.

Phase is important in navigation (e.g. OMEGA System), in signal fading, in Doppler detection radars, and in a variety of techniques for studying ionospheric characteristics such as electron total content, sudden ionospheric disturbances, and traveling ionospheric disturbances. Even on a 'quiet' day, HF waves reflected vertically may have phase fluctuations of the order of a hundred cycles in periods of the order of 10 min. Phase changes diurnally, seasonally, and during disturbances (Davies 1970).

7.2 Power loss

It is convenient to think of attenuation in terms of a loss L (in decibels) defined as the ratio of the actual power received (p_r) to the power that would have been received in the absence of the particular losses under consideration (p_u). The loss is given by

$$L = -10 \log (p_r/p_u) = P_u - P_r \qquad (7.1a)$$

where

$$P_u = 10 \log p_u \qquad (7.1b)$$

and

$$P_r = 10 \log p_r \qquad (7.1c)$$

The advantage of this logarithmic form is that the total loss L is simply the arithmetic sum of its component losses. In terms of electric field amplitudes E_u and E_r the absorption (in decibels) is

$$L = -10 \log (E_r/E_u)^2 = -20 \log (E_r/E_u) \qquad (7.2)$$

The total loss between the output terminals of a transmitter and the input terminals of a receiver include losses in the transmission lines and antennas. Here we confine our attention to the propagation losses incurred outside the transmitting and receiving antennas except for the polarization coupling between the waves and the antennas. We consider distance spreading loss L_d and focusing (or defocusing) loss L_f, ionospheric absorption loss L_a, and polarization coupling loss L_p. The total loss is

$$L = L_d + L_f + L_a + L_p \qquad (7.3)$$

Note that with focusing L_f is negative and that L_p is made up of two parts that involve the transmitting antenna and the receiving antenna.

Now power can reach the receiver along several paths. When we assume that the power input to each path is the same, we can regard the paths as a set of conductors in parallel. The presence of additional signals, although each has suffered losses, increases the net received signal. By analogy with electrical conductors let us define

$$L = -10 \log g \qquad (7.4)$$

where g is analogous to conductivity and constitutes a negative loss. Let $g_1, g_2, g_3, \ldots,$ represent the gains for the individual paths. Then the total gain g is given by

$$g = g_1 + g_2 + g_3 + \ldots \qquad (7.5)$$

and the net loss is found by substitution of (7.5) into (7.4). For example, with three paths, each having a loss of 40 dB, the net power is increased by a factor of 3 compared with that for a single path, so the net loss is $40 - 10 \log 3 = 35{\cdot}2$ dB.

7.3 Spatial losses

7.3.1 Inverse distance loss

In an isotropic medium the power flux from a small source decays as the square of the distance from the source. Let p_s be the power flux at a standard distance s from the source. Then the power flux p at a distance d is given by

$$p/p_s = (s/d)^2 \tag{7.6}$$

The spreading loss L_d is

$$L_d = P_s - P = 20 \log d - 20 \log s \tag{7.7}$$

Putting the reference distance s at unity, we have

$$L_d = 20 \log d \tag{7.8}$$

In the presence of convergence and/or divergence of rays it is convenient to define an effective distance d_e which, in a uniform medium, gives the actual flux.

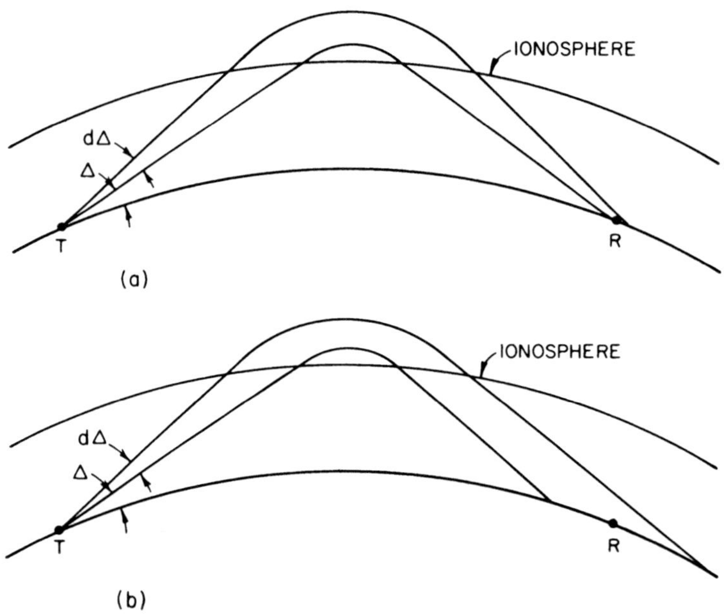

Fig. 7.1 *(a) Focusing (b) defocusing of ionospherically reflected radio signals, caused by curvature of the ionosphere*

7.3.2 Geometrical focusing and defocusing

Curvature of the ionosphere (Fig. 7.1) can lead to focusing and divergence of rays. Such curvature may arise from several causes such as (i) the curvature of the Earth and its atmosphere, (ii) irregularities in electron structure (e.g. field-aligned irregularities), and (iii) traveling wavelike structures.

7.3.3 Skip focusing

Consider a beam, of rectangular cross-section, emitted from a transmitter on the ground (Fig. 7.2). The area A illuminated on the ground is

$$A = a^2 \sin\theta \sin\Delta \, d\eta \, d\theta \tag{7.9}$$

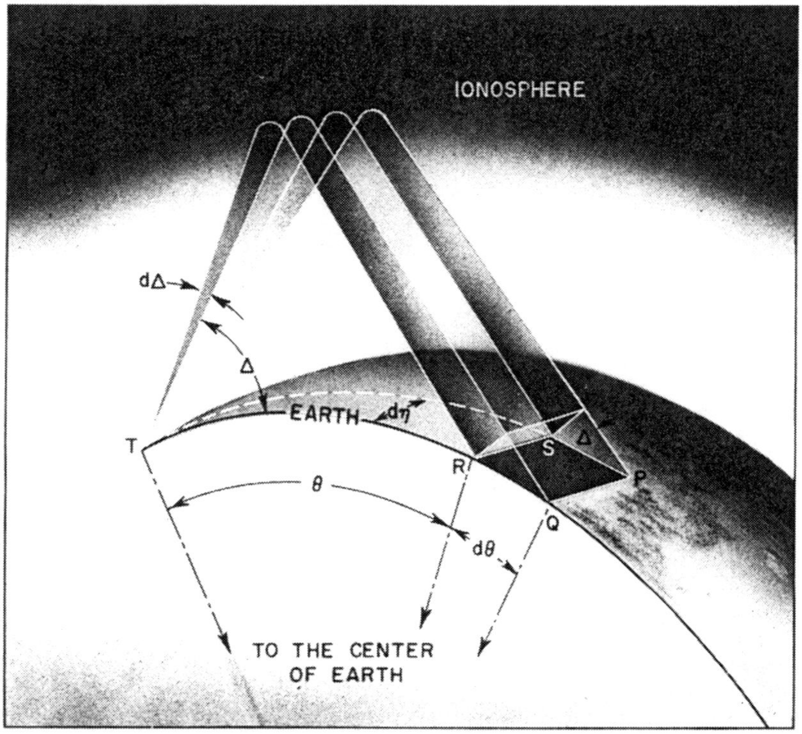

Fig. 7.2 *Divergence of a narrow beam of radio waves reflected from a curved ionospheric layer* The cross-section area perpendicular to Earth's surface at *RS* is $a^2 \sin\Delta \sin\theta \, d\theta \, d\eta$, and the power passing through this area is spread over the area *PQRS* on the ground

where a is the Earth's radius, Δ is the elevation at the sender, $d\eta$ is the azimuthal spread of the beam, and θ is the angle subtended by the path D at the Earth's center. At unit distance from the sender the beam cross-section is

$$A_u = \cos\Delta \, d\Delta \, d\eta \tag{7.10}$$

The effective distance d_e is given by

$$d_e^2 = (A/A_u) = a^2 \sin\theta \tan\Delta \frac{d\theta}{d\Delta} \tag{7.11a}$$

that is

$$d_e^2 = a \sin \theta \tan \Delta \, dD/d\Delta \qquad (7.11b)$$

Since the power loss varies inversely as d_e^2 we have

$$L_f = 10 \log (a \sin \theta \tan \Delta \, dD/d\Delta) \qquad (7.11c)$$

Equation (7.11*c*) is useful in that it indicates the conditions under which focusing (and hence signal strength) is enhanced. The conditions under which L_f is small are these:

(i) $\Delta = 0$, a condition called horizon focusing
(ii) $dD/d\Delta = 0$, a condition called skip focusing because it occurs at the edge of the skip zone
(iii) $\theta = 0$; that is, transmission distance is zero.

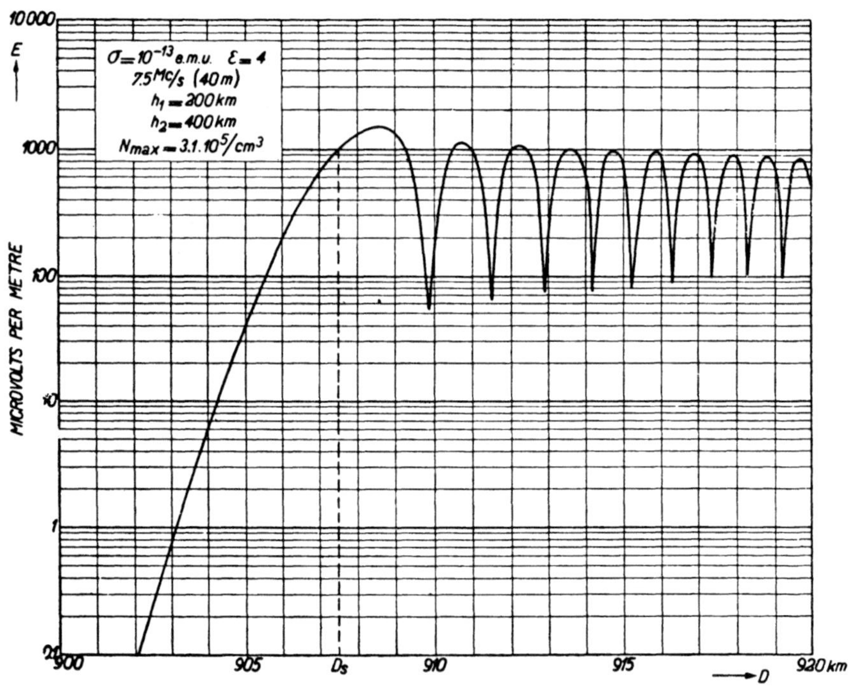

Fig. 7.3 *The variation of the root-mean-square field with distance in the neighborhood of the skip distance D_s (After Bremmer 1949)*
Note that power is received within the skip zone

In practice, d_e is never zero. Furthermore, under conditions of horizon and skip focusing, ray theory breaks down and diffraction effects must be taken into account. In the case of skip focusing, energy is diffracted into the 'shadow' of the skip zone. The calculated variation of amplitude with distance is shown in Fig. 7.3. At the geometrical skip distance the signal amplitude does not fall to zero. Beyond the skip, there are oscillations produced by wave interference between low-angle and high-angle rays. Skip focusing has been observed frequently on short paths, (e.g. Appleton and Beynon 1947; an example is shown in Fig. 7.4 in which the signal fades in as the maximum frequency rises above the signal frequency.

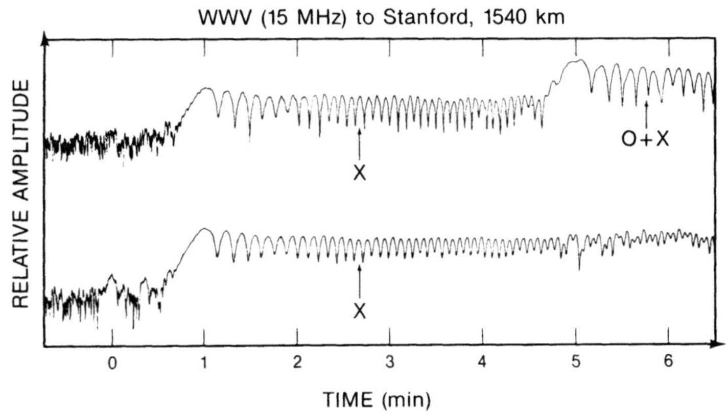

Fig. 7.4 *Variation with time of the received-signal amplitude of WWV 15 MHz signal transmitted from Fort Collins to Stanford near time of sunrise (After Croft 1972)*
The receiver antennas are polarized to reject either the ordinary or extraordinary wave. Note that the extraordinary wave appears first, followed by the ordinary wave some 4 min later

Another type of focusing is antipodal focusing caused by rays emitted in different directions reaching the antipode. Because of the large geographic variations in ionospheric structure, such focusing does not appear to be of great practical importance. The focusing factor $dD/d\Delta$ can be determined by ray tracing or, in the cases of parabolic and quasi-parabolic layers, analytically.

7.3.4 Overhead focusing

The amplitude E of a signal reflected from a cylindrical concave overhead surface at a mean height H is given by

$$E^2 = E_0^2\left(1 - H\frac{d^2z}{dx^2}\right) = E_0^2\left(1 - \frac{H}{2V^2}\frac{d^2P}{dt^2}\right) \tag{7.12}$$

where z and x are vertical and horizontal distances, E_0 is the signal reflected from a plane reflector at height H, V is horizontal speed, and P is the path to and from reflection (Whitehead 1962). When the reflecting surface is sinusoidal with amplitude D and horizontal wavelength L,

$$E^2 = E_0^2(1 + 4\pi^2 HD/L^2) \tag{7.13}$$

Insertion of the values $D = 2$ km, $H = 200$ km, and $L = 200$ km yields an amplitude increase of about 20%. The oblique reflection coefficient of an ionospheric layer containing large-scale irregularities is given by Caldwell *et al.* (1969).

7.4 Ionospheric absorption

7.4.1 Nondeviative and deviative absorption

Absorption is the process by which ordered energy of the radio wave is transformed into heat and electromagnetic noise by electron collisions with neutral molecules and ionized particles. When the absorption is small, and spatial spreading is neglected, absorption is given by the imaginary part κ of the complex propagation function k. The absorption loss L_a (in decibels) is given by

$$L_a = -8{\cdot}68 \int \kappa \, ds \tag{7.14}$$

If p_r is the received power, p_u is the unabsorbed power that would have been received in the absence of absorption, and $\rho = (p_r/p_u)^{1/2}$ is the effective amplitude reflection coefficient,

$$L_a = -10 \log (p_r/p_u) = -20 \log \rho \tag{7.15}$$

In scientific work, absorption is sometimes expressed in nepers. On the basis of the natural logarithm,

$$1 \text{ neper} = 8{\cdot}68 \text{ decibels}$$

The wave amplitude E decays exponentially with distance; that is

$$E = E_0 \exp\left(-\int \kappa \, ds\right) \tag{7.16a}$$

Hence, the absorption in nepers is

$$L_a = \int \kappa \, ds = -\ln \rho \tag{7.16b}$$

In the absence of the Earth's magnetic field and when $\chi^2 \ll \mu^2$, (3.8) gives

$$\kappa = \frac{e^2}{2\varepsilon_0 mc\,\mu} \frac{N\nu}{\omega^2 + \nu^2} = 4{\cdot}6 \times 10^{-2} \frac{N\nu}{\mu(\omega^2 + \nu^2)} \tag{7.17}$$

where κ is in decibels per kilometer.

Equation (7.17) enables us to define two types of absorption:

(i) Nondeviative absorption, which occurs in regions where μ is approximately unity but where the product $N\nu$ is large. This is the type of absorption of HF and VHF waves that occurs in the D region (Fig. 7.5).

(ii) Deviative absorption, which occurs near the top of the trajectory or anywhere along the path where marked ray bending takes place, for example, when $\mu \rightarrow 0$. Deviative absorption is associated with group retardation.

On high and very high frequencies, $\omega^2 \gg \nu^2$, so the nondeviative absorption coefficient

$$\kappa = 1{\cdot}15 \times 10^{-3} N\nu/f^2 \tag{7.18}$$

which shows that nondeviative absorption is inversely proportional to the square of the radio frequency. This is the basis for using as high a frequency as possible for radio communications.

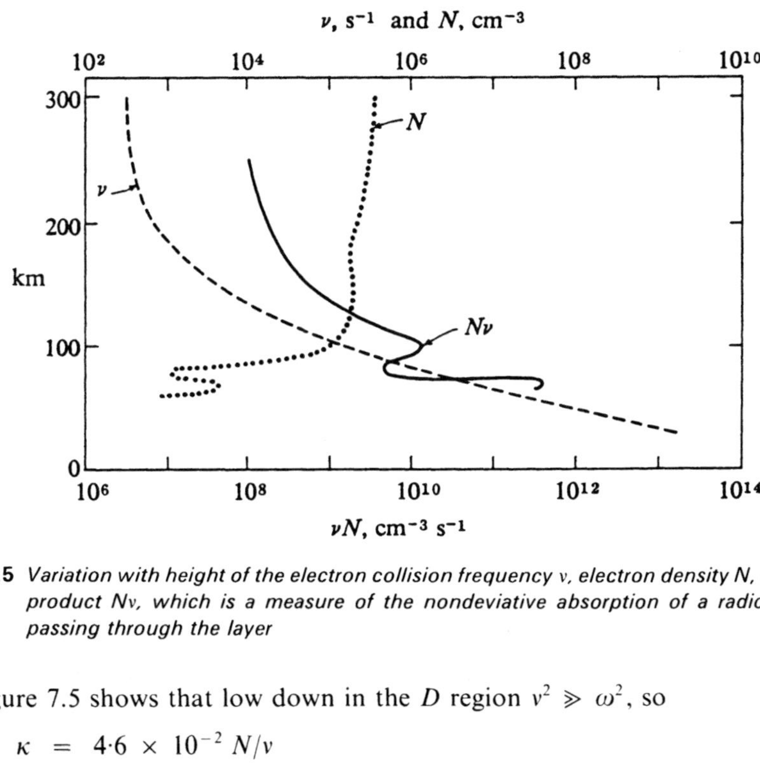

Fig. 7.5 *Variation with height of the electron collision frequency ν, electron density N, and the product Nν, which is a measure of the nondeviative absorption of a radio signal passing through the layer*

Figure 7.5 shows that low down in the D region $\nu^2 \gg \omega^2$, so

$$\kappa = 4{\cdot}6 \times 10^{-2} N/\nu \tag{7.19}$$

and the absorption coefficient maximizes where $\omega = \nu$. The physical reason for the decrease of absorption, where ν is large, is that collisions restrict the motions of electrons and little energy is abstracted from the wave.

Deviative absorption takes place where the wave spends a long time in the absorbing region, e.g. around a critical frequency. The absorption coefficient is given approximately by (Piggott *et al.* 1957)

$$\kappa \approx \frac{\nu M}{2c}\left(\frac{1}{\mu} - \mu\right) = \frac{\nu M}{2c}(\mu' - \mu) \approx \frac{\nu M}{2c}\mu' X \tag{7.20}$$

where μ' is the group refractive index and where M is a number near unity, depending on wave frequency, magnetic field, etc. Since $\mu \simeq 1/\mu' \simeq (1 - X)^{1/2}$, we have

$$\kappa \approx \frac{\nu M}{2c} \frac{X}{(1 - X)^{1/2}} \tag{7.21}$$

It must be realized that close to reflection the wavelength becomes large and ray theory breaks down. To find an expression for the absorption near the reflection level with vertical propagation, the phase integral technique (Budden 1985; Thorpe 1971) has been used. A simple application of this method is as follows.

In the presence of collisions the refractive index $n(=\mu - i\chi)$ is complex and it is necessary to assume that reflection occurs at a complex height z_0 at which $n = 0$. The total absorption L is given by the phase integral

$$L = L_1 + L_2 \tag{7.22a}$$

where

$$L_1 = 2k_0 \int \chi \, dz \tag{7.22b}$$

and

$$L_2 = 2k_0 \, \mathrm{Im} \int n \, dz \tag{7.22c}$$

and $k_0 = \omega/c$.
With a linear gradient of electron density,

$$L_2 = 1{\cdot}34 \times 10^{-10} (\nu f)^{1{\cdot}5} (dN/dz)^{-1} \tag{7.23}$$

where L_2 is in decibels, f is in hertz, N is in m^{-3}, and z is in meters.

Equation (7.20) can be integrated to give the total deviative absorption during normal reflection:

$$L_d = \int \frac{\nu M}{c} (\mu' - \mu) \, dh = \frac{\bar{\nu} M}{c} (h' - h_p) \tag{7.24}$$

where $\bar{\nu}$ is an average collision frequency, h' the group height, h_p the phase height, and M a factor near unity. This equation has been used to measure the effective collision frequency in the F layer from measurements of absorption on closely spaced frequencies near f_oF_2 (e.g. Setty *et al.* 1970).

In the presence of the Earth's magnetic field the nondeviative absorption coefficient is

$$\kappa = 4{\cdot}6 \times 10^{-2} \frac{N\nu}{(\omega \pm \omega_L)^2 + \nu^2} \tag{7.25}$$

where ω_L $(=\omega_H \cos\theta)$ is the angular gyrofrequency corresponding to the longitudinal (parallel) component of the magnetic field and where the + and − signs refer to the ordinary and extraordinary waves, respectively. From this we see that the nondeviative absorption of the extraordinary wave is greater than that of the ordinary wave. This is particularly important on frequencies near the gyrofrequency (i.e. 1 to 2 MHz) where the extraordinary wave is heavily absorbed. When $\nu^2 \ll (\omega \pm \omega_L)^2$ we have

$$L = A(f \pm f_L)^{-2} \tag{7.26}$$

where L is in decibels and A is the absorption on an equivalent frequency of 1 MHz. The observed dependence of absorption on wave frequency is shown in Fig. 7.6, which shows that (7.26) is valid except near critical frequencies (f_oF_1 and f_oE) and near the gyrofrequency.

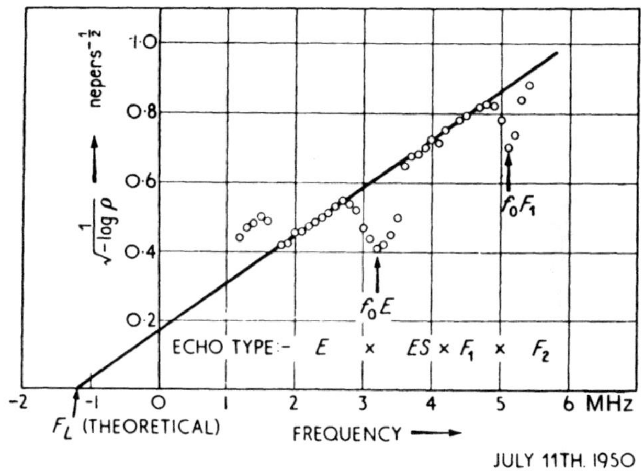

Fig. 7.6 *Variation of ionospheric absorption (−log ρ) of high-frequency (ordinary) waves with frequency at Slough, England (From Appleton and Piggott 1954 © Pergamon Press, Inc.; reprinted with permission)*
The data are plotted to show that the absorption varies as $(f + f_L)^{-2}$. Large deviative absorptions are evident around the critical frequencies f_oE and f_oF_1

7.4.2 Measurement of absorption

The three principal techniques for measuring absorption are called the A1, A2, and A3 methods. These methods are described in detail in Rawer (1976) and Hunsucker (1990).

7.4.2.1 The A1 or pulse-reflection method

This technique consists of the measurement of the amplitudes $E_1, E_2, E_3, \ldots$, etc., of echoes reflected once, twice, etc., from the ionosphere. The method has been discussed by Piggott *et al.* (1957). The effective reflection coefficient ρ is given with one ionospheric reflection by

$$\rho = B\, E_1 h' \tag{7.27a}$$

with two ionospheric reflections and one ground reflection by

$$\rho^2 \rho_g = 2B\, E_2 h' \tag{7.27b}$$

and so on; that is,

$$\rho^n \rho_g^{n-1} = nB\, E_n h' \tag{7.27c}$$

Here ρ_g is the reflection coefficient of the ground, B is a calibration that depends on such factors as transmitter power and antenna gain, and h' is the group height that accounts for the spatial spreading of the wave. Hence

$$\rho\rho_g = \frac{2E_2}{E_1} = \frac{3E_3}{2E_2} = \ldots = \frac{(n+1)\, E_{n+1}}{n\, E_n} \tag{7.28}$$

which eliminates the need to know B. The ground reflection coefficient can be estimated from the ground characteristics. Alternatively, $\rho\rho_g$ can be measured late at night when the absorption is small and $\rho \approx 1$. When ρ is known from (7.28), substitution in (7.27*a*) gives B, which makes possible the measurement of absorption during the day, when only one echo may be present. The effects of distance attenuation on HF absorption measurements have been discussed by Whitehead and Jones (1972).

The measurement of absorption is complicated by echo fading, with amplitude variations on the order of 10 : 1 over fading periods varying from a few seconds to tens of minutes. Amplitudes can be measured visually, photographically or, preferably, by digital sampling of the echo peak. Some additional problems arise because of the superposition of ordinary and extraordinary pulses, by partial reflections, by the spreading of echoes, and by the presence of interfering signals.

7.4.2.2 The A2 or riometer method

The riometer (relative ionospheric opacity meter) (Little and Leinbach 1959) measures the intensity of the wideband noise that impinges on the Earth from deep space. The equipment consists of a stable receiver, a suitable antenna, a recorder, and a noise-diode calibrator. The calibration is carried out periodically by switching from the antenna to the calibrator. The received noise power is a function of sidereal (star) time, since the antenna beam will sample the same strip of sky each day, as the Earth rotates. The absorption is, therefore, the difference betrween the signal actually received and that which would have been received at the same sidereal time in the absence of absorption (Krishnaswamy *et al.* 1985). In using this technique a compromise frequency has to be selected. On the one hand, one would prefer to use as high a frequency as possible to avoid deviative effects but on higher frequencies the galactic noise level decreases and the ionospheric absorption decreases. A typical frequency used with this technique is 30 MHz, with which absorption changes of about 0·1 dB can be measured. The problem of interference can be partially removed by recording the minimum signal received while the receiver frequency is swept over a small range (but many bandwidths). Another difficulty arises during solar radio bursts. Thus a polar-pointing antenna is desirable. For discussion of the limitations of the technique see Little (1957).

7.4.2.3 The A3 or continuous-wave method

In this method, the output of a continuous-wave transmitter is recorded. The method has the advantage of simplicity and sensitivity but it is incapable of discriminating between various echoes. This difficulty can be partially removed by a suitable choice of frequency; for example near the gyrofrequency the contributions from the extraordinary wave and the higher order echoes are small. A CW system can best be calibrated by assuming that the nighttime absorption is zero. This technique is attractive since broadcasting transmitters are available. Also, we are usually interested in changes of absorption so that accurate knowledge of the zero absorption level is not critical. For further discussion of this technique see Schwentek (1966). This technique is subdivided according to frequency range: A3(*a*) on frequencies higher than 2 MHz and A3(*b*) on frequencies lower than 2 MHz. [See Rawer (1976) and Hunsucker (1990) for further details.]

7.4.2.4 The f_{min} method

The minimum frequency observed on an ionogram is a rough measure of the nondeviative absorption. It is useful as an absorption index during large changes of absorption such as those that occur during solar X-ray flares, ionospheric (auroral) storms, and polar cap absorptions. Further details are given in Rawer (1976, Sec. 4.9.8.2).

7.4.2.5 The ionosonde method

One limitation with the techniques A1–A3 is that they give the absorption on a few fixed frequencies. Absorption measurements have been made using ionosondes by displaying the amplitude along with virtual height (e.g. Wakai and Ishizawa 1962). With modern digital ionosondes this is relatively simple (see Sec. 4.4). Calibration is similar to the A1 method.

7.4.3 Absorption in some model layers

To provide some insight into the dependence of absorption on electron-density profiles, solar zenith angle, frequency, etc., the total absorptions in some simple layers are given in Table 7.1. The wave frequently is much greater than the collision frequency, and the magnetic field is ignored. Of particular interest are eqns. (2) and (4) in the Table. Equation (2) has been used, as a model of the *E* region, to separate *D*-layer absorption and *E*-layer absorption. Equation (4) shows that absorption varies as $\cos^{1.5}\chi$. In those cases that include both deviative and nondeviative absorption, that is, (2) and (5), the inverse square dependence on wave frequency does not hold.

Table 7.1 *Total absorption in model layers*

Electron-density profile	Collision profile	Type of absorption	Absorption		Comments
$N = N_0$	$\nu = \nu_0$	Nondeviative	$\int \kappa \, ds = 1{\cdot}2 \times 10^{-7} f^{-2} N_0 \nu_0 T$	(1)	T is slab thickness
Parabolic [eqn. (5.11)]	$\nu = \nu_0$	Deviative and nondeviative	$\int \kappa \, ds = \frac{y_m \nu_0}{2c}\left(\frac{1 + x^2}{2x} \ln \frac{1 + x}{1 - x} - 1\right)$	(2)	$x = \frac{f}{f_c}$, f_c is critical frequency
$N = N_0$	$\nu = \nu_0 e^{-z}$	Nondeviative	$\int \kappa \, ds = 1{\cdot}2 \times 10^{-7} f^{-2} N_0 \nu_0 H(1 - e^{-T/H})$	(3)	$z = \frac{h - h_0}{H}$; H is scale height; T is slab thickness
Chapman [eqn. (2.47)]	$\nu = \nu_0 e^{-z}$	Nondeviative	$\int \kappa \, ds = 4{\cdot}13 \nu_0 H c^{-1} \left(\frac{f_c}{f}\right)^2 \cos^{3/2} \chi$	(4)	Double traverse
$N = a(h - h_0)$	$\nu = \nu_0$	Deviative and nondeviative	$\int \kappa \, ds = \frac{2}{3} \frac{\nu_0 f^2}{cka}$	(5)	Wave reflected, $k = 80{\cdot}5$

7.4.4 *The WKB solutions*

Equation (6.2) can be expressed as

$$\frac{d^2E}{dz^2} + k_0^2 q^2 E = 0 \tag{7.29}$$

An approximate solution of (7.29) for wave amplitude E, valid when q changes very slowly with z, is

$$E = E_0 q^{-1/2} \exp\left(-ik_0 \int q\, dz\right) \tag{7.30}$$

and for the magnetic field

$$H = -H_0\, q^{1/2} \exp\left(-ik_0 \int q\, dz\right) \tag{7.31}$$

These solutions are called the Wentzel, Kramers, and Brillouin, or WKB solutions. The conditions for their validity in ionospheric propagation are discussed by Budden (1985, Ch. 7). In particular these solutions are not accurate at reflection, that is, when $q \to 0$. However, from (7.30) and (7.31) we have

$$H = -q\, E \tag{7.32}$$

Therefore $H \to 0$ near reflection, and the wave is electric. Equations (7.30) and (7.31) show that, as a wave penetrates the ionosphere and q decreases, H will decrease and E will increase.

For a wave reflected at a height z_r and observed at the bottom of the ionosphere z_0,

$$E_r = E_0 q^{-1/2} \exp\left(-2\, ik_0 \int_{z_0}^{z_r} q\, dz\right) \tag{7.33a}$$

and

$$E_i = E_0 q^{-1/2} \tag{7.33b}$$

where E_i and E_r are the incident and emergent amplitudes. Hence, the reflection coefficient ρ is given by

$$\rho = \exp\left(-2ik_0 \int_{z_0}^{z_r} q\, dz\right) \tag{7.33c}$$

7.4.5 *Martyn's absorption theorem*

The absorption L_o of a wave of frequency f incident on a flat ionosphere at an angle ϕ_o is related to the absorption L_v of the equivalent vertical wave, on a frequency $f \cos \phi_o$, by

$$L_o = L_v \cos \phi_o. \tag{7.34}$$

This theorem shows that the additional absorption of the oblique wave, caused by the longer ray path, is more than compensated for by the absorption decrease because of the higher frequency (Martyn 1935).

7.5 Temporal and spatial variations of absorption

7.5.1 Diurnal variations

Ionospheric absorption of radio waves varies greatly during a day and from one day to the next. On an average basis, it has been found that

$$L_a = C \cos^n \chi \tag{7.35}$$

where χ is the solar zenith angle, C depends on season, magnetic disturbance, etc., and n is about 0·75. Equation (7.35) does not apply near sunset or sunrise, and several alternative formulas have been suggested:

$$L_a = L_0 + B \cos \chi \tag{7.36a}$$

where L_0 is the value of L_a when $\cos \chi \rightarrow 0$,

$$L_a = Q (\cos 0{\cdot}881\, \chi)^{1{\cdot}3} \tag{7.36b}$$

where Q is a proportionality factor, and

$$L_a = L_d + L_e \tag{7.36c}$$

In the last case L_d and L_e are contributions from the D and E regions respectively (Piggott 1964). There is residual absorption when $\cos \chi \rightarrow 0$. The exponent n in (7·35) has a latitude variation falling from about 0·85 $\pm$ 0·15 in low latitudes to 0·75 $\pm$ 0·15 in middle latitudes to about 0·2 in the auroral zones. This low value of n in high latitudes results from D-region ionization caused by particle precipitation that maximizes at night rather than during the day (Piggott 1964). In (7.36*c*) the absorption is expressed as the sum of L_d, a D-region contribution, and L_e due to the E layer where L_e is given, for example, by eqn. (2) in Table 7.1. It is sometimes convenient to express the absorption in terms of the absorption $A[=L(f + f_L)^2]$ on an equivalent frequency $(f + f_L)$ of 1 MHz. Contours of constant A show (Fig. 7.7) that the diurnal maximum occurs about 20 min after noon (Appleton and Piggott 1954). This relaxation time lag τ is related to the electron density N and recombination coefficient α by the relation

$$\tau = 1/(2\alpha N) \tag{7.37}$$

Thus it is possible to estimate α when N is known.

7.5.2 Seasonal variations

The seasonal variation of noon D-layer absorption follows a $\cos^n \chi$ law during summer and equinox but during winter there is anomalously high absorption called the winter anomaly. The enhanced absorption occurs on groups of days in middle latitudes and is related to stratospheric warming. It does not appear

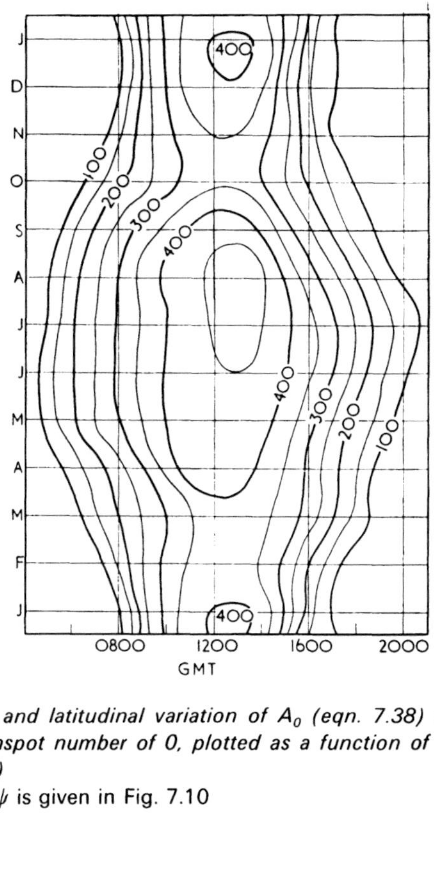

Fig. 7.7 *The seasonal and latitudinal variation of A_0 (eqn. 7.38) for cos χ = 1 and for a smoothed sunspot number of 0, plotted as a function of the modified dip (From Bradley 1976)*
The function ψ is given in Fig. 7.10

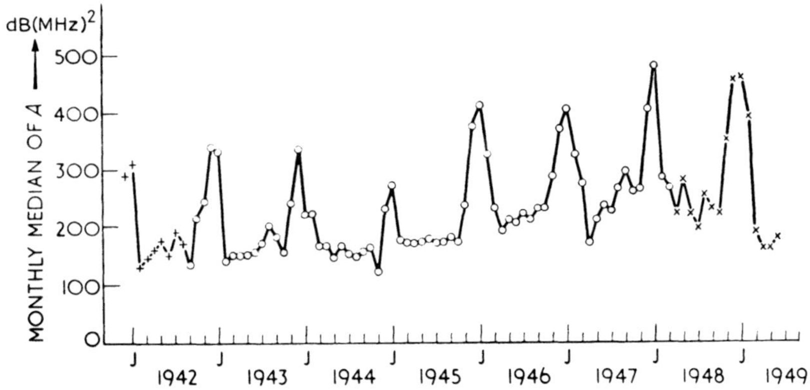

Fig. 7.8 *Variation of ionospheric absorption A at a constant solar zenith angle* (cos χ = 0·25) *at Slough, England (From Appleton and Piggott 1954 © Pergamon Press, Inc; reprinted with permission)*
The values of *A* are the measured values in decibels times MHz^2 on an equivalent frequency of 1 MHz

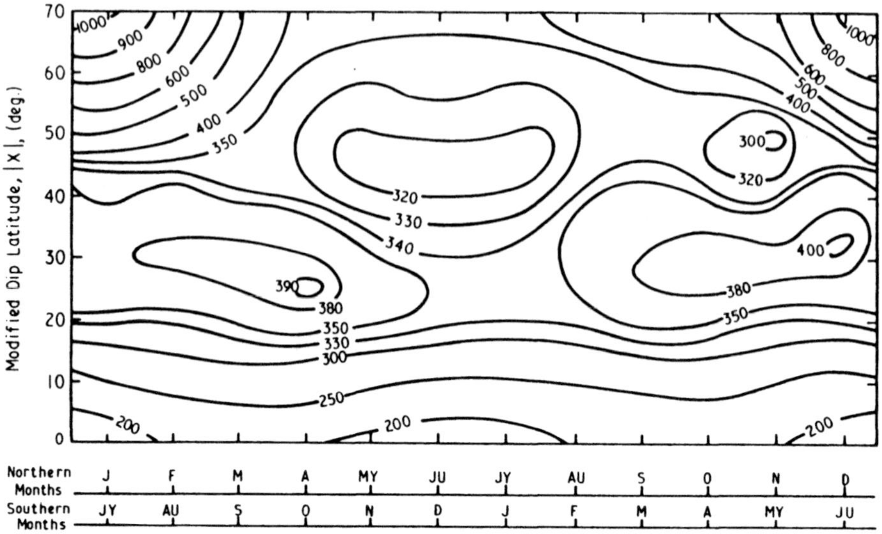

Fig. 7.9 *The seasonal and latitudinal variation of A_o (eqn. 7.38) for* $\cos\chi = 1$ *and for a smoothed sunspot number of 0, plotted as a function of the modified dip (From Bradley 1976)*
The function ψ is given in Fig. 7.10

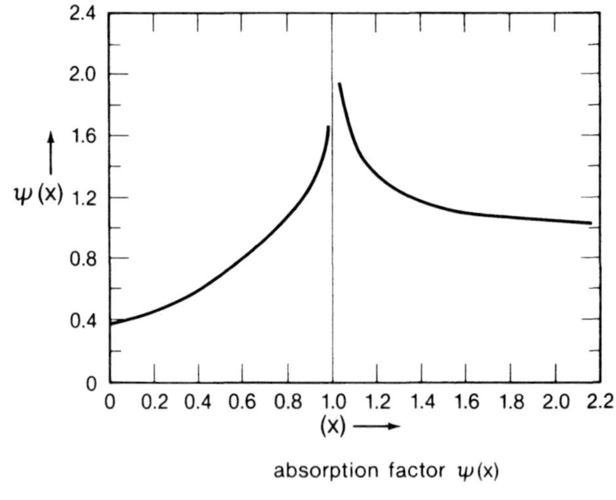

Fig. 7.10 *The variation of the absorption factor $\psi(x)$ with x, where $x = f/f_oE$ (After George 1971)*

in polar regions where, in winter, there is prolonged darkness. The seasonal anomaly is most evident when absorption at a constant value of χ is plotted against time (Fig. 7.8). Days of enhanced absorption in one longitude sector are days of low absorption in other longitude sectors (Thomas 1962), and high absorption is associated with partial reflections (Gregory and Manson 1969). There is little or no correlation between the winter anomaly and magnetic activity (Brasseur and Solomon 1986).

7.5.3 Geographical variations

George (1971) confirmed the existence of a geomagnetic anomaly in the global distribution of absorption during all seasons (Fig. 7.9). Peaks are found in the range 25°–30° of modified dip angle X (see eqn. 7.39) and a trough is centered on the dip equator. In low latitudes, there is a seasonal variation with equinoctial maxima whereas in middle latitudes the winter anomaly dominates the seasonal pattern. There is no evidence for a systematic change in the pattern of Fig. 7.9. It is believed that the anomalous latitudinal and seasonal variations in absorption are due to changes in the electron content of the D region. This in turn is throught to be related to the large-scale circulation in the mesosphere.

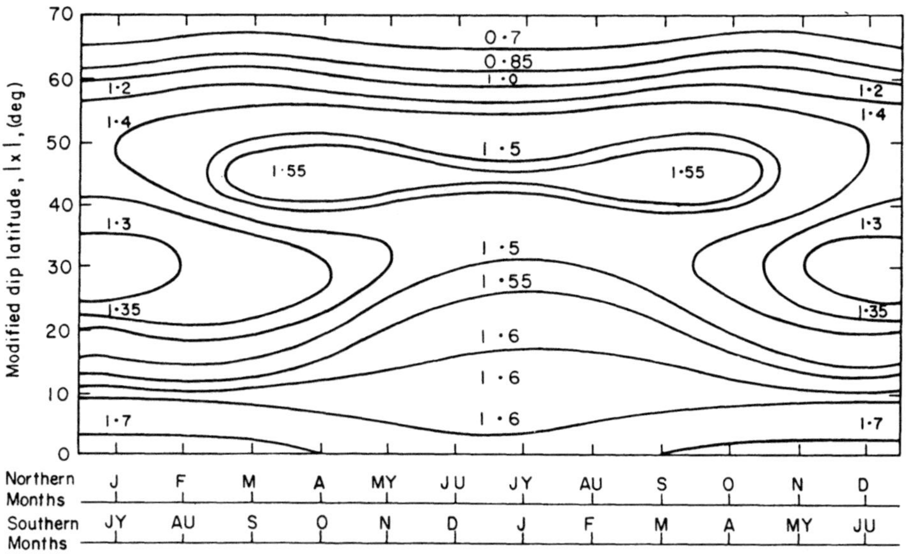

Fig. 7.11 *Variation of exponent p in F(χ) in (7.38) with modified dip (From Bradley 1976)*

For a single-hop path, the absorption L_o of the ordinary wave is given (in dB) by George and Bradley (1974):

$$L_o = A_0(1 + 0{\cdot}0067R) \sec \phi_{100}\psi(x)\, F(\chi)\, (f + f_L)^{-2} \tag{7.38}$$

Here A_0 is the nondeviative absorption of a wave on an equivalent frequency of 1 MHz with vertical propagation, sunspot number = 0 with an overhead sun, and it is given in Fig. 7.9; $F(\chi) = \cos^p(0{\cdot}881\ \chi)$ or 0·1, whichever is the larger; ϕ_{100} is the angle between the ray and the vertical at 100 km; χ is the solar zenith angle; R is the 12-month smoothed sunspot number; $\psi(x)$ is given in Fig. 7.10 where $x = f \cos \phi_{100}/f_o E$. The exponent p is given in Fig. 7.11 as a function of the modified dip X given by

$$X = \arctan (I/\sqrt{\cos \phi}) \tag{7.39}$$

where I is the dip and ϕ is geographic latitude. In middle latitudes some typical values are $X \approx 55°$, $A_0 = 400$ dB, $p \approx 1{\cdot}5$, and $\psi(x) \approx 1$. Note that the lower limit of 0·1 for $F(\chi)$ allows for nighttime residual absorption, and $\psi(x)$ includes an allowance for deviative absorption, especially when $f_v \approx f_o E$. Absorption in low latitudes has been discussed by Piggott (1959).

7.6 Polarization matching

7.6.1 Importance of polarization

The polarization of a radio signal is important because of (i) the interaction with the receiving antenna and (ii) the excitation of the magnetoionic waves by the emitted wave. For example, in high latitudes where the geomagnetic field is nearly vertical, the emergent (ordinary wave) oblique signal will be primarily vertically polarized and, for maximum pickup, the receiving antenna must be vertically polarized. Thus a horizontal dipole or rhombic is inefficient. Similarly for east–west propagation at the magnetic equator a north–south transmitting antenna should be used so that all the power goes into the ordinary wave, and a north–south receiving antenna should be used for maximum response to the reflected wave. These considerations are particularly relevant on medium frequencies, on which the extraordinary wave is heavily absorbed, since the power in the extraordinary wave is lost (see Sec. 11.2).

7.6.2 Limiting polarization

The polarization at the bottom of the ionosphere, where the electron density is zero, is obtained from (3.6) by setting $X = 0$, which gives

$$R_{l\mp} = \frac{i}{2Y_L}\left\{\frac{Y_T^2}{1 - iZ} \mp \left[\frac{Y_T^4}{(1 - iZ)^2} + 4Y_L^2\right]^{1/2}\right\} \tag{7.40}$$

where $R_{l\mp}$ is the limiting polarization of the ordinary and extraordinary waves, respectively. Equation (7.40) shows that these polarizations depend not only on the strength and direction of the magnetic field (through Y_L and Y_T) but also on the collision frequency (through Z). Under these conditions there is tight coupling between the ordinary and extraordinary waves, a subject that has been discussed by Budden (1985, Sec. 17.10). For the present discussion we set $Z = 0$ and obtain

$$R_{l\mp} = \frac{i}{2Y_L}[Y_T^2 \mp (Y_T^4 + 4Y_L^2)^{1/2}] \tag{7.41}$$

Thus the limiting polarizations are, in general, elliptical and will not match the polarization of the transmitting and receiving antennas. When the polarization of the incident wave matches that of the ordinary (o) or extraordinary (x) wave all the incident power will go into the particular matching wave. The division of incident power takes place at the bottom of the ionosphere. Thereafter, the o- and x-waves propagate independently, provided that the ionosphere varies sufficiently slowly to prevent coupling. As the waves are refracted, θ (and hence the polarizations of the waves) will change along the paths.

In general, the polarization of the incident wave will not match the polarization of one of the characteristic waves. Hence, the incident wave must be decomposed into two waves having the polarizations of the characteristic waves. Each of these component waves will then excite its characteristic wave. This matching process makes it possible to calculate the distribution of power between the o-wave and x-wave.

7.6.3 Polarization matching with vertical propagation

Consider two cases with horizontal linearly polarized incident waves:

(i) At a dip pole, $\theta = 0$ and $R_{l\mp} = \pm i$; thus characteristic polarizations are circular and in opposite senses. In this case, the incident power is divided equally between o- and x-waves.
(ii) At the dip equator, the electric vector of a wave is at an angle γ with the dip equator. The propagation is transverse with $R_{l-} = 0$ and $R_{l+} = \infty$. Thus the electric vector E_o of the o-wave is north–south whereas the electric vector E_x of the x-wave is east–west. Therefore $E_o = E_i \sin\gamma$ and $E_x = E_i \cos\gamma$ where E_i is the incident electric field. Hence, the powers in the o-wave P_o and x-wave P_x are

$$P_o = P_i \sin^2\gamma \tag{7.42a}$$

$$P_x = P_i \cos^2\gamma \tag{7.42b}$$

so a north–south antenna excites only an o-wave and an east–west antenna excites only an x-wave. An antenna making an angle of 45° with the dip equator will produce equal o- and x-waves.
(iii) Consider cases with circularly polarized waves propagating vertically upward with their electric fields rotating counter-clockwise when an observer on the ground is looking upward. At the south dip pole this polarization matches the polarization of the o-wave and all the power goes into that wave. At the north dip pole all the power goes into the x-wave.
(iv) Consider cases in which rotation is clockwise: At the north dip pole all the incident power goes into the o-wave, and at the south dip pole all the power goes into the x-wave.
(v) Consider the case in which the rotation is either clockwise or counterclockwise: At the dip equator with either sense of rotation, the incident wave excites o-waves and x-waves equally.

Consider the case of a wave with a north–south electric field incident vertically on the ionosphere in middle latitudes (Fig. 7.12). Since the characteristic polarizations are elliptical, the east–west major axis of the x-wave ellipse must cancel the minor axis of the o-wave ellipse, and the powers in the two waves are given by

$$P_o = \frac{1}{1 + |R_-^2|} P_i \qquad (7.43a)$$

and

$$P_x = \frac{|R_-^2|}{1 + |R_-^2|} P_i \qquad (7.43b)$$

where R_- is given by (7.41). From (7.41) we see that, on high frequencies, Y_T and Y_L are small, so for values of θ not too close to 90° the polarization is essentially circular ($|R_-^2| \approx 1$) and the power divides equally between the o-waves and x-waves.

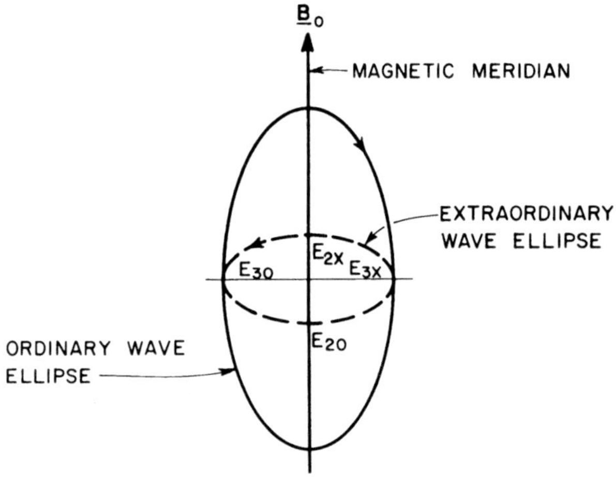

Fig. 7.12 *Matching of a vertically propagating radio wave having a north–south electric field, with ordinary and extraordinary waves on entry into the ionosphere*
Because there is no east–west component of the incident signal the east–west component of the ordinary wave cancels the east–west component of the extraordinary wave

Similarly, when the incident field is east–west, the minor axis of the x-wave ellipse must cancel the major axis of the o-wave ellipse,

$$P_o = \frac{1}{1 + |R_+^2|} P_i = \frac{|R_-^2|}{1 + |R_-^2|} P_i \qquad (7.44a)$$

$$P_x = \frac{|R_+^2|}{1 + |R_+^2|} P_i = \frac{1}{1 + |R_-^2|} P_i \qquad (7.44b)$$

and $|R_+||R_-| = 1$. At the poles, $|R_-| = |R_+| = 1$, and $P_o = P_x = P_i/2$.

7.6.4 Polarization matching with oblique propagation

(i) *East–west propagation at the dip equator:* With the ***E***-vector horizontal and parallel to the magnetic field (north–south) all the power goes into the *o*-wave. With a vertical antenna and a linear east–west antenna, the ***E***-vector is perpendicular to the magnetic field and all the incident power goes into the *x*-wave.

(ii) *Propagation in a magnetic meridian:* Figure 7.13 shows ray paths of a wave incident on a plane ionosphere at an angle of 45° at a latitude where the magnetic dip is 45°. In Fig. 7.13*a* the antenna is in the north–south meridian;

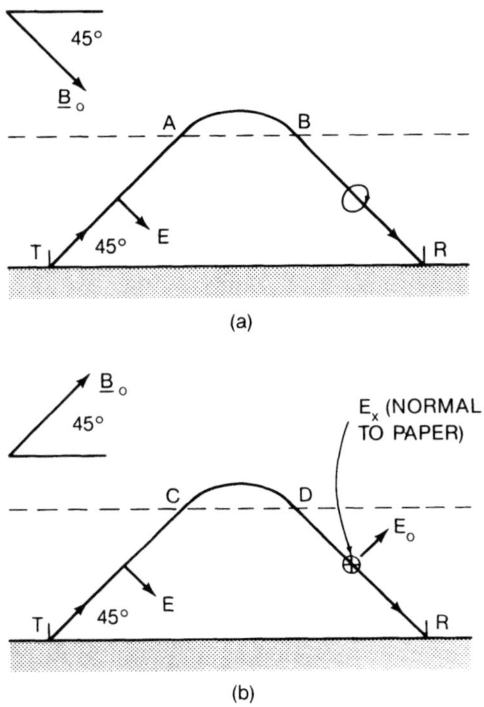

Fig. 7.13 *Polarization matching between a wave incident on the ionosphere at 45°, in a magnetic meridian where the dip is 45°, and the characteristic (o and x) waves. The transmitting antenna is vertical so the E-field lies in the magnetic meridian*

(*a*) The incident wave is perpendicular to the Earth's magnetic field and the emergent ray is parallel to the magnetic field

(*b*) The incident wave is parallel to the Earth's magnetic field and the emergent wave is perpendicular to the magnetic field

the incident electric field is parallel to $\boldsymbol{B}_0$ and all the power goes into the *o*-wave. On exit at B the propagation is parallel to $\boldsymbol{B}_0$ and the polarization is circular. Thus a linear antenna at *R* would pick up half the power in the reflected wave. With an east–west transmitting antenna the incident electric field is perpendicular to $\boldsymbol{B}_0$ and all the power goes into the x-wave. On exit the wave is circularly polarized. In Fig. 7.13b, the antenna is in the magnetic meridian; the incident ray at entry C is parallel to $\boldsymbol{B}_0$, and the characteristic polarizations are circular. The incident power is divided equally between o- and x-waves. At exit D, propagation is transverse, so the $\boldsymbol{E}$-field of the o-wave E_o lies in the meridian plane whereas the field E_x of the x-wave is east–west. Hence, an east–west antenna at R will pick up only the x wave. With a horizontal east–west transmitting antenna, the incident $\boldsymbol{E}$-field is perpendicular to the magnetic field and, since $\theta = 0°$, the power is divided equally between the o- and x-waves; on exit the $\boldsymbol{E}$-field of the x-wave is east–west and the $\boldsymbol{E}$ field of the o-wave lies in the magnetic meridian.

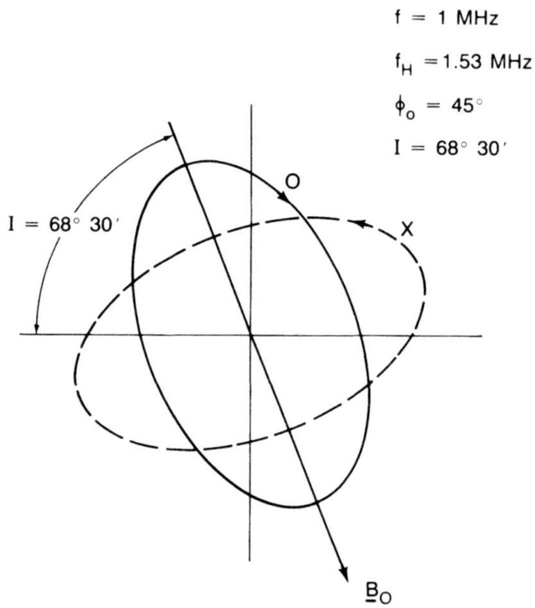

Fig. 7.14 *Limiting polarization ellipses for geographic west-to-east propagation in the USA*

(iii) *East–west propagation*: The characteristic polarizations are elliptical; the major axis of the o-ellipse and the minor axis of the x-ellipse are parallel to the magnetic field (Fig. 7.14). When (7.43) and (7.44) are used and the transmitting antenna is vertical, the division of power is

$$F_{ov} = P_o/P_i = \frac{\sin^2 I + |R_-^2| \cos^2 I}{1 + |R_-^2|} \tag{7.45a}$$

$$F_{xv} = P_x/P_i = \frac{\cos^2 I + |R_-^2| \sin^2 I}{1 + |R_-^2|} \tag{7.45b}$$

where F_{ov}, F_{xv} are the coupling factors, and I is the magnetic dip. Corresponding expressions can be obtained for a horizontal antenna, viz.

$$F_{oH} = P_o/P_i = \frac{\cos^2 I + |R_-^2| \sin^2 I}{1 + |R_-^2|} \tag{7.46a}$$

$$F_{xH} = P_o/P_i = \frac{\sin^2 I + |R_-^2| \cos^2 I}{1 + |R_-^2|} \tag{7.46b}$$

7.6.5 Coupling loss

The term coupling loss is used to describe the loss of power pickup because of the difference between the polarization of the received wave and the polar-

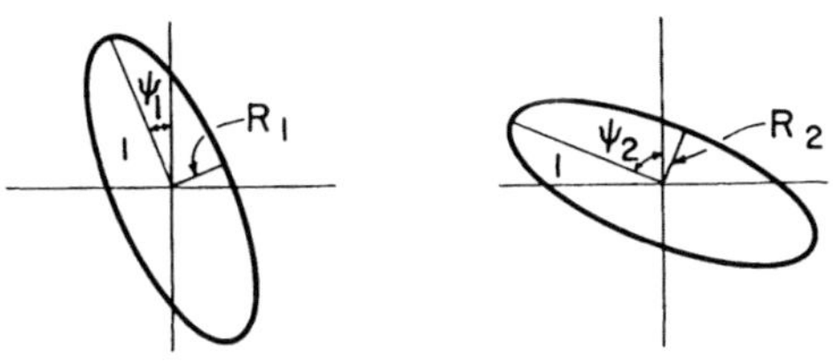

Fig. 7.15 *Coupling between two elliptical polarizations*
(*a*) Polarization of the wave incident on a receiving antenna
(*b*) Polarization of the receiving antenna

ization of the receiving antenna. When an incident wave (Fig. 7.15) with polarization R_1 (ratio of minor axis to major axis) is incident on an antenna of polarization R_2 the coupling factor F, expressed as a power ratio, is (Phillips and Knight 1965)

$$F = \frac{(1 + R_1 R_2)^2 \cos^2 (\psi_1 - \psi_2) + (R_1 + R_2)^2 \sin^2 (\psi_1 - \psi_2)}{(1 + R_1^2)(1 + R_2^2)} \tag{7.47}$$

where ψ_1 and ψ_2 are as shown in Fig. 7.15. Note that when the rotations are in opposite senses ($R_1 R_2 = -1$), the coupling is reduced, whereas when the rotations are in the same senses ($R_1 R_2 = 1$), the coupling is enhanced. Thus when $\psi_1 = \psi_2$ and $R_1 = R_2 = R$, $F = 1$, whereas when $R_1 R_2 = -1$, $F = 0$; there is no coupling between wave and antenna. Note that when the antenna is vertical, $R_2 = 0$ and $\psi_2 = 0$, and (7.47) reduces to (7.45*a*) where $\psi_1 = 90 - I$.

Similarly when the antenna is horizontal, $R_2 = 0$ and $\psi_2 = 90°$, and F reduces to (7.46*a*) where $\psi_1 = 90 - I$. Figure 7.16 shows the variation of the coupling factor with magnetic bearing (from the magnetic meridian) for various dips and for vertical antennas with $f = f_H$ and an angle of incidence of 75°. At the equator, F_{ov} varies between zero for east–west propagation and about 0·5 for north–south propagation. At the pole $I = 90°$, R is small, and F_{ov} is approximately 0·93 and independent of bearing. Polarization is particularly important on medium frequencies since power used to generate the x-wave is wasted because of the high absorption (see Sec. 11.2).

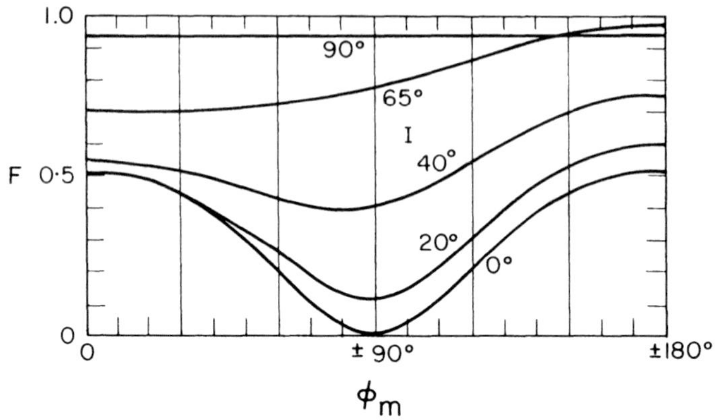

Fig. 7.16 *Variation of coupling factor F with magnetic bearing ϕ_m for vertical transmitting and receiving antennas (After Phillips and Knight 1965)*
Angle of incidence = 75°, $f = f_H$.

7.7 Fading

7.7.1 Meaning of fading

Because the ionosphere changes continuously with time, in general the reflected signal amplitude will fluctuate with time. The fading period can vary from a fraction of a second to months. Fading can result from changes of reflection conditions (e.g. fade-in in the morning and fade-out in the evening) and from changes in absorption. More generally, fading results from the motion of an amplitude pattern on the ground produced by interference of various reflected wavelets (e.g. ordinary and extraordinary waves, E and F echoes, one-hop and two-hop echoes, and so on). The fading can be regular as in the case of polarization, or Faraday, fading or it can be irregular as is generally the case in interference fading.

Fading can be shallow, a decibel or so, or very deep, e.g. 40 dB. It is fading that makes coping with ionospheric signals so difficult. A steady signal, even a weak signal, is often easier to deal with than one that fades below the noise level at one instant and overloads the receiver a few seconds later. Some order-of-magnitude data on fading are given in Table 7.2.

For communication planning, information on the mean value of the received signal is insufficient. Fading has an important influence on the performance of radio systems and on the type of modulation that can be used effectively. It is

Table 7.2 *Fading speeds as a function of frequency (order of magnitude)*

f	Transmission distance (km)	Time of day	Type of fading	Depth of fading (%)	period
20 kHz	7000	Night	Interference	5 to 25	30 to 60 min
20 kHz	0	Night	Interference	50	10 min
20 kHz	Variable	Day	Interference	0	No appreciable fading
100 kHz	0	Night	Interference	50	1 to 15 min
1 MHz	1000	Night	Interference	85	0·1 to 1·0 min
1 MHz	1000	Day	Interference	0	No appreciable fading
10 MHz	0 to 10 000	Night & day	Interference	100	1 to 10 s
10 MHz	3000	Evening	Flutter	?	0·1 to 0·01 s
100 MHz	Oblique	Mostly night	Radio star scintillation	50	10 to 100 s
100 MHz	Moon echoes	Night & day	Libration	Small	1 to 100 s
			Polarization	Large	30 min
20 MHz	Satellite	Night & day	Scintillation	90	0·2 s
20 MHz	Satellite	Night & day	Polarization	Large	1 min

essential to know the severity (magnitude) and rapidity of fading in order to specify (i) the transmitter powers required, (ii) the necessary protection ratio to guard against interference, and (iii) an efficient diversity or coding system.

7.7.2 MUF or skip fading

This is the type of fading encountered when an ionospheric signal is about to penetrate the ionosphere, that is, when it is near the edge of the skip zone. An example is shown in Fig. 7.17 in which focusing occurs as the MUF approaches

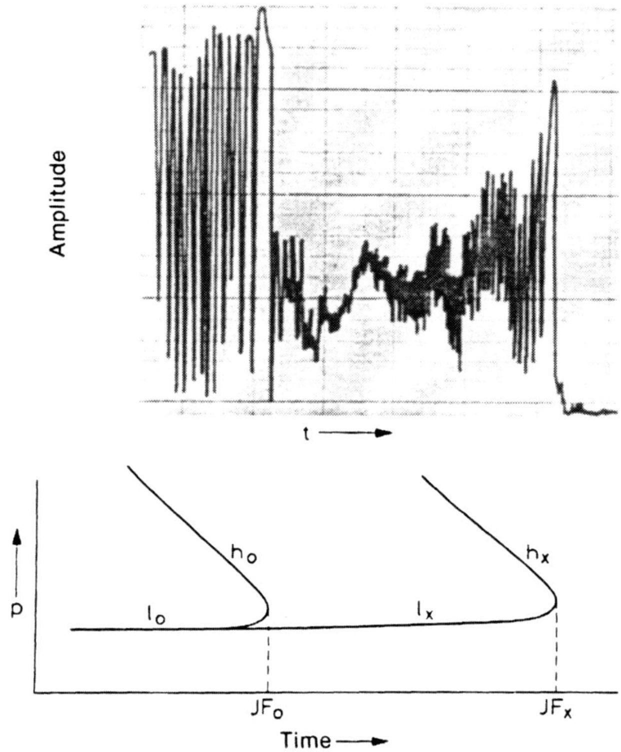

Fig. 7.17 *Skip fading (From Dieminger 1968)*
The ordinary wave penetrates first, followed by penetration of the extraordinary waves. Both waves experience MUF focusing

the signal frequency and rhythmic fading occurs that is due to beating between the high-angle and low-angle waves. The signal is said to fade out. The deep rhythmic fading is superimposed on an irregular, longer period fading. This type of fading can occur as a result of temporal changes in the maximum frequency, over a given ground range, and/or with change of distance with constant ionospheric conditions. Note that the ordinary wave penetrates first; it is followed by the extraordinary wave, after which the signal falls to a very low level (Appleton and Beynon 1947). On a frequency just below the *o*-junction frequency there is interference between four waves: high-angle ordinary and extraordinary, and low-angle ordinary and extraordinary; on a frequency just above the *o*-junction frequency there is interference between the high-angle and low-angle extraordinary waves.

7.7.3 Polarization fading

This type of fading results from changes in the polarization of the downcoming wave brought about by time variations in the phase differences between the ordinary and extraordinary waves. For example, when the amplitudes of these two components are equal and the rotations are opposite, the result is a linearly polarized wave that rotates as the ionosphere and/or the ray path changes (see Sec. 8.3.2). Thus with a linear antenna the induced voltage will vary as

$$V(t) = V_m \cos \Omega(t) \tag{7.48}$$

The receiver output (amplitude versus time) will be a rectified cosine with period π/Ω. In general, the amplitudes of the ordinary and extraordinary waves will differ, because of differential D-region absorption, and the resultant wave will be elliptically polarized. This will have the effect of filling in the deep fades. In a rough approximation the number of fades per second is given by

$$n = \frac{A}{f^2}\frac{dN_T}{dt} \tag{7.49}$$

where N_T is the electron content along the path and A is a proportionality factor. This beating has to be determined by ray tracing, which can be a difficult procedure when the ray paths of the ordinary and extraordinary waves differ appreciably. Polarization fading can be reduced, for example, with vertical propagation, by the use of two orthogonal horizontal antennas. By introducing a phase advance or retardation of $\pi/2$ in one with respect to the other it is possible to select one magnetoionic component (see Sec. 4.2.4).

7.7.4 Absorption fading

Fluctuations in signal amplitude caused by changes in D-region absorption are called absorption fading. Such changes include, for example, short-wave fade-outs, which are associated with X-ray flares and which can last from about 1 minute to about 1 hour. Fading can result from the movement of irregularities (waves) in the D region, from slow changes in solar ionizing radiations, and from the diurnal and 27-day variations of absorption.

7.7.5 Interference fading

By far the most common type of fading is that caused by irregularities in the reflecting surfaces of the ionosphere. These can be caused by (i) clouds of electrons, or depletions, which drift with the neutral winds in the D and E regions, and/or (ii) the passage of acoustic-gravity waves in the neutral atmosphere. A surface of constant electron density (i.e. a reflecting surface) is somewhat similar to the surface of a pool in which waves are traveling in all directions. From such an irregular surface the reflected received signal will come from a relatively broad area (Fig. 7.18). The structure sizes can vary from a few meters up to hundreds and even thousands of kilometers, and they move with speeds varying from about 50 m s^{-1} in the E region to 1000 m s^{-1} in the F region. A speed of 200 m s^{-1} is quite common in the quiet F region. This speed increases markedly during magnetic disturbance, for example, when $K > 4$.

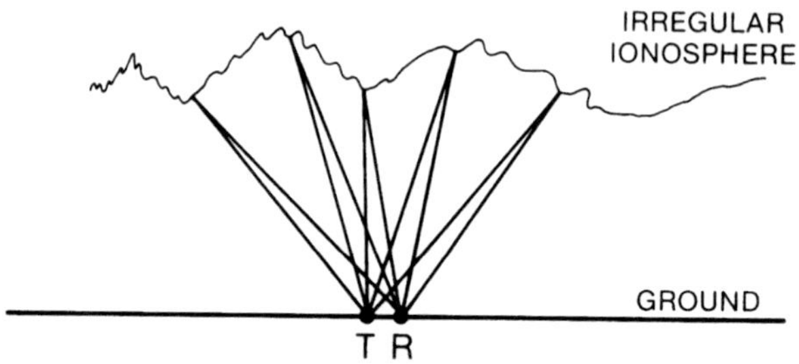

Fig. 7.18 *Reflections from a rough undulating layer*
The received wave is composed of a number of reflected, or scattered, wavelets. T and R represent transmitter and receiver, respectively

7.7.6 *Fading statistics*

In studies of fading time series it is usual to treat the fluctuations of signal strength with time as a random or stochastic process (Brennan 1961). This requires the selection of a sampling interval sufficiently long to give an adequate representation of the fading.

It has been found that the fading of composite ionospheric echoes can be represented as the sum of a steady, specularly reflected signal plus a random component. This is expressed by the Nakagami–Rice distribution:

$$p(v) = (2v/v_n^2) \exp[-(v^2 + v_1^2)/v_n^2]\, I_0(2v_1 v/v_n^2) \tag{7.50}$$

where $p(v)\,dv$ = probability of finding an amplitude between v and $v + dv$,
$I_0(x)$ = modified Bessel function of zero order
v = (received signal envelope voltage)/$\sqrt{2}$
v_1 = steady component
v_n = root-mean-square of the random component.

The probability $P(v_0)$ of finding the amplitude v greater than a value v_0 is given by

$$P(v_0) = \int_{v_0}^{\infty} p(v)\, dv \tag{7.51}$$

when $v_n \gg v_l$ so the random component dominates, the probability function approaches a Rayleigh distribution:

$$p(v) = 2(v/v_n^2) \exp(-v^2/v_n^2) \qquad (7.52)$$

When $v_l \gg v_n$ so the specular component is dominant, the distribution is essentially normal or Gaussian:

$$p(v) = (v_n \sqrt{\pi})^{-1} \exp[-(v - v_l)^2/v_n^2] \qquad (7.53)$$

Under this condition the signal varies about the median value with a standard deviation of $v_n/\sqrt{2}$ (CCIR 1986, p. 123). When the signal strength is expressed in decibels relative to a specified level the distribution in (7.53) is called the log-normal distribution.

A simple way of characterizing empirical signal distributions is to identify the occurrence of specified values v_0. Figure 7.19 shows the distribution of v_0 for various levels of $P(v_0)$ the fraction of the time exceeded in the case of the Nakagami–Rice distribution. The distribution is Gaussian on the left side and Rayleigh on the right.

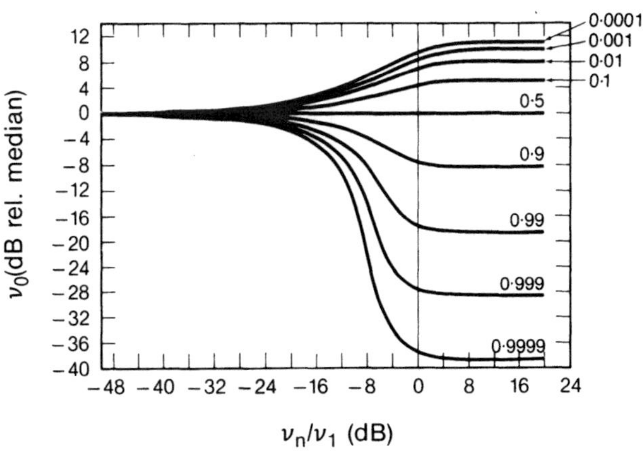

Fig. 7.19 *Variation of v_0 with (v_n/v_1) for various levels of $P(v_0)$ for the Nakagami–Rice distribution (After CCIR 1986)*

With rapid fading, and time samples of 3 to 7 minutes, the distributions are approximately Rayleigh; for 30–60 min intervals the distribution is nearly log-normal. The fading range is defined as the difference between the signal levels exceeded 10% and 90% of the time. Over long paths (1500–6000 km) the fading range is on the order of 15 dB, which is in good agreement with the value 13·4 dB expected for the Rayleigh distribution.

Of equal importance with the depth of fading is its rate or period. A signal fading with, say, a 1 s period is a bigger problem than one fading with a 1 h period. Fading rates are often expressed in terms of the autocorrelation of a time series. The definition of the autocorrelation function (ACF) can be made with reference to Fig. 7.20 where a function $x(t)$ is plotted along with the same function displaced in time by τ seconds. The ACF is obtained by correlating the two curves, viz.,

$$R_{xx}(\tau) = \frac{1}{D}\left\{\frac{1}{n}\sum_{1}^{n}[x(t) - \bar{x}][x(t + \tau) - \bar{x}]\right\}, \qquad (7.54a)$$

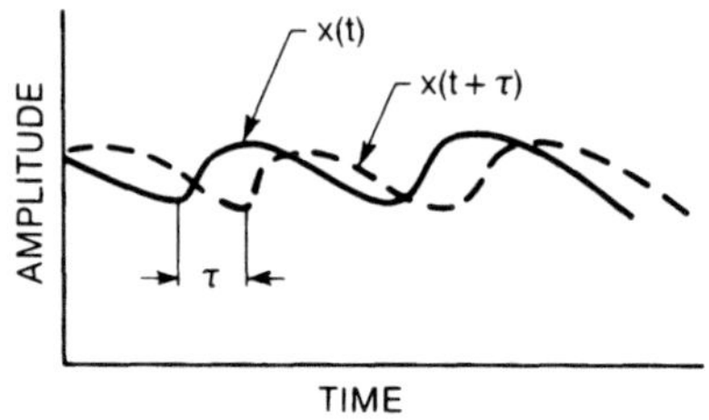

Fig. 7.20 *An example of autocorrelation*
The continuous line and the broken line are similar but displaced in time by an interval τ

where $\bar{x}$ is a mean value of x, and D is the mean square fluctuation, viz.,

$$D = \frac{1}{n}\sum_{1}^{n}[x(t) - \bar{x}]^2 \qquad (7.54b)$$

The ACF is a measure of the rapidity with which a point on the time series is decoupled from the neighboring points. For $\tau = 0$, $R_{xx}(\tau) = 1$ and for large τ, $R_{xx}(\tau)$ will tend to zero. With a Gaussian distribution the ACF is Gaussian:

$$R(\tau) = R(0)\exp(-\tau^2/2\tau_0^2) \qquad (7.55)$$

where τ_0 is called the correlation (or coherence) time of the fading. The quantity $W = \tau_0^{-1}$ is called the correlation bandwidth.

The fading rate can be defined also as the number of positive crossings per unit time through a specified level. With a Rayleigh distribution, the fading rate $N(v)$ through the level v is given by (Rice 1958):

$$N(v) = 2{\cdot}95\,(v/v_m)\,f_n \exp(-0{\cdot}693\,v^2/v_m^2) \qquad (7.56)$$

where v_m is the median and f_n is the root-mean-square of the fading frequency. Putting $v = v_m$ we get $N(v_m) = 1{\cdot}47\,f_n$.

Fading rates over long paths vary between 6 and 16 cycles per minute with a mean value around 11 per minute. On transequatorial paths, flutter fading occurs at times of spread F, near sunset, and especially near the equinoxes (see Sec. 5.6) (Koster 1963). Under these conditions the fading period can be less than 0·1 s. Long-term fading, e.g. days to a solar rotation (27 days), has been fitted to lognormal distributions; these show standard deviations of about 8 dB (CCIR 1986).

7.8 Phase and frequency

7.8.1 Importance of phase and frequency

A wave is specified by its amplitude and phase. For purposes of navigation (i.e. location by radio means) phase comparisons provide a useful tool, especially on very low frequencies. Also, since phase depends on the refractive index integrated along a ray path, phase variations provide a convenient way of monitoring ionospheric changes. The time rate of change of phase gives a (Doppler) frequency shift. Thus ionospheric changes are of vital importance to detection systems, such as over-the-horizon radars, that separate moving from stationary reflectors by the frequency differences in their echoes.

7.8.2 Measurement of phase

Phase is measured by comparison of the echo waveform with a reference that can be derived from the emitted signal or from a local stable reference oscillator (Ranzi 1934; Findlay 1951). The principle is discussed in Sec. 4.2.2.

7.8.3 Measurement of frequency

Frequency changes of ionospheric high-frequency echoes are usually less than 1 Hz. (Methods of measurement of frequency are discussed in Sec. 4.2.3.). An example of frequency fluctuations with time, near sunset and with vertical propagation on 4 MHz, is given in Fig. 7.21. Typically, there are quasi-

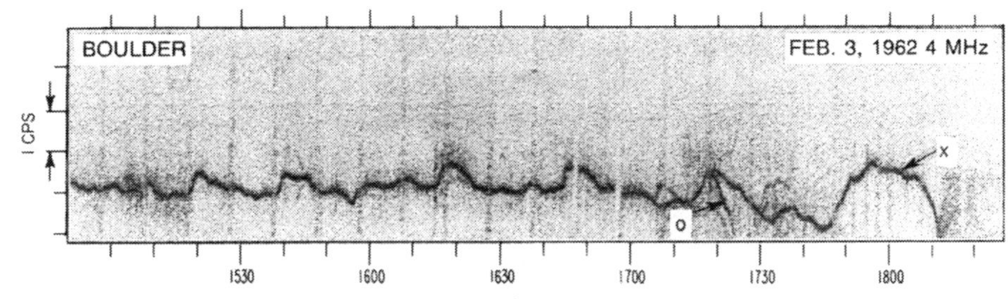

Fig. 7.21 *Frequency variations of a 4 MHz signal incident vertically, near time of sunset*
Note the penetration of the ordinary wave ($\approx$1724) and of the extraordinary wave ($\approx$1811)

sinusoidal frequency deviations with peaks of about 0·1 Hz and with periods ranging from about 5 min to about 50 min. We see in Fig. 7.21 that near sunset the *o*-wave penetrates the ionosphere before the *x*-wave.

7.8.4 Theory of phase and frequency variations

The phase ϕ of a received wave relative to that at the sender is given by

$$\phi = 2\pi \int \frac{ds}{\lambda} = \frac{2\pi f}{c} \int \mu \, ds = \frac{2\pi f}{c} P = \frac{2\pi}{\lambda_0} P \tag{7.57}$$

where P is the phase path and λ_0 is the free-space wavelength. In an anisotropic medium $P = \int \mu \cos \alpha \, ds$, where α is the angle between the direction of phase propagation and the ray. The frequency change Δf is given by

$$\Delta f = -\frac{1}{\lambda_0} \frac{dP}{dt} = -\frac{f}{c} \frac{dP}{dt} \tag{7.58}$$

The change in phase path ΔP is obtained from

$$\Delta P = \lambda_0 \, \Delta\phi / 2\pi = -\lambda_0 \int \Delta f \, dt \tag{7.59}$$

Since, at some time, the ionosphere must return to a previous state the total phase change is zero, so the areas under the $\phi(t)$ and $\Delta f(t)$ curves must be zero.

For the sake of discussion we separate the phase and frequency changes into two types:

(i) Those caused by changes in real path (e.g. height of reflection), in which case

$$\Delta\phi \propto f \tag{7.60a}$$

$$\Delta f \propto f \tag{7.60b}$$

(ii) Those caused by small changes in refractive index in a nondeviating slab (where $\mu \approx 1$), in which case

$$\Delta\phi \propto f^{-1} \tag{7.61a}$$

$$\Delta f \propto f^{-1}. \tag{7.61b}$$

Hence, these two types of changes can be distinguished from measurements on two or more frequencies. When changes in both phase path ΔP and group path $\Delta P'$ are available, we have for a nondeviative change

$$\Delta P' / \Delta P = -1 \tag{7.62}$$

and for a change of reflection height

$$\Delta P' / \Delta P = +1 \tag{7.63}$$

In the presence of irregularities, $\Delta P'$ and ΔP can be in phase quadrature (Bennett and Dyson 1986).

With vertical propagation, we have

$$f \Delta f = \frac{161{\cdot}2}{c} \frac{d\bar{N}}{dt} (h' - h_0) \tag{7.64}$$

where $\bar{N}$ is an average electron density betweeen the bottom of the ionosphere

h_0 and the level of reflection, and h' is the group height. Thus we see that, in general, Δf will not vary as f or f^{-1} since h' is a complicated function of frequency. In the case of nondeviative changes with vertical propagation we have

$$f\,\Delta f = 80{\cdot}6\ \dot{N}_T/c \tag{7.65}$$

where $\dot{N}_T$ is the time derivative of the electron content in a 1 m^2 vertical column in the nondeviating layer.

When electron collisions are introduced, there is a change of radiowave absorption ΔL ($= \int \kappa\, ds$). The ratio of the change in absorption (in nepers) to the change in phase path ΔP is given by

$$\Delta L/\Delta P = \bar{\nu}/c, \tag{7.66}$$

where $\bar{\nu}$ is a mean collision frequency in the absorbing layer.

The effect of the Earth's magnetic field is to split the incident wave into ordinary and extraordinary waves. The expressions for Δf become complicated except in the case of vertical propagation of the ordinary wave at the dip equator (when the no-field formulas apply) and with vertical propagation at the magnetic poles when f (in Eqns. 7.61 and 7.65) is replaced by $(f \pm f_H)$ where f_H is the gyrofrequency.

7.8.5 Phase paths in model layers

Table 7.3 lists some expressions for phase heights in model layers with vertical propagation and for longitudinal propagation ($\theta = 0$). The no-field formulas are derived from the longitudinal formulas by setting $Y = 0$ and/or $f_H = 0$.

7.8.6 Vertical and oblique equivalence

Let $\Delta f(\phi_o, f)$ be the frequency change suffered by a wave of frequency f incident at an angle ϕ_o on a plane-stratified ionosphere having no imposed magnetic field or collisions. Further, let $\Delta f(0, f_v)$ be the frequency change suffered by the equivalent vertical wave on a frequency f_v ($=f\cos\phi_o$). Then

$$\Delta f(\phi_o, f) = \Delta f(0, f_v) \tag{7.67}$$

Hence

$$f\dot{P}_o = f_v \dot{P}_v \tag{7.68}$$

where $\dot{P}_o$ is the time rate of change of the oblique phase path and $\dot{P}_v$ is the time rate of change of the equivalent vertical path (Davies 1969, Ch. 9).

For fixed sender and receiver we have (Bennett 1967)

$$\Delta f = -\frac{f}{c}\int_s \frac{\partial \mu}{\partial t}\, ds \tag{7.69}$$

in which s denotes the ray path.

Table 7.3 *Phase heights in model layers with longitudinal propagation*

Profile	Phase height		
$N = a(h - h_0)$	$h_p = h_0 + \frac{2f^2}{3ka}(1 \pm Y)$	(1)	
$f_N = \beta(h - h_0)$	$h_p = h_0 + \frac{\pi f}{4\beta}(1 \pm Y)^{1/2}$	(2)	
$f_N^2 = F^2 e^{h/H}$	$h_p = 2H\left(\ln\left\{\frac{f}{F}(1 \pm Y)^{1/2} + [(1 \pm Y)(f/F)^2 - 1]^{1/2}\right\} - \{1 + [1/(1 \pm Y)](F/f)^2\}^{1/2}\right)$	(3)	
$f_N^2 = f_c^2\left[\frac{2(h - h_0)}{y_m} - \frac{(h - h_0)^2}{y_m^2}\right]$	$h_p = h_0 + \frac{1}{2}y_m + \frac{1}{4}y_m\left(1 \pm \frac{f_H}{f}\right)^{1/2}\left(\frac{f}{f_c} - \frac{f_c}{f \pm f_H}\right)\ln\frac{f_c + (f^2 \pm ff_H)^{1/2}}{f_c - (f^2 \pm ff_H)^{1/2}}$	(4)	Reflection
$f_N^2 = f_c^2\left[\frac{2(h - h_0)}{y_m} - \frac{(h - h_0)^2}{y_m^2}\right]$	$\Delta h_p = y_m + \frac{y_m}{2}\frac{f^2(1 \pm Y) - f_c^2}{ff_c(1 \pm Y)^{1/2}}\ln\frac{(f^2 \pm ff_H)^{1/2} + f_c}{(f^2 \pm ff_H)^{1/2} - f_c}$	(5)	Penetration

The + sign refers to the z wave
$k = 80{\cdot}6$

7.9 Ionospheric drifts

When a characteristic feature of an ionospheric echo (e.g. amplitude, phase, frequency) is observed at several different points it is generally found that the fading time series are somewhat similar but displaced in time as if the pattern on the ground were moving and changing shape as it drifts. By making some simple assumptions information can be obtained about the magnitude and direction of the horizontal drift.

First, consider the situation (Fig. 7.22) in which a plane front moves over three reflection points (E, S, and U) in the ionosphere with a horizontal velocity vector V_H. Let t_{EU} and t_{ES} be the time delays between the patterns at E and U and at E and S, respectively; then the horizontal apparent or trace speeds along EU and ES are given by

$$V_{EU} = EU/t_{EU} \tag{7.70a}$$

$$V_{ES} = ES/t_{ES} \tag{7.70b}$$

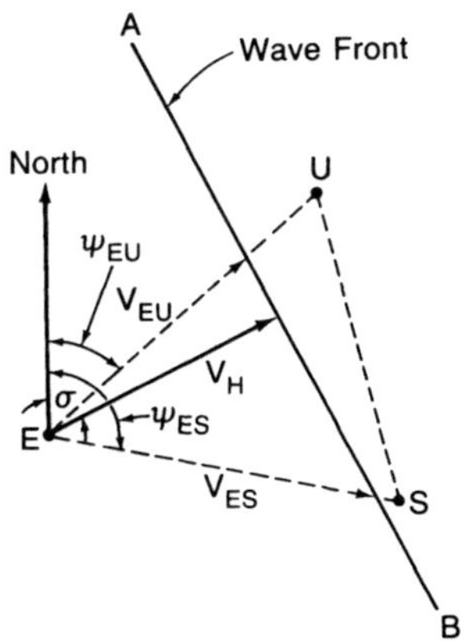

Fig. 7.22 *Movement of plane fronts across three observation sites*
V_{EU} and V_{ES} are the horizontal trace speeds along *EU* and *ES*, and V_H is the resultant horizontal trace speed

where EU and ES are sides of a triangle. The direction of motion σ is given by

$$\tan \sigma = -\frac{V_{EU} \cos \psi_{EU} - V_{ES} \cos \psi_{ES}}{V_{EU} \sin \psi_{EU} - V_{ES} \sin \psi_{ES}} \tag{7.71}$$

where the angles are as shown in Fig. 7.22. The magnitude V_H of the horizontal velocity is given by

$$\frac{\sin^2 (\psi_{ES} - \psi_{EU})}{V_H^2} = \frac{1}{V_{EU}^2} + \frac{1}{V_{ES}^2} - \frac{2 \cos (\psi_{ES} - \psi_{EU})}{V_{EU} V_{ES}} \tag{7.72}$$

These expressions simplify when EU and ES are perpendicular; then (7.72) reduces to

$$\frac{1}{V_H^2} = \frac{1}{V_{EU}^2} + \frac{1}{V_{ES}^2} \tag{7.73}$$

From this we see that V_{EU} and V_{ES} are not components of the velocity vector V_H. With several different frequencies (giving different heights of reflection) it is possible to determine the three-dimensional velocity vector. This technique is restricted to the case in which only one disturbance is present (Hines and Rao 1968).

The time displacements can be obtained in two ways: (i) by the method of similar features (or fades) in which time displacements between similar features (e.g. troughs, peaks, etc.) are averaged over a suitable time interval; (ii) by use of the cross-correlation function. For digitized records of two time series $x(t)$ and $y(t)$ the cross-correlation coefficient is given by

$$R_{xy}(\tau) = \frac{1}{D_x D_y}\frac{1}{n}\sum_{t=1}^{n}[x(t) - \bar{x}][y(t+\tau) - \bar{y}] \tag{7.74}$$

in which τ is the time displacement between the two time series. D_x and D_y are the root-mean-square fluctuations in the two time series; for example,

$$D_x = \left\{\frac{1}{n}\sum_{t=1}^{n}[x(t) - \bar{x}]^2\right\}^{1/2} \tag{7.75}$$

and similarly for D_y. The bar denotes an average, and n is the sample size. When R_{xy} is plotted against τ there is usually a maximum at a value T of τ which gives the appropriate time displacement. The coefficient $R_{xy}(\tau)$ shows how closely the time series $x(t)$ and $y(t)$ are related. When they are identical $R_{xy} = 1$, and when they are uncorrelated $R_{xy} = 0$.

The technique described here is applicable to reflected and transmitted signals. In some versions a single sender is used with spaced antennas, or several transmitters and a single receiver. In the cases of reflection the size of the triangle (or other geometrical figure) in the ionosphere is half that on the ground. In arrangements where the sender is well outside the ionosphere (e.g. geostationary satellite or radio star) and in the case of three (vertical) ionosondes the geometrical figure in the ionosphere is essentially the same size as that on the ground.

In practice it is found that as the pattern moves bodily its shape changes in a random fashion. Thus the total motion consists of a translational speed V and a random or characteristic speed V_c (Briggs 1968). It can be shown that speeds measured from the distance separation ξ and time separation τ' for maximum cross correlation yield an apparent speed V' where

$$V' = \xi/\tau' = V + V_c^2/V \tag{7.76}$$

which is greater than the true speed V. For further information on the analysis of fading records see, for example, Wright *et al.* (1976) and Wright and Pitteway (1978). Observed irregularities are thought to be of two types: (i) clouds of electrons advected by either the neutral wind in the D and E regions, or by electrodynamic ($\boldsymbol{E} \times \boldsymbol{B}$) drift in the F region; (ii) irregularities produced by the action of waves in the neutral atmosphere (D, E, and F) (see JATP 1968).

7.10 Traveling ionospheric disturbances

7.10.1 Observations

Traveling ionospheric disturbances (TIDs) are relatively large-scale, wavelike structures (100 km to 1000 km horizontal wavelengths) that travel with speeds between about 50 m s^{-1} and about 1000 m s^{-1} and have wave periods from a few minutes to more than an hour. Such structures are responsible for the focusing of radio waves (discussed in Sec. 7.3.4). TIDs have been observed by a variety of techniques. For example, Munro (1958) used fixed-frequency $P'(t)$ observations, Bowman (1968) analyzed ionogram traces, Georges (1968) used multi-frequency HF Doppler (Sec. 7.8.3), and Thome (1964) used incoherent scatter.

Observations on several radio frequencies show that these disturbances consist of long fronts that are tilted forward; thus the disturbance appears first at high heights and moves downward (Bowman 1968). TIDs have been classified into very-large-scale and medium-scale disturbances (Georges 1968). Very-large-scale disturbances, which originate in the auroral zone and travel to great distances, are associated with magnetic storms. They have wavelengths of the order of 1000 km and periods of 1 h or more. Medium-scale disturbances with wavelengths of the order of 100 to 200 km, and periods ranging from 5 to 45 min or so, do not appear to travel great distances and may originate in local tropospheric weather (see Sec. 14.2).

7.10.2 Signatures of ionospheric irregularities

An ionospheric irregularity can often be identified by the shape or appearance of an echo signal. For example, consider a cylindrical irregularity with small radius, at a height H, traveling horizontally with speed V (Fig. 7.23*a*). The variation of range s with time is hyperbolic; that is,

$$s^2 - V^2 t^2 = H^2 \tag{7.77}$$

At $t = 0$ the irregularity is overhead. The variation of range with time is represented in Fig. 7.23*b*.

In the case of a cosinusoidal reflecting surface with amplitude D and wavelength L at a mean height H moving with speed V (Fig. 7.24*a*), the range s and the frequency shift Δf are given by

$$s = (H + D \cos\theta)(1 + k^2 D^2 \sin^2\theta)^{1/2} \tag{7.78}$$

where $k = 2\pi/L$, and $\theta = kx$ is the phase of the point of reflection, and

$$\Delta f = \frac{2fDkV \sin\theta}{c(1 + k^2 D^2 \sin^2\theta)^{1/2}} = \frac{2f}{c} V_l \approx \frac{2\,fDkV \sin\theta}{c} \tag{7.79}$$

where V_l is the line-of-sight velocity. Thus, although the reflection point moves along the reflecting surface, the Doppler shift is the same as that produced by a point reflector moving horizontally with the wave speed. This has been called Pfister's theorem (Brownlie *et al.* 1973).

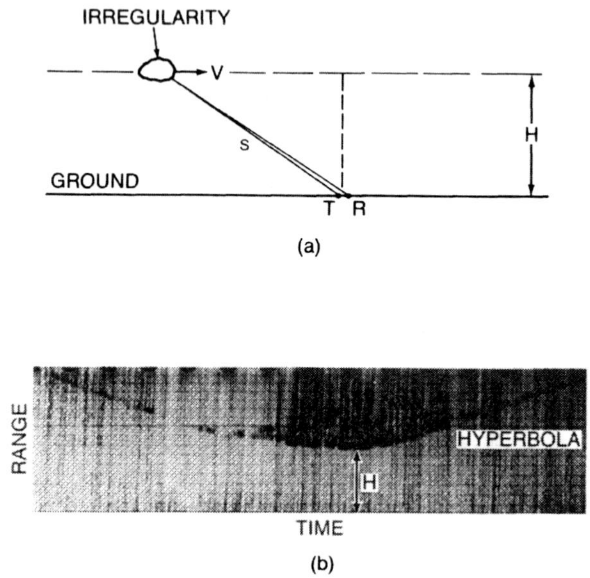

Fig. 7.23 *(a) Signal reflected from a horizontally moving irregularity*
(b) Time variation of the measured range of the echo

The variations, with time, of range and Doppler shift are shown in Figs. 7.24*b* and *c*. It is seen that, for a range of times and for $D > 1$ (approximately), there are three echoes. The maximum frequency deviation $\Delta f_{\max} = 2fkDV/c$ is directly proportional to the amplitude D of the ionospheric perturbation and to its speed V. An example of a switchback event is shown in Fig. 7.25. Using spaced sensor measurements one can determine the speed and the period τ. From the maximum shift $\Delta f_{\max}$ we obtain the approximate amplitude of the ionospheric disturbance:

$$D = c\tau\Delta f_{\max}/4\pi f \tag{7.80}$$

Now $\Delta f_{\max}$ and τ can be measured at a single location and H can be found from a real height analysis of an ionogram. The approximate condition for a

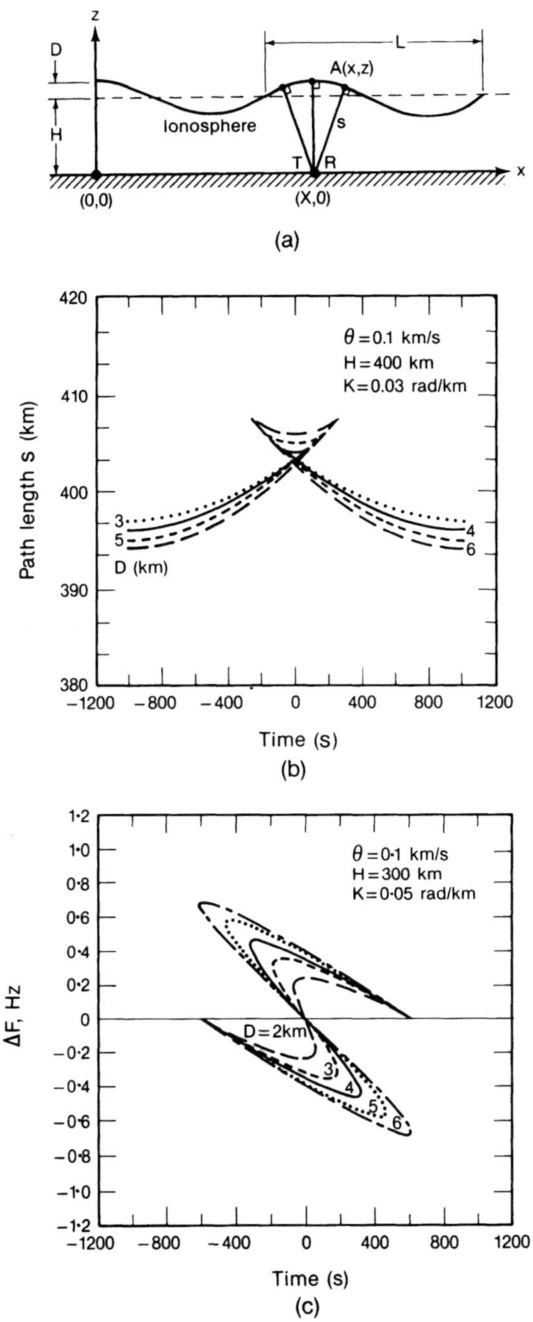

Fig. 7.24 *(a) Moving sinusoidal structure*
(b) Variation of range with time
(c) Variation of Doppler shift with time

switchback to occur is that

$$HDk^2 = 4\pi^2 HD/V^2\tau \geqslant 1 \tag{7.81}$$

Therefore we can measure the horizontal speed V when the Δf-versus-time trace is vertical.

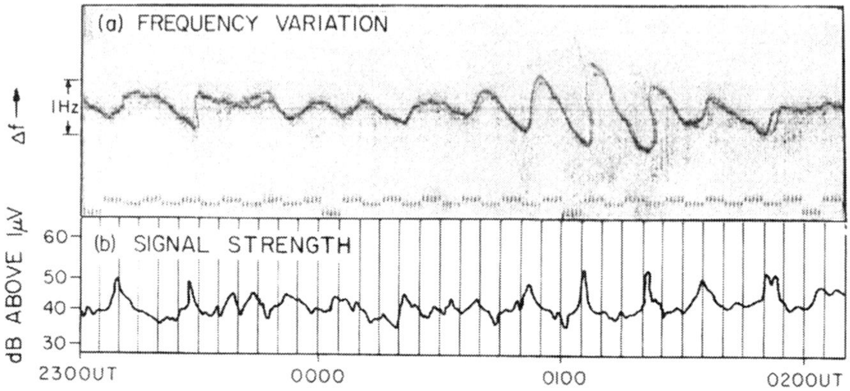

Fig. 7.25 *Record showing (a) frequency record with switchbacks, and (b) amplitude enhancement at onset of switchback conditions*
Sunset to Boulder, Colorado, 40 km; April 30, 1966; 5·054 MHz

In Sec. 7.3.4 we saw that a sinusoidal disturbance produces amplitude focusing. From (7.12) and (7.58) we have

$$E^2 = E_0^2\left[1 + \frac{Hc}{2V^2 f}\frac{d}{dt}(\Delta f)\right] \tag{7.82}$$

Thus focusing should occur at the onset of the switchback condition as is seen in Fig. 7.25*b*.

7.11 Acoustic-gravity waves

Various suggestions have been made as to the origin of wavelike disturbances in the ionosphere (Kent and Wright 1968). The most popular view is that TIDs are the ionospheric manifestation of waves in the neutral atmosphere. We are familiar with longitudinal sound waves in a uniform medium, which are dependent on compression and rarefaction. As the wave period increases to minutes (e.g. $\tau \sim 5$ min) the effects of buoyancy become important because of atmospheric stratification. At these long periods the atmosphere becomes dispersive and anisotropic. Here we consider some simple properties of these waves; for comprehensive discussions see Hines (1974) and Gossard and Hooke (1975).

7.11.1 Characteristic frequencies

Consider a plane-stratified, isothermal atmosphere under gravity that is constant with height. If this atmosphere were compressed and then released it would expand and would oscillate about its equilibrium state at the acoustic-cutoff frequency ω_a given by

$$\omega_a = \frac{C}{2H} \tag{7.83}$$

where $C\ (=\sqrt{\gamma g H})$ is the sound speed, H is the scale height, and γ is the ratio of the specific heats at constant pressure and constant volume.

Next consider a parcel of gas in the same atmosphere. If the parcel is displaced vertically, under adiabatic conditions, it will no longer be in equilibrium but will oscillate about its equilibrium height at the buoyancy, or Brunt–Väisälä frequency, ω_g, where

$$\omega_g^2 = \frac{g}{H}\frac{\gamma - 1}{\gamma} \tag{7.84}$$

and g is the gravitational acceleration. From (7.83) and (7.84) we have

$$\omega_g^2 = \omega_a^2\, 4(\gamma - 1)/\gamma^2 \tag{7.85}$$

Hence for a diatomic gas ($\gamma = 1.4$)

$$\omega_g = 0{\cdot}9\ \omega_a$$

In the absence of viscous absorption the vertical power flux is constant and the amplitude ζ will increase exponentially with height h as

$$\zeta = \zeta_0 \exp\left(\frac{h - h_0}{2H}\right) \tag{7.86}$$

in which ζ_0 is the amplitude at a height h_0. Such waves are often called acoustic-gravity waves.

7.11.2 Dispersion relations

The refractive index for a wave traveling in the above-described atmosphere, at an angle ϕ with the vertical, is given by

$$\mu^2 = \frac{C^2}{v_p^2} = \frac{C^2 k^2}{\omega^2} = \frac{1 - X}{1 - Y^2 \sin^2 \phi} \tag{7.87a}$$

where v_p is the phase speed and

$$X = (\omega_a/\omega)^2 \tag{7.87b}$$

$$Y = \omega_g/\omega \tag{7.87c}$$

We see that ω_a and ω_g are somewhat similar to the plasma frequency and the gyrofrequency, respectively. Equations (7.87) show that the propagation is dispersive and anisotropic and that two families of waves can propagate: (i) acoustic waves on frequencies greater than the acoustic-cutoff frequency, $X < 1$, and (ii) gravity waves on frequencies lower than the buoyancy frequency, $Y > 1$. In terms of the horizontal component $k_x (= k \sin \phi)$ and vertical component k_z $(= k \cos \phi)$ the dispersion relation is

$$\omega^4 - \omega^2 \omega_a^2 - k_x^2 C^2 (\omega^2 - \omega_g^2) - C^2 \omega^2 k_z^2 = 0 \tag{7.88}$$

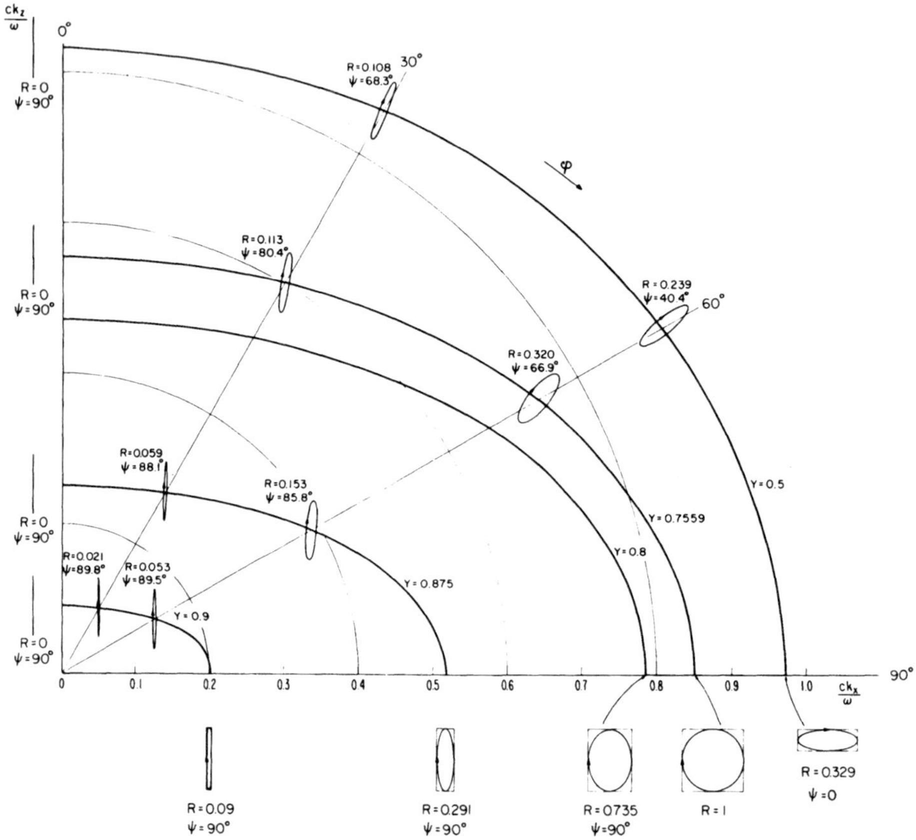

Fig. 7.26 *Refractive index curves and parcel orbits for acoustic waves in an isothermal atmosphere*

R is the axial ratio of the elliptical parcel orbit, and ψ is the inclination of the major axis from the horizontal

The ratio of the horizontal displacement ξ of an air parcel to its vertical displacement ζ is

$$\frac{\xi}{\zeta} = - \frac{\mu \sin \phi}{\mu^2 \sin^2 \phi - 1} [\mu \cos \phi - i(X - Y^2)^{1/2}] \tag{7.89}$$

Thus the motions of the air parcels are, in general, ellipses in the plane of propagation and have components transverse to the direction of wave propagation. Between the acoustic-cutoff frequency and the buoyancy frequency, waves are forbidden. The variations of $\mu(\omega, \phi)$ for acoustic waves and gravity waves are given in Figs. 7.26 and 7.27, respectively, together with sample orbits.

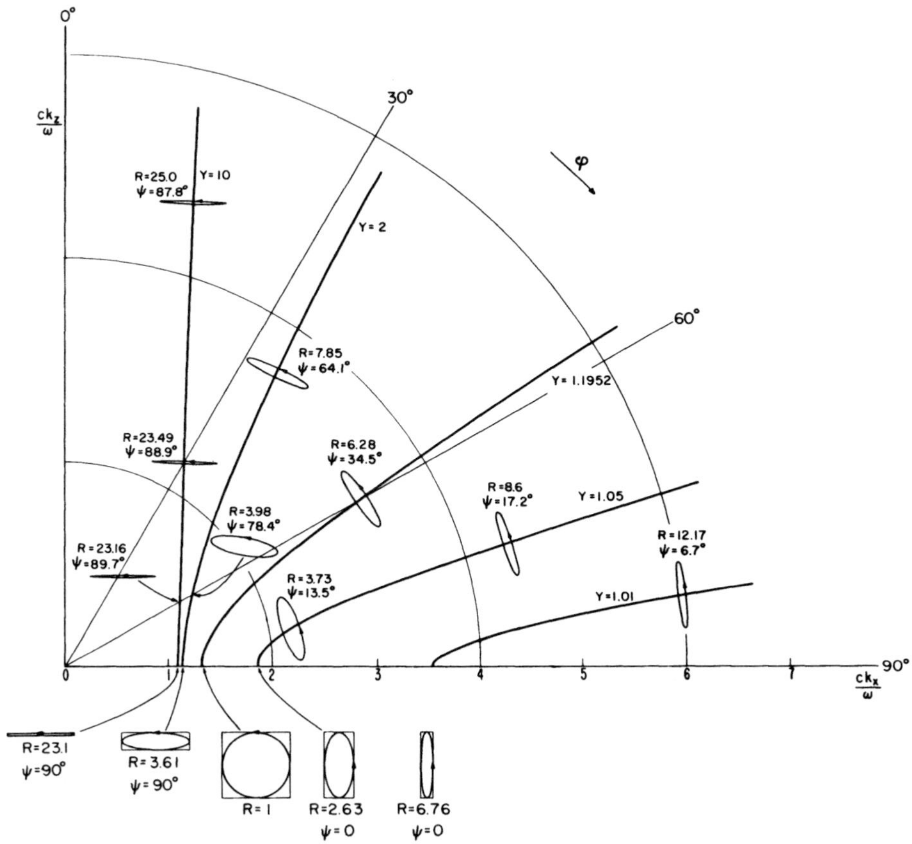

Fig. 7.27 *Refractive index curves and parcel orbits for gravity waves in an isothermal atmosphere*

R is the axial ratio of the elliptical parcel orbit, and ψ is the inclination of the major axis from the horizontal

On frequencies just above the acoustic cutoff the air motion is essentially vertical as it is on frequencies just below the buoyancy frequency. With acoustic waves on high frequencies the motion is radial as in sound waves. The motion is circular with horizontal propagation at a frequency $\omega_a(2/\gamma)^{1/2}$. With gravity waves, propagation is limited to angles between $\phi_{\min} = \sin^{-1}(\omega/\omega_g)$ and

$\phi_{max} = \pi - \sin^{-1}(\omega/\omega_g)$. The sense of rotation of the air for gravity waves is opposite that for acoustic waves. As ϕ approaches its asymptotic values the air motions become linear and transverse to the direction of propagation. Some properties of acoustic and gravity waves are listed in Table 7.4.

Table 7.4 *Some properties of acoustic and gravity waves in an isothermal atmosphere*

Parameter	Acoustic waves	Gravity waves
$X = (\omega_a/\omega)^2$	$X < 1$	$X > 1$
$Y = \omega_g/\omega$	$Y < 1$	$Y \sin\phi > 1$
Refractive index	$\mu < 1$	$\mu > 1$
Phase speed	$v_p > C$	$v_p < C$
Propagation angle	$0 \leqslant \phi \leqslant 2\pi$	$\pi - \sin^{-1}(1/Y) > \phi > \sin^{-1}(1/Y)$
Air parcel rotation	Clockwise	Anticlockwise
Energy vector $\boldsymbol{S}$	Same quadrant as propagation	Energy flows up when phase travels down and vice versa

7.11.3 Phase and ray propagation

The angle α, obtained by substitution of (7.87*a*) in (1.12), is given by

$$\tan\alpha = \frac{Y^2 \sin\phi \cos\phi}{1 - Y^2 \sin^2\phi} \tag{7.90}$$

For acoustic waves, $Y < 1$, and α is positive, but for gravity waves, $Y > 1$, and α is negative; that is, when $\boldsymbol{k}$ points down, the ray $\boldsymbol{S}$ points up. This is an important property since it accounts for the observed downward phase propagation when the source is below the level at which a disturbance is observed. Near the asymptote, $\alpha \to 90°$.

The horizontal u_x and vertical u_z components of the packet velocity $\boldsymbol{u}$ are obtained by substitution of (7.88) into (1.13). This gives

$$u_x = \frac{C^2 k_x(\omega^2 - \omega_g^2)}{\omega(2\omega^2 - \omega_a^2 - C^2k^2)} = C\frac{\mu \sin\phi(1 - Y^2)}{2 - X - \mu^2} \tag{7.91}$$

and

$$u_z = \frac{C^2 k_z \omega^2}{\omega(2\omega^2 - \omega_a^2 - C^2k^2)} = C\frac{\mu\cos\phi}{2 - X - \mu^2}. \tag{7.92}$$

When $Y \to 1$, $u_x \to 0$ and energy can only flow vertically. Note that, in Figs. 7.26 and 7.27, the radius vector is the direction of propagation and the ray direction is tangent to the $\mu(\phi)$ curves. Thus, for low frequencies (i.e. long periods), the $\mu(\phi)$ surface is vertical and the ray direction is horizontal and essentially independent of the direction of phase propagation; although the propagation directions can vary over wide ranges, the energy is 'beamed' into a narrow range.

7.11.4 Effects of neutral wind

The propagation of acoustic-gravity waves is affected by a horizontal wind. The refraction of the waves is expressed by the constancy of the horizontal trace speed V_H:

$$V_H = \frac{v_p}{\sin\phi} + U = \text{a constant} \tag{7.93}$$

where U is the wind component in the plane of propagation. The phase speed v_p can be obtained from (7.87*a*), provided we use the Doppler-shifted (intrinsic) frequency Ω seen by an observer moving with the wind:

$$\Omega = \omega(1 - U/V_H) \tag{7.94}$$

(Hines and Reddy 1967). The angle β between the ray and the vertical is given by

$$\tan\beta = (U + u\sin\psi)/u\cos\psi \tag{7.95a}$$

where

$$\tan\psi = u_x/u_z \tag{7.95b}$$

and $\boldsymbol{u}$ is the packet velocity in the absence of a wind.
Hence, the ray path is given by

$$x = \int_{z_0}^{z} \tan\beta \, dh \tag{7.96}$$

and the time of flight

$$t = \int_{z_0}^{z} \frac{dh}{u\cos\psi} \tag{7.97}$$

When $U/V_H = 1$, the intrinsic frequency $\Omega = 0$ and the wave ceases to propagate. The level at which this occurs is called a critical level (Hines 1968).

7.11.5 Waves in realistic atmospheres

The variations with height of the characteristic periods τ_a and τ_g are shown in Fig. 7.28. We note that τ_a and τ_g vary from about 5 min near sea level, to about 15 min in the F region. Ray paths in a stationary atmosphere are shown in Fig. 7.29 for acoustic waves with a period of 3·0 min emanating from a source at 20 km. Note (i) the ducting in the lower atmosphere, (ii) reflections from the temperature incline above 90 km, and (iii) the iris through which waves can

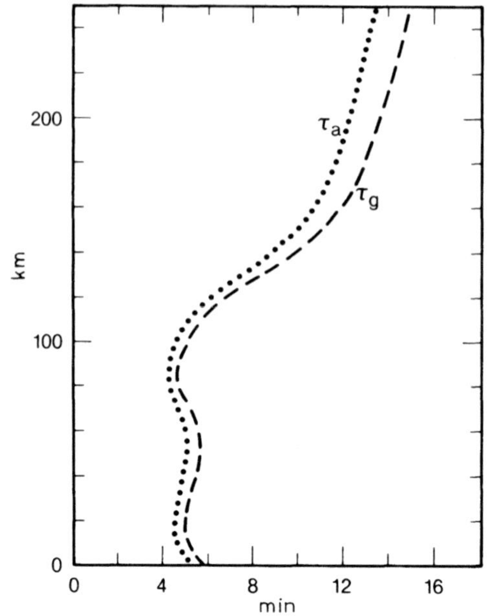

Fig. 7.28 *Height variations of the acoustic-cutoff period τ_a and buoyancy period τ_g in the 1962 US Standard Atmosphere*

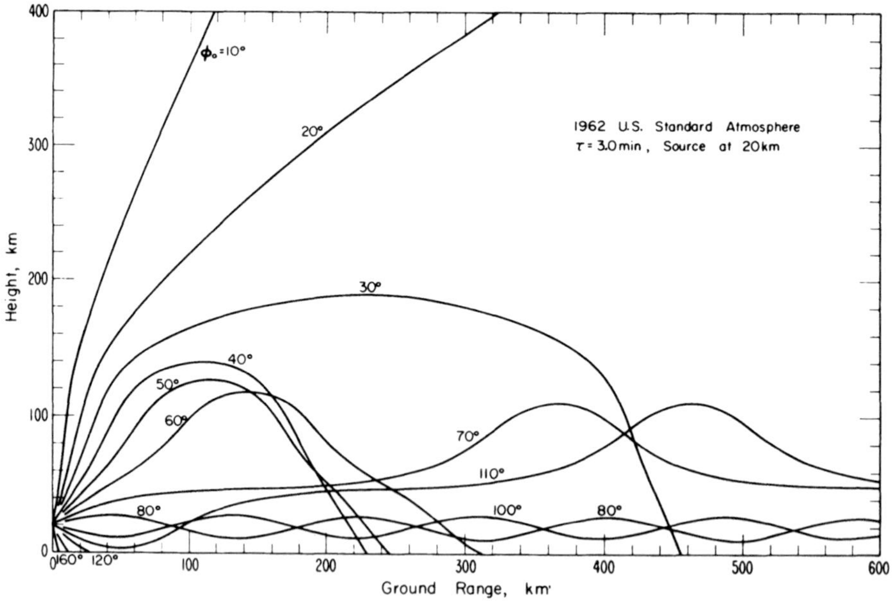

Fig. 7.29 *Acoustic-wave ray paths in the 1962 US Standard Atmosphere*
The source is at a height of 20 km; ϕ is the angle of takeoff measured from the vertical.

reach the upper F region. Gravity-wave propagation is restricted such that waves from the lower atmosphere can reach the F region only through an annulus.

Observations indicate that gravity waves in the upper atmosphere propagate at angles close to the asymptotic limit. Thus waves with periods near the buoyancy period tend to propagate with their phase fronts nearly vertical. Long-period (60 min) waves have phase fronts nearly horizontal, so the vertical wavelength λ_z is much less than the horizontal wavelength λ_x.

Some sources of acoustic-gravity waves are discussed in Chapter 14. For further information on atmospheric motions the reader is referred to books by Chapman and Lindzen (1970), Kato (1980), and Wilkes (1949).

7.11.6 Waves in the ionized media

Disturbances observed with radio techniques are signatures in the ionized medium produced by collisional interaction with neutral air. In the D region and lower E region, ion collisions are sufficiently great that the plasma motion replicates the neutral motion. In the F region, however, the ion motion, gained from the neutral gas, is constrained to move along the geomagnetic field. Figure 7.27 shows that, for long-period waves, the air motion ξ is essentially horizontal, so in the magnetic meridian, the ion motion parallel to the geomagnetic field is $\xi \cos I$, where I = dip, and the vertical component is $\xi \cos I \sin I$ (see Fig. 2.22) which is the disturbance registered by radio probing. For a wave of given period and amplitude the vertical motion is a maximum where the dip is 45° (i.e. near 27° dip latitude). With magnetic east–west propagation the air motion is perpendicular to the geomagnetic field and there will be little ion motion. This is an example of the aspect sensitivity of the ionosphere to the passage of a gravity wave. Thus the fact that few ionospheric waves are seen in certain directions does not necessarily mean that there are few neutral waves. Even when ionospheric perturbations exist, some techniques, such as that of Faraday rotation, are insensitive to them if the contributions from troughs and peaks cancel. For thorough discussions of acoustic-gravity waves and their ionospheric effect, the reader should consult the books by Hines (1974), Gossard and Hooke (1975), and Yeh and Liu (1972).

7.11.7 Summary of observations

Common periods of ionospheric waves are between 5 min and 60 min. Waves of shorter period have been observed under special conditions (e.g. Davies and Jones 1973). Speeds vary from about 50 m s^{-1} in the D and E regions to more than 1 km s^{-1} for long-period waves in the F region. F-layer irregularities induced by acoustic-gravity waves appear to be magnetic-field aligned, especially near the magnetic equator. Kent and Wright (1968) and Hines (1974) summarized experimental results, and Francis (1975) summarized the theory. A considerable amount of information on the effects of atmospheric acoustic gravity waves on electromagnetic wave propagation is contained in AGARD (1972).

7.12 Nonreciprocal phenomena

7.12.1 Reciprocity

The term reciprocity implies that the effects of the medium on a radio wave are the same if we go from *A* to *B* as they are for the reverse direction. In an isotropic medium this reciprocity is always the case, and even in a magnetoplasma reciprocity usually holds for most characteristics. Since the refractive index is unaltered by reversing the direction of propagation it follows that such characteristics as ray path, time delay, phase, absorption, and Doppler shift are independent of the direction of propagation between *A* and *B*.

7.12.2 Nonreciprocity on high frequencies

The nonreciprocity for a path arises only because of the interaction between the magnetoionic waves and the antennas. This can be understood with reference to Fig. 7.30 in which a linear antenna at *A* emits a wave with its electric field at an angle of 45° with the vertical. As the wave propagates parallel to the imposed magnetic field from *A* to *B*, the electric field rotates through an angle of 45°, so at *B* the vertical antenna picks up the entire signal. Now consider the reverse path in which the vertical electric field emitted from *B* is rotated in the right-hand sense with respect to the magnetic field. Therefore, at *A* (see Sec. 8.3.2), the electric field is perpendicular to the antenna that does not pick up the signal. Nonreciprocal polarization fading has been observed; for example, see Jull (1969) and other references therein.

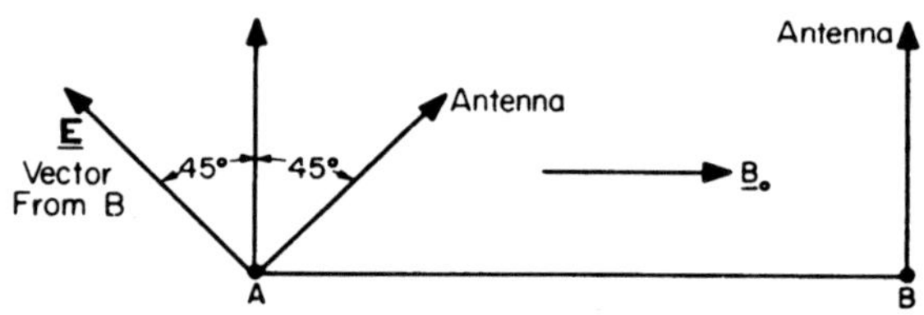

Fig. 7.30 *Example of the nonreciprocity of Faraday rotation, which depends only on the direction of the magnetic field and not on the direction of propagation*

7.12.3 Nonreciprocity on very low frequencies

A different type of nonreciprocity, which occurs on very low frequencies, has been discussed by Barber and Crombie (1959). It involves different responses of the ionospheric plasma to waves traveling in opposite directions such as east-to-west and west-to-east. This subject is discussed further in Sec. 10.3.5 and Sec. 11.5.7.

7.13 References

AGARD (1972): 'Effects of atmosphere gravity waves on electromagnetic wave propagation'. AGARD Conf. Proc. No. 115, Advisory Group for Aerospace Research and Development, North Atlantic Treaty Organisation

APPLETON, E. V., and BEYNON, W. J. G. (1947): 'The application of ionospheric data to radio communications problems: Part II', *Proc. Phys. Soc.*, **59**, p. 58

APPLETON, E. V., and PIGGOTT, W. R. (1954): 'Ionospheric absorption measurements during a sunspot cycle', *J. Atmos. Terr. Phys.*, **5**, p. 141

BARBER, N. F., and CROMBIE, D. D. (1959): 'VLF reflections from the ionosphere in the presence of a transverse magnetic field', *J. Atmos. Terr. Phys.*, **16**, p. 37

BENNETT, J. A. (1967): 'The calculation of Dopper shifts due to a changing ionosphere', *J. Atmos. Terr. Phys.*, **29**, p. 887

BENNETT, J. A., and DYSON, P. L. (1986): 'The effect of small amplitude wave irregularities on radio wave observations of the ionosphere', *Radio Sci.*, **21**, p. 375

BOWMAN, G. G. (1968): 'Movements of ionospheric irregularities and gravity waves', *J. Atmos. Terr. Phys.*, **30**, p. 721

BRADLEY, P. A. (1976): 'A new computer-based method of HF sky-wave signal prediction using vertical-incidence ionosonde measurements' *in* BLACKBAND, W. T. (Ed.): 'Radio systems and the ionosphere'. AGARD Conf. Proc. No. 173, Advisory Group for Aerospace Research and Development, North Atlantic Treaty Organisation, p. 11

BRASSEUR, G., and SOLOMON, S. (1986): 'Aeronomy of the middle atmosphere' (Reidel Publishing Co., Dordrecht) 2nd edn.

BREMMER, H. (1949): 'Terrestrial radio waves' (Elsevier Publishing Co., New York)

BRENNAN, D. G. (1961): 'Probability theory in communication system engineering' *in* BAGHDADY, E. (Ed.): 'Lectures on communications system theory' (McGraw-Hill, New York)

BRIGGS, B. H. (1968): 'On the analysis of moving patterns in geophysics. I: Correlation analysis', *J. Atmos. Terr. Phys.*, **30**, p. 1777

BROWNLIE, G. D., DRYBURGH, L. G., and WHITEHEAD, J. D. (1973): 'Measurement of the velocity of waves in the ionosphere: A comparison of the ray theory approach and diffraction theory', *J. Atmos. Terr. Phys.*, **35**, p. 2147

BUDDEN, K. G. (1985): 'The propagation of radio waves' (Cambridge University Press, Cambridge)

CALDWELL, J. D., STEVENS, E. E., and WARREN, E. S. (1969): 'The reflection coefficient of an ionosphere containing large scale irregularities' *in* JONES, T. B. (Ed.): 'Oblique ionospheric radiowave propagation'. AGARD Conf. Proc. No. 13, Advisory Group for Aerospace Research and Development, North Atlantic Treaty Organisation, p. 285.

CCIR (1986): 'Propagation in ionized media. Recommendations and reports of the CCIR, 1986, VI.' Int. Telecom. Union, Geneva

CHAPMAN, S., and LINDZEN, R. S. (1970): 'Atmospheric tides' (Gordon and Breach, New York)

CROFT, T. A. (1972): 'Sky-wave backscatter: A means for observing our environment at great distances', *Rev. Geophys. Space Phys.*, **10**, p. 73

DAVIES, K. (1969): 'Ionospheric radio waves' (Blaisdell Publishing Co., Waltham, MA)

DAVIES, K. (Ed.) (1970): 'Phase and frequency instabilities in electro-magnetic wave propagation'. AGARD Conference Proceedings No. 33, Advisory Group for Aerospace Research and Development, North Atlantic Treaty Organisation

DAVIES, K., and JONES, J. E. (1973): 'Acoustic waves in the ionospheric F_2 region produced by severe thunderstorms', *J. Atmos. Terr. Phys.*, **35**, p. 1737

DIEMINGER, W. (1968): 'Ionospheric propagation of high frequency radio waves' *in* 'Radio wave propagation'. AGARD Lecture Series XXIX, Advisory Group for Aerospace Research and Development, North Atlantic Treaty Organisation, p. III–1

FINDLEY, J. W. (1951): 'The phase and group paths of radio waves returned from region E of the ionosphere', *J. Atmos. Terr. Phys.*, **1**, p. 353

FRANCIS, S. H. (1975): 'Global propagation of atmospheric gravity waves: A review', *J. Atmos. Terr. Phys.*, **37**, p. 1011

GEORGE, P. L. (1971): 'The global morphology of the quantity $\int N\nu dh$ in the D- and E-regions of the ionosphere', *J. Atmos. Terr. Phys.*, **33**, p. 1893

GEORGE, P. L., and BRADLEY, P. A. (1974): 'A new method of predicting the ionospheric absorption of high frequency waves at oblique incidence', *Telecom. J.*, **41**, p. 307

GEORGES, T. M. (1968): 'HF Doppler studies of traveling ionospheric disturbances', *J. Atmos. Terr. Phys.*, **30**, p. 735

GOSSARD, E. E., and HOOKE, W. H. (1975): 'Waves in the atmosphere' (Elsevier Publishing Co., Inc., New York)

GREGORY, J. B., and MANSON, A. J. (1969): 'Seasonal variations of electron densities below 100 km at mid-latitudes', *J. Atmos. Terr. Phys.*, **31**, Part I, p. 683; Part II, p. 703

HINES, C. O. (1968): 'Some consequencies of gravity-wave critical layers in the upper atmosphere', *J. Atmos. Terr. Phys.*, **30**, p. 837

HINES, C. O. (1974): 'The upper atmosphere in motion' (American Geophysical Union, Washington, DC)

HINES, C. O., and RAO, R. R. (1968): 'Validity of three-station methods of determining ionospheric motions', *J. Atmos. Terr. Phys.*, **30**, p. 979

HINES, C. O., and REDDY, C. A. (1967): 'On the propagation of atmospheric gravity waves through regions of wind shear', *J. Geophys. Res.*, **72**, 1015

HUNSUCKER, R. D. (1990): Radio techniques for probing the terrestrial ionosphere' *in* LANZEROTTI, L. (Ed): 'Series on physics and chemistry in space' (Springer-Verlag, New York)

JATP (1968): 'Special issue on upper atmosphere winds, waves and ionospheric drifts', *J. Atmos. Terr. Phys.*, **30**, pp. 657–1063

JULL, G. W. (1969): 'Short-tem and averaged characteristics of nonreciprocal HF ionospheric paths' *in* JONES, T. B. (Ed.): 'Oblique ionospheric radiowave propagation'. AGARD Conf. Proc. No. 13, Advisory Group for Aerospace Research and Development, North Atlantic Treaty Organisation, p. 303

KATO, S. (1980): 'Dynamics of the upper atmosphere' (Reidel Publishing Co., Dordrecht)

KENT, G. S., and WRIGHT, R. W. H. (1968): 'Movements of ionospheric irregularities and ionospheric winds', *J. Atmos. Terr. Phys.*, **30**, p. 657

KOSTER, J. R. (1963): 'Some measurements of the sunset fading effect', *J. Geophys. Res.*, **68**, p. 2571

KRISHNASWAMY, S., DETRICK, D. L. and ROSENBERG, T. J. (1985): 'The inflection point method of determining riometer quiet day curves', *Radio Sci.*, **20**, p. 123

LITTLE, C. G. (1957): 'The measurement of ionospheric absorption using extra-terrestrial radio waves' *in* (BEYNON, W. J. G., and BROWN, G. M. (Eds.): 'Annals of the IGY, Vol. III', (Pergamon Press, London) p. 207

LITTLE, C. G., and LEINBACH, H. (1959): 'The riometer – a device for the continuous measurement of ionospheric absorption', *Proc. IRE*, **47**, p. 315

MARTYN, D. F. (1935): 'The propagation of medium radio waves in the ionosphere', *Proc. Phys. Soc.*, **47**, p. 323

MUNRO, G. H. (1958): 'Traveling ionospheric disturbances in the F region', *Aust. J. Phys.*, **11**, p. 91

PHILLIPS, G. J., and KNIGHT, P. (1965): 'Effects of polarization on a medium-frequency sky-wave service, including the case of multi-hop paths', *Proc. IEE*, **112**, p. 31

PIGGOTT, W. R. (1959): 'The calculation of median sky wave field strength in tropical regions'. Radio Research Special Report No. 27, Dept. of Sci. and Ind. Res., Slough, England

PIGGOTT, W. R. (1964): 'Studies of ionospheric absorption', *Research in Geophysics:* Vol. 1. Sun, upper atmosphere, and space' (The MIT Press, Cambridge, MA) p. 277

PIGGOTT, W. R., BEYNON, W. J. G., BROWN, G. M., and LITTLE, C. G. (1957): 'The measurement of ionospheric absorption' *in* 'Annals of the IGY, Vol. III, Part II' (Pergamon Press, London) p. 173

RANZI, I. (1934): 'Phase variations of reflected radio waves and a possible connection with the Earth's magnetic field in the ionosphere', *Nature*, **133**, p. 908

RAWER, K. (Ed.) (1976): 'Manual on ionospheric absorption measurements'. Report UAG-57, World Data Center-A, NOAA, Boulder, CO 80303

RICE, S. O. (1958): 'Distribution of the duration of fades in radio transmission: Gaussian noise model', *Bell Syst. Tech. J.*, **37**, p. 581

SCHWENTEK, H. (1966): 'The determination of absorption in the ionosphere by recording the field strength of a distant transmitter', *Ann. Geophys.*, **22**, p. 276

SETTY, C. S. G. K., JAIN, A. R., and VYAWAHARE, M. K. (1970): 'Collision frequency of electron in the F region', *Can. J. Phys.*, **48**, p. 653

THOMAS, L. (1962): 'The winter anomaly in ionospheric absorption', *J. Atmos. Terr. Phys.*, **23**, p. 301

THOME, G. D. (1964): 'Incoherent scatter observations of traveling ionospheric disturbances', *J. Geophys. Res.*, **69**, p. 4047

THORPE, M. R. (1971): 'The phase integral correction for calculations of radiowave absorption in the ionosphere', *J. Atmos. Terr. Phys.*, **33**, p. 1597

WAKAI, N., and ISHIZAWA, K. (1962): 'A method for the simultaneous recording of the echo amplitude with h'-f curves in the ionospheric sounding and some applications to the measurement of ionospheric parameters', *J. Radio Res. Labs.*, **9**, p. 421

WHITEHEAD, J. D. (1962): 'Distribution of echo amplitudes from an undulating surface', *J. Atmos. Terr. Phys.*, **24**, p. 715

WHITEHEAD, J. D., and JONES, T. B. (1972): 'Distance attenuation effects in the absorption of vertically incident radio waves', *J. Atmos. Terr. Phys.*, **34**, p. 165.

WILKES, M. V. (1949): 'Oscillations of the Earth's atmosphere' (The University Press, Cambridge)

WRIGHT, J. W., and PITTEWAY, M. L. V. (1978): 'Computer simulation of ionospheric radio drift measurements, and their analysis by correlation measurements', *Radio Sci.*, **13**, p. 189

WRIGHT, J. W., AMAYENC, P. and VASSEUR, G. (1976): 'The interpretation of ionospheric radio drifts IX', *J. Atmos. Terr. Phys.*, **38**, p. 731

YEH, K. C., and LIU, C. H. (1972): 'Theory of ionospheric waves' (Academic Press, New York)

Chapter 8

Earth–space propagation

8.1 Introduction

Since the launch of Sputnik I in 1957, satellite communications, navigation, and surveillance have increased steadily. Because satellite systems employ very high frequencies, wider channels are available than in the high-frequency band. Satellites are now used regularly for intercontinental telephone and television transmission. For discussion purposes satellite orbits can be classified into two categories: (i) low-inclination synchronous orbits, at approximately 36 000 km above the Earth, which appear to be essentially fixed in the sky to an observer on the Earth and, hence, are called geostationary, and (ii) orbits in which the satellite moves with respect to a ground observer. In the early days of satellite telemetry, very high frequencies were used (e.g. $\approx$ 137 MHz) partly because it was believed that, on these frequencies, ionospheric effects were negligible. Almost immediately two ionospheric effects were evident: (i) the rotation of the electric vector of the wave gave rise to polarization, or Faraday, fading, and (ii) amplitude fluctuations (scintillations) were produced by ionospheric irregularities. Thus the notion that satellite radio systems were immune to ionospheric effects was quickly dispelled. In the 1960s the decision was made to go to still higher frequencies (e.g. 4–6 GHz) but although ionospheric effects were reduced they were still troublesome in the auroral zones and in the equatorial region at certain local times and seasons.

Satellites have been used to study the ionosphere, the most successful being the various topside sounders (IEEE 1969). It is important to remember that more than two-thirds of the ionosphere electron content lies above the peak of the F_2 layer. Thus, for satellite-to-ground transmissions, the topside can be more important than the bottomside and may contain scintillation-producing irregularities that cannot be seen from ionosondes on the ground. This Chapter discusses the use of satellite sounders to study the topside and then considers the use of fixed-frequency transmissions to study total electron content (TEC) and irregularities, and the reverse, namely, the effects of ionospheric electron content and irregularities on radio transmissions.

8.2 Topside sounding

8.2.1 Satellite sounders

These are of two types: (i) those with a set of fixed frequencies and (ii) those with swept-frequency ionosondes. Both use pulse-amplitude modulation and measure the group delay of the echo. Orbits vary widely from the near-circular orbit of Alouette I at about 1000 km to elliptical orbits (e.g., Alouette II; perigee 502 km, apogee 2982 km). These sounders have produced a wealth of information not only about the distribution of electrons in the topside but also about the response of plasma when the transmitter is embedded in it.

Topside sounders have these main advantages:

(i) The topside can be observed; it cannot be observed with ground-based ionosondes.
(ii) Large geographical areas can be sounded in relatively short time intervals
(iii) Equipment characteristics are constant, so ionogram changes are caused by ionospheric, rather than equipment, differences.
(iv) The signals do not suffer D-region absorption and, therefore, can be observed during disturbances.
(v) Plasma effects (resonances) can be observed because the sounder is embedded in the plasma.

Although fixed-frequency sounders have advantages of tuned circuitry and fine time resolution, the most successful sounders have been of the swept frequency type such as the Canadian Alouette I and the Japanese (Ionospheric Sounding Satellite) ISS–b. A disadvantage of the swept-frequency technique is the lack of tuning. Furthermore, all nongeostationary satellites move during a sweep. The joint US–Canadian topside program was called ISIS (International Satellites for Ionospheric Studies) and contained, in addition to a swept- plus fixed-frequency topside sounder, a mass spectrometer, an ion temperature probe, an ion number- density probe, and an untuned VLF receiver (50 Hz to 30 kHz). For further information on the Alouette/ISIS program, see Jackson and Warren (1969).

8.2.2 Ray paths with topside sounding

Some ray paths for waves having a vertical direction of phase propagation are shown in Fig. 8.1 for various heights above the peak of a Chapman layer in the Northern Hemisphere. Note that the ordinary (o) ray is deviated toward the Equator and the extraordinary (x) ray is deviated toward the pole. As the height of the satellite decreases (i.e. as the surrounding electron density increases) first the x-wave and secondly the o-wave get 'imprisoned' in the satellite because the plasma is overdense. In high latitudes, when the satellite is near the reflection level of the ordinary wave, energy slides almost sideways. An additional path (the z path) is possible, which may be understood by reference to Fig. 3.2a. For $Y < 1$ it is seen that there is a branch of the extraordinary curve lying in the range

$$\frac{1 - Y^2}{1 - Y_L^2} < X < 1 + Y$$

Normally this wave can originate only from a source embedded in the plasma. To get some idea of what the ray paths may look like we notice that $\mu^2 > 1$ near $X = (1 - Y^2)/(1 - Y_L^2)$ and μ drops below 1 as it approaches reflection. The quantity $(\mu^2 - 1)$, and hence the angle α between the wave normal and the ray,

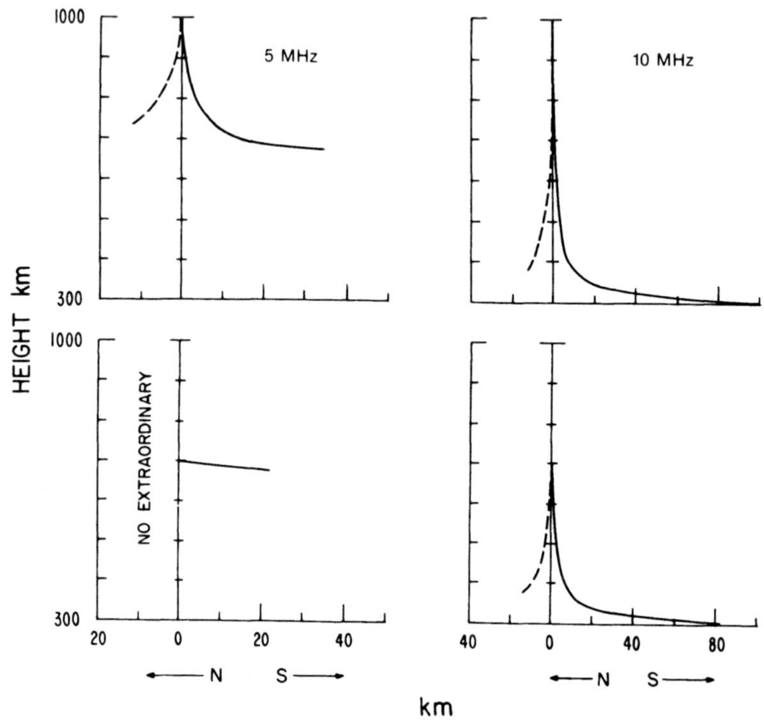

Fig. 8.1 *Ordinary (——) and extraordinary (- - -) ray paths from topside sounders at 1000 km and 700 km in a Chapman layer*
Peak plasma frequency $f_c = 10$ MHz, $H = 100$ km, $h_{max} = 300$ km, $f_H = 0{\cdot}75$ MHz, $I = 72{\cdot}9°$

will change sign. The ray path veers first to one side and then to the other (Fig. 8.2). The existence of the z-wave has been confirmed in soundings of the topside (Fig. 8.3). Figure 8.3 shows another feature common on topside ionograms, namely the petering out of the signals on frequencies just below the F_2 critical frequencies. Attempts to match topside electron-density profiles and bottomside profiles usually lead to overlap; that is, $h_m F_2$ from the topside is at

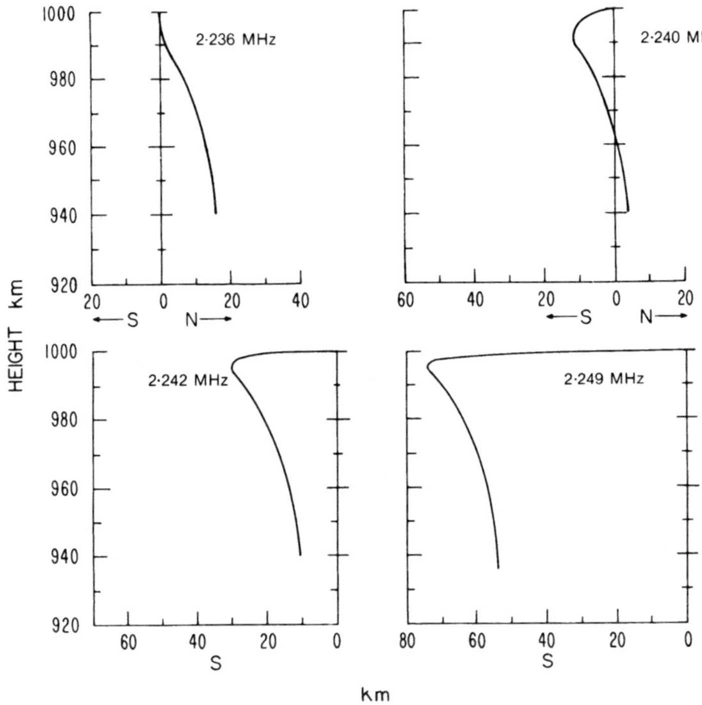

Fig. 8.2 *Z-rays from a topside source at 1000 km in a Chapman layer with constant magnetic field*
Peak plasma frequency f_c = 10 MHz, H = 100 km, h_{max} = 300 km, f_H = 0·75 MHz, dip = 72·9°

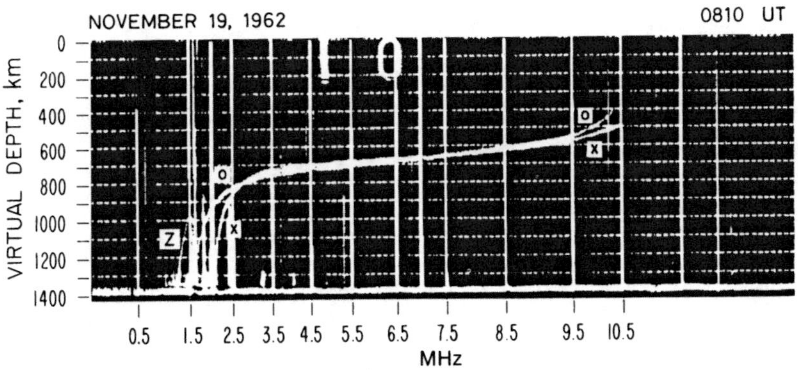

Fig. 8.3 *Alouette I ionogram taken near Singapore*
Note the *o*-, *x*-traces and the *z*-trace over a narrow frequency range

a lower altitude than $h_m F_2$ from the bottomside (Jackson 1969). This would arise from a horizontal wavelike structure in electron density. From the ionogram in Fig. 8.3 it is seen that the echoes start at the source, which is not the case with ground-based sounders. The electron density and magnetic field at the satellite can be determined from the frequencies at which the group delays of the *o*-, *x*-, and *z*-traces go to zero. Because of ionospheric shielding, there is no inteference from broadcast transmitters (0·5 MHz to 1·6 MHz). On frequencies above the F_2 penetration frequencies, echoes may be obtained from strong sporadic E and/or from the Earth's surface as seen in Fig. 8.4.

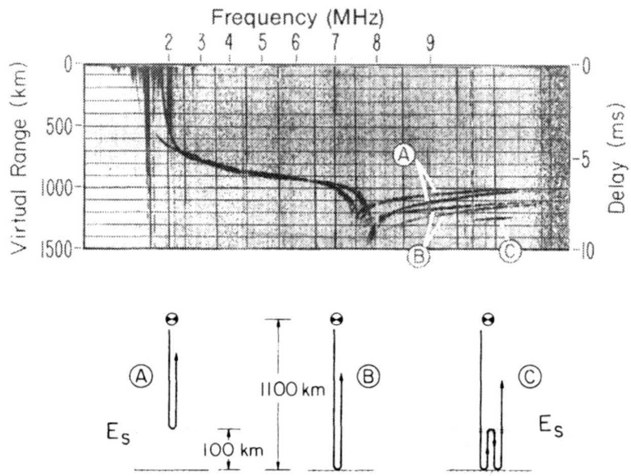

Fig. 8.4 *Alouette I topside ionogram showing echoes from sporadic E (From Calvert and Warnock 1969)*
Schematic diagrams of the (A) E_s, (B) ground, and (C) *W*-mode echo paths are shown. Ionogram was taken at 1811 UT, December 17, 1966

The lowest frequency on which a particular (z, o, x) wave is observed is called the wave cutoff frequency. Let $f_z S$, $f_o S$, and $f_x S$ be the observed cutoffs; then

$$f_o^2 S = f_N^2 = f_x S\,(f_x S - f_H) = f_z S\,(f_z S + f_H) \tag{8.1}$$

The upper frequency limit of the z trace is given by

$$f_z I = \left[\frac{f_T^2 + (f_T^4 + 4 f_N^2 f_H^2 \cos^2\theta)^{1/2}}{2}\right]^{1/2} \tag{8.2a}$$

where

$$f_T = (f_N^2 + f_H^2)^{1/2} \tag{8.2b}$$

is called the upper hybrid frequency or upper hybrid resonance.

Note that the electron density at the satellite can be obtained directly from $f_o S$ and the magnetic field from the difference between $f_z S$ and $f_x S$ or from f_T. Near the apogee of some elliptical orbits (e.g. Alouette II) the electron density falls to values (e.g. less than 0·15 MHz) too low for direct measurement. In these cases, f_N has to be obtained indirectly from (8.2*b*) (Timleck and Nelms 1969).

The above principles can be illustrated by scaling the cutoff frequencies in Fig. 8.3. Because of the nonlinear frequency scale, accurate values of the cutoffs are not possible but visual estimates give the following: $f_x S \approx 1{\cdot}9$ MHz, $f_o S \approx 1{\cdot}6$ MHz, $f_z S \approx 1{\cdot}0$ MHz. From (8.1) we find $f_H \approx 0{\cdot}55$ MHz at the height (1000 km) of the satellite, which is in rough agreement with the inverse cube law for a dipole (see Sec. 2.4.1). Because the ionogram in Fig. 8.3 pertains to low latitudes, $f_z I \approx f_T \approx 1{\cdot}7$ MHz, which is in rough agreement with observation.

8.2.3 Scattering and ducting in the topside

Examples of spread echoes on topside ionograms are shown in Fig. 8.5. This is the topside version of spread F on the bottomside (see Sec. 5.6). Spread echoes

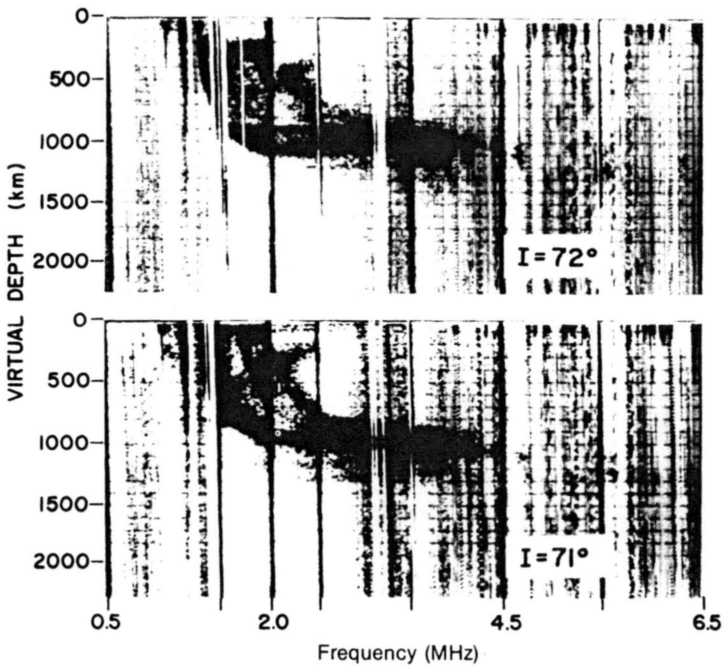

Fig. 8.5 *Topside ionograms with spread echoes*

are produced by ionospheric irregularities that scatter and duct energy. Field-aligned irregularities produce ducts in the topside in which waves can be trapped either when the satellite passes through the duct or when the wave is refracted and so enters the duct with its wave normal parallel to the duct. The wave proceeds along the duct and is reflected at one of the conditions for reflection given in (4.16). Strong echoes are possible because of containment by the duct.

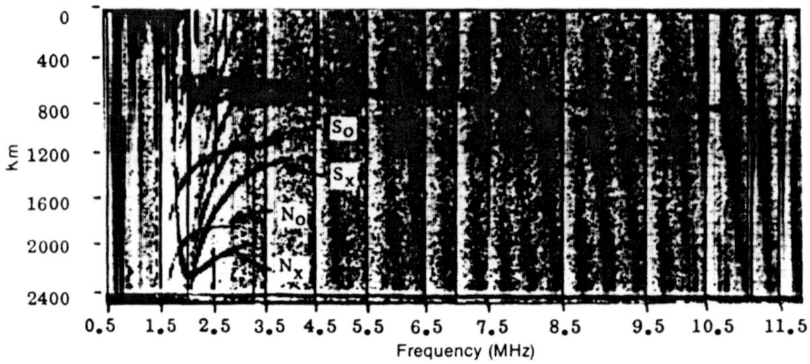

Fig. 8.6 *Alouette I ionogram recorded at Quito, Ecuador (80.1°W, 6.3°S) on December 18, 1962, 1725 UT, showing trace resulting from propagation along the Earth's magnetic field (From Muldrew 1969 © IEEE; reprinted with permission)*
S_o and S_x are the ordinary- and extraordinary-wave echoes reflected at the south end of the path; N_o and N_x are the corresponding echoes from the north end of the path

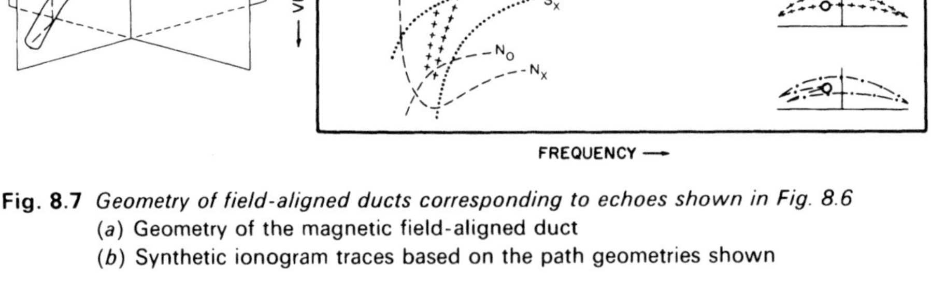

Fig. 8.7 *Geometry of field-aligned ducts corresponding to echoes shown in Fig. 8.6*
(*a*) Geometry of the magnetic field-aligned duct
(*b*) Synthetic ionogram traces based on the path geometries shown

Conjugate echoes are illustrated in Fig. 8.6, obtained with a sounder operating in the Northern Hemisphere. The duct lies in the magnetic meridian, and *o*- and *x*-waves are returned to the sounder after reflection in both the Northern and Southern Hemispheres (Fig. 8.7). For more extensive information on topside sounding see IEEE (1969).

8.2.4 Plasma resonances

These appear as spikes with lifetimes from a fraction of a millisecond to many milliseconds on topside ionograms (Fig. 8.8), and are produced when the transmitter power is confined to the neighborhood of the satellite. Some ionograms have only a few resonances. Others have ten or more; that is, the group

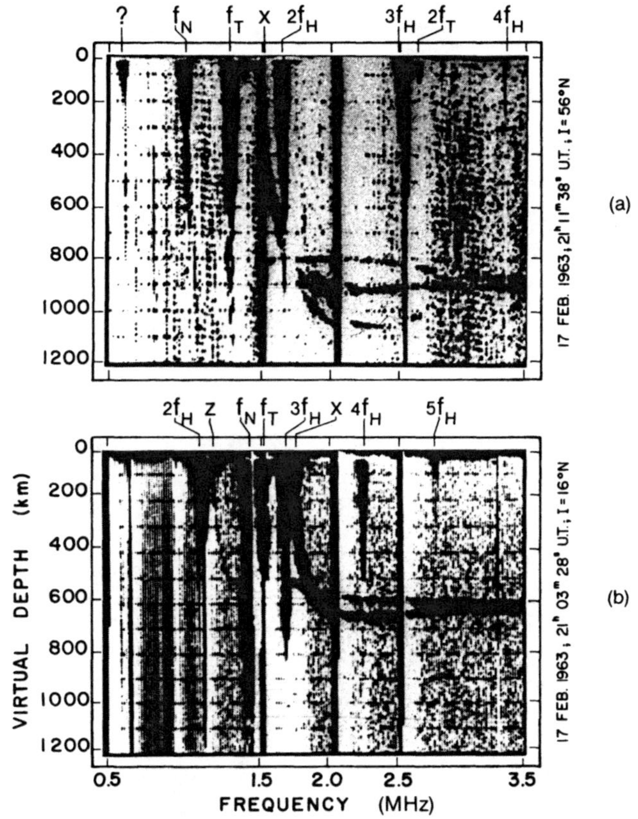

Fig. 8.8 *Alouette I ionograms for February 17, 1963, showing the presence of plasma resonances (From Calvert and Goe 1963)*
f_N is the local plasma frequency at the sounder, nf_H are the harmonics of the electron gyrofrequency, f_T is the upper hybrid resonance
(*a*) for dip = 56°N
(*b*) for dip = 16°N

velocity is near zero. The f_N resonances are caused by electrostatic waves (in which the electron motion is in the direction of propagation) having frequencies within a few kilohertz of f_N. The waves propagate up to several hundred meters from the satellite, and then they are reflected back to the source (McAfee 1974) where interference patterns can be produced (Fig. 8.9) (Calvert and McAfee 1969).

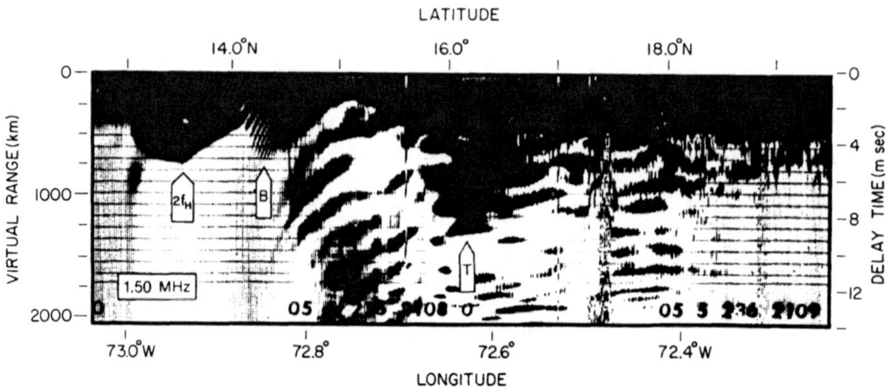

Fig. 8.9 *Examples of the resonance at the upper hybrid frequency (T), showing the fringe pattern due to interference between resonances (From Calvert and McAfee 1969)*

Table 8.1 *Plasma resonances*

Type	Symbol(s)	Frequency
Electron plasma	f_N, N	$\frac{1}{2\pi}\left(\frac{Ne^2}{m\varepsilon_0}\right)^{1/2}$
Cyclotron	f_H, H	$\frac{1}{2\pi}\frac{e}{m}B_0$
Cyclotron harmonics	nf_H	$n = 2, 3, 4, \ldots$, etc.
Upper hybrid	f_T, T	$(f_N^2 + f_H^2)^{1/2}$
Electrostatic resonances	f_{Q_n}, Q	
Diffuse resonance	f_{D_n}	$0{\cdot}95\,(f_N/f_H)^{1/2}\,n^{1/2}$ $n = 1, 2, 3, \ldots$, etc.
Remote resonance	f_c	$\geqq 2f_H$ (at reflection of x-wave).
Subsidiary resonances	f_s	$\frac{n}{m}$ (f_H or f_N or f_T), m, n are small integers
Floating spike		$f_T/2 = f_N - f_H$
Lower hybrid	LHR	$(f_N f_H/f_T)(m_e/m_p)^{1/2}$

The resonance frequencies are related to the local plasma frequency and the gyrofrequencies of electrons and ions (on very low frequencies) as given in Table 8.1. The main resonance spikes (Hagg *et al.* 1969; Muldrew 1972) are f_T, f_N, nf_H, f_{Q_n}, the floating spike, the diffuse resonance, the remote resonance, etc.

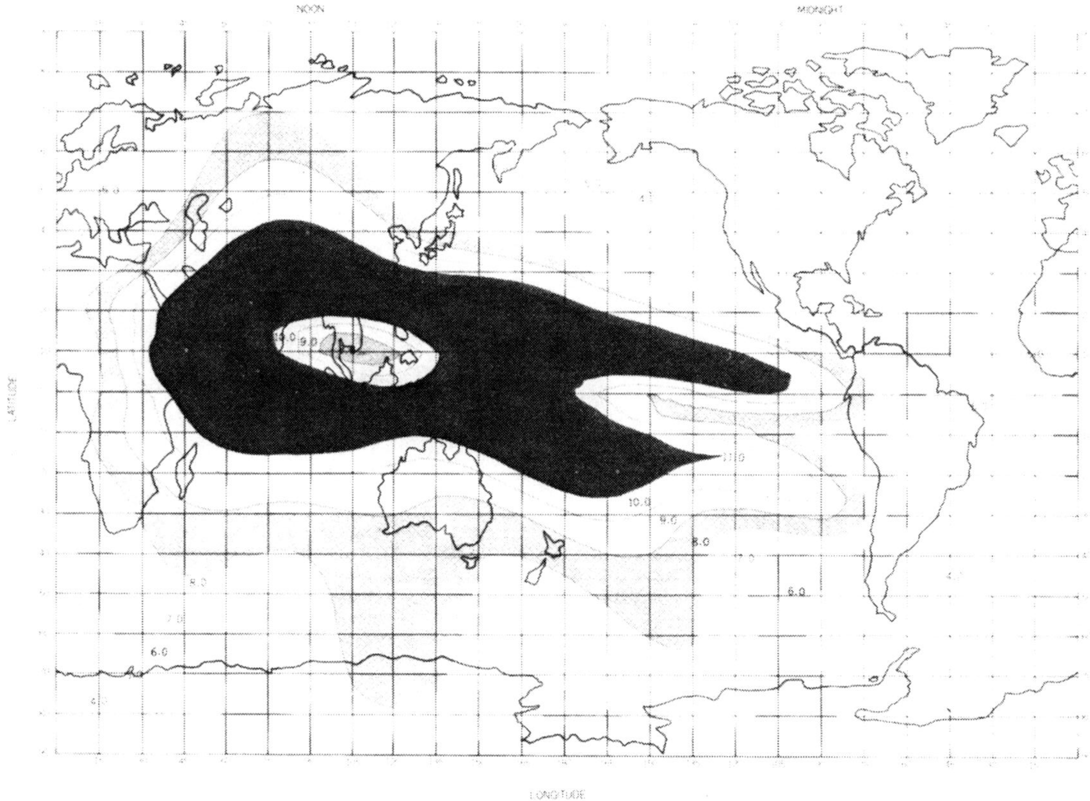

Fig. 8.10 *World map of the critical frequency of the F_2 layer obtained from the ISS-b topside soundings (From Matuura 1979 © Communications Research Laboratory, Japan; reprinted with permission)*
Because of the slow precession of the satellite orbit it takes about 4 months to obtain such a map

Here f_N is the plasma resonance, nf_H is the harmonic gyro-resonances ($n = 2, 3, \ldots$) and f_T is the upper hybrid resonance. Note that the fundamental gyrofrequency, f_H, resonance is not prominent. The diffuse resonance (Benson 1982) is confined between f_N and f_T. The so-called Q resonances are observed on frequencies higher than f_T and between the nf_H harmonics. The diffuse plasma resonances (Osherovich 1987) are related approximately by

$$f_{Dn} \approx 0{\cdot}95\,(f_N f_H)^{1/2} n^{1/2} \tag{8.3}$$

where $n = 1, 2, 3, \ldots$.

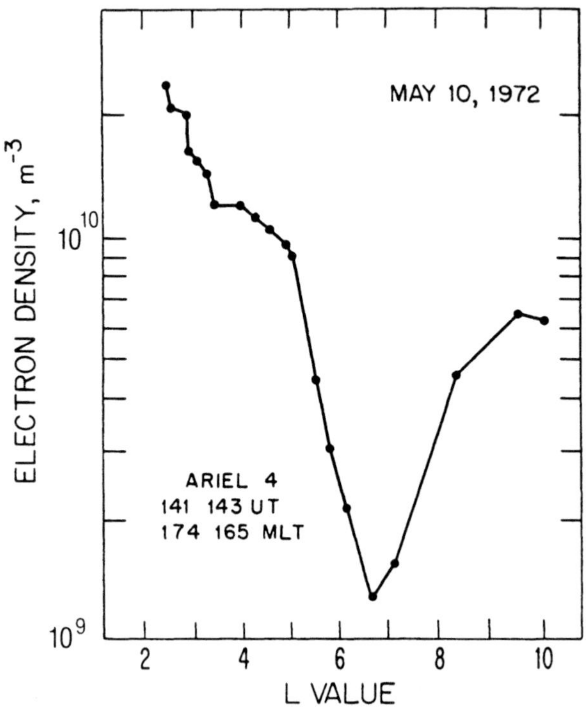

Fig. 8.11 *Electron density variation across a middle-latitude trough at about 500 km (From Grebowsky et al. 1976 © Pergamon Press; reprinted with permission)*

Very-low-frequency noise on Alouette I has been observed (Barrington 1969). It has a sharp lower frequency limit at the lower hybrid resonance (LHR) of the ambient plasma. For an electron–proton gas,

$$f_{\mathrm{LHR}} = \frac{f_N f_H}{f_T}\left(\frac{m_e}{m_p}\right)^{1/2} \tag{8.4}$$

where m_e and m_p are the electron and proton masses. At 1000 km, f_{LHR} is about 20 kHz.

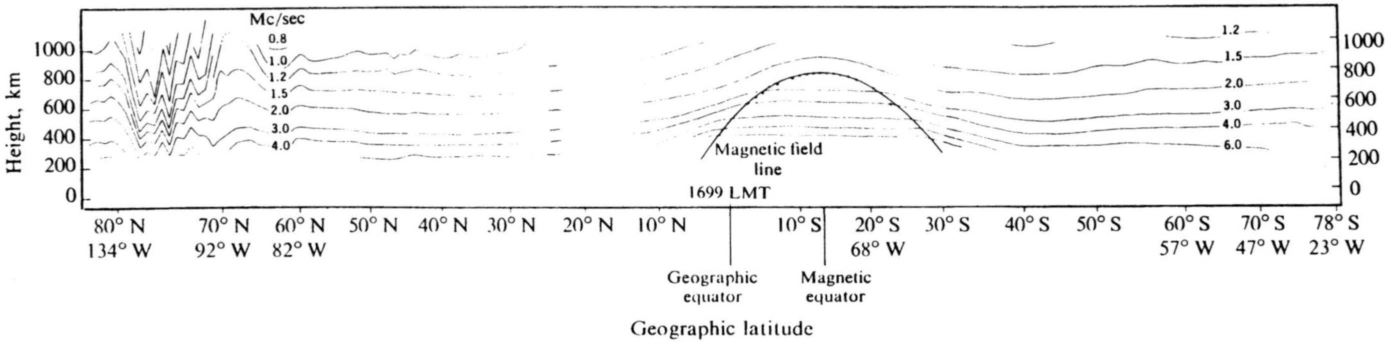

Fig. 8.12 *Contours of constant plasma frequency obtained from Alouette I soundings. (From Warren 1969 © IEEE; reprinted with permission)* Note the distortions near the magnetic equator and in high northern latitudes. November 12, 1962, 2121 UT.

8.2.5 Geophysical results from topside sounding

Besides the interesting radio discoveries made by topside sounders there have been many important geophysical discoveries, of which a few are mentioned here.

Polar-orbiting sounders make possible global measurements of certain ionospheric parameters (e.g. f_oF_2) which, however, are approximately at a fixed local time. Obtaining a global distribution depends on the slow precession of the orbit. For example, Fig. 8.10 shows a map of f_oF_2 for a given universal time, averaged over some 4 months, derived using spherical harmonic analysis (Sec. 5.3.3) of the ISS–b topside sounder (Matuura 1979). These maps provide valuable f_oF_2 data over ocean areas for testing the accuracy of maps based on conventional soundings on land (Rush *et al.* 1984).

Another important contribution was the identification of the ionospheric troughs (Muldrew 1983), which are zones of low electron densities lying between about 50° and 70° of geomagnetic latitude. The main trough lasts from about 1800 to about 0600 magnetic time, and its width is about 500 km to 1000 km (Timleck and Nelms 1969). The electron density inside the trough is drastically reduced (Fig. 8.11) by as much as a factor of 2 at 1000 km altitude and as much as an order of magnitude at the F_2 peak.

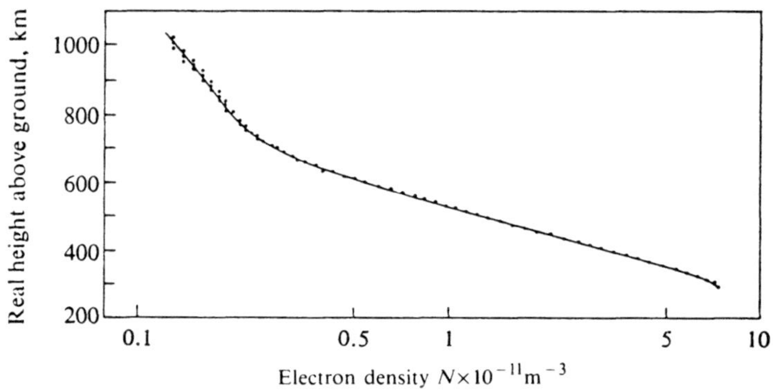

Fig. 8.13 *Electron density distribution in the topside of the ionosphere, for a point 9°N, 88°E recorded at Singapore, 1614 UT, October 1, 1962 (From King 1963 © Macmillan Magazines, Ltd.; reprinted with permission)*

Figure 8.12 shows contours of various plasma frequencies during a passage of Alouette. One of the more distinctive features is the domelike shape around the magnetic dip equator which is the topside version of the equatorial anomaly (Sec. 5.1.1). It develops after sunrise and becomes most evident at about 2000 local time. Note the appearance of a trough just below 70°N latitude and the highly irregular structure near 80°N, which is probably associated with the cleft where magnetospheric particles have direct access to the ionosphere. This region is also characterized by small-scale (tens of meters) irregularities, which produce aspect-sensitive echoes (spread F) on topside ionograms. In the nighttime auroral zone, small-scale irregularities are located above large-scale (tens of kilometers) irregularities (Muldrew 1983).

Topside profiles (Fig. 8.13) often show a sharp change of slope indicating a change in the scale height of the ionosphere. In general, the electron density in the topside decays exponentially with height as

$$N = N_0 \exp\left[-(h - h_0)/H_{pl}\right] \tag{8.5}$$

where H_{pl} is the plasma scale height. This change in scale height indicates a change from O^- to H^+ as the dominant ion. The plasma scale height differs from the neutral scale height because of the electric field E created by the tendency for the electrons to diffuse to great heights and for the heavy positive ions to settle to low heights [see (2.55)]. The plasma scale height is given by (2.50). The height at which the number densities of O^+ and H^+ are equal is called the transition height.

8.3 Radio beacons

8.3.1 Advantages of radio beacons

A radio beacon is a satellite-borne transmitter that can be used to measure propagation effects on radio signals traversing the ionosphere. The first radio beacon was the 20 MHz transmitter on Sputnik I. The motion of the satellite produced, at the ground, a characteristic fading, which was recognized as Faraday rotation or polarization fading with deep nulls (i.e. a full-wave rectified sine wave as seen in Fig. 8.14). Since Sputnik, many satellite telemetry transmitters have been used as radio beacons. Both low-orbit satellites (e.g. Navy Navigation Satellite System NNSS, S-66B, BE-B and BE-C, INTASAT) and geostationary satellites (e.g. ATS-1, ATS-3, ATS-6, San Marco, and ETS-2) have been used. The normal telemetry frequencies were about 137 MHz, but in recent years telemetry signals have been moved to the gigahertz range and are

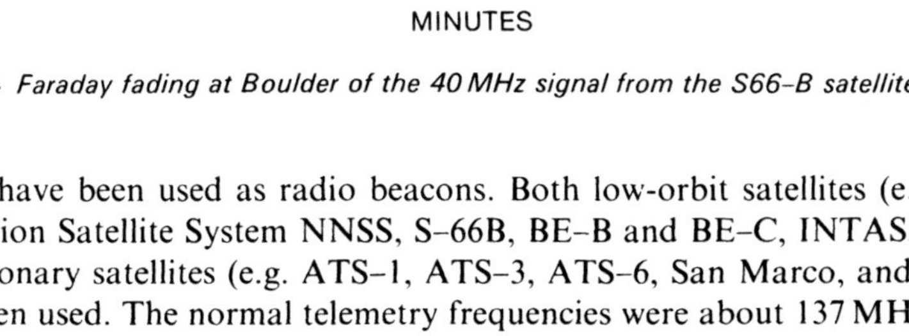

Fig. 8.14 *Faraday fading at Boulder of the 40 MHz signal from the S66-B satellite*

no longer suitable for Faraday rotation measurements. The greatest advantage of a radio beacon, and especially a geostationary beacon, is the simplicity of the receiving equipment. Since the source is provided, a simple linearly polarized antenna, a receiver, and a pen recorder suffice for the basic measurements. An arrangement for the measurement of Faraday rotation is called a polarimeter.

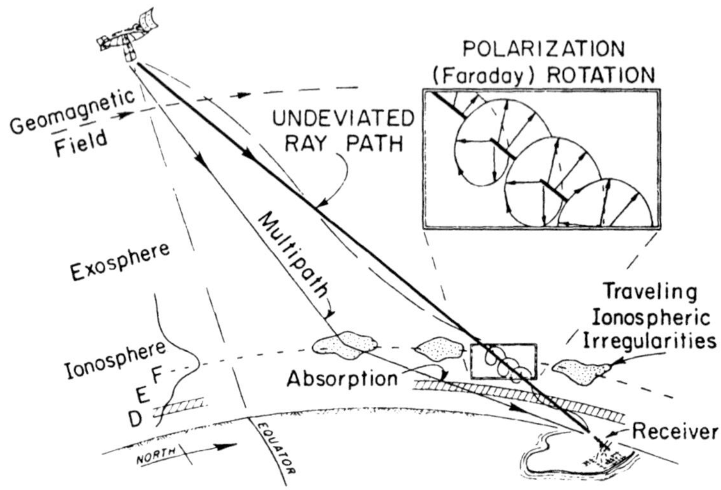

Fig. 8.15 *Major ionospheric propagation effects on satellite-to-ground links*
These include Faraday rotation of the electric field, group delay, scattering and multipath caused by irregularities, phase and amplitude fluctuations caused by traveling ionospheric disturbances, and absorption in the D region

In addition to experiencing Faraday rotation, satellite beacon signals scintillate because of the presence of ionospheric irregularities; hence, amplitude (and phase) measurements can be inverted to determine the occurrence and properties of irregularities. The main ionospheric effects on Earth-space propagation include Faraday rotation of the electric field, scattering by irregularities, absorption in the D region, and time delay (Fig. 8.15).

8.3.2 Faraday rotation

8.3.2.1 Case of negligible electron collisions

The rotation of the electric vector may be thought of as a result of the different characteristics of the ordinary and extraordnary waves. These waves have different wavelengths and phase speeds and opposite senses of rotation, as discussed in Sec. 3.4. Equation (3.24) shows that when a linearly polarized VHF wave is emitted from a satellite ($f \gg f_N$, $f \gg f_H$), $R = \pm i$; the waves are circularly polarized and the emitted power is shared equally by the two waves.

Consider the case of propagation parallel to the magnetic field (Fig. 8.16). In going a distance s from P to Q the o- and x-waves rotate in opposite senses; the resultant rotation Ω is given by

$$\Omega = \tfrac{1}{2}(\Omega_+ - \Omega_-) = \frac{\pi s}{\lambda_0}(\mu_+ - \mu_-) \tag{8.6}$$

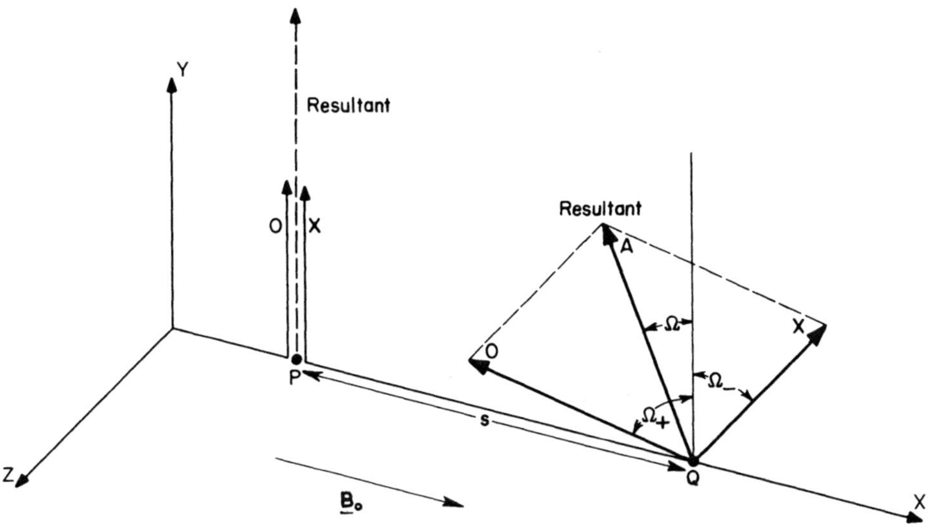

Fig. 8.16 *Faraday rotation as the combination of ordinary and extraordinary waves of equal amplitude propagating parallel to the imposed magnetic field*

where λ_0 is the free space wavelength and μ_+ and μ_- are the refractive indices of the o- and x-waves, respectively. On VHF, $\mu_+ \approx \mu_- \approx 1$ and (3.9) gives

$$\mu_+ - \mu_- \approx \frac{X}{2}\frac{[Y_T^4 + 4(1 - X)^2\, Y_L^2]^{1/2}}{(1 - X)(1 - Y_L^2) - Y_T^2} \tag{8.7}$$

For further discussion we simplify this. When $X \ll 1$, Y_L and $Y_T \ll 1$

$$\Omega = \frac{\pi s}{c}\frac{f_N^2 f_L}{f^2} = 2{\cdot}365 \times 10^4 f^{-2}\, s\, NB_0 \cos\theta \tag{8.8}$$

where N is the electron density (assumed constant along the path s). The quantity Ns is the total electron content in a cylinder of 1 m^2 cross-section and length s. The Faraday rotation is proportional to (i) the total electron content (TEC), (ii) the magnetic field strength and (iii) the inverse frequency squared. Note that (8.8) applies only to quasi-longitudinal propagation which, on a frequency of 138 MHz, holds at angles θ to $> 89°$.

To an observer looking along the direction of propagation, and along the field, the tip of the resultant electric vector traces out a helix with a right-hand sense of rotation (the thumb being in the direction of the magnetic field). Hence, to a fixed observer looking up the magnetic field in the Northern Hemisphere, an increase in electron content produces a counterclockwise rotation of the electric field (Fig. 8.17). To an observer looking up the magnetic field in the Southern Hemisphere an increase in *TEC* will produce a clockwise rotation.

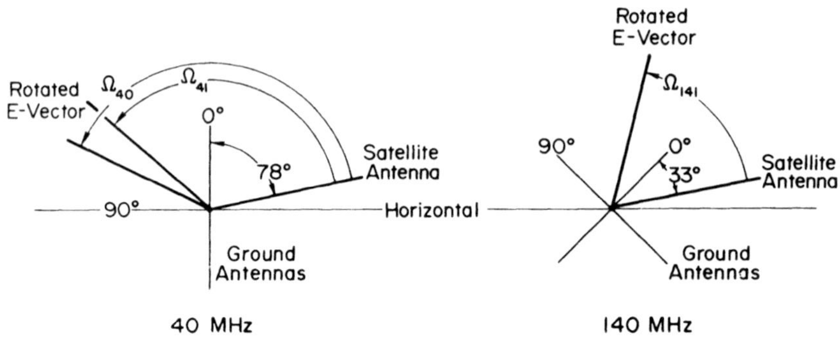

Fig. 8.17 *Faraday rotation, as seen from Boulder to ATS-6*
Note that the rotation increases with increase of electron content in a right-handed sense with the right thumb pointing along the geomagnetic field (i.e. counterclockwise)

Equation (8.8) can be approximated by

$$\Omega = 2{\cdot}365 \times 10^4 f^{-2} \bar{M} N_F. \tag{8.9a}$$

N_F is the Faraday electron content, and

$$\bar{M} = \overline{B_0 \cos\theta \sec\chi} \tag{8.9b}$$

is an average value of $B_0 \cos\theta \sec\chi$ along the path where χ is the angle between the ray path and the vertical. With a geostationary beacon, $\bar{M}$ can be taken as the value of M at 420 km (Titheridge 1972).

Sample values of Faraday rotation for various frequencies and *TEC*s are given in Fig. 8.18. Note that daytime values of *TEC* are of the order of 10^{17}, so Faraday rotation on 137 MHz is of the order of 10 radians (1·6 cycles), and on 40 MHz a typical value is 120 radians or about 20 rotations. To obtain the total Faraday rotation we need to know (i) the initial direction ϕ_0 of the electric field at the satellite, and (ii) the number n of full rotations (cycle ambiguity or ramp number). From the orthogonal amplitudes X and Y the measured angle is given by

$$\phi = 2\Omega = \phi_0 + 2n\pi + \tan^{-1}(Y/X) \tag{8.10}$$

The integer n has to be determined by inspection (e.g. continuity of rotation) and ϕ_0 must be known from the satellite orientation or by other means such as knowledge of the f_oF_2 and slab thickness τ, which is defined as

$$\tau = N_F/N_mF_2 \tag{8.11}$$

A typical value of slab thickness is 300 km.

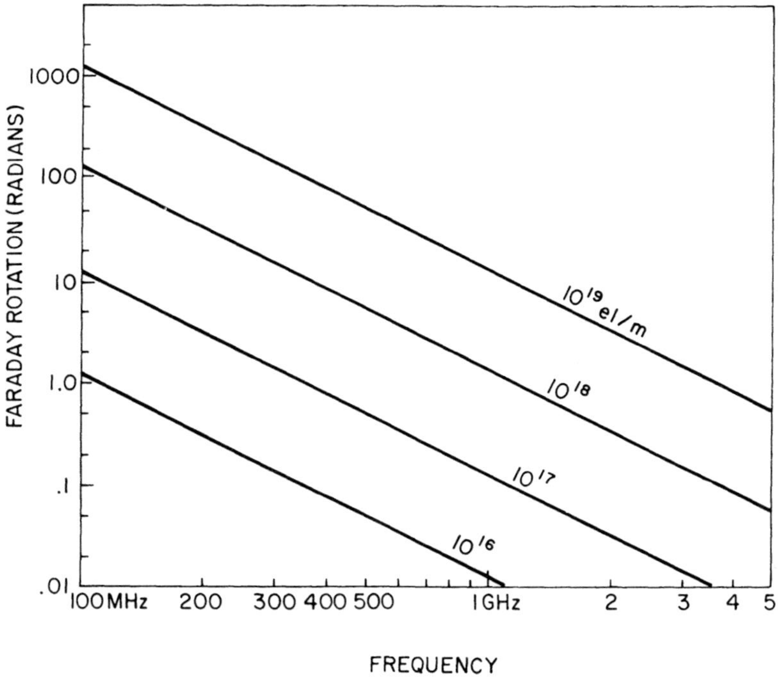

Fig. 8.18 *Approximate Faraday rotation as a function of frequency and TEC for a ray path in a magnetic meridian (From Klobuchar 1978, with permission)*

8.3.2.2 Effect of electron collisions

The main effect of collisions is to produce differential absorption of the ordinary and extraordinary waves and thus change the resultant polarization from linear to elliptical. At a fixed point on the ray path, the wave is elliptically polarized. The rotation of the major axis of the ellipse is given by

$$\Omega = (\pi s/\lambda_0)\ \mathrm{Re}\ (n_+ - n_-)$$

where n_+ and n_- are the complex refractive indices of the *o*- and *x*-waves respectively. For propagation parallel to the magnetic field ($\theta = 0°$), substitution from (3.8) gives

$$\Omega = \frac{\pi s}{\lambda_0}\frac{XY(1 - Z^2 - Y^2)}{(1 - Z^2 - Y^2)^2 + 4Z^2} \tag{8.12}$$

This formula is useful for calculating Faraday rotation in the D region. It is seen that there is no Faraday rotation at the level where $\omega^2 + \nu^2 = \omega_H^2$.

8.3.3 Differential phase

Differential phase (also called dispersive Doppler, carrier phase) measures the time delay by comparing the phases (ϕ_1 and ϕ_2) of two signals on widely separated frequencies (f_1 and f_2) when they are translated to a common reference frequency f_0. On very high frequencies the phases of received signals are obtained from (7.57) and (3.11):

$$\phi_1 \lambda_1 = \int \mu_1 \, ds = \int ds - \frac{40 \cdot 3}{f_1^2} \int N \, ds = S_1 - \frac{40 \cdot 3}{q_1^2 f_0^2} N_T$$

and

$$\phi_2 \lambda_2 = \int \mu_2 \, ds = \int ds - \frac{40 \cdot 3}{f_2^2} \int N \, ds = S_2 - \frac{40 \cdot 3}{q_2^2 f_0^2} N_T$$

where q_1 and q_2 are integers relating f_0 to f_1 and f_2, and λ_1 and λ_2 are the free space wavelengths. On VHF the refraction is small, so we assume that the real paths S_1 and S_2 are equal and, hence, the phase difference $\Delta\phi$ measured on the frequency f_0 is given by

$$\Delta\phi = \frac{\phi_2}{q_2} - \frac{\phi_1}{q_1} = \frac{40 \cdot 3 N_T}{f_0 c} \left(\frac{1}{q_1^2} - \frac{1}{q_2^2} \right) \text{ in cycles} \tag{8.13a}$$

That is,

$$\Delta\phi = \frac{80 \cdot 6\pi \, N_T}{f_0 c} \left(\frac{1}{q_1^2} - \frac{1}{q_2^2} \right) = \frac{8 \cdot 447 \times 10^{-7} \, N_T}{f_0} \left(\frac{1}{q_1^2} - \frac{1}{q_2^2} \right) \text{ in radians} \tag{8.13b}$$

For example, in a beacon with frequencies of 150 MHz and 400 MHz we could have $f_0 = 50$ MHz, $q_1 = 3$, and $q_2 = 8$. One big advantage of this method is that the effect of satellite motion is eliminated by taking the difference of the phases on the reference frequency.

8.3.4 Modulation phase

In this method (also called differential group delay and phase dispersion) the phase of the modulation (of frequency Δf) on a frequency f_1 is compared with the same modulation on a (much higher) frequency f_2. The modulation phase ϕ_m is given by

$$\phi_m = \frac{80 \cdot 6\pi}{c} \Delta f \left(\frac{1}{f_1^2} - \frac{1}{f_2^2} \right) N_T = 8 \cdot 447 \times 10^{-7} \, \Delta f \left(\frac{1}{f_1^2} - \frac{1}{f_2^2} \right) N_T \tag{8.14}$$

When $f_2^2 \gg f_1^2$

$$\phi_m \approx 8{\cdot}447 \times 10^{-7}\,\Delta f/f_1^2 \tag{8.15}$$

8.3.5 Excess time delay

The differential phase and modulation phase both give the time delay T of the signal over and above that for free space:

$$T = \left(\frac{40{\cdot}3\,N_T}{cf^2}\right) = 1{\cdot}343 \times 10^{-7}\,N_T/f^2 \tag{8.16}$$

where T is in seconds, f is in hertz, and N_T is in electrons m^{-2}. A plot of excess time delay versus frequency is shown in Fig. 8.19, from which we see that on a frequency of 100 MHz and with a $TEC = 10^{17}$ electrons m^{-2} the excess time delay is $1{\cdot}34\,\mu s$. Multiplying T by the free space speed (c) gives the range error.

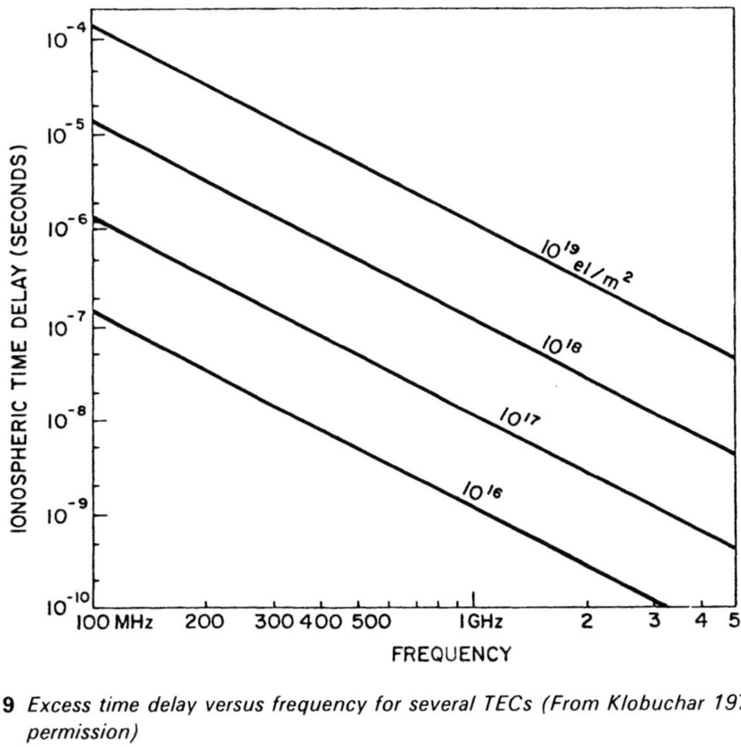

Fig. 8.19 *Excess time delay versus frequency for several TECs (From Klobuchar 1978, with permission)*

8.3.6 Refraction

Figure 8.20 illustrates the refraction of a ray in traversing the ionosphere. The angular error ξ is related to the range error $\rho(=cT)$ by

$$\xi \approx \frac{(R + a\sin\Delta)\,a\cos\Delta}{h_i(2a + h_i) + a^2\sin^2\Delta}\,\frac{\rho}{R} \approx \frac{(R + a\sin\Delta)\cos\Delta}{2h_i + a\sin^2\Delta}\,\frac{\rho}{R} \tag{8.17}$$

where Δ is the true elevation angle, a is the Earth radius, R is the range, and h_i is the height of the centroid of the electron-density profile (Klobuchar 1978), generally between 300 km and 450 km. Substitution of typical values

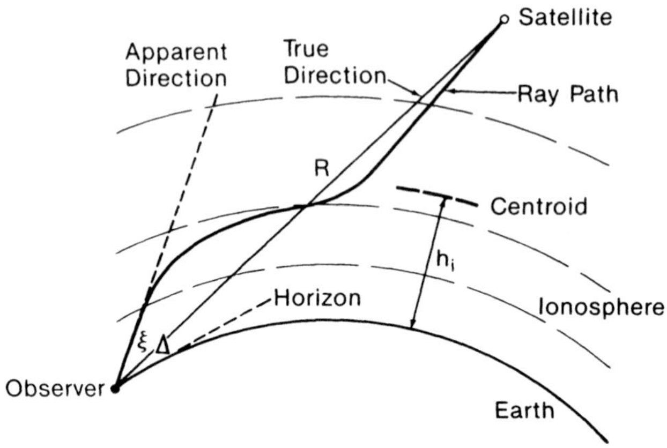

Fig. 8.20 *Ray path from a satellite to ground, showing refractive bending and angular deviation*

(e.g. f = 100 MHz, $N_T = 10^{18}$ electrons m^{-2}, h_i = 400 km, $\Delta = 45°$, and R = 1000 km gives ξ = 5 mrad or 0·3°.

When Δ is small,

$$\xi = \rho \cos \Delta / 2h_i \tag{8.18}$$

and when Δ is large, and R is very much less than $a \sin \Delta$,

$$\xi = \rho \cot \Delta / R \tag{8.19}$$

On 1·6 GHz, with an elevation of 5° and with $N_T = 10^{19}$ electrons m^{-2}, a typical value of ξ is 0·3 mrad; on 400 MHz, ξ is approximately 4 mrad.

In the presence of a horizontal gradient in *TEC*, and for angles of incidence near the zenith, the wedge refraction ζ (Smith 1952) is given by

$$\zeta = 40{\cdot}3\, f^{-2}\, dN_T/dx \tag{8.20}$$

8.3.7 Doppler shift

The Doppler shift is made up of two parts: (i) a part due to the motion of the satellite with respect to the receiver, and (ii) a part due to the rate of change of the total electron content along the path. With a low-orbiting satellite we have

$$\Delta f = f\mu_s v_l / c + \frac{40{\cdot}3}{cf}\frac{dN_T}{dt} \tag{8.21}$$

where μ_s is the refractive index at the satellite and v_l is the line-of-sight component of the satellite velocity. With a geostationary satellite the first term is small and the Doppler shift, caused by temporal changes in electron content such as the drift of ionospheric irregularities across the ray path, is

$$\Delta f = d\phi/dt = 1{\cdot}34 \times 10^{-7} f^{-1}\, dN_T/dt \tag{8.22}$$

This frequency shift is generally small. For example: An upper limit of the rate of change of *TEC* of 0·1 × 10^{16} electrons $m^{-2}s^{-1}$, gives a shift of less than 0·01 Hz on 1·6 GHz and about 3·3 Hz on 40 MHz.

8.3.8 Problems associated with TEC measurements

In the measurement of the absolute value of *TEC* there are two major unknowns: (i) the zero value of the Faraday rotation, and (ii) the number of cycles of rotation (also called cycle ambiguity or ramp number) (Lawrence *et al.* 1964; Garriott *et al.* 1970).

In the case of Faraday rotation the polarization of the emitted signal is unknown except for a few specially designed beacons such as the ATS–6 radio beacon (Davies 1980). Various indirect means have been used such as the simultaneous observations of satellite and moon-reflected signals (Klobuchar 1975). The cycle (or ramp) number can be large (e.g. 50 – 100 half-rotations on 40 MHz), so keeping track is a formidable task when the rotation is changing rapidly as it does after dawn in the winter in middle latitudes. Computer control is desirable but, in the presence of rapid and deep scintillation, loss of lock is common.

One way of estimating the zero calibration is to measure concurrently the total Faraday rotation (or phase) and the critical frequency at night in winter when the number of rotations is zero or a very few. From an estimate of the slab thickness τ we have for ϕ, from (8.9), (8.11), and (8.12):

$$\phi = 5{\cdot}865 \times 10^2 f^{-2} \tau\, (f_oF_2)^2 = \phi_0 + 2n\pi + \tan^{-1}(Y/X) \tag{8.23}$$

where f_oF_2 and f are in hertz and τ is in meters.

The number of cycles can be evaluated by use of the differential rotation $\delta\phi$ (or phase) between two closely spaced frequencies f and $f + \delta f$:

$$\frac{\delta\phi}{\phi} = \frac{\delta\Omega}{\Omega} = -2\frac{\delta f}{f} \tag{8.24}$$

The differential rotation on frequencies of 40·0 MHz and 40·1 MHz corresponding to 75 cycles on 40 MHz is about 0·4 cycles, which is good enough to resolve the ambiguity but is not an accurate measurement of the rotation. Note that the modulation phase technique of Sec. 8.3.4 is of this differential type (Davies 1980). Since Faraday rotation and group delay vary approximately as the inverse square of the wave frequency, a low frequency gives high sensitivity. On the other hand, a low frequency aggravates the cycle problem and, by day, leads to errors caused by ionospheric refraction. Other problems that arise include (i) the group delay in the transmitting and receiving systems, and (ii) ground reflections (Davies *et al.* 1977; Rees *et al.* 1977; Bhonsle *et al.* 1977).

A difficulty arises if and when the ray path passes through the condition of being perpendicular to the magnetic field. In this case the longitudinal components changes direction so the Faraday rotation on one part of the path cancels that on the other part.

8.4 Spatial and temporal structures of total electron content

8.4.1 Total content, Faraday content, and protonospheric content

Before we discuss the geographical and temporal variations of total electron content it is important to distinguish between the true content N_T and the Faraday, or ionospheric, content N_F. This is especially important in the case of a geostationary beacon. First we note that with the phase techniques, the effect of the geomagnetic field is a second-order (or less) effect whereas Faraday rotation is affected directly by the magnitude and direction of the geomagnetic field. Now the geomagnetic field decays approximately as the inverse cube of the radial distance (Sec. 2.4.1), so electrons at high altitudes (e.g. greater than about 2000 km) contribute little or nothing to the Faraday rotation and, conversely, are not detected by the Faraday rotation technique. Calculations with a wide variety of electron-density profiles and geostationary-satellite-to-ground ray paths (Titheridge 1972) show that the Faraday rotation technique measures the content up to an altitude of some 2000 km. This content is called the Faraday content (N_F) or the ionospheric content. The phase techniques measure the total electron content N_T up to the source. The difference between the total content and the Faraday content gives the protonospheric, or plasmaspheric, content N_p from about 2000 km to the satellite; that is,

$$N_p = N_T - N_F = N_T(1 - F/\bar{F}) \tag{8.25}$$

where F and $\bar{F}$ are shape factors given by

$$F\,N_T = \int_R^S f_L N ds = \bar{F} N_F \tag{8.26}$$

(Davies *et al.* 1976; Hargreaves 1978). Because of the simplicity of the Faraday method most *TEC* measurements refer to Faraday content.

8.4.2 Long-term variations

TEC follows the sunspot cycle. For example, at Sagamore Hill (Massachusetts) the Faraday content varies approximately linearly with R_{12}, the 12-month running average sunspot number (Huang 1978). That is,

$$N_F = a(1 + bR_{12}) \tag{8.27}$$

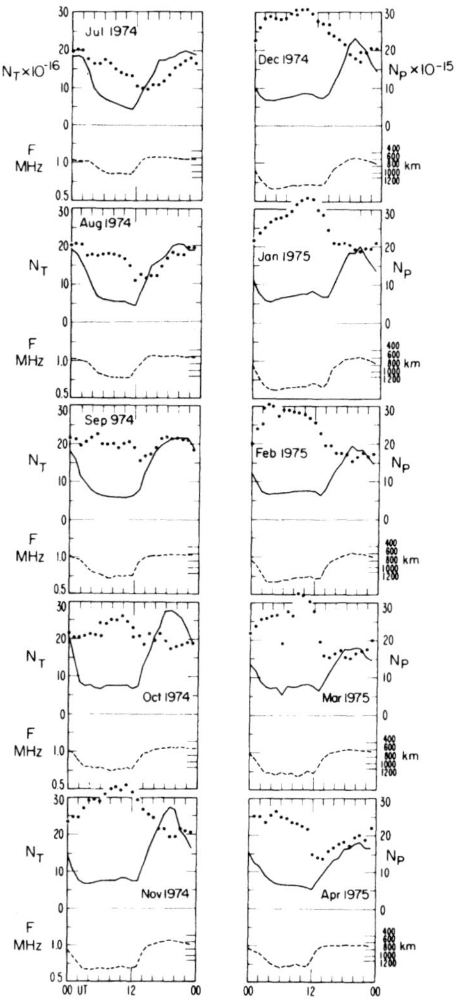

Fig. 8.21 *Monthly median hourly values of total slant electron content N_T (——), slant Faraday content N_F (– – –), and the slant plasmaspheric content N_P (.), July 1974 through May 1975, ATS-6 to Boulder*
The kilometer scale indicates the height at which the geomagnetic field strength corresponds to the value of the shape factor *F*
LT = UT − 7 hours

Both *a* and *b* vary diurnally; *b* is about 0·10 at night and about 0·20 during the day. The *TEC* is about 5 × 10^{16} electrons m^{-2} at night at low sunspot numbers and about 30 × 10^{16} electrons m^{-2} during the day at R_{12} = 100. There is a marked seasonal variation with peaks in February–March and October–November. There is a seasonal variation of the slab thickness (Evans 1977) expressed by

$$\tau = C(1 + 0{\cdot}005R_{12}) \tag{8.28}$$

where C = 270 km, 240 km, and 210 km in summer, equinox, and winter respectively.

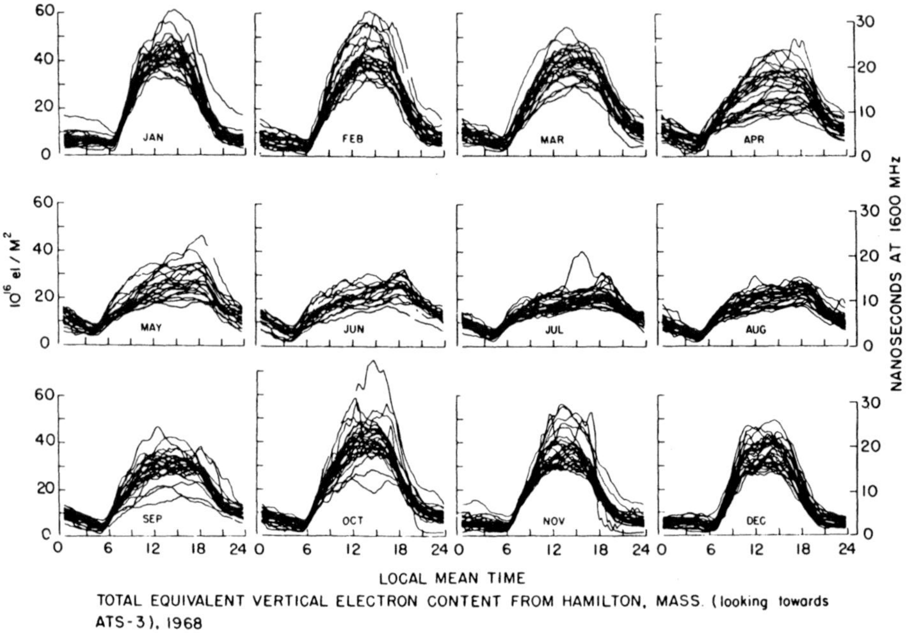

Fig. 8.22 *Day-to-day variations of TEC in middle latitudes (From Klobuchar 1978, with permission)*

8.4.3 Diurnal variations

The diurnal variation of total electron content at Boulder at sunspot minimum is similar to that of N_mF_2. There is a maximum during the day and a minimum just before sunrise (Fig. 8.21). The most noticeable characteristic of total content is the large day-to-day fluctuation (Fig. 8.22). Note the seasonal changes: The lowest *TEC* is in summer, and maxima are at the equinoxes. The distribution of the hourly *TEC* about the monthly median is approximately normal or Gaussian; the standard deviation is about 20% during the day and about 25% at night (Klobuchar 1978).

8.4.4 Geographical variations

Figure 8.23 shows the world distribution of excess time delay, which is proportional to *TEC*. The distribution of *TEC* is similar to that of f_oF_2 (Sec. 5.3.3) with

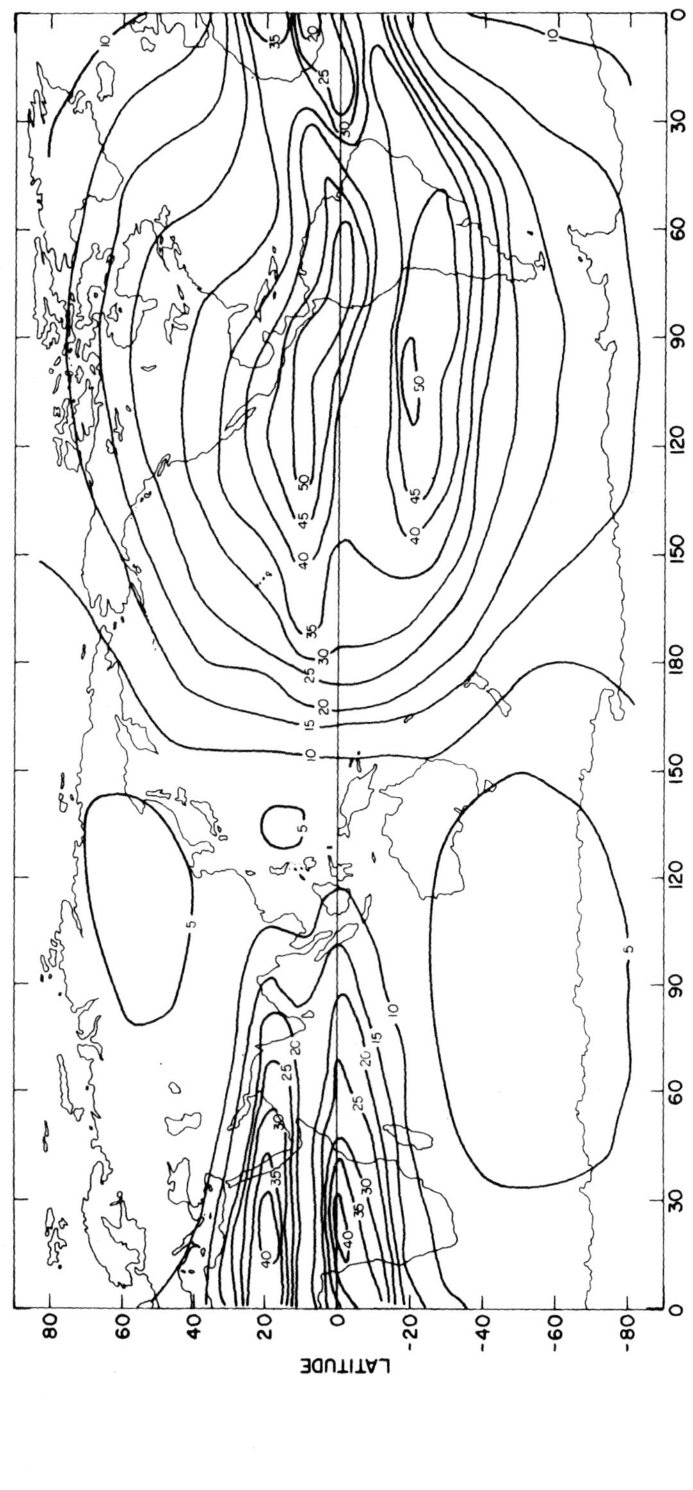

Fig. 8.23 *World contours of excess time delay on 1·575 GHz in nanoseconds for 2000 UT March 1980, $F_{10.7}$ = 167 (Courtesy J. A. Klobuchar)*

regard to the presence of an equatorial anomaly. Contour maps of *TEC* have been constructed on a regional scale by combining geostationary satellite data (which give high time resolution) with polar-orbiting-satellite data (which give latitudinal resolution). The example in Fig. 8.24 shows the presence of large-scale features such as troughs, ridges, and holes in the *TEC* structure.

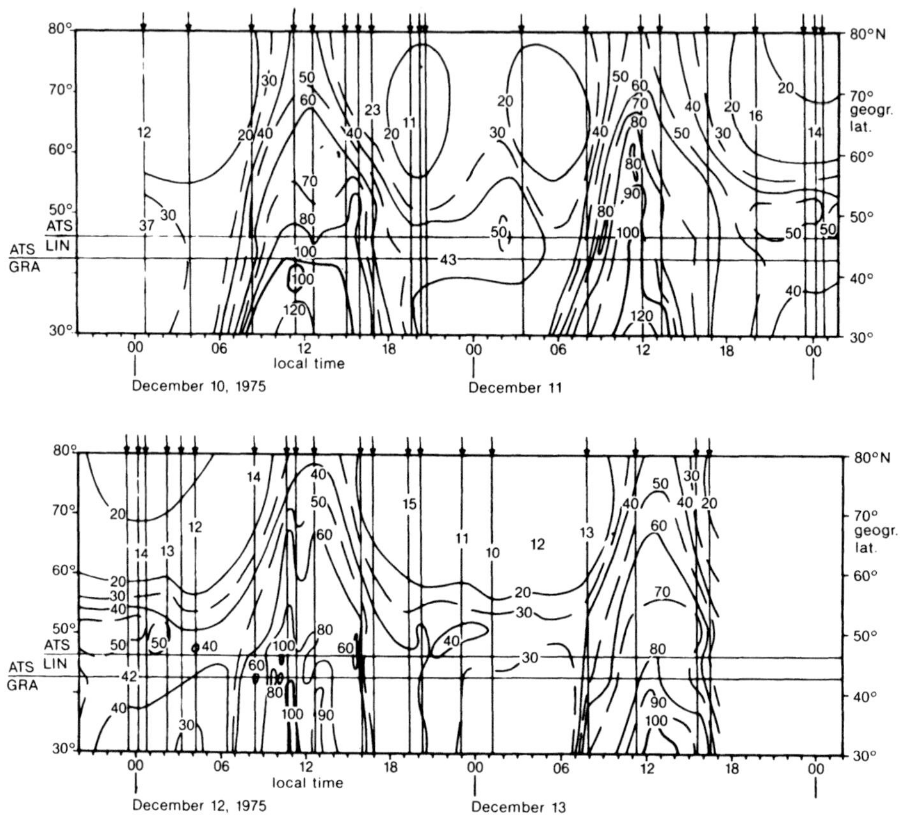

Fig. 8.24 *Contours of total vertical electron content from 30°N to 80°N geographic latitude over Europe in units of 10^{15} electrons m^{-2}, December 10–13, 1975*
The maps are constructed by combining data from the ATS-6 geostationary satellite and the polar-orbiting NNSS satellites.

8.4.5 Modeling TEC

For system design purposes a global model of *TEC* is useful for estimating the time delay, refraction, etc. Empirical models are available such as the numerical maps of $f_o F_2$ (Sec. 5.2.4) that give $N_m F_2$. The slab thickness can be obtained from the Bent *et al.* (1975) model (Sec. 5.3.4) in which the shape of the topside was defined by data from Alouette ionograms. Klobuchar (1987) constructed a simple algorithm designed to remove approximately 50% of the root-mean-square ionospheric time delay error for single-frequency users of the GPS advanced navigation system.

8.4.6 TEC near the magnetic equator

Faraday measurements at and near the magnetic equator are difficult to interpret because the propagation is nearly perpendicular to the geomagnetic field and the propagation varies rapidly with the angle of propagation. Further, with transverse propagation there is no true Faraday rotation (see Lawrence *et al.* 1964; Davies 1969, Sec. 9.4.3). Measurements of *TEC*, from time delay, are shown in Fig. 8.25 for sunspot minimum. Note the deep pre-sunrise low and the secondary maximum in the late afternoon (Space Science Symposium 1977).

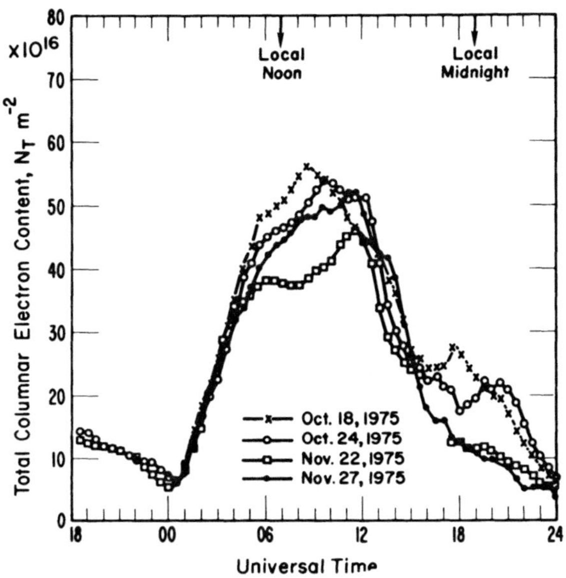

Fig. 8.25 *Examples of the diurnal variations of the total electron content (slant path) for the ATS-6-to-Ootacamund path*
LT = UT + 5 hours

8.4.7 Vertical distribution

In a Chapman layer with scale height H, the layer thickness is given by

$$\tau = 4{\cdot}133H \tag{8.29}$$

In such a layer the ratio of topside content N_a to bottomside content N_b is

$$N_a/N_b = 2{\cdot}15 \tag{8.30}$$

In practice this ratio is greater than 3, largely because of the transition from O^+ ions to H^+ ions in the topside. This is illustrated in the profiles of Fig. 8.26. The height at which the number densities of O^+ and H^+ are equal is called the transition height (h_T). Near sunspot minimum the transition height varies from about 4000 km during a summer day to about 800 km during winter nights. Above the transition height the electron density falls off exponentially up to the plasmapause which, under quiet conditions, occurs at an L-shell of about 6, that is, at about the altitude of a geostationary satellite (see Sec. 2.4.1). At the plasmapause the electron density falls by a factor of 10 to 100.

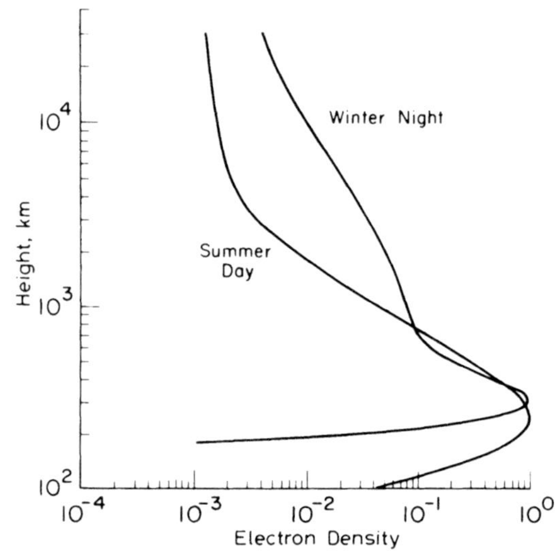

Fig. 8.26 *Electron density profiles fitted to ATS-6-to-Boulder (phase and Faraday) data for average conditions for a summer day (1974) and a winter night (1974)*
The profiles are normalized at the peak. Note the change of slope in the topside at the transition height.

8.5 Scintillation

A radio signal traversing the ionosphere will be modified by irregularities with sizes from a few (3) meters to a few kilometers. Scintillations are variations of amplitude, phase, polarization, and angle of arrival. Scintillations can be severe and present problems to the radio operator. Transionospheric effects of scintillation have been observed on frequencies ranging from 20 MHz to about 10 GHz. Scintillations are particularly severe in tropical regions during evenings around the equinoxes and in the auroral zones (Aarons 1977).

8.5.1 Amplitude and phase scintillations

Amplitude scintillations can be characterized by a depth of fading and a fading period or frequency. A widely used index, S_4, is the root-mean-square of the power P divided by the average power $\bar{P}$ (Briggs and Parkin 1963):

$$S_4 = \frac{[\Sigma(P - \bar{P})^2]^{1/2}}{\bar{P}} \tag{8.31}$$

For simplicity of scaling, a scintillation index SI has been defined:

$$SI = \frac{P_{max} - P_{min}}{P_{max} + P_{min}} \tag{8.32}$$

(Whitney *et al.* 1969). An example of amplitude scintillation is shown in Fig. 8.27.

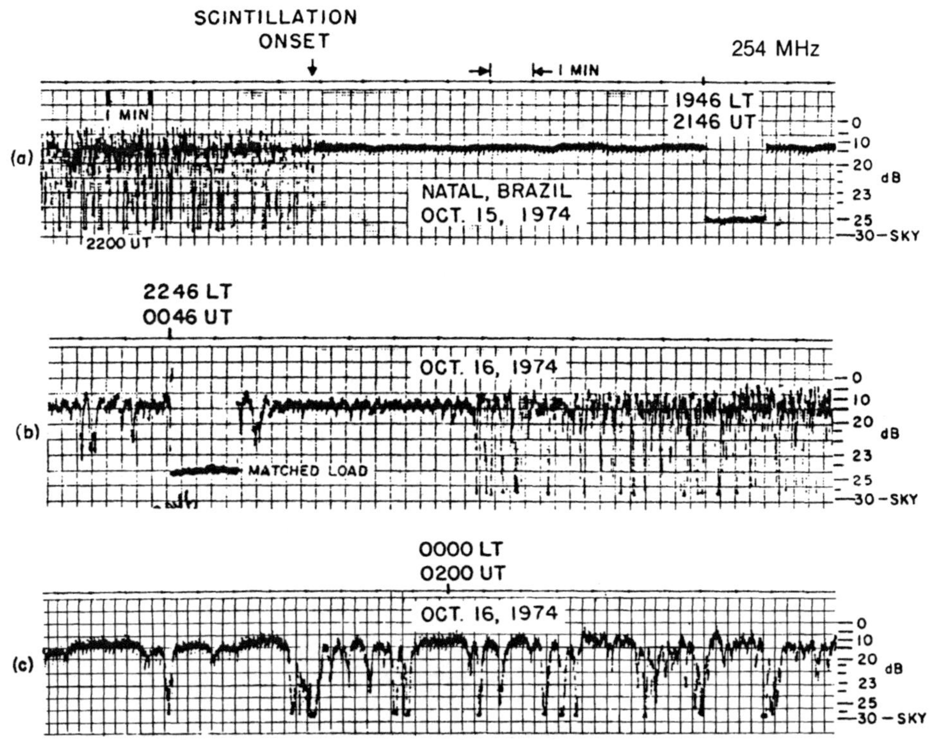

Fig. 8.27 *Observations recorded at Natal, Brazil, of the 254 MHz transmissions of LES–6 on October 15–16, 1974 (From Aarons 1977 © IEEE; reprinted with permission)*
(*a*) Sudden scintillation onset characterized by rapid deep fading
(*b*) Scintillation patches interspersed with sections of low-level fading
(*c*) End of scintillation activity, with long fades below fade margin of 6 dB for 5–30 s

Because there is a spectrum of sizes of ionospheric irregularities there is a spectrum of amplitude scintillation. To a first approximation the fading spectrum follows a power law (Rufenach 1975; Crane 1978; Rino 1976; Ishimaru 1975) as seen in Fig. 8.28. With a geostationary satellite and radio stars, the scintillation spectrum exhibits a 'corner' that represents the change from scattering to specular propagation. The corner frequency ν_f depends on the Fresnel radius and the drift speed of the irregularities. On occasion there are (Fresnel) oscillations in the spectrum in which minima occur at $\sqrt{2}$, $\sqrt{3}$, and $\sqrt{4}$ multiples of the frequency of the first minimum (Fig. 8.29). On frequencies greater than ν_f the scintillation power $P(\nu)$ can be represented by

$$P(\nu) = A\nu^{-p} \tag{8.33}$$

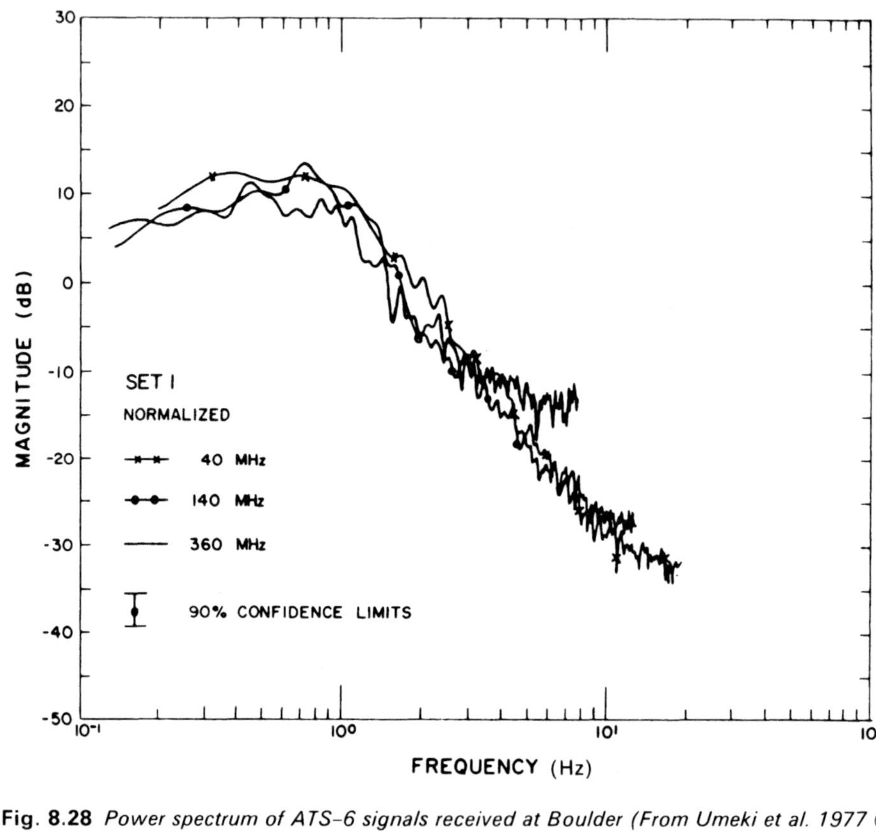

Fig. 8.28 *Power spectrum of ATS-6 signals received at Boulder (From Umeki et al. 1977 © American Geophysical Union)*

where ν is the scintillation frequency, A is a proportionality factor, and p varies between about 1 and about 4. The scintillation index S_4 varies as (Umeki *et al.* 1977)

$$S_4 = \alpha f^{-n} \tag{8.34}$$

where α is a proportionality factor and n varies between about 1·6 for low S_4 and about 0·8 for high S_4. The exponent n is related to p by

$$n = (2 - p)/4 \tag{8.35}$$

The frequency exponent for S_4 holds only for $S_4 < 0{\cdot}5$. With strong scintillation ($S_4 > 0{\cdot}5$) these relationships do not apply.

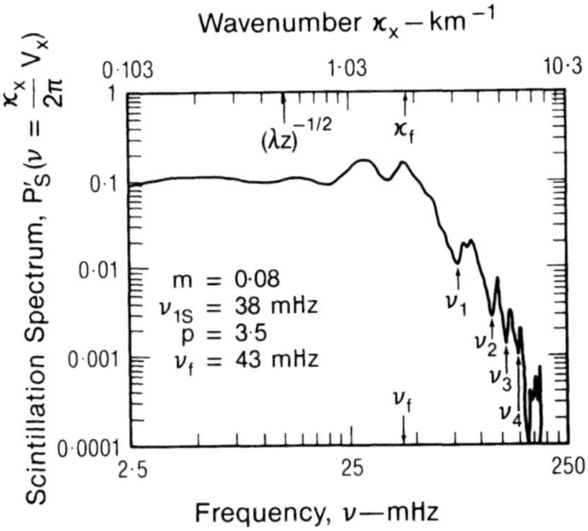

Fig. 8.29 *Power spectrum with Fresnel oscillations, Cygnus A to Boulder, July 21, 1970; 26 MHz (From Rufenach 1972)*
Here p is the spectral index and m is the scintillation index.

Phase scintillations are measured by the root-mean-square fluctuation of phase ϕ. The frequency dependence of ϕ_{rms} is given approximately by

$$\phi_{rms} = Af^{-1} \tag{8.36}$$

where A is a proportionality factor. Both S_4 and ϕ_{rms} depend on sunspot number R_{12} as $1 + 0{\cdot}04R_{12}$ (Singleton 1980). Fluctuations occur also in angle of arrival and Faraday rotation.

8.5.2 Equatorial scintillations

The picture of global scintillation in Fig. 8.30 shows the important, equatorial and auroral/polar, regions. The equatorial scintillations are, on the whole, stronger than those in high latitudes. Scintillations occur in middle latitudes but they are less intense than elsewhere and do not present a serious problem. At night, scintillations are predominantly an F-layer phenomenon with irregularities covering the height range from 200 km to 600 km. However, daytime scintillation is produced by sporadic E (Rastogi *et al.* 1977).

In the equatorial region, the development of scintillation can be abrupt (Fig. 8.27), sometimes occurring within a single fade. Fading of the order of

20 dB below the mean is observed. Rapid and deep fading characterizes the beginning of the activity. As the night progresses, the scintillation may become more sporadic with intervals of shallow fading. The small-scale irregularities are contained in large-scale patches that move across the ray path. Toward the end of the scintillation activity, the scintillations change character in which long fades may stay below fade margins of 6 dB for 5 to 10 s.

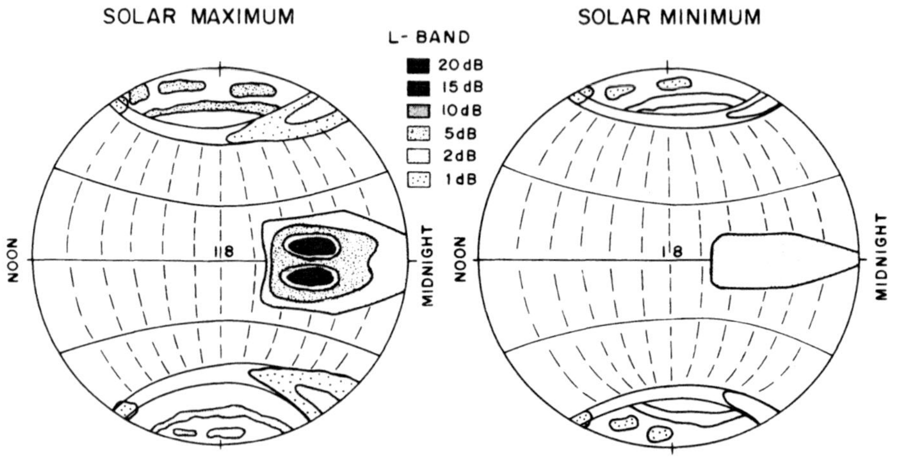

Fig. 8.30 *'Worst-case' fading depths of scintillation at L-band during low and high sunspot numbers (From Basu et al. 1988 © American Geophysical Union)*

Scintillation patterns vary greatly; in a patchy pattern scintillation starts and stops several times in one night, as the patch moves through the ray path in times of the order of 20 min to several hours. Similar lifetimes of scintillations have been noted in HF transequatorial propagation (Röttger 1973). In the equatorial region, there are large-scale irregularity patches (several hundreds of kilometers E–W and many times that in the N–S direction) that contain small-scale irregularities, some as small as 1 m, which produce VHF and microwave scintillation (Basu and Basu 1981).

Observations during solar maximum (1979–81) show that the scintillation maximizes at the peaks of the equatorial anomaly rather than at the geomagnetic equator. Scintillations as strong as 9 dB have been observed on the 4 GHz COMSAT downlink passing through an anomaly crest. Scintillations peak at latitudes (λ_m) $\pm$ 10° of the magnetic equator and decay with a Gaussian shape: $\exp\{-[(\lambda_m - 10)/10]^2\}$ (Goodman 1980).

During evening hours, when scintillation is most intense, incoherent scatter radar observations at Jicamarca, Peru, show the existence of plumes in the topside of the ionosphere (Fig. 8.31). The time history of the plume includes an eastward drift of an irregularity patch. These plumes have been seen up to heights of about 800 km, which is the limit of the radar. A sample time history is as follows (Aarons 1978): The scintillation starts shortly after sunset; 5 dB peak-to-peak values of UHF scintillations are typical, and fading rates are about 10/min. During the buildup of the plumes the scintillation depth increases to about 20 dB peak-to-peak, with fading rates greater than 10/min. It is during

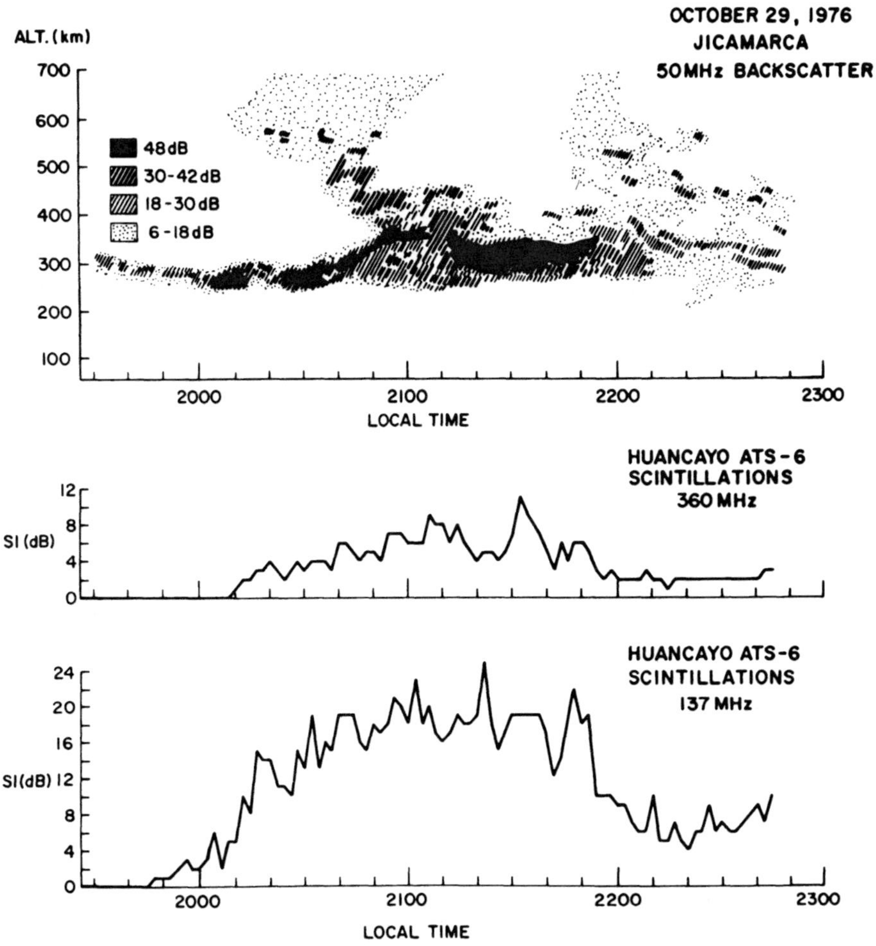

Fig. 8.31 *Incoherent scatter plumes and scintillation (From Woodman and Basu 1978 © American Geophysical Union)*

this stage that microwave scintillations are observed. After formation, an irregularity structure moves east at speeds between 50 and 150 m s^{-1}, and the scintillation depth decreases. Observations suggest that the eastward motion diminishes near midnight. The plume forms a patch of irregularities that extend along the magnetic field like a banana (Fig. 8.32); east–west size is greater than about 100 km, vertical height is 50 to several hundred km, and north–south dimension is about 2000 km. *In situ* (satellite) measurements of ion density in the F region (at 225 km) show severe depletions. Patches show depletions that produce fluctuations in Faraday rotation. Some depletions are more than 90% (Aarons 1982). The occurrence of scintillation shows equinoctial maxima.

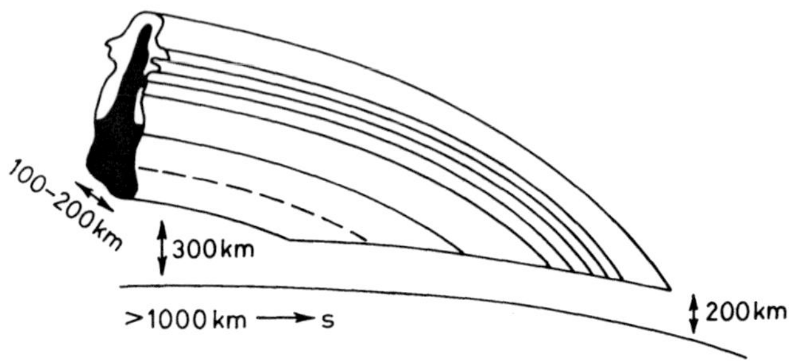

Fig. 8.32 *3-D model of patch of equatorial irregularities (From Aarons 1982 © IEEE; reprinted with permission)*

8.5.3 High-latitude scintillations

Strong scintillations occur near the auroral zones (Aarons 1985*a*; Basu *et al.* 1985). In the auroral and polar regions energetic electron precipitation and current systems (Fig. 2.17) are dominant factors in producing irregularities. The irregularities that are primarily responsible for auroral scintillation are at altitudes of 250 to 500 km but can extend from the E layer to more than 1000 km. Scintillations are most intense near midnight during March, April, and May in the Northern Hemisphere, during both magnetically quiet and disturbed conditions. In polar regions the scintillation is strong but weaker than in the auroral zone. There is a marked increase of auroral scintillation with increase of magnetic activity. Formulas are available for calculating the scintillation index (SI) in terms of day, time, solar flux, etc. (Aarons 1985*b*).

The holographic technique has been used to reconstruct the source structure from spaced receiver measurements of the complex phase fronts (Rogers and Ireland 1980; Tauriainen 1982).

8.5.4 Storm effects on scintillations

In general, increased magnetic activity inhibits equatorial scintillation before midnight. After midnight the scintillation activity, in general, increases slightly during magnetic storms. Scintillations on 136 MHz, 1·7 GHz, 4 GHz, and even on 11·5 GHz were observed in Japan (e.g. Minakoshi *et al.* 1981) (Fig. 8.33). The frequency dependence of S_4 was $f^{-0 \cdot 5}$ for $0 \cdot 136 < f < 1 \cdot 7$ GHz and f^{-2} for $4 < f < 11 \cdot 5$ GHz. The Japanese data show the following relation between

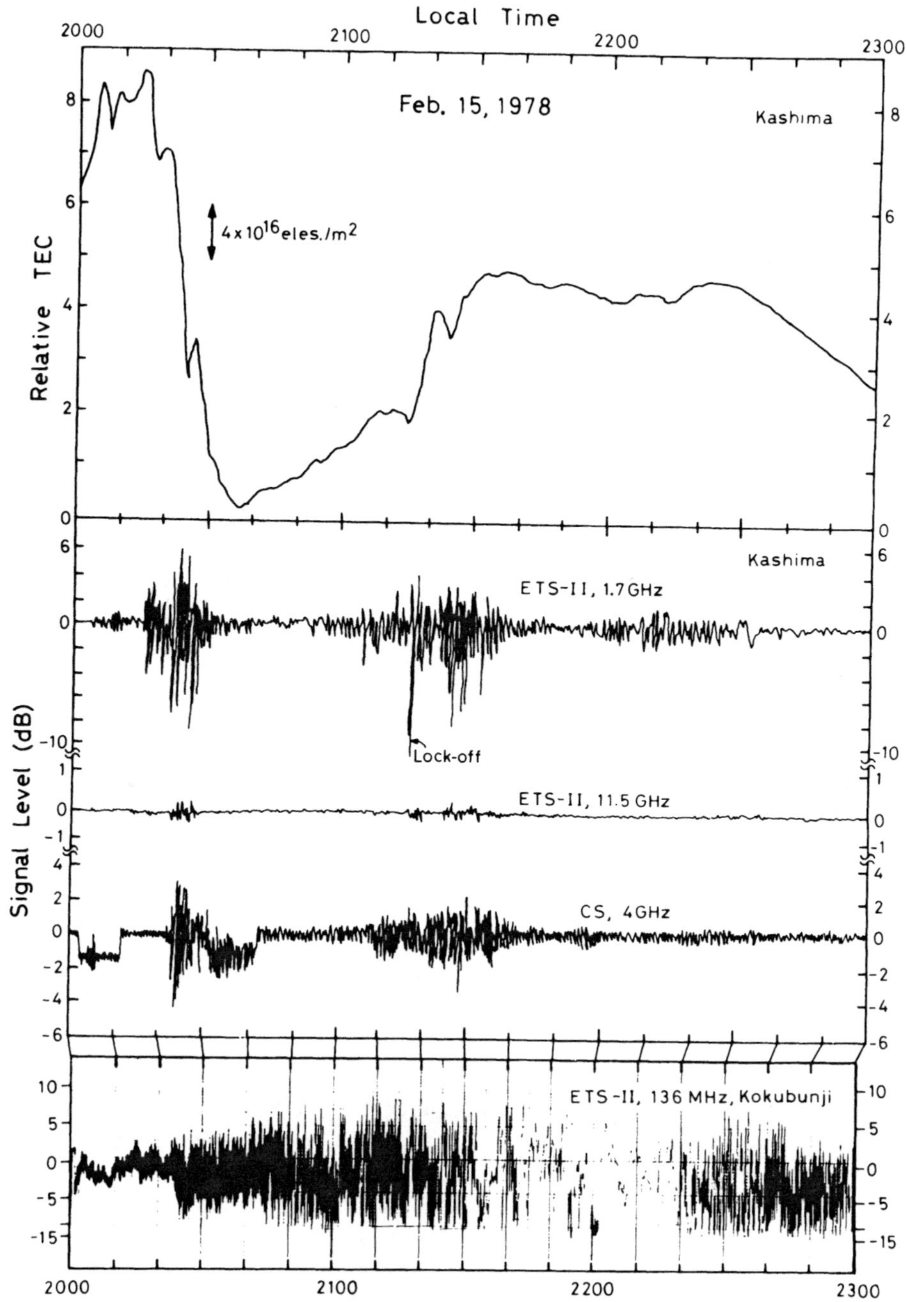

Fig. 8.33 *Scintillation on 136 MHz of ETS-2 at Kokubunji; 1·7 GHz and 11·5 GHz of ETS-2 and 4 GHz of CS at Kashima, Feb. 15, 1978; and total electron content for Kashima (From Ogawa et al. 1980 © Pergamon Press; reprinted with permission)*

the scintillation depth S ($= P_{max} - P_{min}$) dB and fluctuations $\Delta\Omega$ in Faraday rotation:

$$S = 11 \log (0{\cdot}5\ \Delta\Omega) \tag{8.37}$$

Deep scintillation occurs when the ray path is essentially parallel to the geomagnetic field (e.g. ETS–2 to Japan).

8.5.5 Models of scintillation

Various models have been developed to express scintillation data in terms of radio and ionospheric parameters. For a three-dimensional irregularity power spectrum (Rino and Matthews 1980) with an outer scale size K, the mean square phase fluctuation σ_ϕ^2 is given by

$$\sigma_\phi^2 = A\lambda^2\ LK \sec\chi\ \langle\Delta N^2\rangle \tag{8.38a}$$

and the scintillation index S_4 is given by

$$S_4 = B\lambda^3\ ZLK \sec^2\chi\ \langle\Delta N^2\rangle \tag{8.38b}$$

where A and B depend on geometrical factors, fundamental constants, etc. From (8.38) we see how the scintillation depends on radio wavelength λ, layer height Z, layer thickness L, angle χ between the ray and the vertical, and the mean square deviation $\langle\Delta N^2\rangle$ of electron density. Input parameters for such models have been developed (Donnelly 1980, Vol. IV).

8.6 Some uses of satellite data in ionospheric studies

Satellite radio beacon data have proved useful in studies of a number of ionospheric phenomena, for example, traveling ionospheric disturbances, storm behavior, sudden increases in *TEC* (caused by flare X-rays), isolated plasma lenses, pulsations, troughs, protonosphere–ionosphere coupling, night enhancements, and volcanic effects. This Section presents a few illustrative examples.

8.6.1 Protonospheric content

The protonospheric content N_p ($= N_T - N_F$) has been measured using several satellites but particularly using the specially designed ATS–6 radio beacon (Davies 1980). Figure 8.21 shows monthly median hourly values of N_T and N_p at Boulder near sunspot minimum. The striking feature of the diurnal variation of N_p is the nighttime maximum. In Europe, on the other hand, N_p peaks during the daytime; this has been explained on the basis of plasma flow from the local plasmasphere and conjugate ionosphere (Kersley and Klobuchar 1978). It is important to note that there is essentially no plasma production in the protonosphere and, hence, ionization has to diffuse up from the ionosphere. Thus the protonosphere acts like a reservoir; it takes plasma from the ionosphere by day, stores it in a loss-free environment, and returns it to the ionosphere at night, thus helping to maintain the nighttime F layer. The plasma flows along magnetic tubes and, when there are asymmetries at the feet of these tubes, plasma can flow between magnetically conjugate areas. In the USA near sunspot minimum, the monthly mean N_p is approximately 10% of the midday Faraday content, and about 50% of N_F at night. During summer at sunspot minimum, when there is little diurnal variation in N_p, the monthly mean total electron content N_T can be estimated from measurements of the Faraday content N_F:

$$N_T(LT) = N_F(LT) + a\,N_F(\max) \tag{8.39}$$

where LT is local time, N_F (max) is the maximum diurnal value, and a is about 0·1 in the U.S.A. and about 0·15 in Europe.

8.6.2 Traveling ionospheric disturbances

The geostationary beacon technique has been used extensively for observing TIDs. Figure 8.34*a* shows Faraday rotation records for three stations, separated

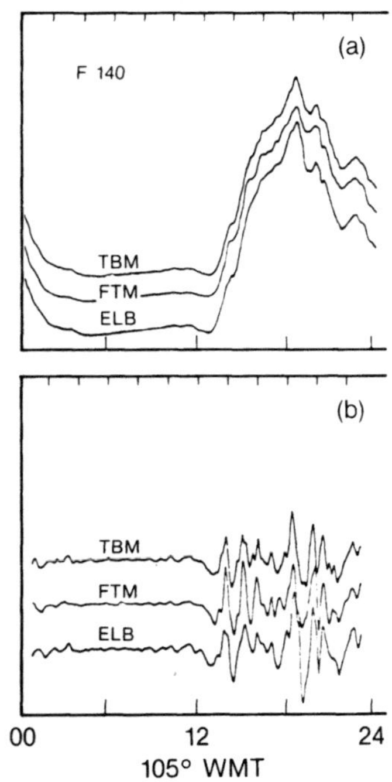

Fig. 8.34 *Faraday rotations at three stations in Colorado (Table Mountain, Fort Morgan, and Elbert) from ATS-6 on Oct. 11, 1974*
(*a*) Measured data
(*b*) High-pass filtered data

by about 120 km, using the ATS–6 radio beacon. The electron content fluctuations have periods ranging from 15 min to more than an hour and, therefore, are probably caused by gravity waves in the atmosphere (see Sec. 7.11). The peak-to-peak amplitude can be more than 10% of the background content. The relevant time displacements are more easily obtained from the high-pass-filtered data of Fig. 8.34*b*, which remove the large diurnal variation, and the calculation of horizontal velocity is similar to that given in Sec. 7.9. Some results (Fig. 8.35)

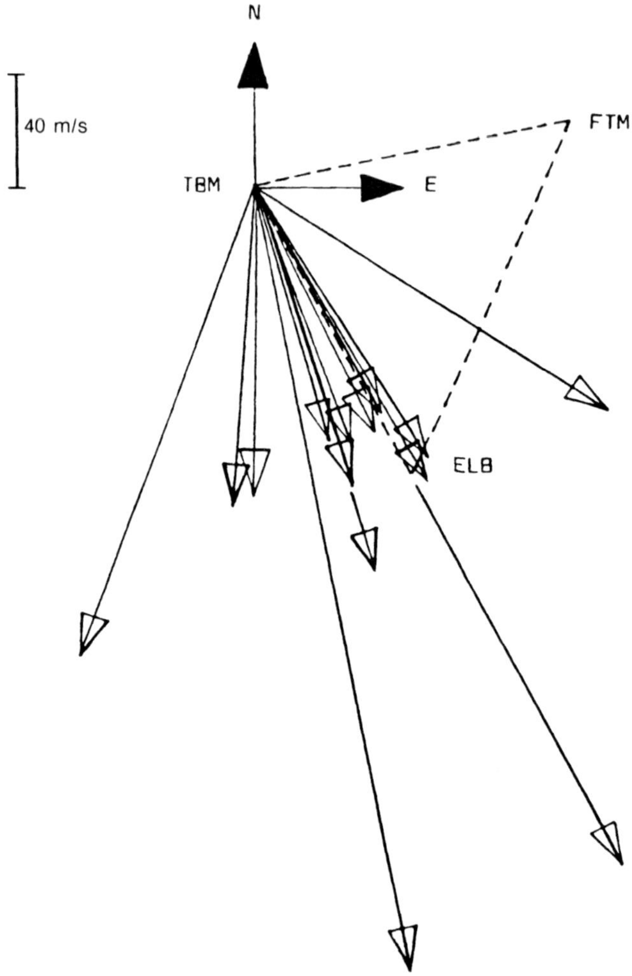

Fig. 8.35 *Horizontal trace speeds of traveling ionospheric disturbances*
Deduced from Faraday measurements at Table Mountain, Fort Morgan, and Elbert on December 6, 9, and 14, 1974, between 0700 and 1400, 105° west meridian time

for three stations near Boulder show a mean speed of 138 m s^{-1} and a mean azimuth of 163°. Similar measurements have been made in many other parts of the world. For example, in India the speeds are about the same, namely, 140 m s^{-1} by day, 110 m s^{-1} by night (Singh *et al.* 1977) but the directions are different. Large-scale TIDs with speeds of 500 m s^{-1} have been identified with auroral substorms (Bowman 1978; Davis 1971; Leitinger 1972; Al'pert 1975; Mendillo 1976; Checcacci 1978; Tyagi 1983; Tauriainen 1986).

8.6.3 Isolated irregularities

Occasionally an isolated rhythmic fading event occurs that is caused by diffraction, or interference, between waves at the edge of an irregularity or by a strong irregularity. Such events may last from a few seconds to several minutes and have been observed in middle and low latitudes. The example in Fig. 8.36 is characterized by a deep central minimum and amplitude fluctuations on both sides of the minimum. This event has been modeled in terms of diffraction around a non-absorbing cylindrical irregularity that produces a Gaussian-shaped phase advance ϕ (Fig. 8.37), where

$$\phi = \phi_0 \exp\,[-(x - x_0)^2/d^2] \tag{8.40}$$

The scale size d of these ringing irregularities is of the order of 100 – 1000 m.

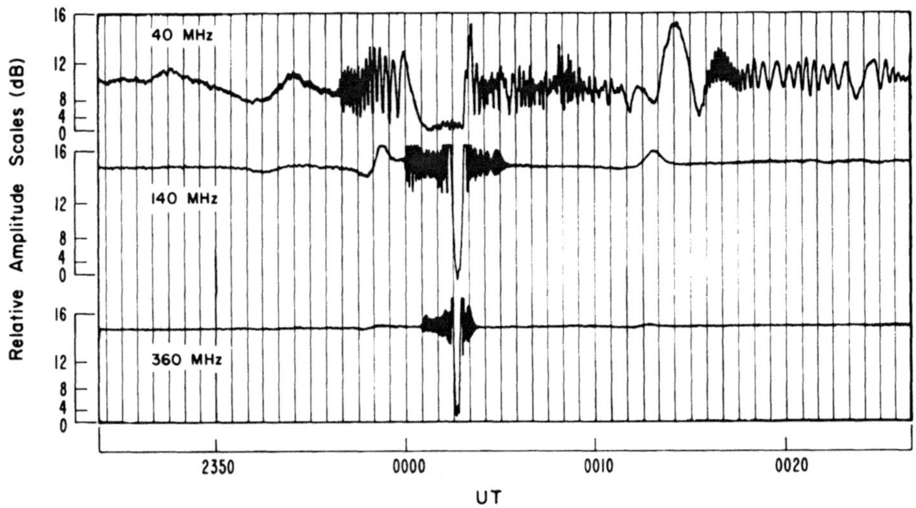

Fig. 8.36 *Amplitude fading on 40, 140, and 360 MHz, ATS-6 to Boulder, showing modulation of fading pattern*

Here x is the horizontal distance from the center x_0. Over the emergent wave front the amplitude will be constant since there is no absorption in the cylinder. However, because the phase is perturbed, the re-radiated Huygens wavelets will interfere to produce a diffraction pattern on the ground. This is a simple example of a phase-changing screen (Booker *et al.* 1950; Ratcliffe 1956). For such diffraction, with vertical propagation, the wave components (X) in phase with and (Y) in quadrature with the undisturbed wave are

$$X = 1 - 2\int \sin(P_0 + \phi/2)\sin(\phi/2)\, dx/(r\lambda)^{1/2} \tag{8.41a}$$

and

$$Y = 2\int \cos(P_0 + \phi/2)\sin(\phi/2)\, dx/(r\lambda)^{1/2} \tag{8.41b}$$

where

$$P_0 = 2\pi(r - h)/\lambda - \pi/4 \tag{8.41c}$$

Here h is the height of the irregularity and r joins the element dx on the emergent wave front to a point R on the ground. The integrals are taken over the irregularity in the plane of the wave front (Heron 1979; Titheridge 1971).

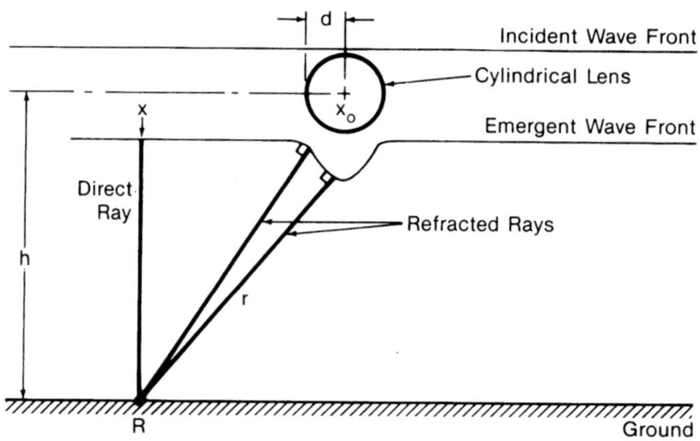

Fig. 8.37 *Geometry of diffraction of a plane wave by a cylindrical lens*

Ionospheric plasma lenses can be of the focusing type or defocusing type, both of which are present in Fig. 8.36. In a defocusing lens the plasma density is greater than the background value whereas in a focusing lens the plasma density is less than the background value.

Now irregularities can occur singly (Davies and Whitehead 1977) or in tandem (Basu *et al.* 1976) or superimposed (Trivedi *et al.* 1978). Figure 8.38 shows how the diffraction pattern on the ground changes with the radio wavelength for a set of closely spaced cylindrical lenses. It is seen that, on the shortest waves, the diffraction is weak and there is a close correspondence between the structure in the ionosphere and that on the ground. As the wavelength increases, the depth of the amplitude fading increases and the pattern spreads out so there is little relationship between the ionospheric structure and the ground pattern.

This is true also for the phase ψ variations where

$$\psi = \tan^{-1}(Y/X) \tag{8.42a}$$

and the power P is given by

$$P = X^2 + Y^2 \tag{8.42b}$$

Thus diffraction patterns, especially of the ringing sort, should not be (mis)-interpreted from observations obtained by using a single long wave.

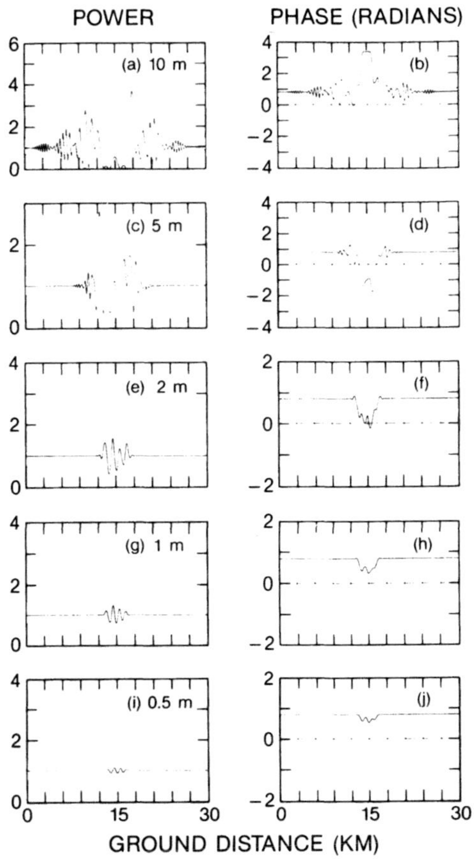

Fig. 8.38 *Power and phase diffraction patterns on various wavelengths produced by three closely spaced irregularities at 400 km height*
Sizes: 400, 410, and 390 m
Respective strengths: −0·467, −0·560, and −0·373 rad
Respective positions: 13·9, 15·0, and 15·9 km

8.6.4 Pulsations in total content

These consist of trains of weak quasi-sinusoidal oscillations (Fig. 8.39) observed with the sensitive carrier phase technique (Sec. 8.3.3). These oscillations have periods of 30 – 50 s and can last for several hours near noon on geomagnetically quiet days. They are, therefore, analogous to the Pc4 type of magnetic pulsations. The amplitudes of these pulsations in N_T are of the order of 0·01% to 0·1% of the background (Davies and Hartmann 1976; Okuzawa and Davies 1981).

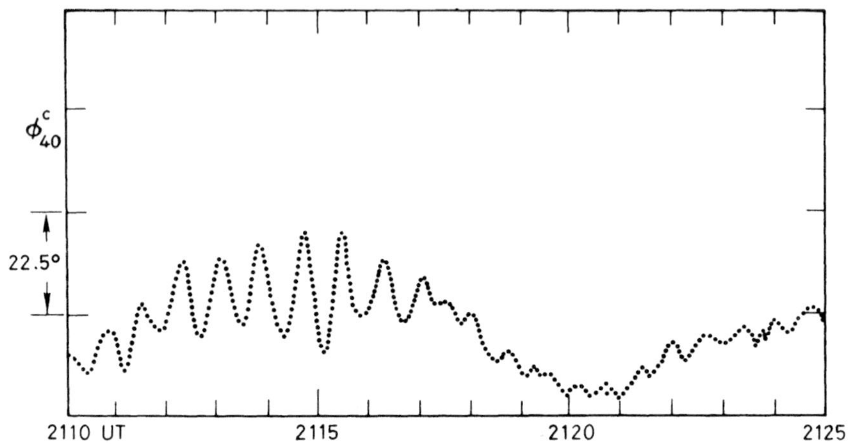

Fig. 8.39 *Filtered output of 40 MHz carrier phase of ATS–6-to-Boulder, showing pulsations*

8.6.5 Troughs

Polar-orbiting-satellite beacons provide excellent means of locating the sub-auroral trough because of the good latitudinal resolution provided by a few meridionally aligned receivers. This technique has been used with the Navy Navigation Satellite System (NNSS) to study the structure and motion of the trough over Scandinavia and central Europe (Leitinger *et al.* 1983) (Fig. 8.40). Since the vertical *TEC* is obtained by projecting an oblique measurement onto the vertical, a low-pass spatial filtering effect is introduced. The spatial cutoff size is greater the more distant a given structure is from the receiver. Overhead even small-scale structures are undistorted whereas for large zenith angles only the large-scale structures are reliable.

8.6.6 Auroral ionosphere

Orbiting beacons have been used in conjunction with incoherent scatter sounders (Chandra and Kersley 1985) to study the irregularity formation and dynamics of the high-latitude ionosphere. Measurements indicate that auroral irregularities responsible for radio scintillations are field-aligned sheets rather than rods (Fremouw and Lansinger 1981).

8.6.7 Winter-night enhancements

A secondary diurnal maximum in total electron content occurs during the night in winter in middle latitudes. This occurs on magnetically quiet days and covers wide geographical areas (Mendillo *et al.* 1977; Huang 1983). The ATS-6 radio beacon has provided evidence (Fig. 8.41) that, although the night enhancement is most evident during winter, it is present, over both the USA and Europe, in other seasons, but occurs earlier in the day when it is obscured by the diurnal

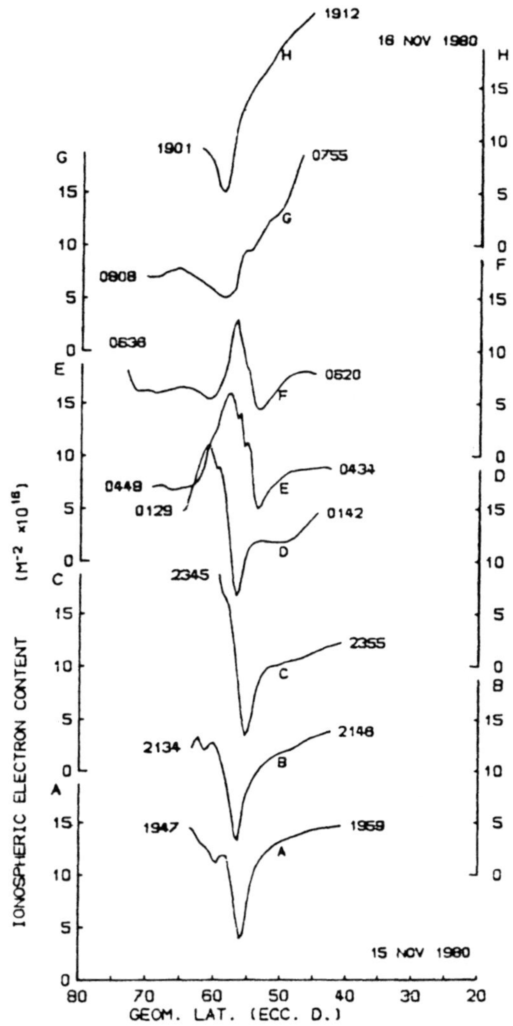

Fig. 8.40 *Troughs in electron content over Scandinavia from measurements of NNSS (150 MHz and 400 MHz) at several ground stations (From Leitinger et al. 1983)*

decline (Davies 1980). In winter there appear to be two peaks, one near midnight and the other near 0300 LT. Increases in *TEC* of 30% over the background have been observed, and the peak(s) tend to move westward at fixed local time. However, there is no evidence of depletion in the protonospheric content.

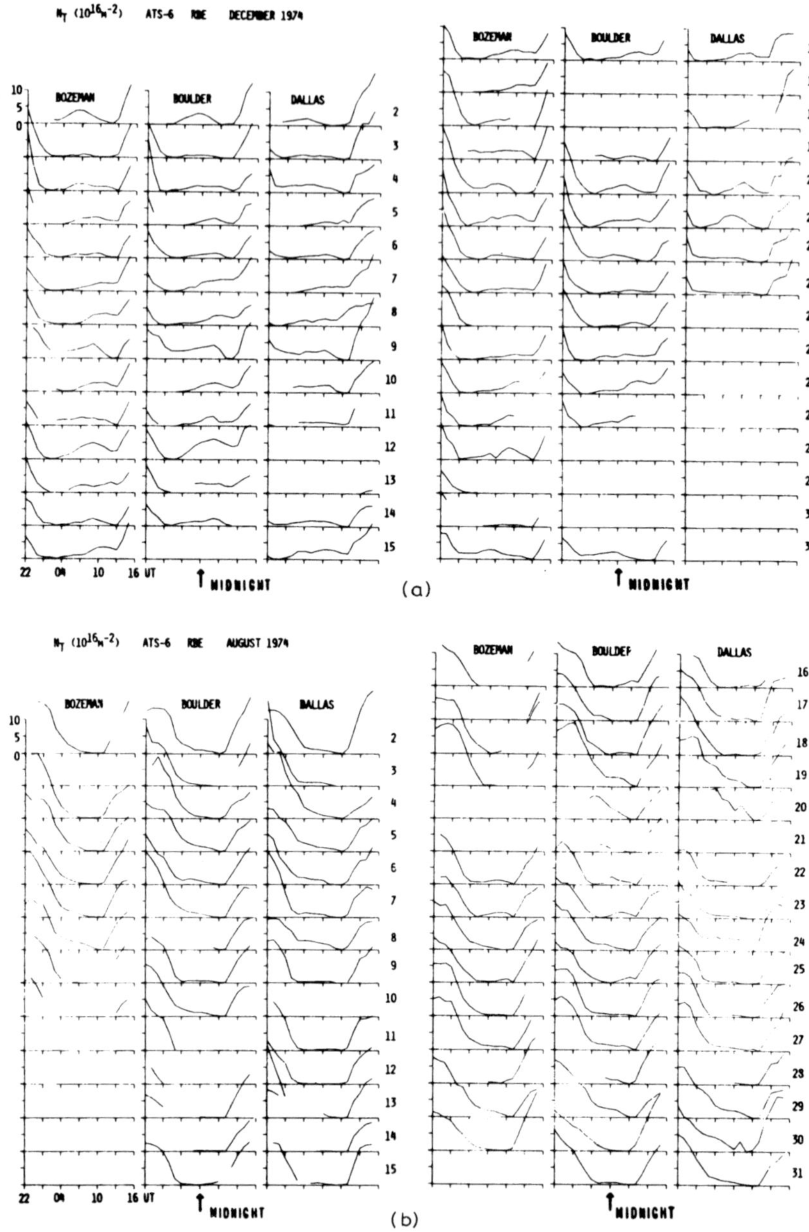

Fig. 8.41 *Nighttime variations of total electron content, ATS-6 to Bozeman, Montana; Boulder, Colorado; and Dallas, Texas*

07 UT corresponds to 00 LT for 105°W meridian. The curves are normalized to pre-dawn minimum values

(*a*) December 1974

(*b*) August 1974

8.7 Space systems applications

8.7.1 Transionospheric propagation systems

There are numerous radio systems in operation that involve propagation between satellite and ground, satellite and satellite, or a distant space probe and ground. Thus a satellite system may be defined as one having a satellite or spacecraft at one of its terminals. Satellite systems normally operate on frequencies higher than 200 MHz because VHF telemetry on 137 MHz is being phased out. The exact frequencies used for the various satellite systems are in accord with the internationally agreed frequency allocation rules and regulations. Satellite systems are used for navigation, communication, surveillance, etc. Magnitudes of the possible ionospheric effects, and their frequency dependencies, are given in Table 8.2. Some examples of systems involving transionospheric propagation are discussed below.

Table 8.2 *Estimated maximum ionospheric effects on 100 MHz for elevation angles of about 30° one-way traverse*

Effect	Magnitude	Frequency dependence
Faraday rotation	30 rotations	f^{-2}
Propagation delay	25 μs	f^{-2}
Refraction	< 1°	f^{-2}
Variation in the direction of arrival	20 min of arc	f^{-2}
Dispersion	0·4 ps/Hz	f^{-3}
Scintillation	10 dB	

8.7.2 The global positioning system (GPS)

The GPS is a highly accurate space-based navigation system that provides users with precise position (15 m), velocity (0·10 m s^{-1}) and time (0·1 μs) information. The system operates on two L-band frequencies – f_1 = 1575·42 MHz and f_2 = 1227·6 MHz – and it is composed of 18 satellites, three in each of six orbital planes, with circular (19 652 km) orbits of 63° inclination and 12 h periods. Let δT_1 be the needed correction to the measured time on f_1 to allow for the ionospheric delay. Then

$$\delta T_1 = \Delta(\delta T) f_2^2/(f_1^2 - f_2^2) = F\Delta(\delta T) \qquad (8.43)$$

where $\Delta(\delta T)$ is the difference between the measured time delays on the two frequencies f_1 and f_2. The factor F, called the ionospheric scaling factor, is 1·55. The actual ionospheric time delay is measured from the difference in arrival times of 10 MHz modulation envelopes on f_1 and f_2. For a 10 ns time delay on f_1 a differential measurement of 6·5 ns is required.

8.7.3 Satellite communication systems

Satellite communication systems provide large bandwidths, permitting many data to be transmitted rapidly. As discussed above, satellite systems are affected by the ionosphere, especially by scintillation. Many systems employ synchronous satellites (INTELSAT, COMSAT, DSCS, NATO, MARISAT) but orbiting satellite systems still operate (e.g. GPS). Satellite systems have experienced fading on frequencies of 6 GHz; the peak-to-peak values are several decibels (Fang and Liu 1983). Fading of more than 20 dB has been seen on a 1200 MHz signal (Fremouw *et al.* 1978). A high occurrence of fading greater than 20 dB peak-to-peak has been seen on a 250 MHz signal (Aarons 1978). Both in equatorial areas and in high latitudes, scintillation effects are seen on UHF (100–600 MHz) communication systems as well as on L-band (>1 GHz) systems. The operational difficulty introduced by fading is that for a certain fraction of time the signal may drop below an acceptable level; for example, with a Rayleigh distribution (Sec. 7.7.6) the signal fades to $<18{\cdot}4$ dB, 1% of the time.

Diversity schemes attempt to reduce the effects of fading by combining two signals that fade independently. For two independent Rayleigh processes, the larger of the two signals at any point fades to 9·5 dB, 1% of the time, yielding a gain in protection of 9 dB. In general, the diversity gain is greater the lower the correlation coefficient of the two signals. When the correlation is $<0{\cdot}6$ the diversity gain is 8 dB or better. A correlation coefficient of 0·6 may be used as a threshold for diversity action although it applies strictly to Rayleigh fading only (strong scintillation). With weak scintillation, a correlation coefficient less than 0·6 occurs on two frequencies having the same paths, only if the frequencies are separated by more than a factor of 3, which shows that frequency diversity is not effective against scintillation. Simultaneous fluctuations on orthogonally polarized channels are also highly correlated on frequencies greater than 100 MHz.

Other diversity systems employ time and space. Time diversity requires that the same message be sent at a time difference sufficient for the correlation coefficient to be less than 0·6. Space diversity requires that the signal be received on two antennas separated sufficiently for the correlation coefficient to be less than 0·6 (Aarons 1978). Diversity schemes are normally required for strong scintillation only because weak scintillation can be offset by a reasonable fade margin. Spatial correlation coefficients of 0·6 and 0·3 correspond to separations of 300 m and 1000 m respectively for an ionospheric drift velocity of 50 m s^{-1}. Hence space diversity is feasible only on land.

8.7.4 Satellite surveillance systems

In addition to navigation and communication, satellites are used for surveillance, for example, SEASAT for monitoring activities over oceans, and LANDSAT for obtaining information about the Earth's natural resources. These systems operate on frequencies between 2·0 and 3·0 GHz. It is assumed that the propagation medium is 'frozen' during a sounding. The moving satellite makes the antenna appear large, and such antenna systems are called synthetic-aperture antennas. Irregularities in the ionosphere produce amplitude and phase fluctuations in the signal and, since synthetic-aperture antennas rely on the principle of coherent integration, phase scintillation degrades the coherence across the aperture (Szuszczewicz *et al.* 1983).

8.7.5 Time transfer systems

Satellite systems, and the GPS system in particular, are used for the accurate synchronization and comparison of clocks and frequency standards having stabilities of a few parts in 10^{14}. Propagation effects (noise) on the transmitted signal place a limit on the capability of such systems. Such effects limit the ability to transfer time with stabilities of a few nanoseconds and time accuracies on the order of 10 ns (Allan and Ping-ping 1987).

8.7.6 Geodetic systems

The ionosphere degrades geodetic and radio astronomical measurements. High accuracy is required in geodetic work such as the measurement of continental drift. For such measurements a geodetic user can average the data over relatively long time intervals. Averaging results in high accuracy because the principal errors in ionospheric delay are random rather than systematic. Furthermore, high accuracy is obtained by interferometric phase measurements on signals from the GPS system. With baselines of 100 km to 200 km, accuracies of 0·06 m have been obtained (Spoelstra 1986). In radio astronomy, ionospheric refraction is significant for single-dish antennas and for connected-element radio interferometers on frequencies up to about 2 GHz and for very-long-baseline interferometers (VLBI) on frequencies higher than 2 GHz (Spoelstra 1986).

8.8 References

AARONS, J. (1977): 'Equatorial scintillations: A review', *IEEE Trans.*, **AP-25**, p. 729

AARONS, J. (1978): 'Ionospheric scintillation – An introduction'. AGARD Lecture Series No. 93

AARONS, J. (1982): 'Global morphology of ionospheric scintillations', *Proc. IEEE*, **70**, p. 360

AARONS, J. (1985*a*): 'Equatorial F-layer irregularity patches at anomaly latitudes', *J. Atmos. Terr. Phys.*, **47**, p. 875

AARONS, J. (1985*b*): 'Construction of a model of equatorial scintillation intensity', *Radio Sci.*, **20**, p. 397

ALLAN, D. W., and PING-PING, L. (1987): 'Estimating combined errors due to propagation and ephemeris and their effect on time and frequency transfer'. Proc. 41st Annual Frequency Symposium, IEEE Document No. 87CH2427-3, p. 144

AL'PERT, Ya. L. (Ed.) (1975): 'Proc. Beacon Satellite Investigations of the Ionosphere Structure and ATS-6 Data: Vols. I and II. Inst. of Terr. Mag. Iono. Radio Wave Prop., Moscow

BARRINGTON, R. E. (1969): 'Ionospheric ion composition deduced from VLF observations'. *Proc. IEEE*, **57**, p. 1036

BASU, Sa., and BASU, Su. (1981): 'Equatorial scintillations – A review', *J. Atmos. Terr. Phys.*, **43**, p. 473

BASU, S., GUHATHAKURTE, B. K., BHATTACHARYYA, G. N., and DAS GUPTA, A. (1976): 'VHF scintillation observations at Calcutta with orbiting and geostationary satellites' *in* MENDILLO, M. (Ed.): 'The geophysical use of satellite beacon observations'. Boston University, Boston, MA

BASU, Su., BASU, Sa., MACKENZIE, E., and WHITNEY, H. E. (1985): 'Morphology of phase and intensity scintillations in the auroral oval and polar cap', *Radio Sci*, **20**, p. 347

BASU, Sa., MACKENZIE, E., and BASU, Su. (1988): 'Ionospheric constraints on VHF/UHF communications links during solar maximum and minimum periods', *Radio Sci.*, **23**, p. 363

BENSON, R. F. (1982): 'Stimulated plasma instability and nonlinear phenomena in the ionosphere', *Radio Sci.*, **17**, p. 1637

BENT, R. B., LLEWELLYN, S. K., NESTERCZUK, G., and SCHMID, P. E. (1975): 'The development of a highly-successful worldwide empirical ionospheric model and its use in certain aspects of space communications and worldwide total electron content investigations' *in* GOODMAN, J. M. (Ed.): 'Effects of the ionosphere on space systems and communications'. US Govt. Printing Office, Washington, DC

BHONSLE, R. V., ALURKAR, S. K., and SAWANT, H. S. (1977): 'Effects of ground reflection on the polarization of beacon transmissions from ATS–6 satellite', *Indian J. Radio Space Phys.*, **6**, p. 5

BOOKER, H. G., RATCLIFFE, J. A., and SHINN, D. H. (1950): 'Diffraction from an irregular screen with applications to ionospheric problems', *Phil. Trans. Roy. Soc.*, **242A**, p. 579

BOWMAN, G. G. (1978): 'A relationship between polar magnetic substorms, ionospheric height rises and the occurrence of spread-F', *J. Atmos. Terr. Phys.*, **40**, p. 713

BRIGGS, B. H., and PARKIN, J. A. (1963): 'On the variation of radio star and satellite scintillation with zenith angle', *J. Atmos. Terr. Phys.*, **25**, p. 339

CALVERT, W., and GOE, G. B. (1963): 'Plasma resonances in the upper ionosphere', *J. Geophys. Res.*, **68**, p. 6113

CALVERT, W., and McAFEE, J. R. (1969): 'Topside sounder resonances', *Proc. IEEE*, **57**, p. 1089

CALVERT, W., and WARNOCK, J. M. (1969): 'Ionospheric irregularities observed by topside sounders', *Proc. IEEE*, **57**, p. 1019

CHANDRA, H., and KERSLEY, L. (1985): 'Irregularity heights near the high-latitude scintillation boundary', *Indian J. Radio Space Phys.*, **14**, p. 96

CHECCACCI, P. (Ed.) (1978): 'Beacon satellite measurements of plasmaspheric and ionospheric properties'. Instituto di Ricerca Sulle Onde Elettromagnetiche, Florence, Italy

CRANE, R. K. (1978): 'Variance and spectra of angle-of-arrival and Doppler fluctuations caused by ionosphere scintillations', *J. Geophys. Res.*, **83**, p. 2091

DAVIES, K. (1969): 'Ionospheric radio waves' (Ginn-Blaisdell, Waltham, MA)

DAVIES, K. (1980): 'Recent progress in satellite radio beacon studies with particular emphasis on the ATS–6 Radio Beacon Experiment', *Space Sci. Rev.*, **25**, p. 357

DAVIES, K., and HARTMANN, G. K. (1976): 'Short-period fluctuations in total columnar electron content', *J. Geophys. Res.*, **81**, p. 3431

DAVIES, K., and WHITEHEAD, J. D. (1977): 'A radio lens in the ionosphere', *J. Atmos. Terr. Phys.*, **39**, p. 383

DAVIES, K., FRITZ, R. B., and GRAY, T. B. (1976): 'Measurements of the columnar electron contents of the ionosphere and the plasmasphere', *J. Geophys. Res.*, **81**, p. 2825

DAVIES, K., DEGENHARDT, W., HARTMANN, G. K., and LEITINGER, R. (1977): 'Electron content measurements over the US', Max Planck Institute for Aeronomy, Katlenburg-Lindau 3, Germany

DAVIS, M. J. (1971): 'On polar substorms as the source of large-scale traveling ionospheric disturbances', *J. Geophys. Res.*, **76**, p. 4525

DONNELLY, R. F. (Ed.) (1980): 'Solar–terrestrial predictions proceedings, IV, D1'. US Govt. Printing Office, Washington, DC 20402

EVANS, J. V. (1977): 'Satellite beacon contributions to studies of the structure of the ionosphere', *Rev. Geophys. Space Phys.*, **15**, p. 325

FANG, D. J., and LIU, C. H. (1983): 'A morphological study of gigahertz equatorial scintillations in the Asian sector', *Radio Sci.*, **18**, p. 241

FREMOUW, E. J., and LANSINGER, J. M. (1981): 'Dominant configurations of scintillation-producing irregularities in the auroral zone', *J. Geophys. Res.*, **86**, p. 10087

FREMOUW, E. J., LEADABRAND, R. L., LIVINGSTON, R. C., COUSINS, M. D., RINO, C. L., FAIR, B. C., and LONG, R. A. (1978): 'Early results from the DNA wideband satellite experiment – Complex-signal scintillation', *Radio Sci.*, **13**, p. 167

GARRIOTT, O. K., DA ROSA, A. V., and ROSS, W. J. (1970): 'Electron content obtained from Faraday rotation and phase path length variations', *J. Atmos. Terr. Phys.*, **32**, p. 705

GOODMAN, J. M. (1980): 'A resume of anticipated FLEETSATCOM and GAPFILLER scintillation effects during the peak of solar cycle 21 (1980–1982)' *in* DONNELLY, R. F (Ed.): 'Solar-terrestrial predictions proceedings IV'. US Govt. Printing Office, Washington, DC 20402

GREBOWSKY, J. M., MAYNARD, N. C., TULUNAY, Y. K., and LANZEROTTI, L. J. (1976): 'Coincident observations of ionospheric troughs and the equatorial plasmapause, *Planet. Space Sci.*, **24**, p. 1177

HAGG, E. L., HEWENS, E. J., and NELMS, G. L. (1969): The interpretation of topside sounder ionograms', *Proc. IEEE*, **57**, p. 949

HARGREAVES, J. K. (1978): 'On the presentation of ATS-6 electron content data', *J. Atmos. Terr. Phys.*, **40**, p. 493

HERON, M. L. (1979): 'Diffraction from discrete and homogeneously structured ionospheric irregularities', *Radio Sci.*, **14**, p. 97

HUANG, Y.-N. (1978): 'Solar cycle variation in the total electron content at Sagamore Hill', *J. Atmos. Terr. Phys.*, **40**, p. 733

HUANG, T.-X. (1983): 'Studies of winter nighttime increases in total electron content of the ionosphere at mid-latitudes', *Acta Geophys. Sinica* **26**, 101

IEEE (1969): 'Special issue on topside sounding and the ionosphere', *Proc. IEEE*, **57**, p. 859

ISHIMARU, A. (Ed.) (1975): 'Special issue on random media', *Radio Sci.*, **10**, pp. 1–137

JACKSON, J. E. (1969): 'The reduction of topside ionograms to electron density profiles', *Proc. IEEE*, **57**, p. 960

JACKSON, J. E., and WARREN, E. S. (1969): 'Objectives, history and principal achievements of the topside sounder and ISIS programs', *Proc. IEEE*, **57**, p. 861

KERSLEY, L., and KLOBUCHAR, J. A. (1978): 'Comparison of protonospheric electron content measurements from the American and European sectors', *Geophys. Res. Lett.*, **5**, p. 123

KING, J. W. (1963): 'Investigations of the upper ionosphere deduced from topside sounding data', *Nature*, **197**, p. 639

KLOBUCHAR, J. A. (1975): 'Polarization of VHF waves emitted from geostationary satellites', *J. Geophys. Res.*, **80**, p. 4387

KLOBUCHAR, J. A. (1978): 'Ionospheric effects on satellite navigation and air traffic control systems'. AGARD Lecture Series No. 93, Advisory Group for Aerospace Research and Development, North Atlantic Treaty Organisation

KLOBUCHAR, J. A. (1987): 'Ionospheric time delay algorithm for single frequency GPS users', *IEEE Trans*, **AES-23**, p. 325

LAWRENCE, R. S., LITTLE, C. G., and CHIVERS, H. J. A. (1964): 'A survey of ionospheric effects upon earth-space radio propagation', *Proc. IEEE*, **52**, p. 4

LEITINGER, R. (Ed.) (1972): 'Proc. Symp. on the future application of satellite beacon measurements'. University of Graz, Austria

LEITINGER, R., HARTMANN, G. K., LOHMAR, F. J., and PUTZ, E. (1983): 'Experiences with geophysical and geodetic campaigns used for collecting differential Doppler data' *in* TYAGI, T. R. (Ed.): 'Beacon satellite studies of the earth's environment'. National Physical Laboratory, New Delhi, India

MATUURA, N. (Ed.) (1979): 'Atlas of ionospheric critical frequency (f_oF_2) obtained from ionosphere sounding satellite-*b* observation'. Radio Res. Lab., Tokyo.

McAFEE, J. R. (1974): 'Electron plasma resonances in the topside ionosphere', *Fundamentals of Cosmic Phys.*, **1**, p. 71

MENDILLO, M. (Ed.) (1976): 'The geophysical use of satellite beacon observations'. Boston University, Boston, MA

MENDILLO, M., BUANSANTO, M. J., and KLOBUCHAR, J. A. (1977): 'Distortions of the winter nighttime ionosphere at $L = 4$', *J. Geophys. Res.*, **82**, p. 3223

MINAKOSHI, H., SINNO, K., KAN, M., KURIHARA, H., and FURUTA, O. (1981): 'Severe ionospheric scintillation associated with magnetic storm on March 22, 1979', *J. Radio Res. Labs.*, **28**, p. 1

MULDREW, D. B. (1969): 'Nonvertical propagation and delayed-echo generation observed by the topside sounders', *Proc. IEEE*, **57**, p. 1097

MULDREW, D. B. (1972): 'Electron resonances observed with topside sounders', *Radio Sci.*, **7**, p. 779

MULDREW, D. B., (1983): 'Alouette-ISIS radio wave studies of the cleft, the auroral zone, and the main trough and of their associated irregularities', *Radio Sci.*, **18**, p. 1140

OGAWA, T., SINNO, K., FUJITA, M., and AWAKA, J. (1980): 'Severe disturbances of VHF and GHz waves from geostationary satellites during a magnetic storm', *J. Atmos. Terr. Phys.*, **42**, p. 637

OKUZAWA, T., and DAVIES, K. (1981): 'Pulsations in total columnar electron content', *J. Geophys. Res.*, **86**, p. 1355

OSHEROVICH, V. (1987): 'Physical nature of the diffuse plasma resonances in the ionosphere', *J. Geophys. Res.*, **92**, p. 316

RASTOGI, R. G., DESHPANDE, M. R., VATS, H. O., DAVIES, K., GRUBB, R. N., and JONES, J. E. (1977): 'Amplitude scintillations of ATS–6 radio beacon signals within the equatorial electrojet region (Ootacamund, dip 4°N)', *Pramana*, **8**, p. 1

RATCLIFFE, J. A. (1956): 'Some aspects of diffraction theory and their applications to the ionosphere', *Rep. Progr. Phys.*, **19**, p. 188 (Physical Society, London)

REES, P. R., KERSLEY, L., and EDWARDS, K. J. (1977): 'Ground reflections and the ATS–6 radio beacon experiment', *J. Atmos. Terr. Phys.*, **39**, p. 545

RINO, C. L. (1976): 'Some new results on the statistics of radio wave scintillation: 2. Scattering from a random ensemble of locally homogeneous patches', *J. Geophys. Res.*, **81**, p. 2059

RINO, C. L., and MATTHEWS, S. J. (1980): 'On the morphology of auroral zone radio wave scintillation', *J. Geophys. Res.*, **85**, p. 4139

ROGERS, G. L., and IRELAND, W. (1980): 'Ionospheric holography: II. The analysis of a set of ionospheric holograms', *J. Atmos. Terr. Phys.*, **42**, p. 397

RÖTTGER, J. (1973): 'Some properties of large-scale equatorial spread-F irregularities interpreted by influences of atmospheric gravity waves', *Z. Geophys.*, **39**, p. 799

RUFENACH, C. L. (1972): 'Power-law wavenumber spectrum deduced from ionospheric scintillation observations', *J. Geophys. Res.*, **77**, p. 4761

RUFENACH, C. L. (1975): 'Power spectra of large scintillation signals', *J. Atmos. Terr. Phys.*, **37**, p. 569

RUSH, C. M., POKEMPNER, M., ANDERSON, D. N., PERRY, J., STEWART, F. G., and REASONER, R. (1984): 'Maps of f_oF_2 derived from observations and theoretical data', *Radio Sci.*, **19**, p. 1083

SINGH, L., VIJAYAKUMAR, P. N., GARG, S. C., TYAGI, T. R., BHARDWAJ, R. K., DABAS, R. S., NAGPAL, O. P., and SEN GUPTA, A. (1977): 'Study of large and medium scale irregularities using 140 MHz ATS–6 Faraday rotation records from three stations', *Indian J. Radio Space Phys.*, **6**, p. 245

SINGLETON, D. G. (1980): 'An improved ionospheric irregularity model' *in* DONNELLY, R. F. (Ed.): 'Solar–terrestrial predictions proceedings, IV, D1-1', US Govt. Printing Office, Washington, DC 20402

SMITH, F. G. (1952): 'Ionospheric refraction of 81·5 Mc/s radio waves from radio stars', *J. Atmos. Terr. Phys.*, **2**, p. 350

Space Science Symposium (1977): *Indian J. Radio Space Phys.*, **6**, p. 209–260

SPOELSTRA, T. A. Th. (1986): 'Correcting ionospheric effects in radio astronomy and satellite geodesy' *in* TAURIAINEN, A. (Ed.): 'Radio beacon contribution to the study of ionization and dynamics of the ionosphere and to corrections to geodesy: II', University of Oulu, Finland, p. 253

SZUSZCZEWICZ, E. P., RODRIGUEZ, P., SINGH, M., and MANGO, S. (1983): 'Ionospheric irregularities and their potential impact on synthetic aperture radars', *Radio Sci.*, **18**, p. 765

TAURIAINEN, A. (1982): 'Application of wave field reconstruction of VHF radio waves to investigate single, isolated ionospheric irregularities', *Radio Sci.*, **17**, p. 684

TAURIAINEN, A. (Ed.) (1986): 'Radio beacon contribution to the study of ionization and dynamics of the ionosphere and corrections to geodesy'. University of Oulu, Finland

TIMLECK, P. L., and NELMS, G. L. (1969): 'Electron densities less than 100 electrons cm^{-3} in the topside ionosphere', *Proc. IEEE*, **57**, p. 1164

TITHERIDGE, J. E. (1971): 'The diffraction of satellite signals by isolated ionospheric irregularities', *J. Atmos. Terr. Phys.*, **33**, p. 47

TITHERIDGE, J. E. (1972): 'Determination of ionospheric electron content from the Faraday rotation of geostationary satellite signals', *Planet. Space Sci.*, **20**, p. 353

TRIVEDI, A. I., DESHPANDE, M. R., and VATS, H. O. (1978): 'Diffraction of ATS–6 signals by isolated ionospheric irregularities: Some new aspects', *Proc. Indian Acad. Sci.*, **87A**, p. 165

TYAGI, T. R. (Ed.) (1983): 'Beacon satellite studies of the Earth's environment'. National Physical Laboratory, New Delhi

UMEKI, R., LIU, C. H., and YEH, K. C. (1977): 'Multifrequency spectra of ionospheric amplitude scintillation', *J. Geophys. Res.*, **82**, p. 2752

WARREN, E. S. (1969): 'The topside ionosphere during geomagnetic storms', *Proc. IEEE*, **57**, p. 1029

WHITNEY, H. E., AARONS, J., and MALIK, C. (1969): 'A proposed index for measuring ionospheric scintillations', *Planet. Space Sci.*, **17**, p. 1069

WOODMAN, R. F., and BASU, S. (1978): 'Comparison between in-situ spectral measurements of equatorial F-region irregularities and backscatter observations at 3 m wavelength', *Geophys. Res. Lett.*, **5**, p. 869

Chapter 9

Ionospheric disturbances and their effects on radio communication

9.1 Types of disturbances

The term 'ionospheric disturbance' is used to cover a wide variety of ionospheric conditions that depart from the usual or quiet state. However, it must be recognized that even in the quiet ionosphere things are not static. For example: sporadic E may appear, fluctuate wildly, and disappear; small ripples are almost always present; the F_2 layer is marked by fluctuations in f_oF_2, h_mF_2, and TEC from day to day and within a day. Because the causes of these quiet-time variations cannot always be identified they are called 'geophysical noise'.

There is a set of disturbances that are associated, either directly or indirectly, with events on the Sun. These include the following:

(i) Sudden ionospheric disturbances
(ii) Ionospheric storms
(iii) Polar cap absorption (PCA) events
(iv) Traveling ionospheric disturbances (large scale)
(v) Associated disturbances in the geomagnetic field (SFE = solar flare effect), aurora, and magnetospheric substorms.

These disturbances have important, usually adverse, effects on radio communications over the entire radio spectrum (ELF to SHF, see Table 1.1). Of course, what is adverse to one user may be advantageous to another. For example an increase in D-region absorption can reduce the needed signal of one operator and concurrently remove interference for another. Further, the decrease of TEC during a storm is beneficial to satellite navigation since it reduces the bothersome excess time delay. In general, however, storms adversely affect propagation and there are many instances. For example: (i) phase changes on VLF navigation systems, (ii) shortwave fadeouts, (iii) MUF failure in the F_2 layer (iv) increase of scintillation near the auroral zones. Of the various disturbances, ionospheric F_2 region storms are the most disruptive because (i) they extend into the highly populated middle latitudes, and (ii) they last for several days.

The various effects – simultaneous and delayed – are noted in Fig. 9.1. Disturbance effects are most pronounced in the D and F_2 layers, although sporadic E appears in some storms. This Chapter discusses some aspects of these effects on radio signals and some remedial actions, which include the operation of forecasting and warning centers. Forecasts and warnings are of value to operators, both civilian and military, because they enable the operators to (i) provide for alternative means of communication, (ii) pass priority material before the radio circuit closes, (iii) initiate or delay rescue operations, and (iv) identify the cause as a natural event rather than equipment failure. Short-term forecasts are of particular value in high latitudes in which ionospheric communications may be the only contact with the outside world. For further information concerning the effects of solar disturbances on radio propagation the reader is referred to the books by Gassmann (1963), Agy (1970), and Mitra (1974) and to the *SESC Glossary of Solar-Terrestrial Terms* (SESC 1988).

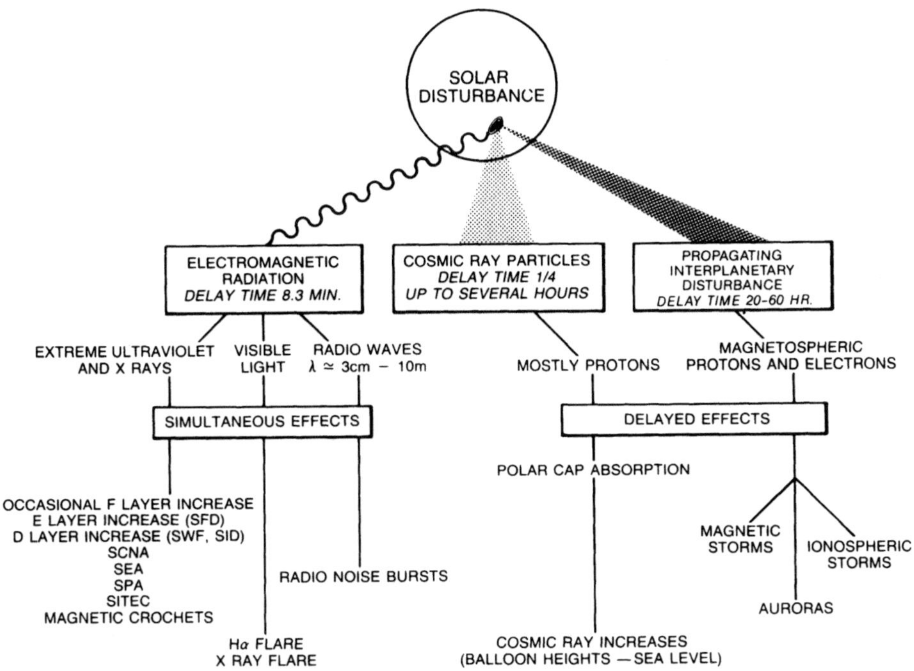

Fig. 9.1 *Terrestrial effects of solar disturbances*

9.2 Solar disturbances

9.2.1 Solar flares

One of the more spectacular solar disturbances is a solar flare, which is a sudden brightening of the Hα line (6563 Å) in an active region (*Solar Geophysical Data*

1985, No. 489). Its brightness increases, and the area expands to a peak in a few minutes. The duration can last from 3 to 4 minutes to several hours; the average lifetime is about 30 minutes. Flares are assigned an 'importance' based on area at maximum brightness (0 to 4) and brightness (faint F, normal N, and bright B). An optical flare may be accompanied by X-ray and EUV emissions and radio bursts.

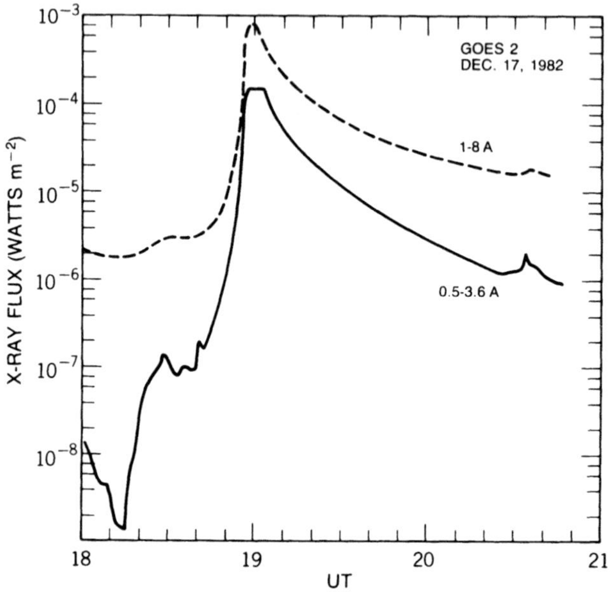

Fig. 9.2 *Observed flux of an X-ray flare in two wavelength ranges (Courtesy H. Garcia)*

X-ray flares are increased emissions in the wavelength range 1–8 Å. They are particularly important from the point of view of ionospheric radio because these wavelengths ionize the D region, which results in radio absorption. The peak X-ray flares are rated C, M, and X according to the fluxes (E) in the 1–8 Å range as measured by instruments on satellites in earth orbits where for C, $10^{-6} < E < 10^{-5}\,\mathrm{W\,m^{-2}}$; for M, $10^{-5} \leqslant E \leqslant 10^{-4}\,\mathrm{W\,m^{-2}}$, and for X, $E > 10^{-4}\,\mathrm{W\,m^{-2}}$. An additional digit may be added to the C, M, X to indicate the coefficient of the flux; e.g. $C8 = 8 \times 10^{-6}$, $M3 = 3 \times 10^{-5}$. Most class-X and some class-M flares produce shortwave fadeouts that affect radio communications. From 1969 through 1976 there were 1344 class-M and 141 class-X flares observed (Kildahl, 1980). From June 1976 through September 1986 (sunspot cycle 22) there were 2634 class-M and 172 class-X flares observed (Hirman 1987, private communication). The X-ray detectors operated on the Geostationary Operational Environmental Satellite (GOES) by the National Oceanic and Atmospheric Administration are simple ion chambers covering the two channels 0·5–4 Å and 1–8 Å. These wavebands were selected because they are directly applicable for computing D-region absorption. Since solar X-rays cannot be measured when the satellite is eclipsed by the Earth or by the Moon, observations by at least two satellites are required for continuous data. Figure 9.2 gives an example of the observed flux of a class-X solar flare. The C, M, and X classification is based on full-disk emission; a more satisfactory measurement can be obtained from a solar X-ray image from which one can locate the flare and, therefore, separate the flare emission from the whole disk emission. Some large flares have an 'explosive phase' during which they reach peak brightness in a minute or two, and then slowly decline. Both the X-ray and EUV flares occur higher in the solar atmosphere than the optical flare; this shows that a flare has a large vertical structure extending into the corona.

Other disturbance features related to terrestrial effects are prominences, which are clouds of glowing gas near the limb suspended (at about 50 000 km) in magnetic fields in the corona. On the disk a prominence appears dark and is called a filament. Active filaments live a few hours and appear as surges and loops (McIntosh and Dryer 1972). Delayed ionospheric effects are sometimes associated with disappearing filaments (Joselyn and McIntosh 1981).

A comprehensive flare index (SESC 1988) is given by

$$\text{CFI} = A + B + C + D + E \tag{9.1}$$

where

A = importance of ionizing radiation as indicated by the importance of the associated SID; scale 1–3

B = importance of H-alpha flare, scale 1–3 (3 includes classes 3 and 4)

C = log (peak $F_{10\cdot7}$)

D = effects associated with the dynamic radio spectrum: Type II burst = 1, Continuum storm = 2, Type IV burst = 3

E = log (200 MHz flux in units of $10^{-22}\ \text{W m}^{-2}\ \text{Hz}^{-1}$)

The CFI ranges from 1 to 17; values > 10 indicate flares with unusually strong electromagnetic radiation. A TENFLARE is a solar flare accompanied by a 10·7 cm noise burst of intensity greater than 100% of the background flux.

9.2.2 Coronal holes

Coronal holes are large regions of the solar corona that are exceptionally low in density, associated with unipolar photospheric regions having open magnetic field topology. Coronal holes are sources of high-speed solar wind that can cause ionospheric and geomagnetic storms. They can be long-lived, some surviving several 27-day solar rotations, and they give rise to recurrent storms. Coronal holes can be seen in X-ray through radio wavelengths (Bohlin 1977).

9.2.3 Coronal mass ejections (CMEs)

CMEs are observed in white light by coronagraphs and imply a large-scale disturbance propagating from the outer corona into the solar wind. CMEs are associated with flares and prominences (Wagner 1984).

9.2.4 Radio bursts

Flares associated with particle emissions generally have strong microwave bursts as well as Type II and IV radio emission. Microwave bursts are generally broadband, often extending into the millimeter and decimeter domains.

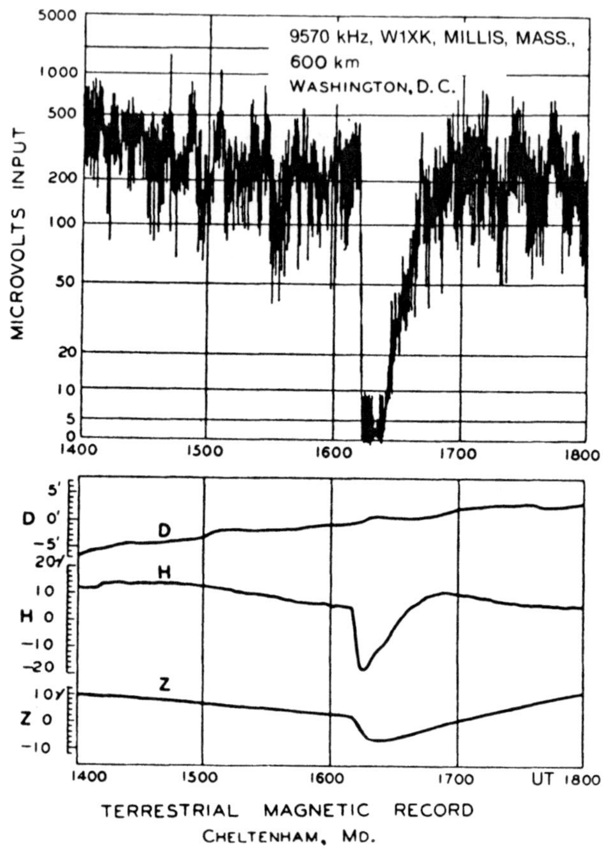

Fig. 9.3 *Example of (top) a shortwave fadeout and (bottom) a geomagnetic crochet (November 26, 1936)*

9.3 Sudden ionospheric disturbances

9.3.1 Shortwave fadeout

Sometimes called a Mögel-Dellinger effect, the shortwave fadeout (SWF) was the first of the various sudden ionospheric disturbances to be associated with a solar flare. An SWF is illustrated in Fig. 9.3. It consists of a sudden decrease in a high-frequency signal, which may last from a few minutes to several hours. Onset of this absorption is usually (but not always) abrupt, and it is followed by a relatively slow recovery. As shown in the figure, SWFs are sometimes accompanied by a transient variation in the geomagnetic field (called a magnetic crochet or solar flare effect) indicating the existence of electric currents (Ohshio 1964). The magnitude of a shortwave fadeout depends on (i) the size of the X-ray flare, (ii) the solar zenith angle χ, and (iii) to some extent, the background ionospheric conditions. Essentially all solar X-ray flares on or near the front side of the Sun are detected by the Space Environment Monitors on the GOES satellites. The high time resolution (≈ 3 s) and high sensitivity of satellite detectors enable the early stage of a flare to be detected minutes before the X-ray intensity is high enough to produce significant ionospheric effects (e.g. a class-X flare). The absorption in an SWF depends on the solar zenith angle as $\cos^n \chi$ where n is about 0·75. Table 9.1 demonstrates the dependence of the absorption and duration of an SWF on the optical importance of flares.

Table 9.1 *Average magnitudes of attenuation and duration, on 5 MHz and for an overhead Sun, of shortwave fadeouts*

Flare category	Importance	Attenuation (dB)	Duration (hours: minutes)
SF	0·1	007·0	0 : 17
SN	0·2	010·0	0 : 19
SB	0·3	020·0	0 : 21
1F	0·7	008·0	0 : 24
1N	1·0	018·0	0 : 28
1B	1·3	025·0	0 : 31
2F	1·7	012·0	0 : 37
2N	2·0	025·0	0 : 45
2B	2·3	032·0	0 : 54
3F	2·7	015·0	1 : 06
3N	3·0	030·0	1 : 18
3B	3·3	040·0	1 : 32
4F	3·7	018·0	1 : 40
4N	4·0	035·0	2 : 00
4B	4·3	045·0	2 : 15

* From Stonehocker (1970)

The behavior of a radio circuit during an X-ray burst is illustrated in Fig. 9.4 where we see that, almost concurrent with the X-ray flare, the lowest observed frequency (LOF) increases from 9 MHz to about 17 MHz and, to a first approximation, follows the time profile of the X-rays. Observations indicate that the LOF depends on solar zenith angle χ according to

$$\mathrm{LOF} = A \cos^{0.25} \chi \tag{9.2}$$

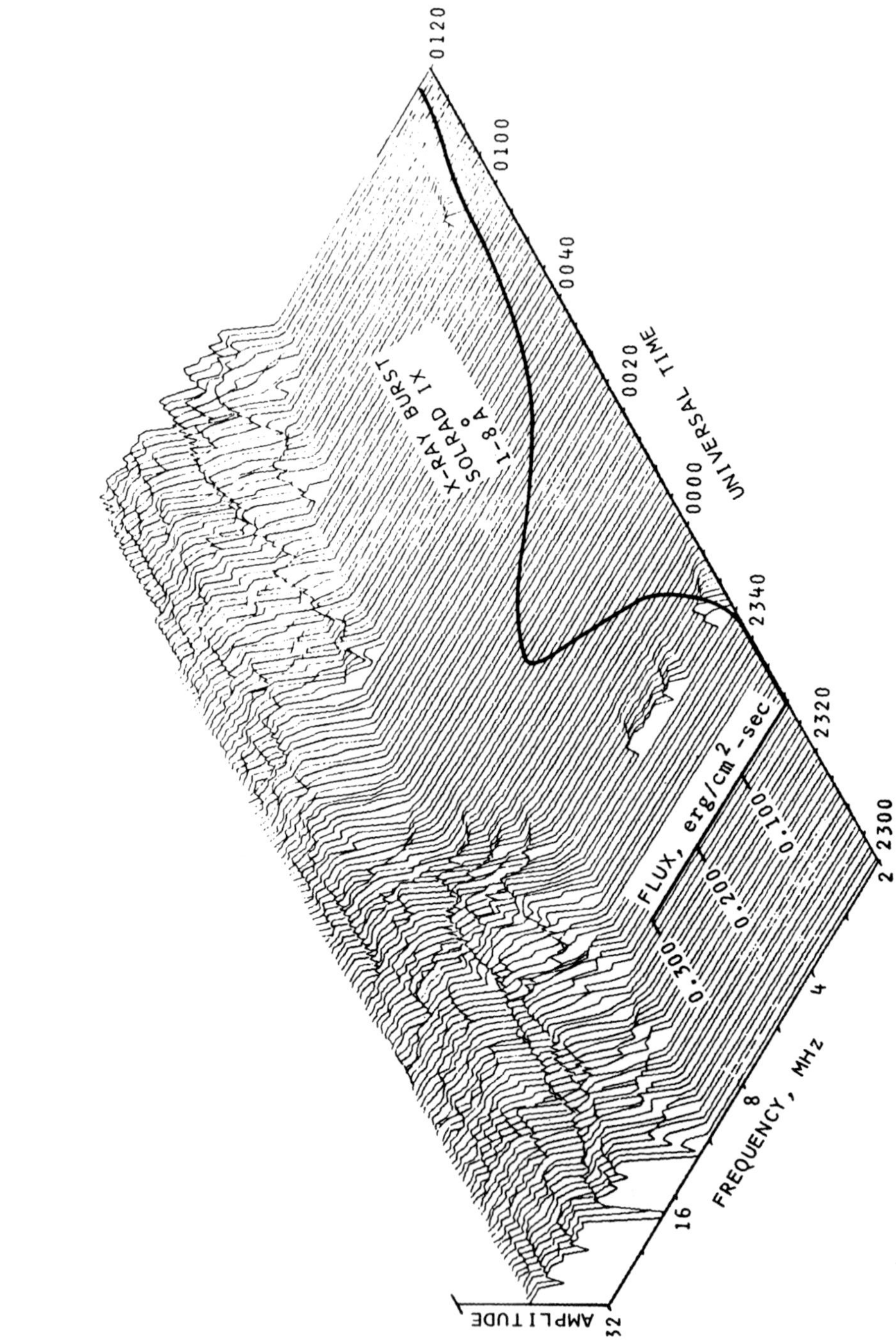

Fig. 9.4 *HF oblique sounder data for the 4000 km path (Hawaii to Corona, California) during a shortwave fadeout and the associated solar X-ray flare (From Bleiweiss et al. 1973)*

where A depends on equipment, size of the X-ray burst, and circuit characteristics. There is evidence (Ohshio 1964, p. 391) of solar flare effects in the night hemisphere.

Not all solar X-ray bursts commence rapidly, and there is a category called gradual ionospheric disturbances (GID). Monthly tables of the start, maximum, and end times (in UTC) together with importance of the associated flare, X-ray class, and NOAA/SESC region are published in *Solar Geophysical Data*. The following sub-categories of SWFs are used:

SWF(S): sudden dropout and slow recovery
SWF(SL): slow dropout taking 5–15 min, and gradual recovery
SWF(G): gradual disturbance; fade irregular in either dropout or recovery or in both

Ionospheric disturbances have importance ratings that depend on (i) the degree to which the normal field strength is reduced – allowing for the cos χ dependence, (ii) the duration of the fade, especially at the minimum (the longer the duration the greater the importance), and (iii) the definiteness of the event. For importance rating 1 – to 3+, we have roughly as follows: 1–, for duration of 14 min or less; 1, for 15 to 29 min; 1+, for 30 to 44 min; 2–, for 45 to 59 min; 2, for 60 to 74 min; 2+, for 75 to 90 min; 3–, for 90 to 104 min; 3, for 105 to 119 min; and 3+, for more than 120 min. The importance is increased or decreased according to whether the fading depth is different from the quiet variations about the median or diurnal trend by more or less than a factor of 2. The definiteness ranges from 5 = definite to 0 = questionable, depending on frequency, local time, interference, calibration break, etc. (Lincoln 1964).

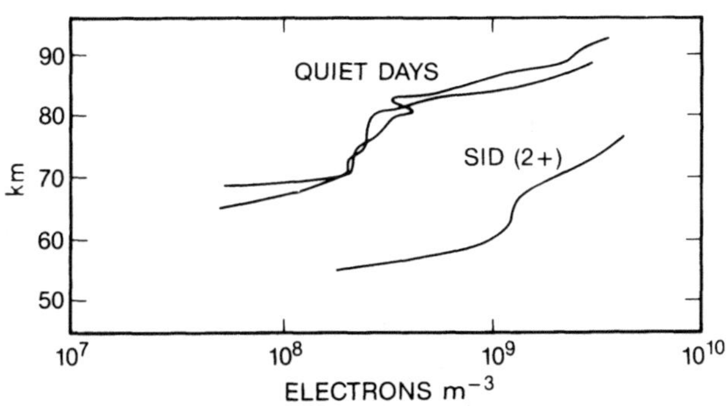

Fig. 9.5 *Electron-density profile on quiet days and during a (2+ flare) sudden ionospheric disturbance, measured by the partial reflection technique at Ottawa, Canada (From Belrose 1964; with permission)*

In some very large SWFs the electron density in the D region may increase by a factor of about 10 (Fig. 9.5). From the point of view of a radio operator, the occurrence of a shortwave fadeout may not be too serious because, although a nuisance, it lasts a relatively short time. For example, during the sunspot maximum year 1937 at Washington, DC, SWFs were observed on 84 days. On 66 days the fadeout was intense; on 39 days there was more than one SWF and on 33 days the SWF lasted for more than 1 hour. On the other hand, in the sunspot minimum year of 1944, the corresponding numbers were only 5, 3, 0, and 2. In sunspot-maximum 1947 they were high again: 121, 104, 54, and 33.

9.3.2 Sudden phase anomaly

A sudden phase anomaly (SPA) is a sudden change of phase $\Delta\phi$ caused by changes in reflection heights (Δh) of VLF (3–30 kHz) signals. Over short (a few hundred kilometers) paths we have

$$\Delta h = \lambda\Delta\phi\,(d^2 + h^2)^{1/2}/4\pi h \tag{9.3}$$

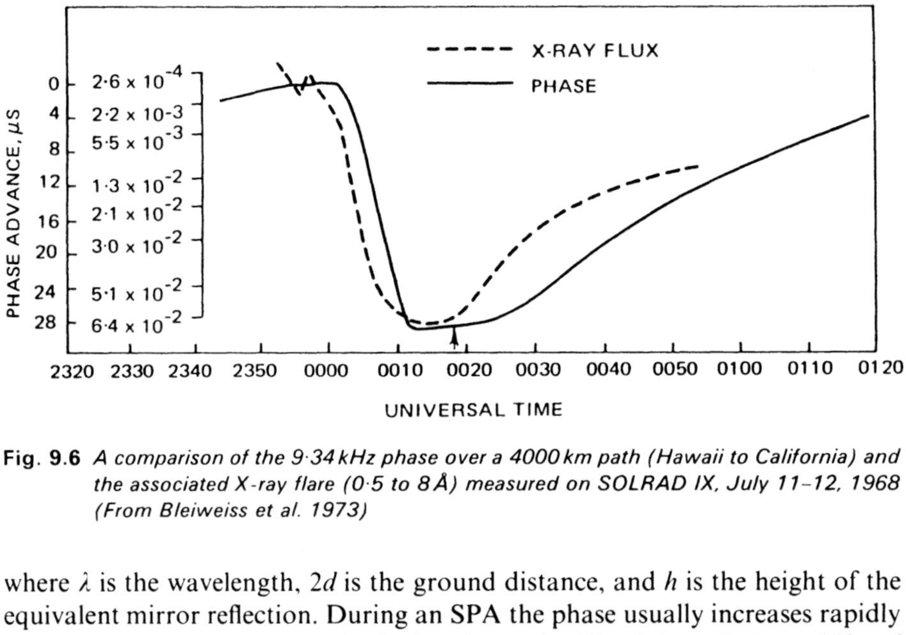

Fig. 9.6 *A comparison of the 9·34 kHz phase over a 4000 km path (Hawaii to California) and the associated X-ray flare (0·5 to 8 Å) measured on SOLRAD IX, July 11–12, 1968 (From Bleiweiss et al. 1973)*

where λ is the wavelength, $2d$ is the ground distance, and h is the height of the equivalent mirror reflection. During an SPA the phase usually increases rapidly (in 1–5 min) and recovers slowly (in 30 min to 3 h) (Fig. 9.6), and phase shifts of 30° to 60° are common for moderate flares. Certain large flares produce large phase perturbations. For example, on May 4, 1960, an SPA of 240° occurred on the Rugby(GBR)-to-Boulder path; that corresponds to a change in effective height of about 18 km. There appears to be a threshold value of the reflection height of about 6 km above which SPAs are accompanied by SWFs. SPAs are usually accompanied by sudden enhancements of [VLF] signals (SESs). A similar sudden enhancement of atmospherics (SEA) is an increase in the field strength of VLF noise near 27 kHz. The phase variations follow fairly closely the time profile of the solar X-ray inputs (Jones 1971). Not all X-ray flares are accompanied by phase advances; indeed there are cases of negative SPAs

(Ohshio 1971) observed on the Fort Collins (WWVL 20 kHz)-to-Japan path and no (or null) SPAs, which occur for solar zenith angles greater than 75°.

9.3.3 *Sudden frequency deviation*

The phenomena discussed above (SWF, SPA, SEA) occur in the D region. Solar flare effects have also been observed in the F_2 region (Munro *et al.* 1961; Donnelly 1971) and in the E layer (Bibl 1963). An example is the sudden frequency deviation (SFD), which is a sudden change in the frequency of a reflected HF signal. SFDs can be simple or complex (Fig. 9.7). The simple SFD (Fig. 9.7*a*) shows an increase of frequency (positive stage) followed by a recovery (negative stage). Since the ionosphere eventually returns to its pre-SFD state the total area under the $\Delta f(t)$ curve must be zero (see Sec. 7.8.4). Observations

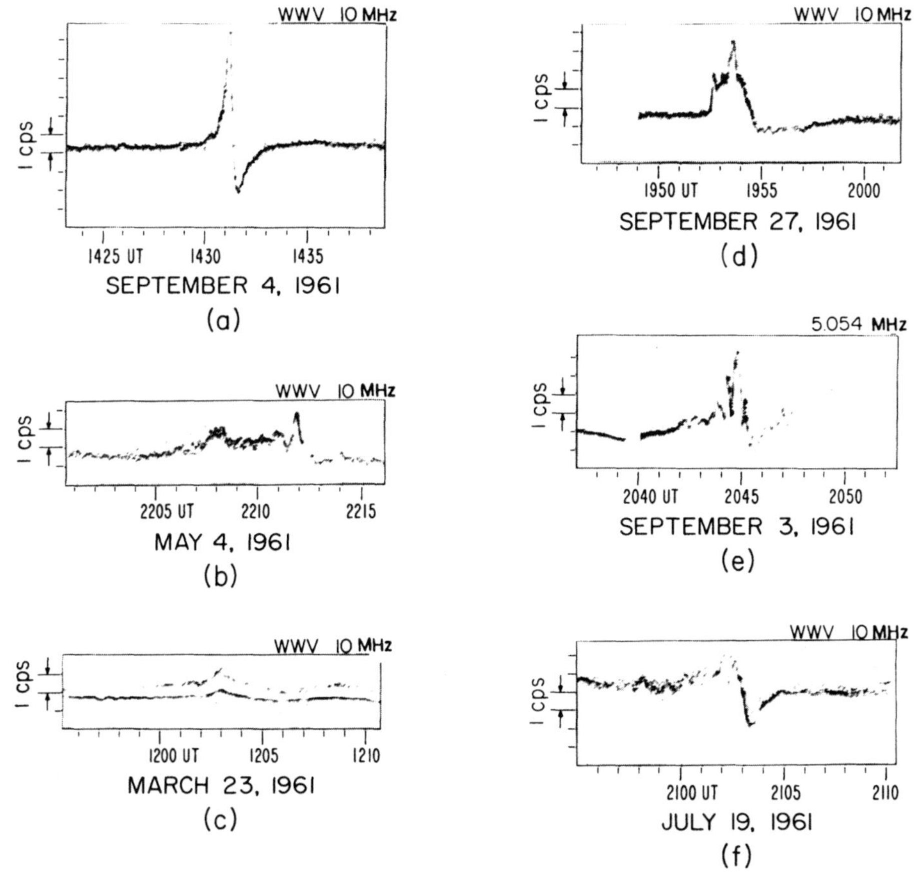

Fig. 9.7 *Examples of sudden frequency deviations recorded at Boulder, Colorado*

indicate that many SFDs occur without excessive absorption, indicating that the additional ionization is produced, by solar extreme ultraviolet (EUV) radiation, in the E region where the electron collision frequency is small (Donnelly 1971). The rate of decay enables us to estimate the linear loss rate from the continuity equation:

$$\Delta f(t) \propto \Delta \dot{N} = \Delta q - \Delta N/\tau \tag{9.4}$$

The enhancement in electron density ΔN is caused by an increase Δq in production rate. The dot denotes the time rate of change,

$$\tau = 1/\beta \tag{9.5}$$

is the 'relaxation time or time constant' and β is the loss coefficient. In the case of quadratic (recombination) loss for which Δq and ΔN are small compared with the undisturbed values (q_0, N_0),

$$\tau = 1/2\alpha N_0 \tag{9.6}$$

where α is the recombination coefficient.

In the case of a highly structured SFD in which the enhanced production is much greater than the additional losses we have

$$\Delta f(t) = 27 \times 10^{-8} f_v^{-1} \Delta q_T \tag{9.7}$$

where f_v is the equivalent vertical frequency in hertz and Δq_T is the total production rate in a vertical column of the region of enhanced production. Note that (9.7) shows that for short-lived enhancements the time profile of the SFD follows the electron production and, hence, the time profile of the EUV flare. When this method is used, the relaxation time of the upper E region turns out to be about 40 s (Baker and Davies 1966). From simultaneous measures of SWF and SFD we find that electrons may be produced over a range of heights; thus SFDs occur having various characteristics:

(i) Large absorption changes, small phase changes – due to D-region enhancements
(ii) Large absorption changes, large phase changes – due to enhanced production in and above the D region
(iii) Small absorption changes, large phase changes – due to enhanced production above the D region.

The majority of SFDs appear to belong to type 3, in which the bulk of the enhanced electron content lies between heights of about 110 km and 200 km. Many of the larger SFDs occur near the explosive phases of flares (Donnelly 1969) and are often accompanied by impulsive bursts of radio noise.

9.3.4 Sudden cosmic noise absorption

Sudden cosmic noise absorption (SCNA) is a sudden descrease in the strength of cosmic noise followed by a gradual recovery (Fig. 9.8). As in the case of an

SWF, an SCNA is assigned an importance: 1 – small intensity change and short duration; 2 – moderate intensity change and relatively long duration; 3 – large intensity change and long duration. The size S of an SCNA can also be expressed in terms of the diode current I_n in the absence of the flare to that I_f at the peak of the flare. That is,

$$S = (I_n - I_f)I_n \tag{9.8}$$

The ratio of the peak absorptions in Fig. 9.8 of the extraordinary wave and the ordinary wave ($14/8 = 1{\cdot}75$) is approximately $(f + f_H)^2/(f - f_H)^2 \approx (10 + 1{\cdot}5)^2/(10 - 1{\cdot}5)^2 = 1{\cdot}83$. From this, one concludes that the enhanced ionization lies at heights where $(\omega \pm \omega_H)^2 \gg \nu^2$ (see Sec. 7.4.1). In most X-ray flares the ionization is produced between about 70 km and 85 km and may be accompanied by changes in ion chemistry.

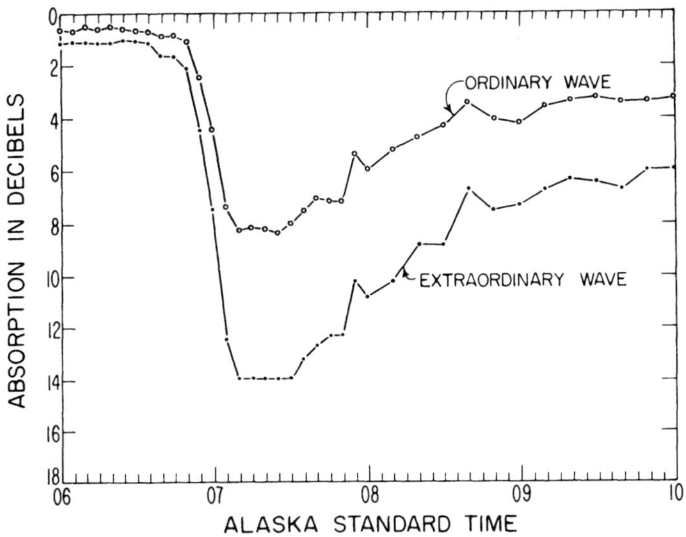

Fig. 9.8 *Sudden cosmic noise absorption, on 10 MHz, at College, Alaska, June 11, 1961 (From Lawrence et al. 1964)*

9.3.5 Sudden enhancement/decrease of atmospherics

The field intensity of atmospherics changes during an X-ray flare. The signal may increase or decrease depending on the wave frequency. Sudden enhancements of atmospherics (SEAs) occur approximately between 10 kHz and 75 kHz, the upper limit depending on path length (Sao *et al.* 1970). Enhancement also occurs on extremely low frequencies (ELF), below approximately 1 kHz, whereas between 1 and 10 kHz a sudden decrease of atmospherics (SDA) is observed.

9.3.6 Sudden increase in total electron content

Sudden increases in total electron content (SITECs) indicate sudden enhancements in ionospheric electron content produced by extreme ultraviolet (EUV) flares. These increases can amount to about 5% of the background electron content. Figure 9.9 is an example of a particularly large SITEC in which the rise time was about 2 min (Davies 1980). SITECs give information on the increases of ionization in the upper E and F regions.

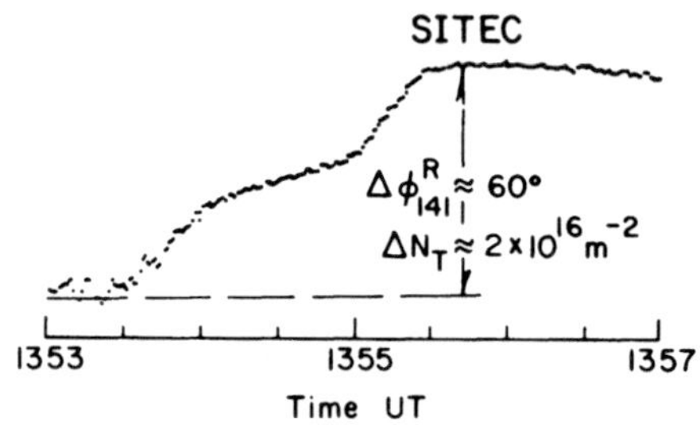

Fig. 9.9 *Sudden increase in total electron content on 141 MHz, ATS-6 to Boulder, July 4, 1974*

9.4 Polar cap disturbances

9.4.1 Solar proton events

During some large flares, highly energetic protons are emitted, which enter and ionize the polar ionosphere. Flares also emit energetic nuclei, relativistic electrons, gamma rays, etc. Proton energies are in the range 1 to 400 megaelectronvolts (MeV) where 1 eV is the energy acquired by an electron falling through a potential of 1 V. In general, the proton density decreases exponentially with energy. The most energetic events, detected at the ground by cosmic ray detectors, are called ground-level events (GLE) (Hultqvist 1969). Thirty-nine GLE events were observed in the 45 years from 1942 through 1987 (Shea 1988, private communication).

Since the early 1960s protons have been monitored continuously from spacecraft such as the NOAA–GOES satellites. For low energies ($\leqslant$ 10 MeV) the solar proton energy flux far exceeds that of the galactic cosmic rays. A proton flare is one that produces significant fluxes of greater-than-10 MeV protons in the vicinity of the Earth (Fig. 9.10). Generally, only the largest flares produce protons with energies and fluxes large enough to affect the ionosphere. Protons with energies of about 30 MeV can penetrate to about 50 km and produce polar cap radio absorption (PCA) whereas the energy needed for a GLE is about 500 MeV. The precipitation of protons into the Earth's atmosphere is characterized by an invariant cutoff latitude. Above the cutoff latitude the proton fluxes appear to be quite uniform, but asymmetries are present, mainly during the initial stages, over the central polar areas and near the cutoff regions. The influx near the cutoff latitudes is time and energy dependent, especially at lower energies ($\leqslant 20$ MeV). The cutoff latitude λ_c is related to the proton energy E (in MeV) through the cutoff rigidity R_c by

$$R_c^2 = E^2 + EE_0 = 222 \times 10^6 \cos^8 \lambda_c \tag{9.9a}$$

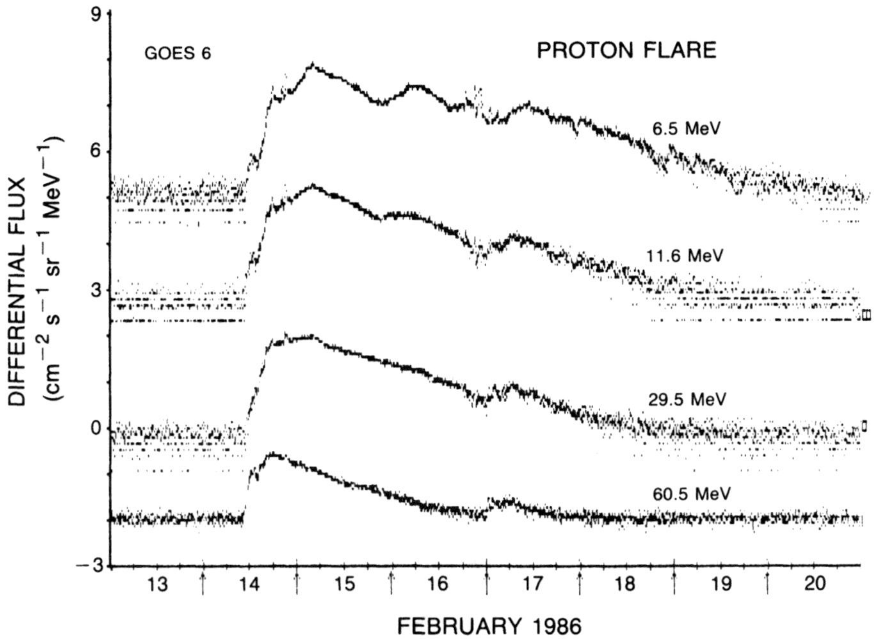

Fig. 9.10 *Proton flare measured on GOES 6 in February 1986 (Courtesy H. H. Sauer)*

where

$$E_0 = \tfrac{1}{2} mc^2 \tag{9.9b}$$

is the rest mass energy of the particle. Observations show that protons of medium and low energies can penetrate to latitudes lower than those given by theory (Reid 1986). The variation of λ_c with proton energy is shown in Fig. 9.11 from which we see that, with a model magnetic field, protons with energies less than about 10 MeV are excluded from latitudes below about 65°. For many practical purposes the polar cap is defined as the area at latitudes higher than about 65°.

When a proton enters the atmosphere it produces ionization by collision, expending 35 eV per ion pair. Because the air density increases with penetration depth the ion production is concentrated near the end of the path. The penetration of protons and electrons, their energies, and cutoff latitudes as functions of speed are shown in Fig. 9.12. Pronounced north–south asymmetries have been observed (Reid 1986); one polar cap may have uniform fluxes while the other is structured. The ion chemistry associated with a solar proton event (SPE) has been discussed by Reid (1986).

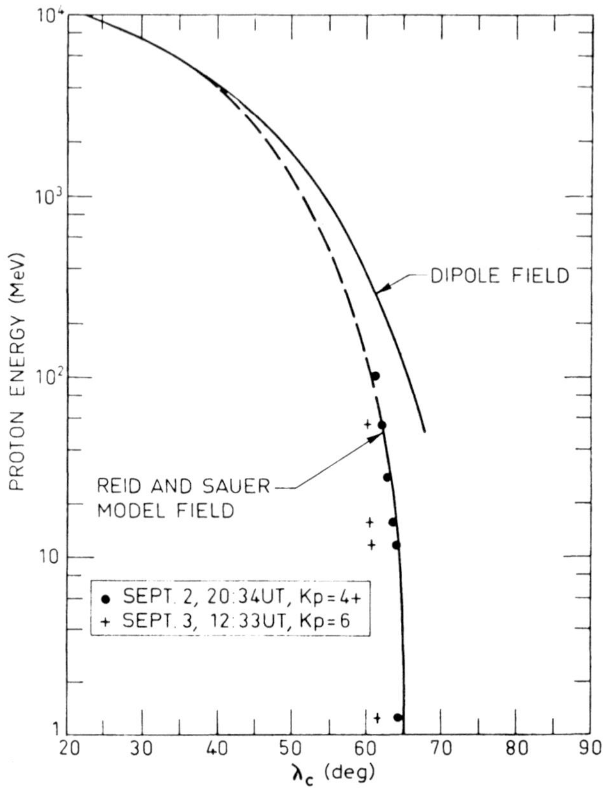

Fig. 9.11 *Comparison of geomagnetic cutoff latitudes (λ_c) in a realistic field with those observed during the solar proton event of September 2, 1966 (From Hofmann and Sauer 1968)*

9.4.2 Polar cap absorption

Polar cap absorption (PCA) is the result of the intense ionization created by energetic protons. PCAs last from an hour or so to 60 hours or more, depending on the latitude and on the duration of the source. They are associated with the polar cap rather than the auroral zone, and they are not closely correlated with magnetic disturbance except during the later stages. PCAs have been studied mostly using riometers, VHF forward scatter, partial reflections, and ionosondes. Figure 9.13 gives an idea of the seasonal variations of occurrence and delay times. There is a winter minimum in occurrence, and a winter maximum of time delay that varies from about 2 h to about 10 h, depending on the size of the event. The occurrence of PCAs tends to follow the sunspot cycle (Larsen

1979). A list of SPEs is given by Svestka and Simon (1975); for 1955 through 1969, 352 events are 'confirmed'. The variation of PCA occurrence with sunspot number is shown in Fig. 9.14. The number of PCAs reported depends on the technique used and, therefore, intercomparisons of the different cycles are not particularly reliable.

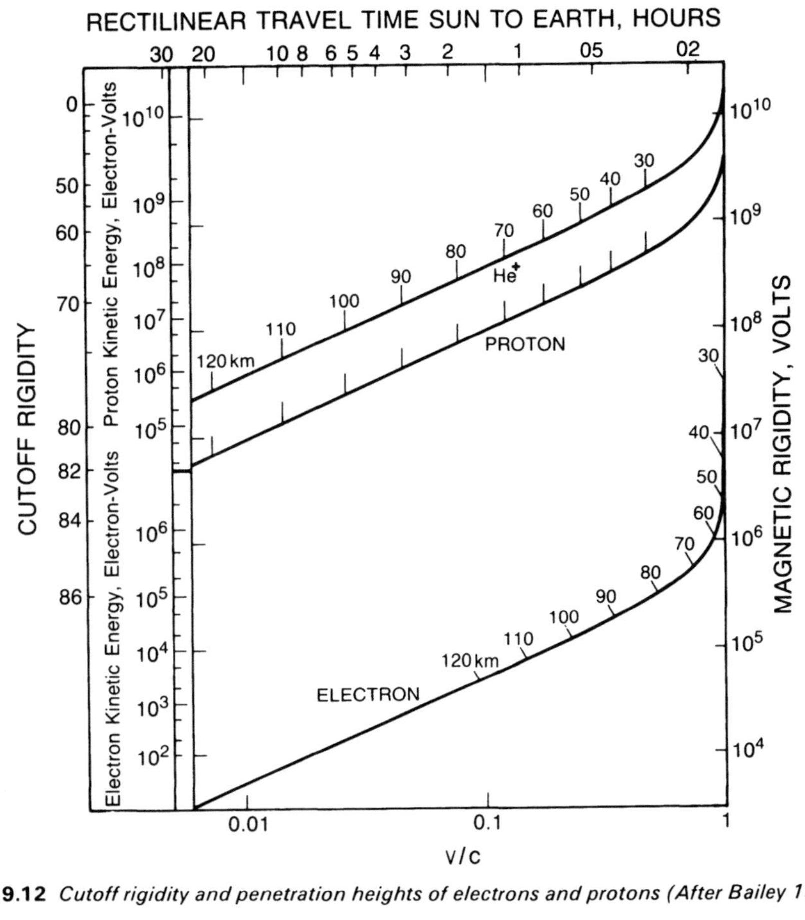

Fig. 9.12 *Cutoff rigidity and penetration heights of electrons and protons (After Bailey 1959)*

The history of a PCA (Fig. 9.15) shows that the absorption increases with latitude and is high during daylight and low at night, owing to electron attachment to O_2 molecules. The absorption appears first in high latitudes and expands equatorward. The maximum absorption on 27 MHz is about 13 dB in the high latitudes. For a 10 MHz ordinary wave reflected from the F layer this would lead to absorption of about 170 dB, which would produce circuit blackout for low power transmissions. There is a short strong absorption associated with the sudden commencement (SC) of the geomagnetic storm followed by a temporary decrease. To summarize:

(i) PCAs are almost always preceded by a major solar flare. The time delay may be as short as $\frac{1}{2}$ hour; average delay is 3 to 4 hours.
(ii) The duration varies from less than 1 day to about 10 days.
(iii) Magnetic activity is usually absent during the first day or so of the event, after which an SC magnetic storm may occur. Most, but not all, PCAs have their maxima before the start of the magnetic storm.
(iv) Nearly every PCA is preceded by a broadband centimeter Type IV radio burst (Svestka 1972).

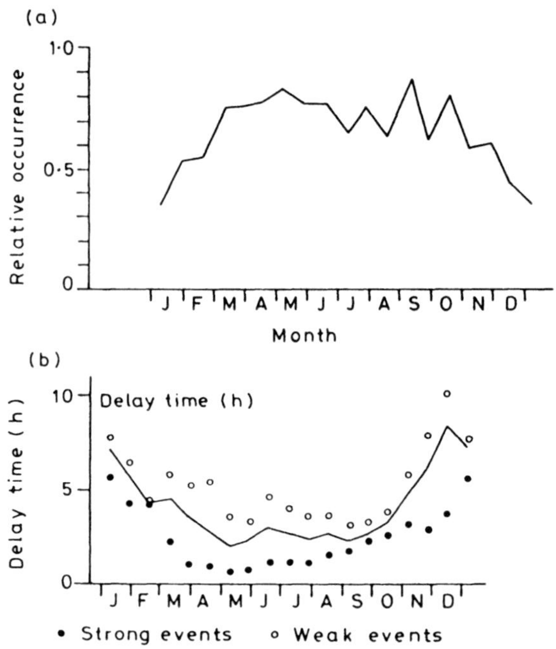

Fig. 9.13 *Seasonal effects in polar cap absorption (From Hargreaves 1979 © Van Nostrand Reinhold Co. Ltd.; reprinted with permission)*
(*a*) Fraction of the flares with Type IV radio bursts that produce PCA
(*b*) Seasonal variation of delay time for strong and weak PCA events

Below 40 km the collision frequency is so high that the absorption coefficient (eqn. 7.19) is small. Hence, the main part of a PCA is due to ionization by protons of less than ≈ 100 MeV, and the cutoff latitude for these protons determines the extent of the PCA. An empirical relationship between riometer absorption L (in dB) on 30 MHz and the omnidirectional proton flux (F) above 20 MeV is

$$L = 0{\cdot}13\, F^{1/2} \tag{9.10}$$

Thus, during a large PCA ($L > 20$ dB) $F > 20\,000$ protons $\mathrm{cm^{-2}\, s^{-1}}$.

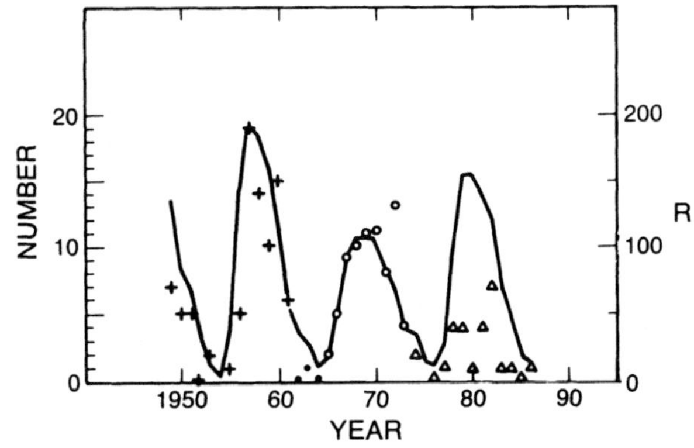

Fig. 9.14 *Annual occurrences of polar cap absorption events and sunspot numbers (R) 1949–1986*
+ from Davies 1965, Fig. 6.14
○ from Larsen 1979, Fig. 5.2
△ from solar proton data

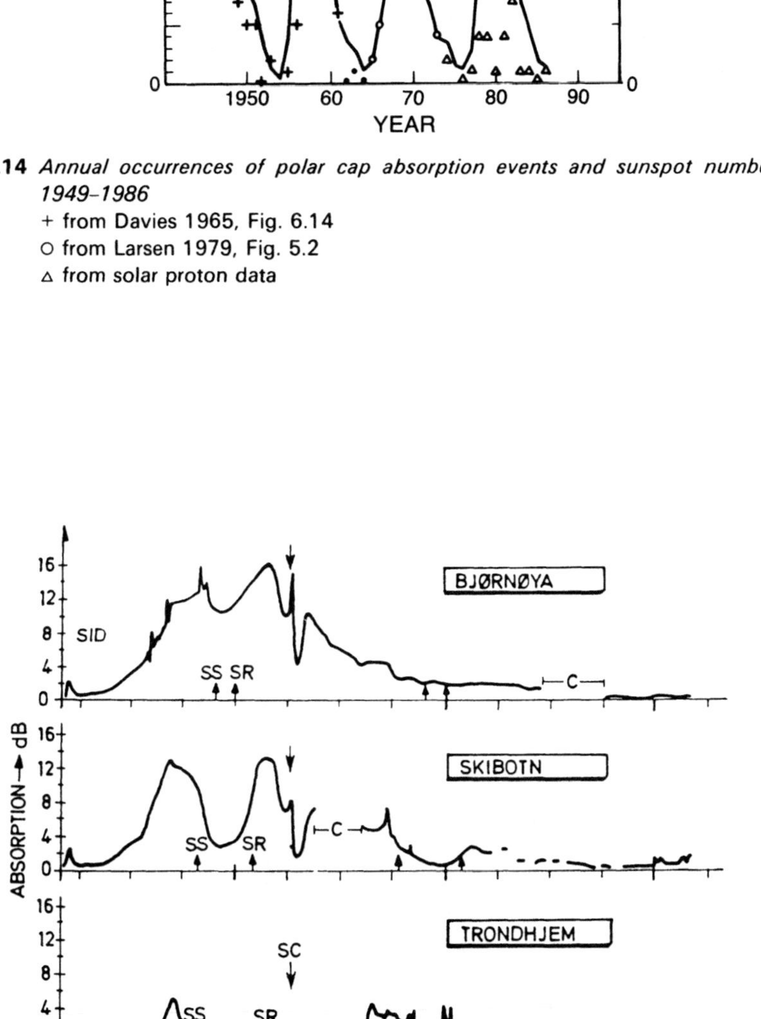

Fig. 9.15 *Absorption on 27·6 MHz riometers during a polar cap absorption event (From Holt 1968 © Plenum Press; reprinted with permission)*

9.4.3 *Effects on very low frequencies*

Transpolar VLF paths are particularly sensitive to solar proton events (Larsen 1979). The following features have been reported:

(i) Phase (on 18·6 kHz) is advanced (by cycles) and amplitude is attenuated (by 30 dB and more).
(ii) Field strength is sensitive to proton fluxes of about 1 proton $cm^{-2} s^{-1} sr^{-1}$ for protons of energy greater than 20 MeV.
(iii) Signals are 10–40 dB stronger at night than by day.
(iv) Signals are 10–20 dB stronger in the winter than in summer.
(v) Heavy attenuation occurs on paths that cross the southern part of the (north) polar cap where the auroral zone dips below the Arctic circle (Oelbermann 1970).

Very low frequencies are quite sensitive to the presence of the Greenland ice cap. The phase variations ($\Delta\phi$) normalized to a frequency of 26·1 kHz and an effective path of 5300 km (north of 62·5° corrected geomagnetic latitude) are as follows:
For a path across the Greenland icecap:

$$\Delta\phi = 12{\cdot}9 + 24{\cdot}5 \ln (F/F_0),\ F \geqslant 5{\cdot}4 \qquad (9.11a)$$

For a path over sea water:

$$\Delta\phi = -10{\cdot}9 + 16{\cdot}5 \ln (F/F_0),\ F \geqslant 7 \qquad (9.11b)$$

where F is the omnidirectional flux ($E > 25$ MeV) and $F_0 = 1$ proton $cm^{-2} s^{-1}$. The VLF reflection height is essentially the same over the whole polar cap during the peak of a PCA (Westerlund *et al.* 1969). The reflection height for a path over sea water is given by

$$h = 75 - 7{\cdot}9 \ln [\ln(F/F_0)] \qquad (9.12)$$

Values of F range from 10 to $> 10^4$ protons $cm^{-2} s^{-1}$. PCA events advance the phases of VLF signals from some 20 μs (≈ 1 cycle) to > 100 μs (> 6 cycles). Polar cap events can occur singly or overlap.

9.4.4 *Relativistic electron precipitation*

Relativistic electron precipitation (REP) was first identified on forward scatter circuits on very high frequencies (30–300 MHz) in the auroral zone ($4{\cdot}5 < L < 6$) (Bailey 1968). REP events have a large latitudinal extent (≈ 2000 km), and most events last from 1 to 6 hours but can last longer. The seasonal variation has equinoctial peaks. Depending on the relationships between the height of scattering and the height of peak ion production, REP can be positive (signal enhancement, usually by day) or negative (signal decrease, by day and by night). REP has been observed also as absorption increases on riometers (Fig. 9.16). REP events are associated with enhanced geomagnetic disturbance and are essentially daytime phenomena having maximum occurrence before noon. They are associated with bursts of *bremsstrahlung* (X-rays produced by the stoppage of electrons in the atmosphere). Electron energies in REP events are about 500 keV, and the energy spectrum is exponential. REP events are also observed as VLF phase anomalies (Larsen 1979).

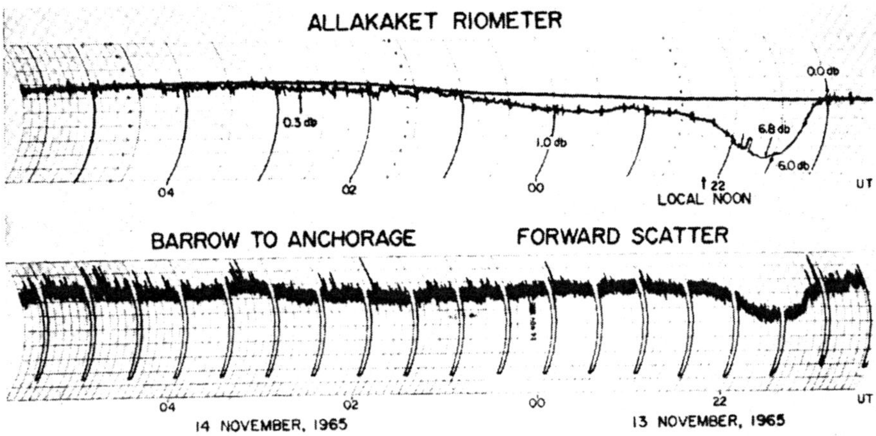

Fig. 9.16 *Example of absorption and forward scatter during a relativistic electron precipitation event in Alaska (From Bailey 1968)*

9.5 Storms

9.5.1 Importance in radio propagation

Ionospheric storms are by far the most important disturbances from the point of view of their societal impact because of their durations (several days), their adverse effects on the radio spectrum, and their geographical locations. Although PCAs are severe and long lasting they are rare and confined to relatively sparsely populated areas; SIDs affect populated regions but are relatively short lived. On the other hand, ionospheric storms affect middle and low latitudes where most ionospheric radio communication is conducted. Ionospheric storms are associated with geomagnetic storms, auroral and magnetospheric storms, etc. These storms affect human systems in a variety of ways. For example, (i) they can result in outages of electrical power systems (Albertson and Van Baelon 1970), (ii) they produce increase of atmospheric drag on satellites, (iii) they affect the electrical discharging of satellites (Schmerling 1987), and they induce large electric voltages in big conductors such as the 1100 km Alaskan pipeline (Campbell 1979). Storms can also affect other human activities such as geophysical exploration (Joselyn 1986*a*) and search and rescue.

9.5.2 Magnetic storms

Magnetic storms are disturbances of the geomagnetic field lasting from a few hours to several days during which time the various components of the Earth's field fluctuate over much wider limits (several percent of the total field in middle latitudes) than the limits on quiet days (a few tenths of 1%). In spite of the irregularity of the field the overall pattern can generally be classified into certain types such as the sudden commencement (SC) storm, discussed in Sec. 2.4.4.2, in which the initial phase is followed by a large decrease (the main phase), and by a recovery phase in which the field returns to normal after 2 to 3 days. The sudden commencement is essentially simultaneous over the globe. Magnetic storms are usually most intense near the auroral zones, which move equatorward in severe storms. Other storms develop gradually and so do not have sudden commencements.

The intensity of a storm is measured by several indices, e.g. the 3 h planetary index K_p and the daily 'equivalent amplitude' A_p (see Sec. 2.5). Magnetic storm data published monthly in *Solar Geophysical Data* include various K indices; daily average A_p; the Bartels diagram; and a list of the commencement and end times (when K drops to 2), SC amplitudes, and maximum K indices. In any given month the international quiet and disturbed days are tabulated. The K figures usually represent the magnetic effects of electric currents flowing in high latitudes. In low latitudes the major magnetic disturbance is associated with the ring current and is represented by the equatorial D_{st}. Hourly D_{st} data are based on values of the horizontal component at four stations: San Juan, Honolulu, Kakioka, and Hermanus. At the Space Environment Services Center in Boulder, Colorado, 'minor storms' indicates that the daily A_p index is between 30 and 49, and 'major storms' indicates that A_p equals or exceeds 50 (Joselyn 1986*b*).

Even during quiet periods, the daily variations of each of the geomagnetic components show remarkable day-to-day changes of amplitude and phase, which indicate day-to-day variations of the ionospheric current intensities and locations (Matsushita 1967). Geomagnetic storms may or may not be accompanied by several ionospheric phenomena such as auroral D-region absorption, D-region sudden-commencement absorption, auroral sporadic E, F-region storms, scintillation, radio noise emission, optical aurora, and substorm phenomena (Hultqvist 1967; Reid 1967).

9.5.3 Electron precipitation

Magnetic storms occur when solar wind with a sufficiently large southward magnetic field impacts the magnetosphere (Baker 1986). The magnetic field of the solar wind 'reconnects' with the geomagnetic field. The convection of these magnetic field lines over the polar caps constitutes a gigantic dynamo that generates up to 10^5 V across the magnetosphere from dawn to dusk; electric currents of 10^7 A generate 10^{12} W of power. It is thought that during quiet periods ions from the solar wind, and especially from the ionosphere, fill the magnetosphere with plasma. Magnetospheric substorms accelerate ions and electrons on the nightside from the plasma sheet, some along magnetic field lines into the auroral zones, causing visible aurora, auroral ionization, radio absorption, neutral atmosphere heating, and so forth. Thus ionospheric ions and electric fields affect the magnetosphere which, in turn, modifies the ionosphere and atmosphere. This is called atmosphere–ionosphere–magnetosphere (AIM) coupling or interaction.

Auroral storm ionization and heating are caused mostly by the precipitation of electrons with energies in the range of ~ 1 keV to 20–30 keV. Auroral heating of the neutral gas can completely change the global pattern of winds in the thermosphere. Large equatorward winds are established on the nightside, and large-amplitude gravity waves are generated that help to transport energy to lower latitudes; both of these modify the ionosphere on a worldwide scale.

Statistical patterns of energy deposition into the upper atmosphere, based on data from the total energy detector on the polar orbiting NOAA–TIROS satellites, have been compiled by Fuller-Rowell and Evans (1987) for different levels of activity. Figure 9.17 shows the extent of such energy deposition. During quiet periods the weak precipitation is well north of the radio circuits from the USA to Japan and Europe, whereas in a major disturbance the precipitation zone intensifies and moves to lower, and more highly populated, latitudes engulfing the circuits. Note that the energy input is concentrated in the midnight sector.

Electrons with energies of 1–20 keV produce most ionization above 100 km. Thus, in the disturbed auroral ionosphere, the peak electron density may be in the E region (Fig. 9.18).

9.5.4 D-region storms

9.5.4.1 Auroral absorption

Auroral absorption (AA) can be large (Fig. 9.19), but its temporal and spatial characteristics are different from those of PCAs. Whereas a PCA varies smoothly with time and location, AA fluctuates (Reid 1967) widely both in time and in space; the spatial correlation can fall to one-half in a distance of about 300 km, and the autocorrelation (Sec. 7.7.6) can fall to 0·65 in 10 min. The maximum occurrence of auroral absorption appears to be just equatorward of the maximum of visible aurora. It is infrequent in high geomagnetic latitudes but can extend into middle latitudes in a major storm. Figure 9.20 shows that AA maximizes in midmorning. At a fixed location the occurrence of auroral absorption may be higher at sunspot minimum than at sunspot maximum.

In middle latitudes the disturbed D region is characterized by a storm 'aftereffect' or 'poststorm' effect in which the D-region disturbance (particularly the VLF phase) persists for some days after the magnetic storm is over. This is believed to be caused by electron precipitation from the magnetosphere.

9.5.4.2 Sudden commencement absorption

At times of some SC storms there are brief absorption enhancements with durations of about 5 to 10 min. These enhancements are essentially auroral zone phenomena and appear on both the dayside and the nightside.

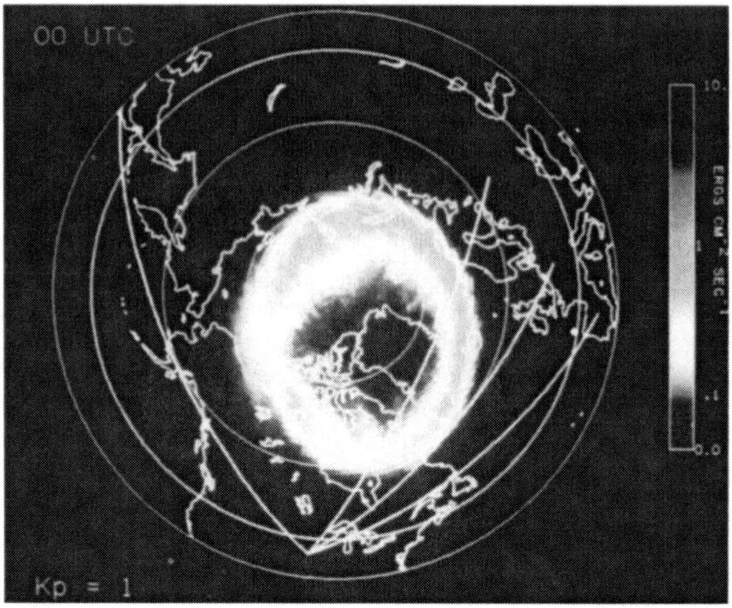

(a)

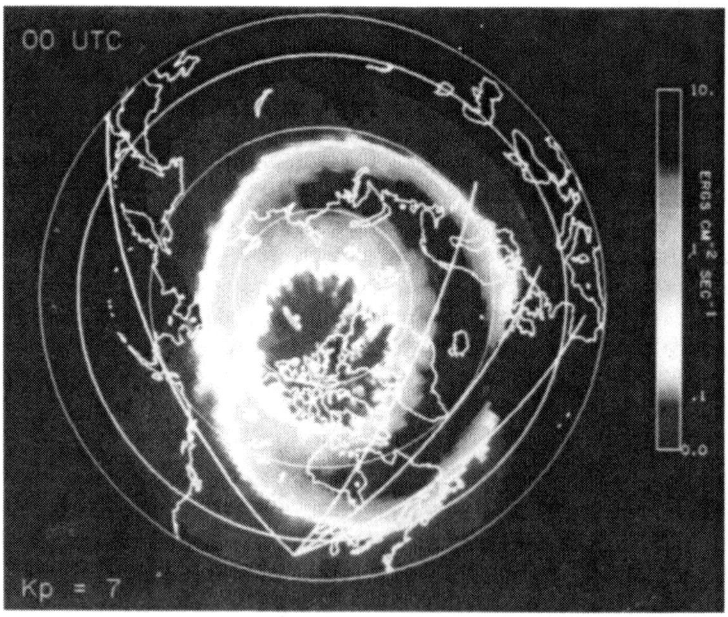

(b)

Fig. 9.17 *Energy precipitation zones, for different levels of activity, in relation to great circle paths between the central USA, Tokyo, Moscow, Frankfurt, and Madrid (Courtesy D. S. Evans)*

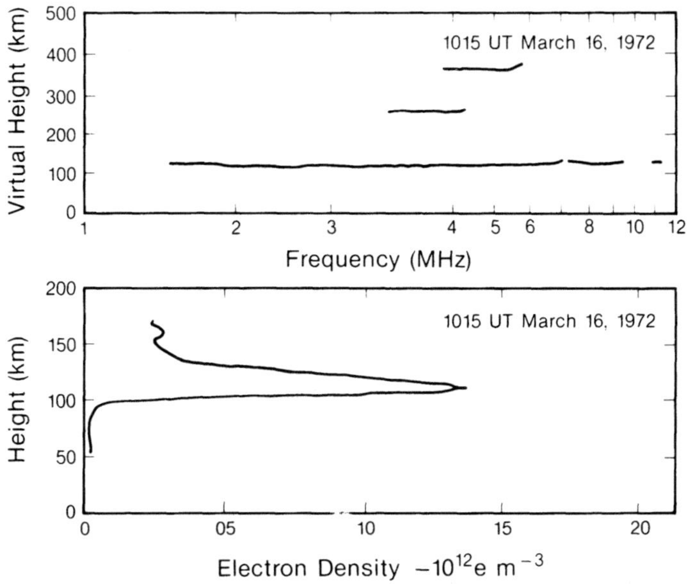

Fig. 9.18 *Examples of E-region phenomena in high latitudes during electron precipitation events (From Bates and Hunsucker 1974 © American Geophysical Union)*
Top: Ionogram traces indicating high values of E critical frequency
Bottom: Electron-density profile showing high electron densities

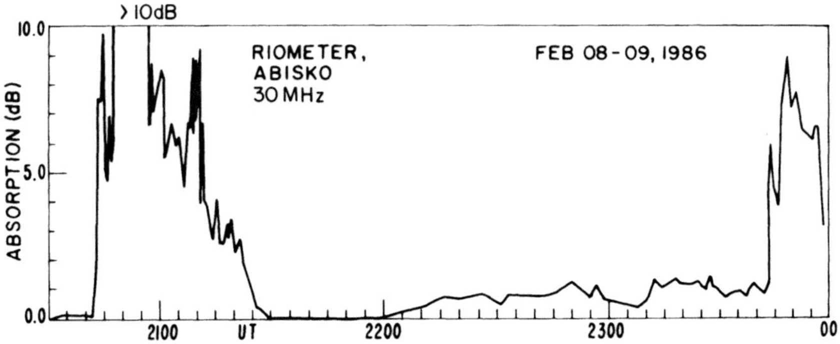

Fig. 9.19 *Auroral absorption in northern Europe during the storm of February 1986 (Courtesy J. Hargreaves)*

9.5.5 Auroral sporadic E

The most conspicuous feature of the disturbed E region is the appearance of auroral or storm E_s, particularly during nighttime (Reid 1967; Smith and Matsushita 1962). In the auroral zone the occurrence of E_s is more frequent over North America than over Siberia (Leighton et al. 1962). A thick E layer associated with magnetic disturbance is called night-E (King 1962; Wakai 1971). Near the auroral zone the critical frequency of night-E can be as high as 7 MHz. Auroral E_s is produced by energetic electron bombardment.

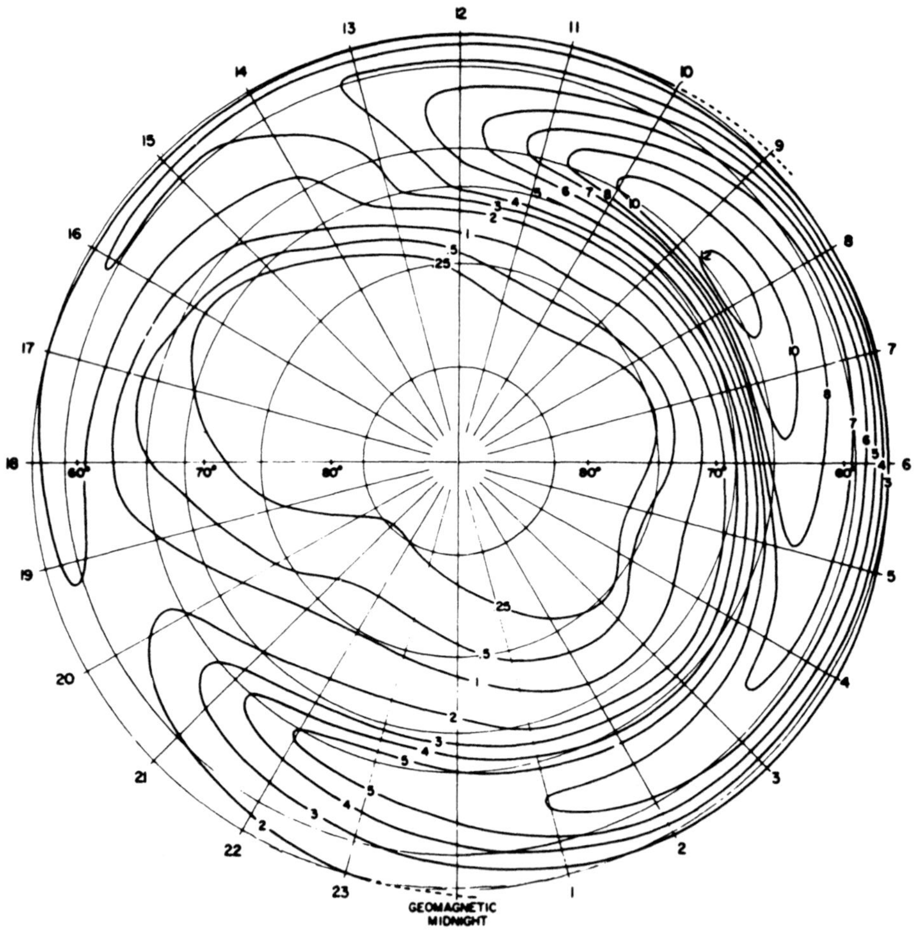

Fig. 9.20 *The percentages of times of occurrence of auroral radio absorption greater than or equal to 1·0 dB on 30 MHz (From Hartz et al. 1968 © Canad. J. Phys.; reprinted with permission)*

9.5.6 Polar radio blackout

Extensive studies of polar radio blackouts have been made by counting the occurrences of the qualifying letter B on tabulated ionogram data (Agy 1954).

This condition indicates an absorption exceeding 50 dB on frequencies near the critical frequency. Alternatively, periods of high absorption have been identified from the lowest frequency f_{min}. The most probable time of occurrence of blackout is in the morning, and the least probable time is in the evening. The time of maximum occurrence of blackout increases with latitude in North America, from about 0700 LT at 60° geomagnetic latitude to about 1300 LT at the magnetic pole. These statistical results are a superposition of polar cap absorption and auroral absorption.

9.5.7 F-region storms – critical frequencies

The most prominent feature of an F-region storm is the reduction in F_2 critical frequency and, hence, of the MUF. When such a reduction is combined with increased absorption, which increases the lower frequency limit, the net effect is a narrowing of the usable frequency spectrum. The effects of a storm on f_oF_2 at a given location depend, in a complicated way, on the local time, season, latitude, and storm time (which is time measured from the time of the sudden commencement).

The effect on the F_2 critical frequencies in middle latitudes is evident in the f-plots of Fig. 9.21. The associated magnetic storm started near 0400 LT on March 26, maximized with K = 8+ on the morning of March 27. and ended on March 29. The quiet-day noon f_oF_2's (March 25) are near 12·5 MHz whereas on March 28 at noon they are about 4·5 MHz, which corresponds to a decrease of N_mF_2 by a factor of nearly 8. The statistical variations of f_oF_2 departures from their quiet values are shown in Fig. 9.22 for 51 strong storms and 58 weak storms that occurred from 1946 through 1955. The most severe depressions (negative storm) of N_mF_2 (and hence f_oF_2) occur between about 20 and 40 hours storm time in geomagnetic latitudes of 45° and higher (zones 1 through 3). In intermediate latitudes (zones 4 through 6) there is a mild depression; in equatorial regions (zones 7 and 8) the storm is positive (increase of N_mF_2). It is seen also that, in general, there is an initial positive stage in which N_mF_2 is enhanced, followed by a negative stage in which N_mF_2 is depressed. However, individual storms can depart markedly from this statistical pattern.

9.5.8 F-region storms – total electron content

The enhancements and depressions are also evident in the departures (from quiet values) of total electron content at Hamilton, Massachusetts (Fig. 9.23). The positive stage is most pronounced on the afternoon of day 1, on which the geomagnetic storm started, and the negative stage lasts for some three days during which conditions return slowly to normal. The percentage of depression is largest near 0400 LT. The maximum density follows a somewhat similar pattern. The ratio of these characteristics (i.e. the slab thickness) shows marked positive departures, the largest occurring on day 2.

In middle latitudes it appears that, although a sudden commencement during daylight generally produces a TEC increase, a storm in which the SC occurs in

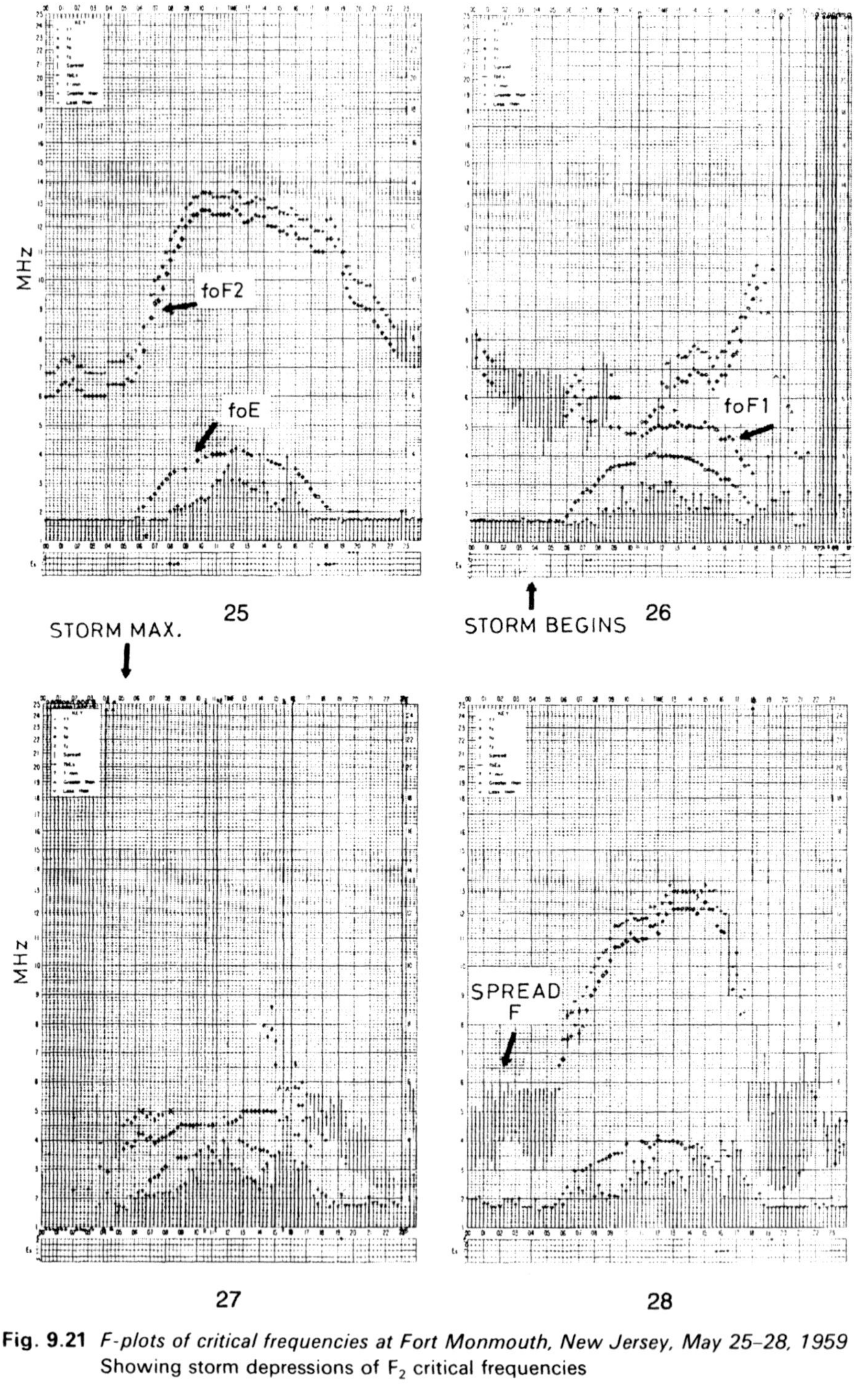

Fig. 9.21 *F-plots of critical frequencies at Fort Monmouth, New Jersey, May 25–28, 1959*
Showing storm depressions of F_2 critical frequencies

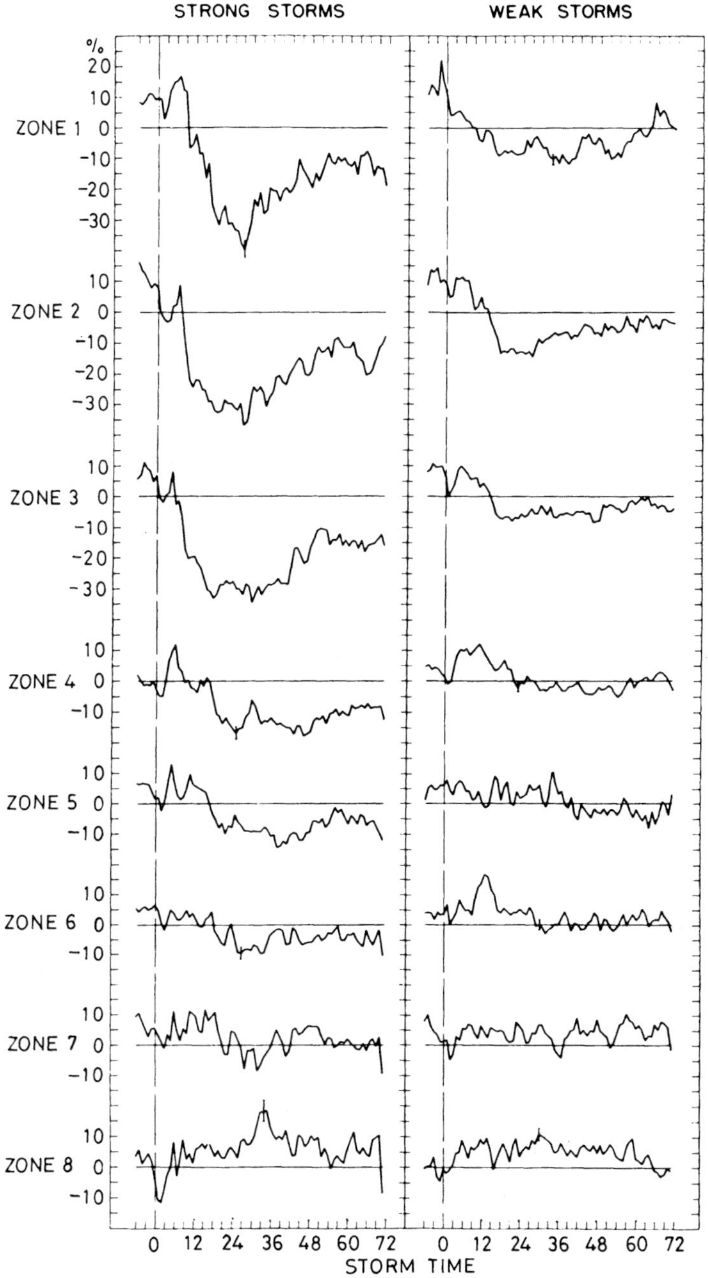

Fig. 9.22 *Storm-time variations of the F_2 maximum electron densities in different latitudes (From Matsushita 1959 © American Geophysical Union)*
The world is divided into eight zones of geomagnetic latitude as follows: Zone 1, 60°–55°; Zone 2, 55°–50°; Zone 3, 50°–45°; Zone 4, 45°–40°; Zone 5, 40°–29°; Zone 6, 29°–20°; Zone 7, 20°–9°; and Zone 8, +9° to −9°

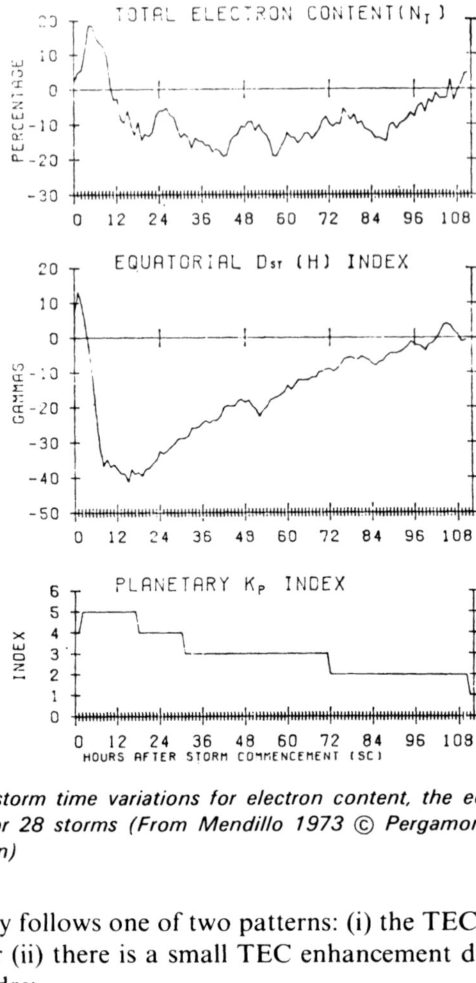

Fig. 9.23 *Average storm time variations for electron content, the equatorial $D_{st}(H)$ index, and K_p for 28 storms (From Mendillo 1973 © Pergamon Press; reprinted with permission)*

darkness generally follows one of two patterns: (i) the TEC goes directly to the negative stage, or (ii) there is a small TEC enhancement during the afternoon of the following day.

9.5.9 F-region storms — layer heights

Along with the depression of f_oF_2 there is usually a marked increase in the virtual height $h'F_2$. In some large storms f_oF_2 is depressed until it is near or below f_oF_1, giving a so-called G condition. When f_oF_2 is only slightly greater than f_oF_1, the $h'F_2$ is large because of retardation in the underlying layer. Therefore changes in $h'F_2$ are misleading indicators of changes in the real height of the F_2 layer (King 1971). Figure 9.24 shows the effect of a storm on the electron-density profile. An explanation for the negative F-region storm is that auroral heating causes atmospheric circulation that changes the composition of the neutral atmosphere, which increases the effective electron-loss rate. Since this heating occurs below the F layer the bottom side of the F_2 layer erodes, increasing the height and decreasing the electron densities.

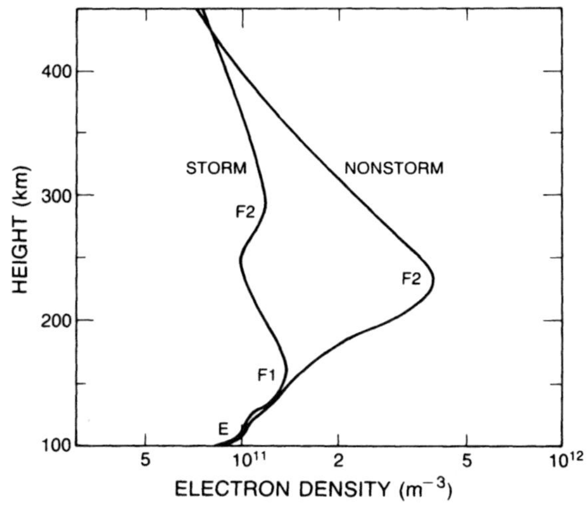

Fig. 9.24 *Electron-density profiles for quiet and storm conditions (From Norton 1969)*

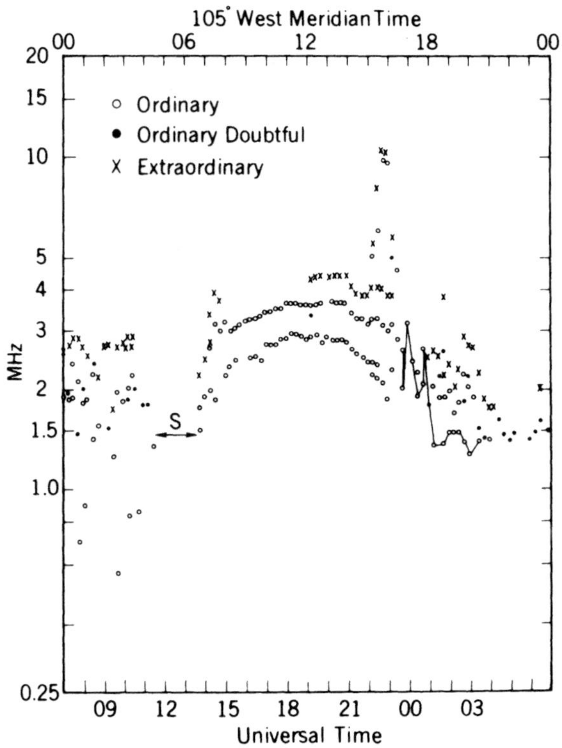

Fig. 9.25 *Fifteen-minute f-plot at Boulder, showing short-lived increase in F_2 critical frequency during a large storm on February 8, 1986*

9.5.10 Individual F-region storms

The above picture of an F-region storm is based on gross averaging, which can be misleading when applied to an individual storm. In general, it can be said that in a given storm the response is not identical at any two locations and that, at a given location, the responses are different for similar (magnetic) storms.

The fact that the statistical picture in Fig. 9.22 can be a little misleading is shown by the seasonal variation of storm pattern in zone 6. The data shown would suggest that there is relatively little storm-time variation in the F_2 layer in these latitudes. The results of a seasonal analysis, however, show that the electron density is depressed during the summer but enhanced in the winter. The overall average is, therefore, not representative of a single storm in either season.

Individual storms vary greatly. For example, some have only a positive phase; others have only a negative phase. There is no direct relationship between the size of an ionospheric storm and the size of the associated magnetic storm. The relationship of the onset of a storm to the storm sudden commencement (SSC) is highly variable, and the time of the main phase onset (MPO) may be a better indication of the start of a storm (Thomas and Venables 1966). Individual storms can have peculiar features such as short-lived decreases or increases. For example, Fig. 9.25 shows that during one of the largest magnetic storms in recent years, at the bottom of the sunspot cycle, the maximum electron density increased by a factor of 10 and decayed again within about 2 hours. The total electron content also went up by a factor of 10, and the strip of enhanced TEC swept across the USA with the Sun. There are marked Northern–Southern Hemisphere asymmetries in storm response of maximum electron-densities as seen in Fig. 9.26.

9.5.11 Protonospheric storms

During magnetic storms there are profound changes in the residual or protonospheric content N_p. A marked depletion of the protonospheric content reaches a minimum on day 3, after which the content increases with a superposed diurnal filling and emptying (Fig. 9.27). In that case the recovery took some 14 days at sunspot minimum (Kersley and Klobuchar 1978). It is believed that disturbances are sufficiently frequent near the maximum of the sunspot cycle that protonospheric filling (from the ionosphere) is never achieved before depletion by a subsequent storm. This depletion of the protonosphere is accompanied by an inward motion of the plasmapause L_{pp} (invariant latitude in degrees), which is related to magnetic index K_p and local time t by (Rycroft and Burnell 1970)

$$L_{pp} = 62{\cdot}0 - 1{\cdot}0\,\mathrm{K_p} - 0{\cdot}4\mathrm{t} \tag{9.13}$$

The latitude L_T of the middle-latitude trough, which also moves equatorward with magnetic disturbance, is given by

$$L_T = 60{\cdot}4 - 1{\cdot}4\,\mathrm{K_p} - 0{\cdot}7t \tag{9.14}$$

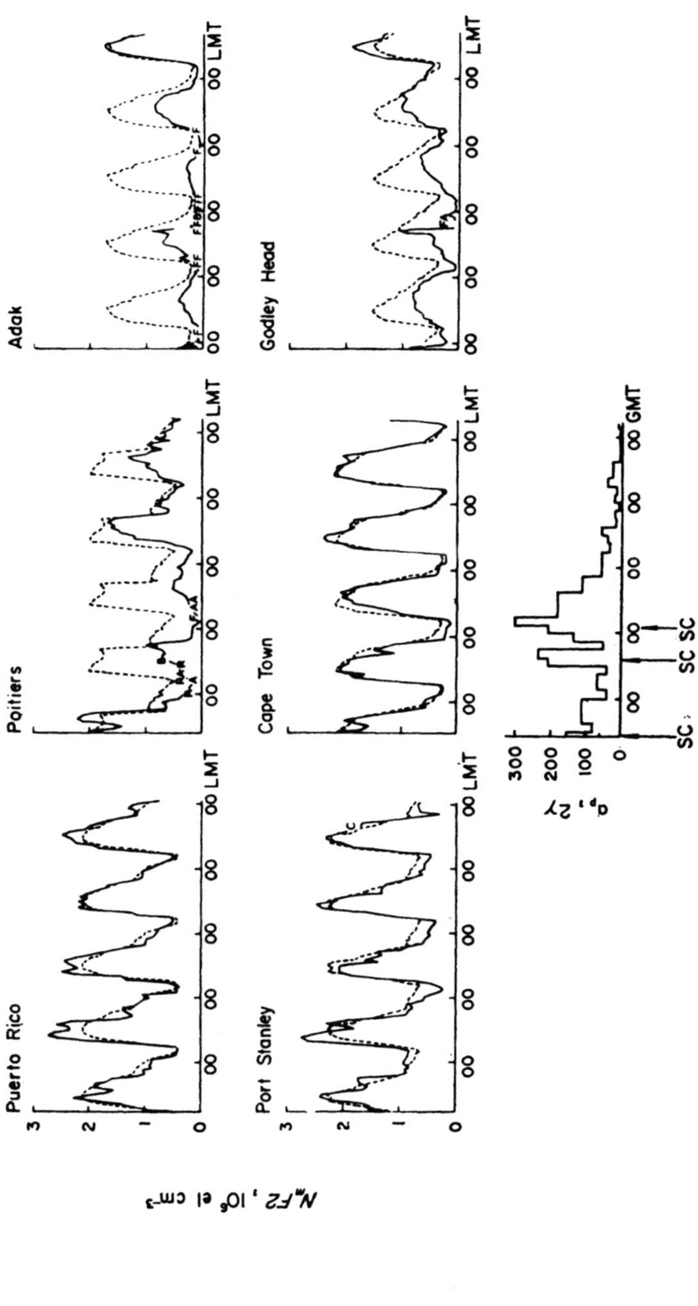

Fig. 9.26 *Storm-time variations of N_mF_2 at three pairs of magnetically conjugate stations, and of the magnetic disturbance index a_p (From Thomas 1970 © Pergamon Press; reprinted with permission)*

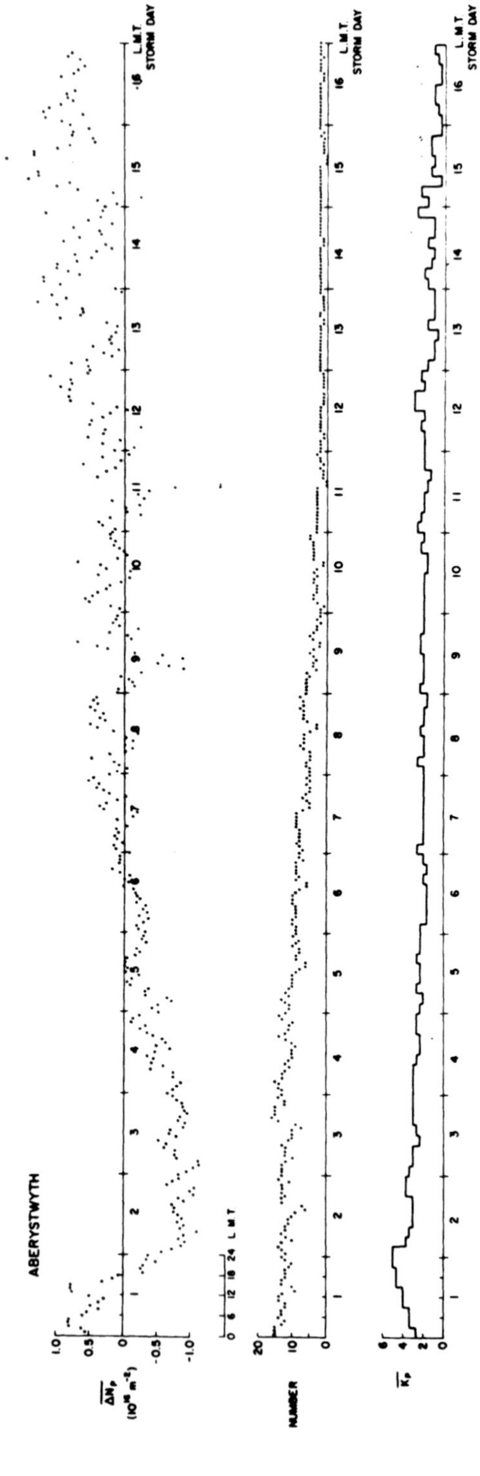

Fig. 9.27 *Storm depletion and replenishment of the protonosphere columnar electron content during storms; ATS-6 to Aberystwyth, Wales (From Kersley and Klobuchar 1978)*

At the poleward edge of the trough there is a sharp increase in density caused by auroral particle precipitation.

9.5.12 Thermospheric storms

During disturbed periods, thermospheric heating due to particle precipitation and Joule (frictional) dissipation is typically of the same magnitude as heating from EUV. This heating causes the thermosphere to expand upward and equatorward, driving thermospheric winds and increasing the neutral density at F-region heights, thus leading to increased satellite drag (Straus and Hickman 1979).

9.6 Global picture of storms

The relationships of physical processes associated with an F-region storm are depicted in Fig. 9.28. Particle precipitation into the auroral atmosphere and Joule heating by auroral currents heat the atmosphere. The heating results in equatorward neutral winds, which move the plasma up the geomagnetic field where the electron loss is smaller; the result is a positive storm. The neutral wind carries the disturbance equatorward. The changes in loss rate produce the

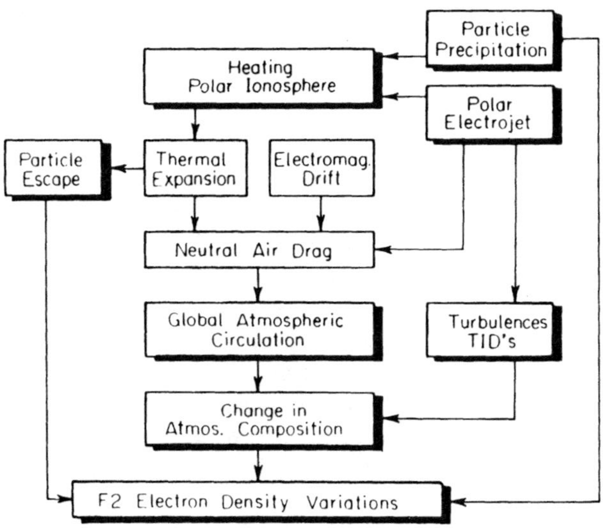

Fig. 9.28 *Physical processes in an F-region disturbance*

negative phase. Thus there are two aspects of the movement of storms: (i) the equatorward motion of the auroral zone, and (ii) the movement of the disturbance away from the auroral zone. It appears that a disturbance moves from about 70° geomagnetic latitude to about 50° geomagnetic latitude in about 10 hours. The departures of f_oF_2 from the quiet values for summer and winter in the Northern Hemisphere are shown in Fig. 9.29. The largest percentage decrease in summer is somewhat larger than that in winter, and the maximum depressions occur during the morning. The negative phase rarely extends to equatorial latitudes in which the positive storm dominates (Matuura 1972).

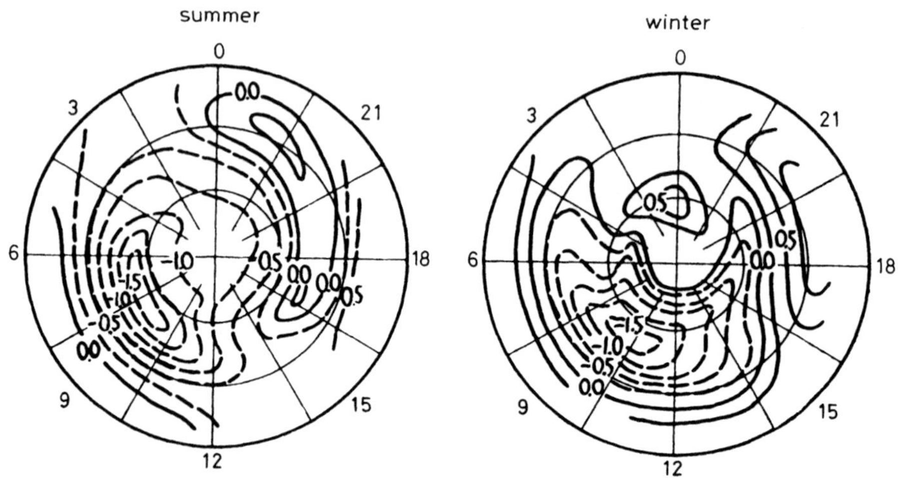

Fig. 9.29 *Patterns of average storm departures $D(f_oF_2)$ from quiet values for northern summer and winter (From Obayashi 1954 © Terra Scientific Publishing Company)*

9.7 Radio communications in high latitudes

Because of the relatively large solar zenith angles and because of their short durations, shortwave fadeouts are relatively unimportant in high latitudes. On the other hand polar cap absorption events and ionospheric storms can be catastrophic. A lowering of the maximum usable frequency (MUF) and an increase in the lowest usable frequency (LUF) can occur separately or concurrently. If they occur separately, a judicious choice of operating frequency can cure the problem but if they are concurrent the usable frequency range may completely close, destroying HF radio contact. An increase in absorption may not be as devastating as it sounds because, in addition to reduction of the strength of the desired signal, there is a decrease in radio interference from distant sources.

More debilitating to circuit operations than signal decrease during a disturbance is the great enhancement in the depth and speed of fading, and during storms, 'auroral flutter' can be disastrous to some forms of modulation (e.g. voice). The combination of these effects makes HF communication involving auroral zone paths rather unreliable. Thus, although HF communication may be suitable for leisure activities, it may be unsuitable when high reliability is required, for example, when lives are in jeopardy.

9.7.1 WWV monitoring

Some useful information on the reliability of HF communications for paths into the Arctic from outside has been obtained from the analyses of monitoring data on the reception of WWV Standard Frequency transmissions in Canada. These consist of aural estimates of the strength of WWV, on frequencies of 2·5, 5·0, 10·0, 15·0, 20·0, and 25·0 MHz. The estimates have a range of 0 to 9 where 0 denotes no audible signal and 9 denotes excellent reception. The monitoring stations are Ottawa, Winnipeg, Churchill, Baker Lake, and Resolute Bay. The analyses show that at least one of the frequencies is received at least 90% of the total time at Resolute Bay, Baker Lake, and Churchill. It appears that the reception is slightly better during sunspot maximum than during sunspot minimum, which implies that the increased outage caused by storms is more than offset by the ability to use higher frequencies resulting from higher MUFs at higher sunspot numbers. This appears to be true on a global scale (Herman and Penndorf 1964). Monitoring data for a disturbed period from September 12–14, 1957, are shown in Fig. 9.30. During disturbed conditions, there are periods of time in which no high frequency is usable, because of high absorption and/or MUF failure. During this storm the WWV–Resolute Bay circuit was the most disturbed, as would be expected. The lowest frequencies (2·5 and 5 MHz) are seldom received at the highest latitudes, and the highest frequency (25 MHz) is seldom received at the lowest latitudes even on quiet days. The middle frequencies, 10 and 15 MHz, are the most reliable. For further information on propagation in high latitudes the reader should consult Landmark (1964). It may be noted that WWV emits a vertically polarized signal that matches the *o*-wave in high latitudes and minimizes the polarization coupling loss (see Sec. 7.6).

9.7.2 Signal distortion

In assessing the effect of a disturbance on a communication system in high latitudes it is not sufficient to consider signal strength alone. The nature of the signal, such as the depth and rate of fading and the time and frequency spreading (distortion) of the signal, must also be considered. For example, Fig. 9.31 shows the time spread of reflected pulses over a middle-latitude path and an auroral path. The extent of the spread limits the speed of transmission of information and the nature of the fading, because the signal consists of a relatively large number of individual interfering wavelets. Indeed, disturbed periods are often characterized by rapid fading called flutter, which seriously affects certain radio systems.

There are two reasons for pulse spreading: (i) the frequency dispersion of the medium, and (ii) scattering by irregularities. Frequency dispersion occurs, for example, with vertical pulse sounding near a penetration frequency and arises because there are large differences in the group delays of the wave components that constitute the pulse. For example, if the center frequency of a pulse coincides with the penetration frequency, only the lower half of the pulse spectrum is reflected. For further discussion of dispersive broadening see Budden (1985, Sec. 11.16) and Paul (1979). The effects of irregularities on pulse broadening have been discussed by Booker *et al.* (1986).

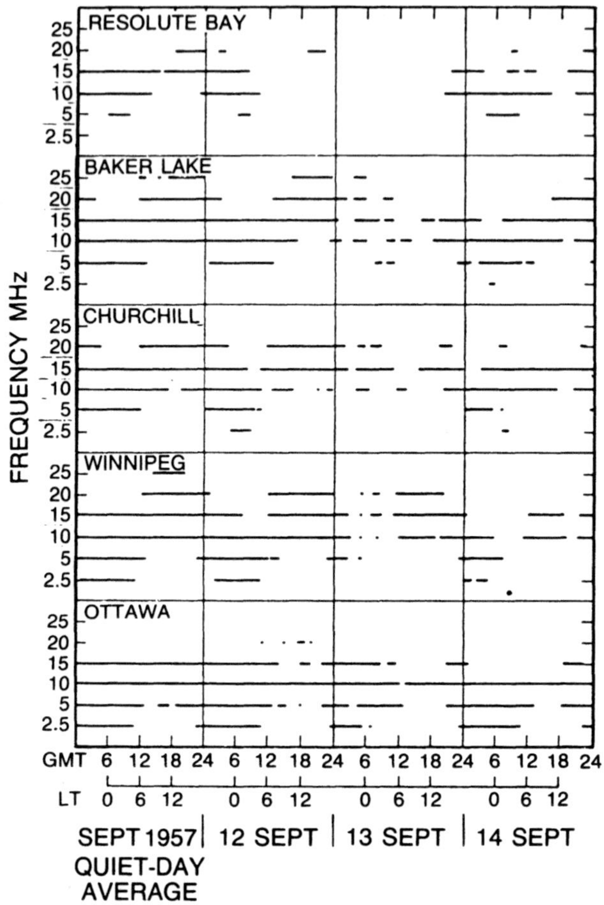

Fig. 9.30 *Hours of WWV reception for a disturbed period from September 12 to 14, 1957 (From Herman and Penndorf 1964: with permission)*
Left-hand column gives quiet-day averages for comparison

9.7.3 Remedial measures

The WWV monitoring results suggest that one remedy for impaired radio communication in high latitudes is the availability of a relatively large number of operating frequencies. Owing to spectrum congestion and the methods of allocating frequencies, this approach is not practical except in a few cases.

Oblique sounders have been used to determine the usable spectrum and such characteristics as range spread (e.g. Jull 1964; Petrie and Warren 1968;

Folkestad 1968; Goodman 1987, Session 1). Oblique sounding is useful in certain ionospheric storms; the usable parts of the frequency band can be identified. For example, enhancement of sporadic E allows propagation on frequencies high enough to overcome the increased absorption (Stevens 1968*a*). This requires the allocation of a number of frequencies. Another procedure that has been used is the relaying of messages around the disturbed area. This is feasible particularly during auroral absorption when blackouts are most frequent along the auroral ring and are limited in longitude. Furthermore, owing to local time patterns of blackouts, it is possible to forecast their occurrence to some extent and hence take remedial action. Such action is, however, of only limited value during polar cap events because of their wide extents.

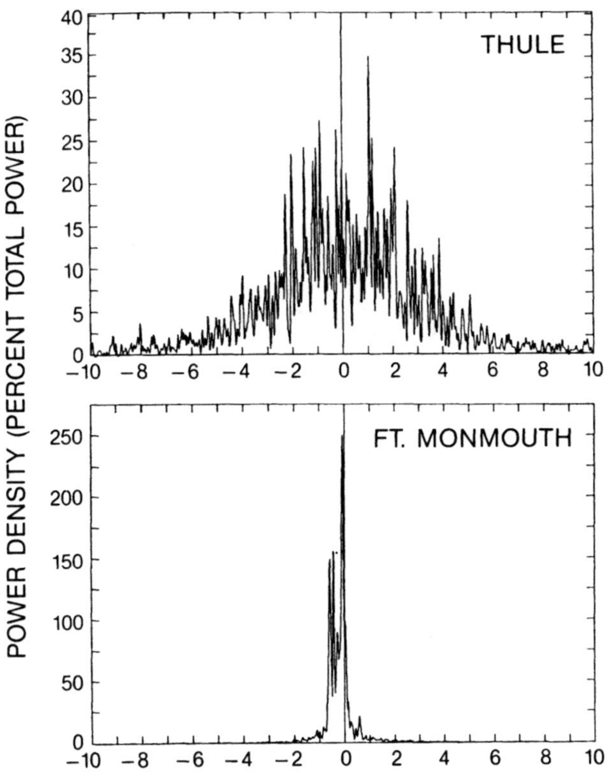

Fig. 9.31 *Power spectra of transmissions from Thule and Fort Monmouth, monitored at Palo Alto (From Hunsucker and Bates 1969 © American Geophysical Union)*

Other remedial measures include the modeling of the particular communications circuit (Vincent *et al.* 1968). This includes taking account of the modulation used; the use of adaptive receivers for multipath channels (Coll 1968);

the operation of channel evaluation sounding (Stevens 1968*b*); and the use of network frequency management by frequency assignment from a central computer (Probst 1968). Although oblique sounders provide a comprehensive picture of propagation conditions over a wide frequency range the method is slow and wasteful of power and spectrum since it produces interference. The ultimate remedial action is to go to (if available) alternative systems that are not affected by the particular disturbance (e.g. satellite systems, microwave systems [troposphere], or cables).

An appropriate response to the relatively harsh ionospheric environment in high latitudes requires the combined efforts of the communication user, the systems engineer, and the scientist. A user needs to define the problem for the systems engineer, and a scientist has to provide the required environmental information (Hatton 1968). Factors that have to be considered in circuit design include effect of failure, tolerable time delay, cost, and accuracy. Of particular importance is user (operator) education so that users will appreciate the nature of the communication environment and will be aware of possible remedial actions.

9.8 Forecasting

9.8.1 Definitions

We need to distinguish between the terms prediction and forecast. According to general usage, prediction of ionospheric propagation conditions refers to the specification of optimum frequencies, signal strengths, angles of elevation, etc., for circuit conditions for months to years ahead. This specification is usually based on the prediction of a parameter such as the sunspot number or its equivalent. On the other hand, we define ionospheric forecasting in terms of disturbances (SIDs, PCAs, storms, etc.) for periods up to one solar rotation (≈ 27 days). These forecasts depend primarily on solar observations, particularly in X-ray, visible light, infrared, and radio. These are supplemented by observations of the geomagnetic field and of the ionosphere, both on the ground and on satellites, and the charged-particle environment.

In forecasting work it is necessary to consider two types of storms: (i) sporadic storms and (ii) recurrent storms. Sporadic storms can be associated with flares and disappearing filaments and are more prevalent around the peaks of sunspot cycles. Recurrent storms are usually associated with coronal holes and are more conspicuous in the years of declining sunspot numbers. They recur at 27-day intervals and are caused by the (solar) rotation of long-lived features.

9.8.2 Solar forecasting

There is a general connection between flares and terrestrial disturbances but this is by no means a one-to-one relationship. Some flares, even big flares, are not followed by ionospheric and magnetic disturbances. This is illustrated in Fig. 9.32. Of 165 class-X flares only 33 were followed by major geomagnetic storms and 34

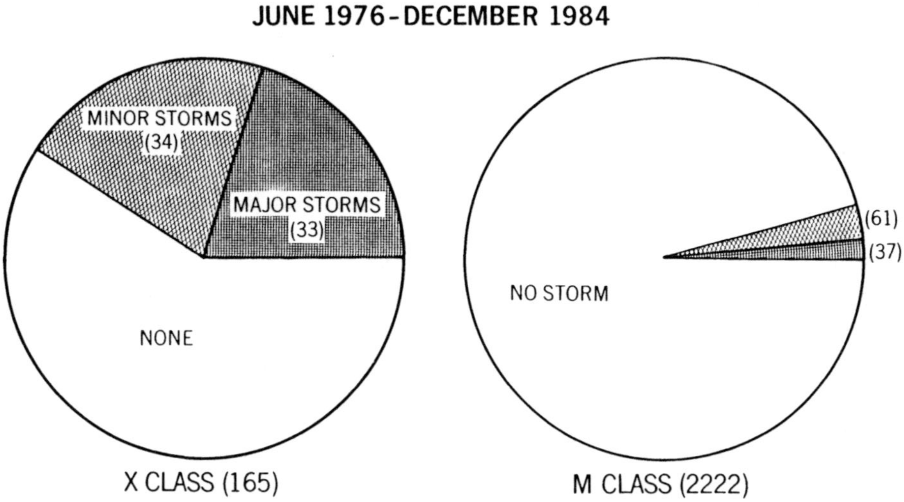

Fig. 9.32 *Data from Sunspot Cycle 21, illustrating the poor correspondence between major flares and subsequent geomagnetic storms (From Joselyn 1986a)*

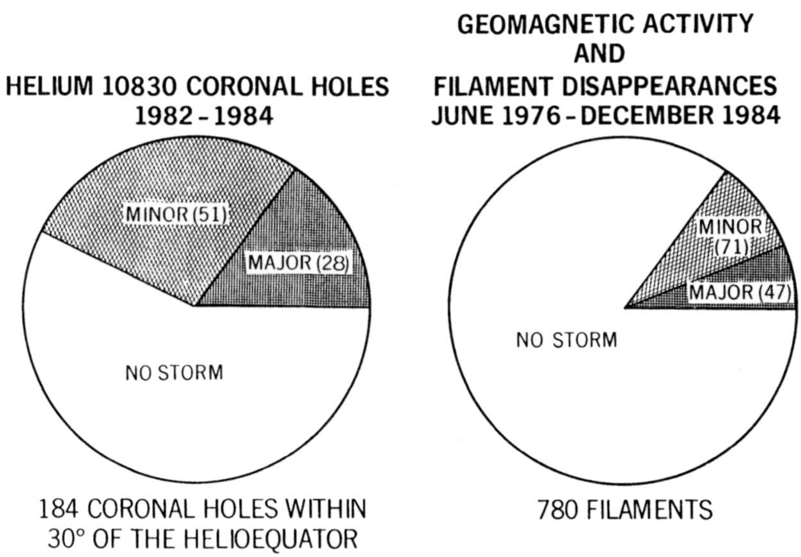

Fig. 9.33 *Data from Sunspot Cycle 21 illustrating the poor correspondence between coronal holes near the helioequator and subsequent geomagnetic storms, and between filament disappearance and subsequent geomagnetic storms (From Joselyn 1986a)*

were followed by minor storms. The corresponding numbers for class-M flares are 2222, 61, and 37. The number of class-X and class-M flares is only about 10% of the total number of observed X-ray flares (see Sec. 2.2.2.1). The number of class-X flares is less than 1% of the total number of X-ray flares. There was only a 1-in-5 chance of a major storm after a class-X flare and only a 1-in-60 chance following a class-M flare. Further, during sunspot cycle 21, for 60% of the class-X flares there was no storm. The geophysical effect of a flare does not appear to be particularly sensitive to the position of the flare on the solar disk (Joselyn 1986*a*). Somewhat similar remarks can be made about the two other solar features associated with storms, coronal holes, and disappearing filaments, as seen in Fig. 9.33. The data for flares, coronal holes, and disappearing filaments are summarized in Fig. 9.34, which shows that out of a sample of 238 geomagnetic storms some 20% (45 storms) are unambiguously associated with flares, about 17% are associated with coronal holes, and about 13% are associated with disappearing filaments. Some 8% of the storms are not associated with these causes, and the rest may be associated with two or more of these solar features.

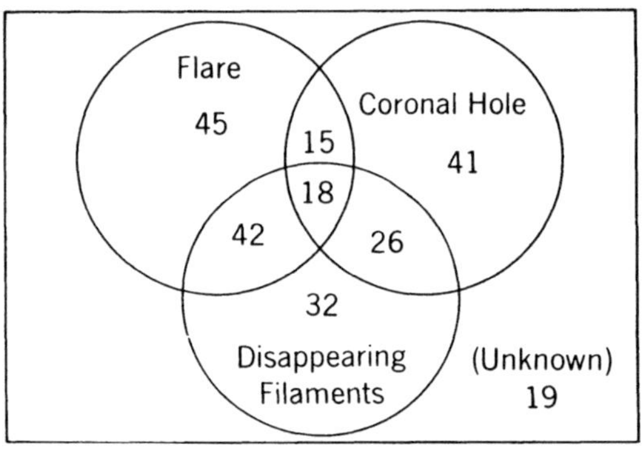

Fig. 9.34 *Numbers of major geomagnetic storms ($A_p \geqslant 5$) associated with flares, coronal holes, or disappearing filaments between June 1976 and December 31, 1984 (From Joselyn 1986a)*

Recurrent (27-day) geomagnetic activity is associated with high-speed streams that emanate from coronal holes. However, recurrence tends to be reliable only for a few years in the declining phases of a sunspot cycle. Coronal holes can be identified on H-alpha synoptic charts such as Fig. 2.4. Again only a small fraction of coronal holes produce geomagnetic disturbances. Thus, forecasting centers keep as close as possible to a 24 h/day watch on the Sun and attempt, by pattern recognition, to forecast the relevant solar phenomena.

As already noted, solar features with similar appearances do not necessarily produce the same effects at Earth, and features that look different can produce similar effects. The amount of effect a solar feature will have at Earth is called its geoeffectiveness.

Although recurrence by itself has been a useful forecasting tool, the presence of coronal holes can be inferred from observations of daily helium 10830 Å spectroheliograms (see *Solar-Geophysical Data*); 10·7 cm east–west solar radio drifts (from Ottawa, Canada); H-alpha and the daily solar noise scans on 692 and 1415 MHz from Fleurs, Australia (Ferguson 1979). For further information the reader is referred to the book by Simon *et al.* (1986).

9.8.3 Application of artificial intelligence

Over the past few decades forecasting solar flares has depended largely on the skill of the forecaster in interpreting the changing patterns on the Sun and the temporal variations of solar radiations. Recently, at the Space Environment Laboratory in Boulder, Colorado, computer technology has been invoked whereby the expertise of a forecaster has been incorporated into a computer decision-making process. The program, called THEO (for Theophrastus, who first recorded observations of sunspots in 325 BC), is based on optical observations of sunspots (Lewis and Dennett 1986). THEO employs decision trees, that is, branching structures in which flare probabilities (the leaves of a tree) are found by tracing a path from the root and selecting branches according to the properties of the active region(s) being analyzed. The idea behind this expert system is that a novice forecaster will have available the knowledge of an expert.

9.8.4 Forecasting centers

Forecasts of solar–terrestrial conditions are made by a number of organizations, including (i) the Space Environment Services Center (SESC), Boulder, Colorado (Heckman 1979); (ii) the Tokyo Regional Warning Center (Maeda and Inuki 1979); (iii) the forecasting center at Meudon, France (Simon 1979); (iv) the Research Institute of the Deutche Bundespost (Dambolt 1979); (v) the Forecasting Center of the Institute of Applied Geophysics, Moscow (Avdyushin *et al.* 1979); (vi) the National Physical Laboratory, New Delhi (Reddy *et al.* 1979); and (vii) the Regional Warning Centre, Sydney (Cook and Davies 1979). These centers are part of the International URSIgram and World Days Service (IUWDS). SESC is a national and international focal point for real-time solar and geophysical data. An elaborate communication system links a global network of observatories with SESC so it is aware immediately of solar activity. In any forecast system it is important not only to acquire timely data but also to provide rapid dissemination of forecasts and warnings before the customers learn of solar activity through circuit interruption. Forecasting goes on continuously, and two forecasts are issued daily. Event notification (warnings) involves the alerting of individuals and organizations of the occurrence of prevailing conditions. Dissemination is by means of satellite, computer bulletin board, telephone, etc., as appropriate. Table 9.2 gives an idea of the types of users of SESC forecasts and the solar–terrestrial phenomena that affect the users. Predictions of the following

Table 9.2 *Summary of uses of the Space Environment Services Center and types of activity affecting systems**

Customer	Types of activity affecting use
Civilian satellite communication	Magnetic storms
Commercial aviation: middle-latitude communication (VHF)	Solar radio emissions
polar cap communication (HF)	PCA, magnetic storms
Commercial aviation navigation (VLF)	PCA, magnetic storms
Commercial electric power	Magnetic storms
Long-line telephone communication	Magnetic storms
High-altitude polar flights (subject to radiation hazards)	Solar proton events
Civilian HF communication: Coast Guard, GSA, commercial companies, Voice of America	X-ray emission, UV emission, magnetic storms
Geophysical exploration	Magnetic storms
Satellite orbital variation, military and civilian	UV emission, magnetic storms
DoD SATCOM communication	Magnetic storms
DoD HF communication	X-ray emission, UV emission, PCA, magnetic storms
DoD reconnaissance	PCA, magnetic storms
DoD navigation	X-ray emission, UV emission
ERDA communication to prospective customers	X-ray emission, UV emission, magnetic storms
Scientific satellite studies: IMS, Solar Maximum Mission Shuttle, solar constant measurements, stratospheric ozone variation, interplanetary missions	Optical solar flares, magnetic storms, X-ray emission, UV emission, solar proton events, solar features
Scientific rocket studies: IMS, magnetosphere, ionosphere, upper atmosphere, Sun	Optical solar flares, solar features, magnetic storms, solar proton emission, X-ray emission
Scientific ground studies: IMS, Sun, interplanetary, magnetosphere, ionosphere, upper atmosphere, stratosphere, troposphere, seismological/geomagnetic	Optical solar flare, magnetic storms, solar proton emission, X-ray emission, UV emission, solar features

* From Heckman (1979)

are made: geomagnetic indices, proton events (PCAs), radiation levels, solar flares (onsets, peak intensities, durations), and storm commencements. The warnings include a MAGNIL indicating the end of a disturbance (i.e. the onset of a quiet interval). It has been known to radio operators for many years that propagation conditions improve before the onset of a storm. Increases of signal strength (10–20 dB for $\approx$ 10 hours) and increases in f_oF_2 prior to a storm are used to predict communication disturbances on polar paths (Ondoh and Obu 1980). The outputs of the Sydney Regional Warning Centre (Cook and Davies 1979) include the probabilities of flares and storms in three grades – possible, probable, or expected – and warnings to HF communicators (concerning MUF and SID effects) and to users of geomagnetic data. In the USSR (Avdyushin *et al.* 1979), the following predictions are issued: (i) short-term disturbances, (ii) long-term maps of MUF, and (iii) long-term predictions for specific paths (Kuleshova *et al.* 1980; Zevakina and Kisileva 1980).

9.8.5 *Some technological developments*

The major advances in solar-terrestrial forecasting over the past 20 years or so have been brought about largely through technological improvements in data acquisition (e.g. satellite sensors) and data processing (computers) and display systems, and more efficient dissemination. Major observing improvements include, for example, (i) the Solar Observing Optical Network (SOON) and (ii) the Radio Solar Telescope Network (RSTN). Another observing improvement is the use of a satellite, ISEE–3, which orbited the Sun-Earth libration point (Fig. 9.35) some 1·5 million km in front of the Earth. In that orbit the satellite intersects the solar wind about 30 to 60 min before the wind reaches the Earth (for example, the electric field signature in Fig. 9.36 precedes the signature observed near the Earth) and therefore provides advance warning of the onset of a geomagnetic disturbance. Because of the relatively large orbit of ISEE–3 and the localized nature of the solar wind it is still possible for a libration point sensor to miss a disturbance that will affect Earth. Alternatively the satellite may detect a disturbance that misses the Earth. Propagation forecasting principles are applied in the PROPHET system (Argo and Rothmuller 1979), which converts real-time solar/geophysical data into propagation predictions for specific systems. Any such Environmental Prediction and Assessment System (EPAS) has three components: (i) sources of solar-geophysical data, (ii) data reduction to usable form, and (iii) an evaluation system in terms of the factors necessary to optimize the system performance (e.g. frequency, antenna, transmitter, power).

Some further improvements have included the development of new indices (e.g. Donnelly 1979) but these have met with limited success. Joselyn (1986*b*) introduced a geomagnetic probability forecast for two geomagnetic zones: a middle-latitude zone (20° to 50° geomagnetic) and a high-latitude zone (50° to 80°). The degree of disturbance is specified in terms of the magnetic A range: 0 to 7, 8 to 15, 16 to 29, 30 to 49, 50 to 100. Climatology (the average conditions expected solely on the basis of historical data) and persistence (the expectation

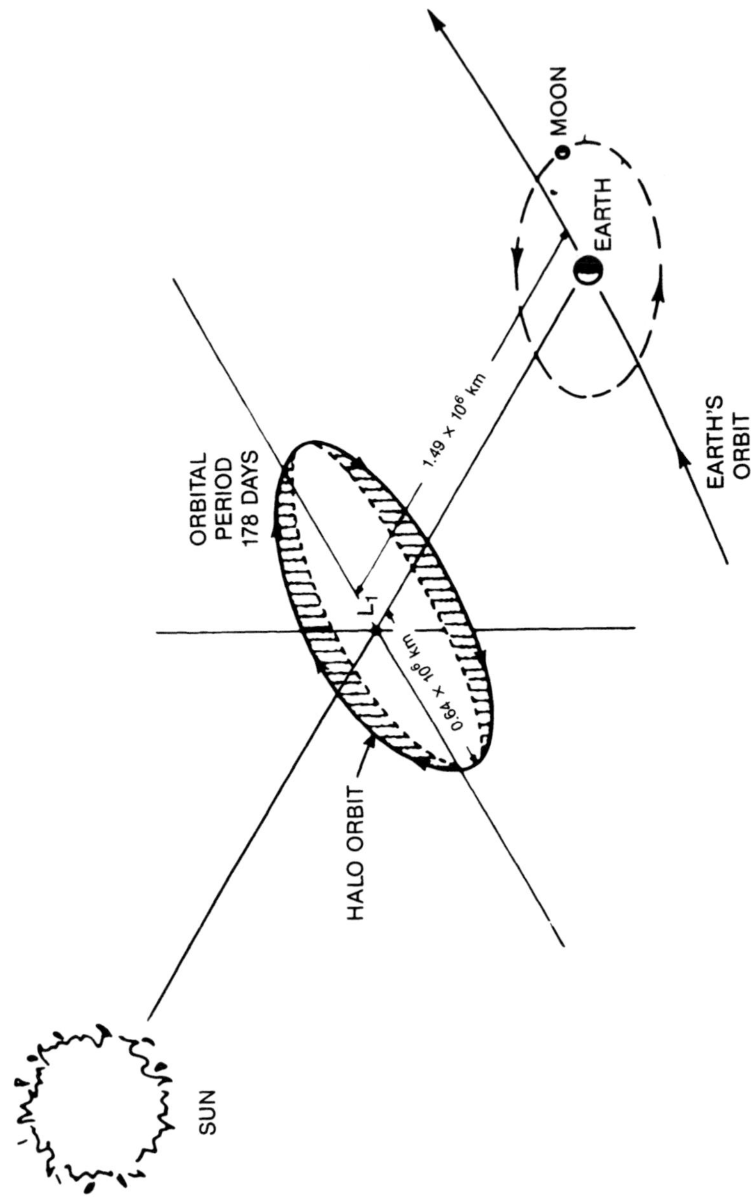

Fig. 9.35 *The orbit of International Sun-Earth Explorer-3 about the Sun-Earth libration point (From Tsurutani and Baker 1979 © American Geophysical Union)*

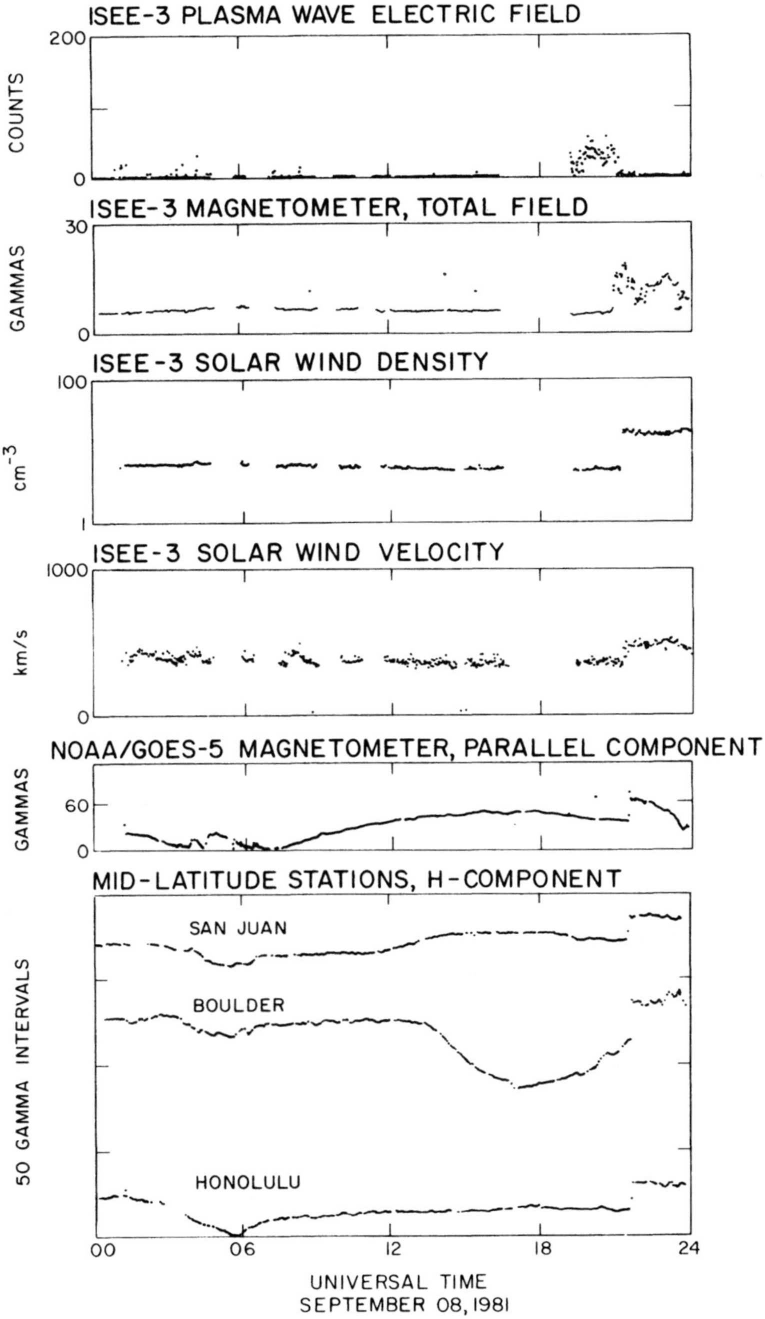

Fig. 9.36 *Real-time data received at SESC, showing a shock observed by ISEE–3 at 2100 UT and at Earth at 2146 UT on September 8, 1981 (From Joselyn 1986a)*

that tomorrow will be like today) are augmented by forecaster skill. An example of a forecast is given in Table 9.3. This forecast is comparable with the forecasts of the US National Weather Service and is available from the Space Environment Laboratory.

Table 9.3 *Sample forecast of geomagnetic conditions, in percentage of probability (after Joselyn 1986b)*

	Geomagnetic conditions (A range)					
	Quiet	Unsettled	Active storm	Minor storm	Major storm	Severe storm
	(0) (7)	(15)	(30)	(50)	(100)	(400)
Middle latitude	10	80	10	0	0	0
High latitude	10	70	20	0	0	0

It should be remembered that solar-terrestrial disturbance forecasting is still largely an art in which the forecaster's experience and skill in recognizing and interpreting observations are dominant. This expertise has been developed to such a level that a forecaster can predict a disturbance within 1 to 3 days but cannot be precise as to (i) when it will start, (ii) its magnitude, or (iii) its duration.

Of particular value to users of solar-terrestrial data is the 'Preliminary report and forecast of solar-geophysical data' published weekly by the Joint NOAA–USAF Space Environment Services Center in Boulder. This booklet contains highlights of solar and geomagnetic activity for the week preceding publication and a forecast for the coming week. The information includes, for example, occurrences of flares (X-ray, visible, radio) and, where appropriate, their locations; proton enhancements; geomagnetic storms; daily sunspot numbers; Ottawa 10·7 cm radio flux; K and A magnetic indices; and a list of alerts issued. Magnetic indices and radio fluxes are forecast 27 days in advance.

9.8.6 Radio propagation forecasting

The Space Environment Services Center (SESC) issues forecasts of radio propagation conditions in the Northern Hemisphere. These forecasts apply to sky-wave propagation (usually in the frequency range 2–30 MHz). The Hemisphere is divided into five latitude zones (0° to 10°, 10° to 30°, 30° to 55°, 55° to 70°, 70° to 90°) and four longitude zones (0° to 90°W; 90°W to 180°, 180° to 90°E, and 90°E to 0°).

A simple letter/number scheme is used to describe the observed and forecast propagation conditions within each area (block) of the zone grid. The letters W, U, and N represent observed worthless (poor), usable (fair), and normal propagation conditions, respectively. Observed conditions (over the preceding 6 hours) are specified on a relative scale. The same three letters (W, U, and N) identify overall propagation conditions *in comparison with* normal conditions for the latitude, local time, season, and phase of sunspot cycle. Thus, an N in auroral latitudes in the sunrise sector *does not* have the same absolute meaning as an N in the afternoon sector in low latitudes. Conditions are normally much better in the latter block than in the former, but N may apply to both blocks. Overall propagation conditions are a weighted combination of MUF, LUF, noise, F-layer stability, presence of sporadic E, etc. Typically, all blocks would be N, since conditions are typically normal.

```
LATEST_SOLAR_REPORT

FROM SPACE ENVIRONMENT SERVICES CENTER, BOULDER, COLORADO
SDF NUMBER 067
JOINT USAF/NOAA REPORT OF SOLAR AND GEOPHYSICAL ACTIVITY.
ISSUED 2200Z 08 MAR 1989
IA.  ANALYSIS OF SOLAR ACTIVE REGIONS AND ACTIVITY FROM 07/2100Z
TO 08/2100Z:  SOLAR ACTIVITY WAS HIGH. REGION 5395 (N35E46)
REMAINED LARGE AND MAGNETICALLY COMPLEX. IT PRODUCED SEVERAL
M-CLASS FLARES INCLUDING AN M5/1B AT 08/0837Z, WHICH WAS
ASSOCIATED WITH MINOR RADIO EMISSIONS, AN M1/2B AT 08/1450Z
ASSOCIATED WITH A TYPE IV RADIO SWEEP BURST, AND AN M4
(OBSERVED BY THE SOLAR MAXIMUM MISSION SATELLITE) AT 08/1857Z.
ONE OPTICALLY UNCORRELATED M2 X-RAY BURST OCCURRED AT 08/0833Z.
TWO NEW REGIONS WERE ASSIGNED TODAY: 5397 (N35E61) AND 5398
(S16E68). BOTH WERE SMALL BIPOLES.
IB.  SOLAR ACTIVITY FORECAST:  SOLAR ACTIVITY IS EXPECTED TO BE
HIGH. THERE IS A GOOD CHANCE FOR MAJOR FLARE ACTIVITY FROM
REGION 5395.
IIA.  GEOPHYSICAL ACTIVITY SUMMARY FROM 07/2100Z TO 08/2100Z:
THE GEOMAGNETIC FIELD HAS BEEN AT QUIET TO ACTIVE LEVELS FOR
THE PAST 24 HOURS. A SUDDEN IMPULSE OF 38 NANOTESLAS WAS
RECORDED AT BOULDER AT 08/1755Z. A SATELLITE PROTON EVENT AT
GREATER THAN 10 MEV BEGAN AT 08/1735Z AND REMAINS IN
PROGRESS. THUS FAR, THE MAXIMUM OBSERVED FLUX HAS BEEN
55 PFU AT 08/1755Z.
IIB.  GEOPHYSICAL ACTIVITY FORECAST:  THE GEOMAGNETIC FIELD IS
EXPECTED TO BE AT ACTIVE TO MINOR STORM LEVELS WITH MOST OF
THE STORMING EXPECTED TOMORROW. THIS ACTIVITY WILL BE IN
RESPONSE TO THE X15/3B FLARE OF 6 MARCH. THE SATELLITE PROTON
EVENT AT GREATER THAN 10 MEV IS EXPECTED TO END WITHIN THE
NEXT 12 TO 24 HOURS.
III.  EVENT PROBABILITIES 09 MAR-11 MAR
CLASS M     90/90/90
CLASS X     50/50/50
PROTON      20/20/30
PCAF        RED
IV.  OTTAWA 10.7 CM FLUX
OBSERVED                08 MAR   206
PREDICTED    09 MAR-11 MAR   208/210/212
90 DAY MEAN             08 MAR   224
V.  GEOMAGNETIC A INDICES
OBSERVED   AFR/AP 07 MAR   012/018
ESTIMATED  AFR/AP 08 MAR   012/012
PREDICTED  AFR/AP 09 MAR-11 MAR   040/045-020/030-015/020
```

a

Fig. 9.37 *Examples of (a) the primary report of solar and geophysical activity, and (b) the daily primary HF radio propagation report, issued jointly by the USAF Global Weather Central and the Space Environment Services Center [see continuation, next page]*

PRIMARY HF RADIO PROPAGATION REPORT ISSUED AT 090600Z MAR 89.
PART I. SUMMARY 090000Z TO 090600Z MAR 89/
FORECAST 090600Z TO 091200Z MAR 89.

		QUADRANT			
		I	II	III	IV
		0 TO 90W	90W TO 180	180 TO 90E	90E TO 0
REGION	POLAR	W2/M20	W2/M20	N3/M20	N3/M20
	AURORAL	N3/M20	N3/M20	N4/M20	N3/M20
	MIDDLE	N6	N7	N7	N7
	LOW	N8	N8	N8	N8
	EQUATORIAL	N7	N8	N8	N8

PART II. GENERAL DESCRIPTION OF HF RADIO PROPAGATION CONDITIONS OBSERVED DURING THE 24 HOUR PERIOD ENDING 08/2400Z, AND FORECAST CONDITIONS FOR THE NEXT 24 HOURS: ALL LATITUDES EXPERIENCED GENERALLY NORMAL CONDITIONS UNTIL ABOUT 18Z, WHEN SOME STRONG GEOMAGNETIC ACTIVITY BEGAN. POLAR STATIONS BEGAN REPORTING SEVERE FADING AND ABSORPTION, RENDERING SOME FREQUENCIES UNUSEABLE. POOR CONDITIONS CONTINUED THROUGH 08/2400Z.
FORECAST: GEOMAGNETIC ACTIVITY IS EXPECTED TO CONTINUE AND MAY INTENSIFY, FURTHER DEGRADING PROPAGATION CONDITIONS. THE SUN IS ALSO VERY ACTIVE, WITH FREQUENT FLARE ACTIVITY CAUSING SHORT WAVE FADING ON SUNLIT PATHS. EXPECT THIS TO CONTINUE THROUGH THE NEXT 24 HOURS.
PART III. SUMMARY OF SOLAR FLARE INDUCED IONOSPHERIC DISTURBANCES WHICH MAY HAVE CAUSED SHORT WAVE FADES IN THE SUNLIT HEMISPHERE DURING THE 24 HOUR PERIOD ENDING 08/2400Z MAR 89 . . .

START	END	CONFIRMED	FREQS AFFECTED
08/0825Z	0959Z	YES	EST. THRU 15 MHZ
1429Z	1503Z	NO	UNKN
1850Z	1908Z	NO	UNKN

PROBABILITY FOR THE NEXT 24 HOURS . . . . STRONG
PART IV. OBSERVED/FORECAST 10.7 CM FLUX AND K/AP.
THE OBSERVED 10.7 CM FLUX FOR 08 MAR 89 WAS 206.
THE FORECAST 10.7 CM FLUX FOR 09, 10, AND 11 MAR 89 ARE 208, 210, AND 212.
THE OBSERVED K/AP VALUE FOR 08 MAR 89 WAS 03/19.
THE FORECAST K/AP VALUES FOR 09, 10, AND 11 MAR 89 ARE 05/45, 04/30, AND 03/20.

SECONDARY HF RADIO PROPAGATION REPORT ISSUED AT 121200Z MAR 89.
PART I. SUMMARY 120600Z TO 121200Z MAR 89/
FORECAST 121200Z TO 121800Z MAR 89.

		QUADRANT			
		I	II	III	IV
		0 TO 90W	90W TO 180	180 TO 90E	90E TO 0
REGION	POLAR	N3/M20	U3/M20	U3/M20	N3/M20
	AURORAL	N4	U3/M20	U3/M20	N4
	MIDDLE	N7	N7	N6	N7
	LOW	N8	N7	N7	N8
	EQUATORIAL	N8	N7	N8	N8

PART II. GENERAL DESCRIPTION OF HF PROPAGATION CONDITIONS.
GEOMAGNETIC STORMING EARLIER IN THE DAY HAS CONTINUED TO CAUSE DEGRADATIONS IN THE POLAR AND AURORAL LATITUDES AS NOTED ABOVE. BECAUSE STORMING CONDITIONS ARE EXPECTED TO RETURN, THESE DEGRADATIONS MAY CONTINUE.

b

Fig. 9.37 *Continued*

Numbers from 1 through 9 represent absolute forecast conditions:

1 – Useless	6 – Fair to good
2 – Very poor	7 – Good
3 – Poor	8 – Very good
4 – Poor to fair	9 – Excellent
5 – Fair	

Two types of forecasts are issued: (i) a primary report of solar and geophysical activity, and (ii) a daily HF radio propagation report. A solar and geophysical activity report (Fig. 9.37*a*) consists of information on solar activity during the past universal day and a plain-language forecast for the coming day. These reports are issued at 0600 UT. An HF propagation forecast (Fig. 9.37*b*) consists of three parts: (i) a description of general HF radio propagation conditions for a 24 h period prior to the forecast, (ii) a summary of possible HF propagation disturbances (SIDs) on sunlit paths for the coming 24 h, and (iii) outlook for general HF radio propagation and geomagnetic field for the coming 24 h. These forecasts are available on the SESC computer bulletin board.

9.9 Accuracy of forecasts

The question of accuracy of forecasts cannot be answered simply. Various statistical studies (e.g. Agy 1970, Ch. 5) have shown that solar activity can be forecast with high accuracy (80% to 90%) and that some 50% to 70% of the more outstanding events have been announced in advance. Also, although statistical improvements in forecasting ionospheric characteristics can be achieved by short-term modifications, the forecasts may be of little benefit to radio operators unless they have rapid access to the information and the ability to use the information (e.g. to change transmitter frequency). Further, although the climatological (average) picture of ionospheric storms has been established with reasonable confidence it may be of little value in forecasting the detailed history of an individual storm.

The value of a forecast is, therefore, much more important than its accuracy in a purely numerical sense. Most users are more concerned with catastrophic events than with minor disturbances, so statistics may be misleading. Thus a particular forecaster may err in forecasting 98 small events, but if he correctly forecasts the two major storms his statistics may be poor but his reputation good. The latter argument has been cited to support unconventional methods such as that based on planetary conjunctions (e.g. Davies 1981 and references cited therein).

Predictions and forecasts should be used mainly as a guide in determining a course of remedial actions. Educated operators are essential for effective utilization of radio circuits, and a small expenditure in training may be a far better investment than expensive equipment. It should also be emphasized that the value of a forecast depends largely on how quickly the information can be communicated to the user. The speed of dissemination has improved greatly in recent years with the use of computers and satellite communication circuits.

9.10 References

AGY, V. (1954): 'Geographical and temporal distribution of polar blackouts', *J. Geophys. Res.*, **59**, p. 499

AGY, V. (Ed.) (1970): 'Ionospheric forecasting'. AGARD Conf. Proc. No. 49, Advisory Group for Aerospace Research and Development, North Atlantic Treaty Organisation

ALBERTSON, V. D. and VAN BAELON, J. A. (1970): 'Electric and magnetic fields at the earth's surface due to auroral currents', *IEEE Trans.*, **PAS-89**, p. 578

ARGO, P. E. and ROTHMULLER, I. J. (1979): 'PROPHET: An application of propagation forecasting principles' *in* DONNELLY, R. F. (Ed.): 'Solar-terrestrial predictions proceedings I', US GPO, Washington, DC 20402, p. 312

AVDYUSHIN, S. I., DANILOV, A. D., MALISHEV, A. B., NOVIKOVA, G. N., and SVIDSKY, P. N. (1979): 'Forecasting ionospheric and geomagnetic conditions at the IAG Forecasting Center' *in* DONNELLY, R. F. (Ed.): 'Solar-terrestrial predictions proceedings I.' US GPO., Washington, DC 20402, p. 104

BAILEY, D. K. (1959): 'Abnormal ionization in the lower ionosphere associated with cosmic-ray flux measurements', *Proc. IRE*, **47**, p. 255

BAILEY, D. K. (1968): 'Some quantitative aspects of electron precipitation in and near the auroral zone', *Rev. Geophys.*, **6**, p. 289

BAKER, D. M. and DAVIES, K. (1966): 'Solar flare effects and the relaxation time of the ionosphere', *J. Geophys. Res.*, **71**, p. 2840

BAKER, D. N. (1986): 'Statistical analyses in the study of solar wind–magnetosphere coupling' *in* KAMIDE, Y. and SLAVIN, J. A. (Eds.): 'Solar-wind–magnetosphere coupling', (Terra Scientific Publishing Co., Tokyo) p. 17

BATES, H. F. and HUNSUCKER, R. D. (1974): 'Quiet and disturbed electron-density profiles in the auroral zone ionosphere', *Radio Sci.*, **9**, p. 455

BELROSE, J. S. (1964): 'Present knowledge of the lowest ionosphere' *in* BLACKBAND, W. T. (Ed.) 'Propagation of radio waves at frequencies below 300 kc/s', AGARDograph 74 (Macmillan Company, New York) p. 3

BIBL, K. (1963): 'Comparison of solar flare effects in the D and E regions of the ionosphere' *in* GASSMANN, G. J. (Ed.): 'The effect of disturbances of solar origin on communications', AGARDograph 59, (Macmillan Company, New York) p. 109

BLEIWEISS, M. P., HILDEBRAND, V. E., and HILL, J. R. (1973): 'A D-region model which accounts for quiet and disturbed VLF propagation phenomena'. Tech. Report TR1868, Naval Electronics Lab. Center, San Diego, CA 92152

BOHLIN, J. D. (1977): 'An observational definition of coronal holes' *in* ZIRKER, J. B. (Ed.): 'Coronal holes and high speed wind streams' (Colorado Associated University Press, Boulder, CO) p. 27

BOOKER, H. G., PASRICKA, P. K., and POWERS, W. J. (1986): 'Use of scintillation theory to explain frequency-spread on F-region ionograms', *J. Atmos. Terr. Phys.*, **48**, p. 327

BUDDEN, K. G. (1985): 'The propagation of radio waves' (Cambridge University Press, Cambridge)

CAMPBELL, W. H. (1979): 'Working group report on geomagnetic applications' *in* DONNELLY, R. F. (Ed.) 'Solar-terrestrial predictions proceedings II', US GPO, Washington, DC 20402, p. 133

COLL, D. C. (1968): 'Adaptive receivers for multipath channels' *in* FOLKESTAD, K. (Ed.): 'Ionospheric radio communications' (Plenum Press, New York) p. 341

COOK, F. E. and DAVIES, P. (1979): 'A review of the operation of the IUWDS Regional Warning Center at the Ionospheric Prediction Service, Sydney' *in* DONNELLY, R. F. (Ed.): 'Solar-terrestrial predictions proceedings I', US GPO, Washington, DC 20402, p. 239

DAMBOLT, T. H. (1979): 'Propagation predictions for the HF-range by the Research Institute of the Deutche Bundespost' *in* DONNELLY, R. F. (Ed.): 'Solar-terrestrial predictions proceedings I'. US GPO Washington, DC 20402, p. 25

DAVIES, K. (1965): 'Ionospheric radio propagation'. NBS Monograph 80, US GPO, Washington, DC 20402

DAVIES, K. (1980): 'Recent progress in satellite radio beacon studies with particular emphasis on the ATS-6 radio beacon experiment', *Space Sci. Rev.*, **25**, p. 357

DAVIES, K. (1981): 'Review of recent progress in ionospheric predictions,' *Radio Sci.*, **16**, p. 1407

DONNELLY, R. F. (1969): 'Energetic X-ray and extreme ultraviolet flashes of solar flares', *Astrophys. J.*, **158**, p. L165

DONNELLY, R. F. (1971): 'Extreme ultraviolet flashes of solar flares observed via sudden frequency deviations: Experimental results', *Solar Physics*, **20**, p. 188

DONNELLY, R. F. (Ed.) (1979): 'Solar-terrestrial predictions proceedings, I'. US GPO, Washington, DC 20402

FERGUSON, B. G. (1979): 'Coronal holes inferred from the "Fleurs" east-west solar scans' *in* DONNELLY, R. F. (Ed.): 'Solar-terrestrial predictions proceedings I', US GPO, Washington, DC 20402, p. 288

FOLKESTAD, K. (Ed.) (1968): 'Ionospheric radio communications' (Plenum Press, New York)

FULLER-ROWELL, T. J., and EVANS, D. S. (1987): 'Height-integrated Pedersen and Hall conductivity patterns inferred from the TIROS–NOAA satellite data', *J. Geophys. Res.*, **92**, p. 7606

GASSMANN, G. J. (Ed.) (1963): 'The effect of disturbances of solar origin on telecommunications'. AGARDograph 59 (Macmillan Co., New York)

GOODMAN, J. M. (Ed.) (1987): 'The effect of the ionosphere on communication, navigation, and surveillance systems.' Naval Research Laboratory, Washington, DC

HARTZ, T. R., MONTBRIAND, L. E., and VOGAN, E. L. (1968): 'A study of auroral absorption at 30 mc/s, *Canad. J. Phys.*, **41**, p. 581

HATTON, W. L. (1968): 'Relationship between the user, system designer and radio physicist' *in* FOLKESTAD, K. (Ed.): 'Ionospheric radio communications' (Plenum Press, New York) p. 439

HARGREAVES, J. K. (1979): 'The upper atmosphere and solar-terrestrial relations' (Van Nostrand Reinhold Company, New York)

HECKMAN, G. R. (1979): 'Predictions of the Space Environment Services Center' *in* DONNELLY, R. F. (Ed.): 'Solar-terrestrial predictions proceedings I'. US GPO, Washington, DC 20402, p. 323

HERMAN, J. R. and PENNDORF, R. B. (1964): 'Reception of mid-latitude transmissions in northern Canada' *in* LANDMARK, B. (Ed.): 'Arctic communications' (Macmillan Company, New York) p. 97

HOFMANN, D. J. and SAUER, H. H. (1968): 'Magnetospheric cosmic-ray cutoffs and their variations', *Space Sci. Rev.*, **8**, p. 750

HOLT, O. (1968): 'Characteristics of polar cap absorption' *in* FOLKESTAD, K. (Ed.): 'Ionospheric radio communications' (Plenum Press, New York) p. 33

HULTQVIST, B. (1967): 'Aurora' *in* MATSUSHITA, S. and CAMPBELL, W. H. (Eds): 'Physics of geomagnetic phenomena II' (Academic Press, New York) p. 633

HULTQVIST, B. (1969): 'Polar cap absorption and ground level effects' *in* DEJAGER, C. and SVESTKA, Z. (Eds.): 'Solar flares and space research' (North-Holland, Amsterdam) p. 202

HUNSUCKER, R. D. and BATES, H. F. (1969): 'Survey of polar and auroral effects on HF propagation', *Radio Sci.*, **4**, p. 347

JONES, T. B. (1971): 'VLF phase anomalies due to a solar X-ray flare, *J. Atmos. Terr. Phys.*, **33**, p. 963

JOSELYN, J. A. (1986*a*): 'Real-time prediction of global geomagnetic activity' *in* KAMIDE, Y.

and SLAVIN, J. A. (Eds): 'Solar wind–magnetosphere coupling' (Terra Scientific Publishing Co., Tokyo) p. 127

JOSELYN, J. A. (1986*b*): 'SESC methods for short-term geomagnetic predictions' *in* SIMON, P. A., HECKMAN, G., and SHEA, M. A. (Eds.): 'Solar-terrestrial predictions', NOAA, Boulder, Co. 80303, p. 404

JOSELYN, J. A. and MCINTOSH, P. (1981): 'Disappearing solar filaments: A useful predictor of geomagnetic activity', *J. Geophys. Res.*, **86**, p. 4555

JULL, G. W. (1964): 'HF propagation in the Arctic' *in* LANDMARK, B. (Ed.): 'Arctic communications' (Macmillan Company, New York) p. 157

KERSLEY, L., and KLOBUCHAR, J. A. (1978): 'Average response of protonospheric electron content to geomagnetic storm activity' *in* CHECCACCI, P. F. (Ed.): 'Beacon satellite measurements of plasmaspheric and ionospheric properties'. Institute di Ricerca sulle Onde Elettromagnetiche, Florence, Italy, p. 18-1

KILDAHL, K. (1980): 'Frequency of class M and X flares by sunspot class' *in* DONNELLY, R. F. (Ed.): 'Solar-terrestrial predictions proceedings III'. US GPO., Washington, DC 20402, p. C–166

KING, G. A. M. (1962): 'The night-E layer' *in* MATSUSHITA, S. and SMITH, E. K. (Eds): 'Ionospheric sporadic E' (Macmillan Company, New York) p. 219

KING, G. A. M. (1971): 'The ionospheric F-region storm', *J. Atmos. Terr. Phys.*, **33**, p. 1223

KULESHOVA, V. P., LAVROVA, E. V., and LYAKHOVA, L. N. (1980): 'Forecasting of $\delta f_o F_2$-variations for ionospheric disturbances' *in* DONNELLY, R. F. (Ed.): 'Solar-terrestrial predictions proceedings IV'. US GPO, Washington, DC 20402, p. C–37

LANDMARK, B. (Ed.) (1964): 'Arctic communications'. AGARDograph 78 (Macmillan Company, New York)

LARSEN, T. R. (1979): 'Effects of D-region ionization on radio wave propagation' *in* DONNELLY, R. F. (Ed.): 'Solar-terrestrial predictions proceedings II', US GPO, Washington, DC 20402, p. 617

LAWRENCE, R. S., LITTLE, C. G., and CHIVERS, H. J. A. (1964): 'A survey of ionospheric effects upon earth–space radio propagation', *Proc. IEEE* **52**, p. 4

LEIGHTON, H. I., SHAPLEY, A. H., and SMITH, E. K. (1962): 'The occurrence of sporadic E during the IGY' *in* MATSUSHITA, S. and SMITH, E. K., 'Ionospheric sporadic E' (Macmillan Company, New York) p. 166

LEWIS, C. and DENNETT, J. (1986): 'Artificial intelligence technology'. Colorado University Dept. of Engineering, Boulder, CO

LINCOLN, J. V. (1964): 'The listing of sudden ionospheric disturbances', *Planet. Space Sci.*, **12**, p. 419

MAEDA, R. and INUKI, H. (1979): 'Radio disturbance warning issuance system' *in* DONNELLY, R. F. (Ed.): 'Solar-terrestrial predictions proceedings I', US GPO, Washington, DC 20402, p. 223

MATSUSHITA, S. (1959): 'A study of the morphology of ionospheric storms', *J. Geophys. Res.*, **64**, p. 305

MATSUSHITA, S. (1967): 'Geomagnetic disturbances and storms' *in* MATSUSHITA, S. and CAMPBELL, W. H. (Eds.): 'Physics of geomagnetic phenomena II' (Academic Press, New York) p. 793

MATUURA, N. (1972): 'Theoretical models of ionospheric storms', *Space Sci. Rev.*, **13**, p. 124

MCINTOSH, P. and DRYER, M. (Eds.): (1972): 'Solar activity observations and predictions' (MIT Press, Cambridge, MA)

MENDILLO, M. (1973): 'A study of the relationship between geomagnetic storms and ionospheric disturbances at mid-latitudes', *Planet. Space Sci.*, **21**, p. 349

MITRA, A. P. (1974): 'Ionospheric effects of solar flares' (Reidel Publishing Company, Dordrecht, Holland)

MUNRO, G. H., KNECHT, R. W., and DAVIES, K. (1961): 'Possible solar flare effects in the F-region of the ionosphere', *Nature*, **192**, p. 347

NORTON, R. B. (1969): 'The middle-latitude F-region during some severe ionospheric storms', *Proc. IEEE*, **57**, p. 1147

OBAYASHI, T. (1954): 'On the world-wide disturbance in F2 region', *J. Geomag. Goelectr.*, **6**, p. 57

OELBERMANN, E. J., Jr. (1970): 'Solar particle effects on polar cap VLF propagation', *J. Franklin Inst.*, **290**, p. 281

OHSHIO, M. (1964): 'Solar flare effect on geomagnetic variation', *J. Radio Res. Labs.*, **11**, p. 377

OHSHIO, M. (1971): 'Negative sudden phase anomaly', *Nature*, **229**, p. 239

ONDOH, T. and OBU, K. (1980): 'Prediction of HF communications disturbance by pre-sc HF field increases on polar paths crossing the auroral zone' *in* DONNELLY, R. F. (Ed.): 'Solar-terrestrial predictions proceedings IV', US GPO, Washington, DC 20402, p. D2

PAUL, A. K. (1979): 'Radio pulse dispersion in the ionosphere' *J. Geophys.*, **46**, p. 15

PETRIE, L. E. and WARREN, E. S. (1968): 'The propagation of high frequency waves on the Winnipeg-Resolute Bay oblique sounding circuit' *in* FOLKESTAD, K. (Ed.): 'Ionospheric radio communications' (Plenum Press, New York) p. 242

PROBST, S. E. (1968): 'The CURTS concept and current status of development' *in* FOLKESTAD, K. (Ed.): 'Ionospheric radio communications' (Plenum Press, New York) p. 370

REDDY, B. M., AGGARWAL, S., LAKSHMI, D. R., SHASTRI, S., and MITRA, A P. (1979): 'Long term solar activity and ionospheric prediction services rendered by the National Physical Laboratory, New Delhi' *in* DONNELLY, R. F. (Ed.): 'Solar-terrestrial predictions proceedings I', US GPO, Washington, DC 20402, p. 188

REID, G. C. (1967): 'Ionospheric disturbances' *in* MATSUSHITA, S. and CAMPBELL, W. H. (Eds.): 'Physics of geomagnetic phenomena II' (Academic Press, New York) p. 627

REID, G. C. (1986): 'Solar energetic particles and their effects on the terrestrial environment' *in* STURROCK, P. A., HOLZER, T. E., MIHALAS, D. M. and ULRICH, R. K. (Eds.): 'Physics of the sun' D. Reidel Publishing Company, Dordrecht, p. 252

RYCROFT, M. J. and BURNELL, S. J. (1970): 'Statistical analysis of movements of the ionospheric trough and the plasmapause', *J. Geophys. Res.*, **75**, p. 5600

SAO, K., YAMASHITA, M., TANAHASHI, S., JINDOH, H., and OHTA, K. (1970): 'Sudden enhancements (SEA) and decreases (SDA) of atmospherics', *J. Atmos. Terr. Phys.*, **32**, p. 1567

SCHMERLING, E. R. (Ed.) (1987): 'The aerospace environment at high altitudes and its implications for spacecraft charging and communications'. AGARD Conf. Proc. No. 406, Advisory Group for Aerospace Research and Development, North Atlantic Treaty Organisation

SESC (1988): 'SESC glossary of solar-terrestrial terms'. Space Environment Services Center, NOAA, Boulder, CO 80303

SIMON, P. (1979): 'The forecasting center of Meudon, France' *in* DONNELLY, R. F. (Ed.): 'Solar–terrestrial predictions proceedings I'. US GPO, Washington, DC 20402, p. 1

SIMON, P. A., HECKMAN, G., and SHEA, M. A. (Eds.) (1986): 'Solar-terrestrial predictions'. NOAA, Boulder, CO 80303

SMITH, E. K. and MATSUSHITA, S. (1962): 'Ionospheric sporadic E' (Macmillan Company, New York)

STEVENS, E. E. (1968*a*): 'The significance of sporadic E propagation in determining the MUF' *in* FOLKESTAD, K. (Ed.): 'Ionospheric radio communications' (Plenum Press, New York) p. 289

STEVENS, E. E. (1968*b*): 'The CHEC sounding system' *in* FOLKESTAD, K. (Ed.): 'Ionospheric radio communications' (Plenum Press, New York) p. 359

STONEHOCKER, G. H. (1970): 'Advanced telecommunication forecasting technique' *in* AGY, V. (Ed.): 'Ionospheric forecasting'. AGARD Conf. Proc. No. 49, Advisory Group for Aerospace Research and Development, North Atlantic Treaty Organisation, p. 27–1

STRAUS, J. M. and HICKMAN, D. R. (1979): 'Predictability of upper-atmospheric density and composition' *in* DONNELLY, R. F. (Ed.): 'Solar-terrestrial predictions proceedings II', US GPO, Washington, DC 20402, p. 71

SVESTKA, Z. (1972): 'Several solar aspects of flare-associated particle events' *in* MCIN-

TOSH, P. S. and DRYER, M. (Eds.): 'Solar activity observations and predictions' (MIT Press, Cambridge, MA) p. 141

SVESTKA, Z. and SIMON, P. (1975): 'Catalog of solar particle events 1955–1969' (Reidel Publishing Company, Dordrecht)

THOMAS, L. (1970): 'F2-region disturbances associated with major magnetic storms', *J. Atmos. Terr. Phys.*, **18**, p. 917

THOMAS, L. and VENABLES, F. H. (1966): 'The onset of the F-region disturbance at middle latitudes during magnetic storms', *J. Atmos. Terr. Phys.*, **28**, p. 599

TSURUTANI, B. T., and BAKER, D. N. (1979): 'Substorm warnings: an ISEE–3 real-time data system', *EOS*, **60**, p. 702

VINCENT, W. R., DALY, R. F., and SIFFORD, B. M. (1968): 'Modeling communications systems' *in* FOLKESTAD, K. (Ed.): 'Ionospheric radio communications', (Plenum Press, New York) p. 321

WAGNER, W. J. (1984): 'Coronal mass ejections', *Ann. Rev. Astron. Astrophys.*, **22**, p. 267

WAKAI, N. (1971): 'Study of the nighttime E region and its effects on the radiowave propagation', *J. Radio Res. Labs.*, **18**, p. 245

WESTERLUND, S., REDER, F. H., and ÅBOM, C. (1969): 'Effects of polar cap absorption events on VLF transmissions', *Planet. Space Sci.*, **17**, p. 1329

ZEVAKINA, R. A., and KISILEVA, M. V. (1980): 'On the short-term prediction of the space-time distribution of auroral absorption' *in* DONNELLY, R. F. (Ed.): 'Solar-terrestrial predictions proceedings IV'. US GPO, Washington, DC 20402, p. D3

Chapter 10

Low, very low, and extremely low frequencies

10.1 Characteristics of LF, VLF, and ELF waves

We have seen that although HF waves provide a convenient and economical means of communication they are highly variable and subject to disruptions. Extremely low frequencies (30–3000 Hz), very low frequencies (3–30 kHz) and low frequencies (30–300 kHz) provide alternatives to the high-frequency band. These frequencies are reflected from the lower side of the D layer and are less affected by D-region disturbances than higher frequency waves that penetrate the D layer, and they are characterized by relatively slow fading rates. The chief disadvantages are the small available bandwidths ($\approx$ 100 Hz) and the relatively large and, hence, expensive transmitter installations (Watt 1967). Further, the use of VLF and LF signals for long-distance navigation has an historical basis in the Austin-Cohen formula (Davies 1969, Sec. 1.1): the longer the wavelength the stronger the signal.

The propagation of VLF signals is characterized by relatively low path attenuation which is relatively stable with time and by the guiding features of the Earth–ionosphere waveguide, which render propagation possible over long distances (5000–20 000 km). Thus VLF is attractive for long distances when reliability and stability are important. The atmospheric noise level on VLF is high, so high transmitter powers and large antennas are required. Hence VLF systems are confined to narrow bandwidths, about 20–150 Hz. Radiation efficiencies of VLF antennas are about 10–20%.

In the OMEGA navigation system, use is made of the lower end of the VLF band (10–14 kHz) and phase comparisons of signals from eight transmitters. The accuracy of the system and the ability to identify lines of position (LOP) unambiguously depend ultimately on the propagation characteristics of the ionosphere–Earth system.

VLF signals propagate over long distances with small attenuation rates of about 2–3 dB per megameter (Mm = 1000 km). Further, they have the remarkable property that the phase delay suffered during propagation follows a predictable diurnal pattern, which makes possible a navigation accuracy of about 1–2 km.

Extremely low frequencies are important in communications with submarines because their relatively large skin depths enable ELF signals to penetrate sea water. The excitation of such long waves ($\lambda \approx 1000\,\text{km}$) requires large antenna systems or natural phenomena, such as lightning, that radiate on these wavelengths.

Low-frequency waves are characterized by higher path attenuation (compared with VLF), lower background noise, and more stable time delays. LF systems are used mostly over intermediate distances (1000–5000 km). Terminal equipment is usually cheaper than in the VLF case, and the available bandwidths are larger.

From the point of view of radio propagation an interesting phenomenon is the propagation in the plasmasphere (protonosphere) of audiofrequency electromagnetic waves, called 'whistlers', generated by lightning. ELF noise can be generated in the plasmasphere by a variety of human activities and natural sources.

10.2 Whistlers

10.2.1 Sources and propagation

Whistlers are bursts of ELF/VLF waves produced by lightning discharges. They travel up through the ionosphere into the plasmasphere where they are guided (ducted) approximately along the geomagnetic field (Fig. 10.1). When they reach the conjugate ionosphere, where there is an abrupt and large change in the refractive index, some of the energy is reflected back and continues to propagate back and forth along the duct. Whistlers are identified as radio signals in the audiofrequency range in which the frequency usually descends in a second or so. For a more complete description of this and allied phenomena the reader is referred to the books by Helliwell (1965) and Al'pert (1983).

10.2.2 Refractive index

Since the frequencies of whistlers are very much less than both the plasma frequency and the electron gyrofrequency, the terms X and Y in the Appleton equation (Sec. 3.1) are much larger than unity and, to a first approximation, the refractive index μ is given by the quasi-longitudinal approximation:

$$\mu \approx f_N/(ff_L)^{1/2} \tag{10.1}$$

In the ionosphere we have $f_N \approx f_L \approx 10^6\,\text{Hz}$; therefore, with $f = 10^3\,\text{Hz}$, $\mu \approx 30$. Thus there is a large change in refractive index as the whistler enters the ionosphere. Application of Snell's Law shows that the $\boldsymbol{k}$ vector in the ionosphere is essentially vertical for a wide range of incident directions. Once within the ionosphere the whistler may be trapped in a magnetic field-aligned irregularity, and it can propagate efficiently to the conjugate ionosphere. The whistler wave corresponds to the upper (o) curve in Fig. 3.2b and substitution of $\mu^2 \gg 1$ in (3.6) shows that $R \approx i/Y_L$; as a result, the wave vectors are elliptically polarized in a clockwise (or right-handed) sense to an observer looking along the magnetic field.

A second approximation to the refractive index formula is

$$\mu^2 = f_N^2/(f|f_L| - f^2) \tag{10.2}$$

which shows that the frequency range is limited to $0 < f < f_H \cos\theta$. Alternatively, for a given frequency less than f_H, the maximum angle θ_m is given by

$$\cos\theta_m = f/f_H \tag{10.3}$$

and, since $f_H \gg f$, propagation is confined to directions close to the magnetic field. Substitution of these values into (3.37) shows that the energy is confined to a narrow cone of semiangle less than 19°28′, which results in beaming.

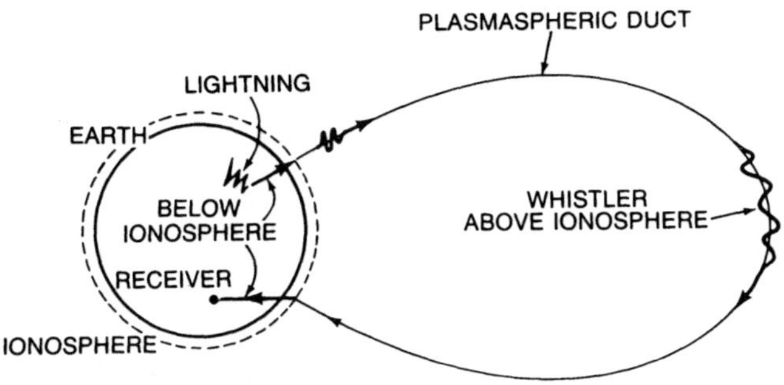

Fig. 10.1 *Generation and propagation of whistlers*
Whistlers are generated by lightning strokes in the troposphere. They propagate in the Earth–ionosphere waveguide and penetrate the plasmasphere. They are ducted along magnetic field lines to the opposite hemisphere where some energy penetrates the ionosphere and propagates to the receiver via the waveguide

10.2.3 Dispersion

The time of flight T of a pulse traveling once along a magnetic line of force from one hemisphere to its conjugate point is given by

$$T = \int \frac{ds}{u} = Df^{-1/2} \tag{10.4a}$$

where

$$D = \frac{1}{2c}\int \frac{f_N\, ds}{f_L^{1/2}} \tag{10.4b}$$

is called the dispersion and u is the group speed. Equation (10.4*a*) shows that whistler propagation is highly dispersive. Thus, the original wideband 'click' of the lightning source is converted into descending tones because the high-frequency components arrive before the low-frequency components (Fig. 10.2). Examples of the guiding of whistlers in the plasmasphere are given by Yabroff (1961), and the trapping of whistlers in ducts is discussed by Helliwell (1965, Ch. 3). More detailed analyses show that there is a minimum frequency ($f_{\min}$)

at which the propagation time is a minimum; therefore 'nose whistlers' are observed (Fig. 10.2) where

$$f_{\min} \approx f_H/4 \tag{10.5}$$

Thus the nose frequency gives the minimum value of f_H along the path; this identifies the approximate path. Whistlers have been detected by satellite-borne receivers on frequencies as high as 0·9 of the local gyrofrequency (Dunckel and Helliwell 1977).

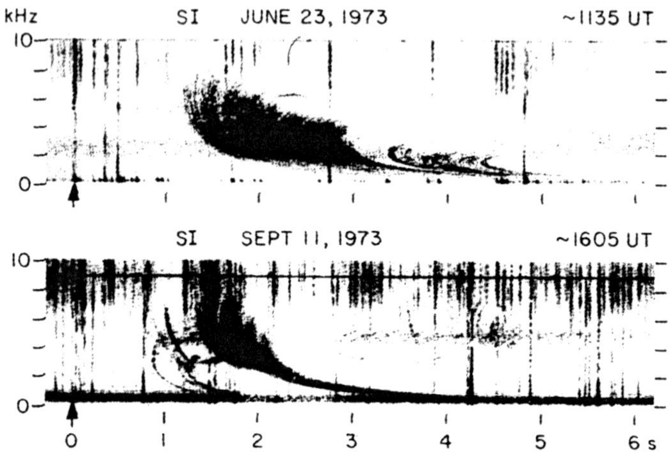

Fig. 10.2 *Variation of frequency with time, showing nose whistlers (Courtesy D. L. Carpenter)*

10.2.4 Allied phenomena

In addition to whistlers there is a set of allied phenomena called VLF emissions. These emissions are classified as hiss, chorus, discrete emissions, periodic emissions, and triggered emissions (Helliwell 1965). Triggered-emission sources include both natural sources (Rosenberg and Dudeney 1986) and VLF transmitters. Triggered emissions are also called stimulated emissions and are caused by wave-particle interaction in the plasmasphere.

Signals from VLF transmitters have been received on a satellite (Rorden *et al.* 1964) having a small loop antenna and a bandwidth of 25 Hz. Signals are generally stronger at night than during the day, presumably because of differences of absorption in the D-region. Time delays range from a few milliseconds to hundreds of milliseconds, and large Doppler shifts (e.g. 7 Hz) have been observed.

10.3 Theoretical aspects of propagation on VLF/ELF

10.3.1 Rays and modes

In some cases it is convenient to describe VLF phenomena in terms of geometrical rays as, for example, in the case of the diurnal variation of the phase, which can be viewed as resulting from the diurnal variation of the height of a fictitious mirror-like reflecting layer.

The propagation of VLF waves between the Earth and the ionosphere, over large distances, can be treated also by waveguide theory. The 'mode theory' of VLF propagation has been considered extensively, for example, by Budden (1961) and by Wait (1970, 1986). The 'mode' of propagation may be defined as 'the form of propagation of waves that is characterized by a particular field pattern in a plane transverse to the direction of propagation, which field pattern is independent of the position along the axis of the waveguide'.

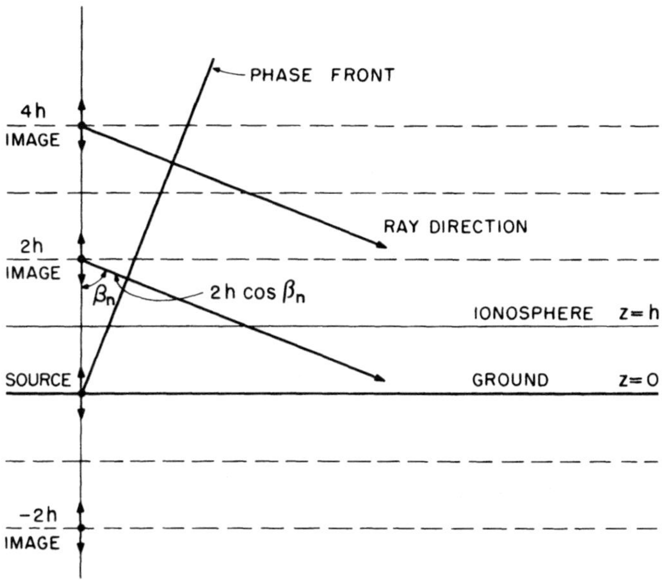

Fig. 10.3 *Ray geometry corresponding to the first-order and second-order modes between two conducting planes*

In the development of mode theory one can consider the signal at a point as consisting of component wavelets emanating from images of the source (Fig. 10.3). The wavelets emanating from the images will reinforce in a direction making an angle β_n with the vertical such that

$$2\text{h} \cos \beta_n = 2hC_n = n\lambda \tag{10.6a}$$

where

$$C_n = n\lambda/2h \tag{10.6b}$$

The ground images will add up at angles $(\pi - \beta_n)$ with the vertical. Thus for a given n, there are two families of waves that have the same horizontal velocities and opposite vertical velocities. We can imagine the superposition of these two sets of waves as a single wave traveling parallel to the Earth with a phase speed

$$v_p = c/S_n, \tag{10.7a}$$

where

$$S_n^2 = 1 - C_n^2 \tag{10.7b}$$

and $n = 0, 1, 2, \ldots$ is called the mode order.

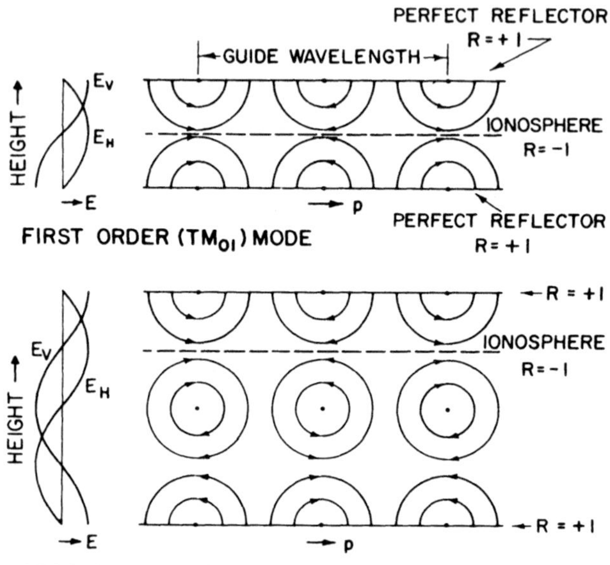

Fig. 10.4 *E-field in an ideal Earth–ionosphere waveguide*

For long-distance propagation the ionosphere behaves approximately like a magnetic conductor with a reflection coefficient of -1, while the ground has a reflection coefficient of $+1$. In this case (Fig. 10.3) the images are located at $z = \pm 2h, \pm 4h$, and so on, but now they alternate in sign, which is equivalent to a change of phase of π. The resonance condition is now

$$2hC_n = (n - \tfrac{1}{2})\pi \tag{10.8}$$

where $n = +1, +2, \ldots$, etc. In this case there is no zero-order mode. The E-field pattern in the plane of propagation in an ideal Earth–ionosphere waveguide is shown in Fig. 10.4 for the first- and second-order modes. Here we are dealing with waves polarized with their electric fields in the (vertical) plane of propagation, and with their magnetic fields transverse to the plane of propagation; such modes are called transverse magnetic (TM) modes.

The vertical wavelength $\lambda_g = \lambda/S_n$ is obtained from (10.6):

$$\left(\frac{1}{\lambda_g}\right)^2 = \left(\frac{1}{\lambda}\right)^2 - \left(\frac{n}{2h}\right)^2 \tag{10.9}$$

which shows that for $\lambda > 2h/n$, λ_g is imaginary and hence the mode is evanescent. Thus there is a minimum cutoff frequency f_n below which waves will not propagate, where

$$f_n = nc/2h \tag{10.10}$$

The cutoff frequency for the first-order mode during daytime ($h = 75$ km) is ≈ 2 kHz.

For the case of a magnetically conducting ionosphere the cutoff frequency is given by

$$f_n = (n - \tfrac{1}{2})c/2h \tag{10.11}$$

It is seen that the change of reflection coefficient R from $+1$ to -1 changes the cutoff frequency (for $n = 1$) from about 2 kHz to about 1 kHz. To be consistent with the mode-numbering system for perfectly conducting electrical walls, the ideal Earth–ionosphere modes should be denoted by $n - \frac{1}{2}$ rather than by n.

The above discussion merely illustrates a simple picture of mode propagation. In no way should it be interpreted as a quantitative theory of VLF propagation. Among others, it neglects the following: (i) the curvature of the Earth and the ionosphere, (ii) the effects of the Earth's magnetic field, (iii) effects of electron collisions, and (iv) irregularities in the Earth–ionosphere waveguide. More detailed treatments require extensive use of large computer programs.

The question often arises as to what is the relationship between modes and rays and whether the use of one is more correct than use of the other. For the idealized horizontally homogeneous Earth–ionosphere waveguide, provided that all modes and all rays are properly taken into account with appropriate corrections for losses, the results from the two methods are equivalent. Sometimes a ray model of VLF propagation is used in which just one ray is reflected several times between Earth and ionosphere. A single-ray model is satisfactory only over short distances. For the realistic horizontally inhomogeneous Earth–ionosphere waveguide, such equivalence has never been demonstrated.

It is appropriate to ask, then, which method of calculation is simplest. For VLF, the wavelength and height of the ionosphere are such that only a few modes are important since the higher-order modes are attenuated at a rate much greater than that of the lowest mode. At great distance only one mode need be considered. In such a case, however, many rays have to be considered. On the other hand, over short distances all propagating modes are comparable in amplitude and must be considered. Only a few rays need to be considered, however, since the higher-order rays are incident at smaller angles of incidence where the reflection coefficient is decreasing rapidly. On higher frequencies (in the LF band), and over long distances, many modes may be important, and the attenuation coefficients of many of the less attenuated modes may be comparable. Thus there is less advantage in using mode theory.

If a waveguide mode formulation is employed, the behavior of the field as a function of height is described by the height–gain function. That function has been discussed by Wait (1970) who pointed out that, over sea water, if the phase speed is greater than c the field initially decreases with height, and vice versa. Some curves showing the variation of phase speed v_p with frequency are shown in Fig. 10.5. Note that two types of signal enhancement are possible for VLF waves: (i) conjugate point enhancement because of whistler propagation, and (ii) antipodal enhancement due to geometry.

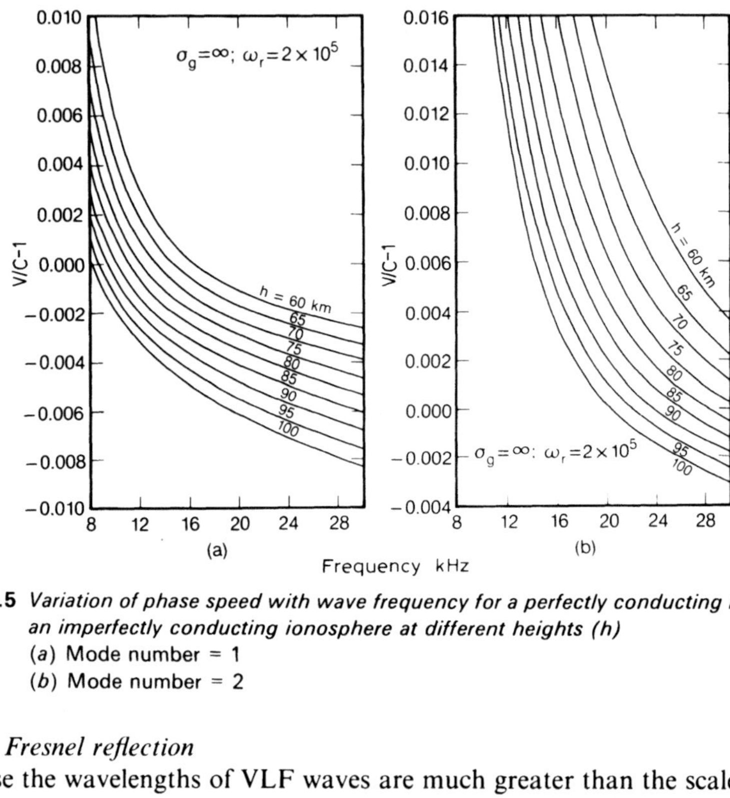

Fig. 10.5 *Variation of phase speed with wave frequency for a perfectly conducting Earth and an imperfectly conducting ionosphere at different heights (h)*
(a) Mode number = 1
(b) Mode number = 2

10.3.2 Fresnel reflection

Because the wavelengths of VLF waves are much greater than the scale size of the reflecting D layer the propagation can be treated approximately by Fresnel reflection at a sharp boundary (Budden 1985). At such a boundary the incident wave is converted into a reflected wave and a transmitted wave that obeys Snell's Law. In the case of an isotropic medium of refractive index μ with a wave having a vertical E-field incident from free space, the reflection coefficient R_{11} (see Sec. 10.3.3) is given by

$$R_{11} = \frac{\mu^2 C - (\mu^2 - S^2)^{1/2}}{\mu^2 C + (\mu^2 - S^2)^{1/2}} \tag{10.12a}$$

and the transmission coefficient T_{11} is

$$T_{11} = 1 - R_{11} = \frac{2(\mu^2 - S^2)^{1/2}}{\mu^2 C + (\mu^2 - S^2)^{1/2}} \tag{10.12b}$$

In the absence of the geomagnetic field we have

$$\mu^2 = 1 - i\omega_N^2/\omega\nu = 1 - i/L \tag{10.13a}$$

where

$$L = \omega\nu/\omega_N^2 = \omega/\omega_r \tag{10.13b}$$

A model of ω_r has been given by Ferguson *et al.* (1985):

For the nth-order mode

$$R_{11}(C_n) = \frac{(L - i)C_n - (C_n^2 L^2 - i\mathrm{L})^{1/2}}{(L - i)C_n + (C_n^2 L^2 - iL)^{1/2}} \tag{10.14}$$

and

$$T_{11}(C_n) = \frac{2(L^2 C_n^2 - iL)^{1/2}}{(L - i)C_n + (C_n^2 L^2 - iL)^{1/2}} \tag{10.15}$$

Reflection from a realistic ionosphere has been simulated by a number of sharply bounded slabs (Fig. 10.6), and the phases and amplitudes of the individual reflections are combined to determine the resulting reflected wave (Johler and Harper 1962).

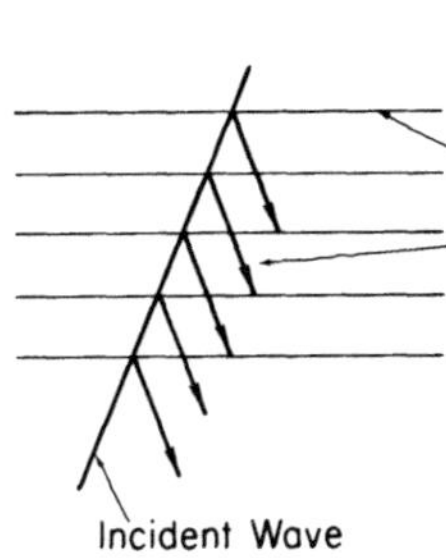

Fig. 10.6 *Simulation of VLF reflection by the addition of reflections from many thin layers*

The reflection coefficient is zero when the incident angle β is given by

$$\tan\beta_B = \mu \tag{10.16}$$

which is called the Brewster angle. For real n, when $\beta \approx \beta_B$ the phase of the reflected wave changes discontinuously by π as β passes through β_B.

When the incident wave is polarized with its E-field perpendicular to the plane of incidence, the reflection coefficient R_{22} and transmission coefficient T_{22} are

given by

$$R_{22} = \frac{C - (\mu^2 - S^2)^{1/2}}{C + (\mu^2 - S^2)^{1/2}} \qquad (10.17)$$

and

$$T_{22} = 2C/[C + (\mu^2 - S^2)^{1/2}] \qquad (10.18)$$

In this case R_{22} cannot be zero (unless $\mu = 1$), so there is no Brewster angle.

With vertical propagation we have for reflection

$$R_{11} = -R_{22} = -\frac{\mu - 1}{\mu + 1} \qquad (10.19)$$

and for transmission

$$T_{11} = 2\mu/(1 + \mu) \qquad (10.20a)$$

$$T_{22} = 2/(1 + \mu) \qquad (10.20b)$$

10.3.3 Effect of the Earth's magnetic field

The effect of the Earth's magnetic field is to generate characteristic waves that rotate the polarizations. Thus an incident wave having its electric vector in the vertical plane of incidence will produce a reflected wave having components both in the plane of incidence and perpendicular to that plane. Similarly an incident wave having its electric vector perpendicular to the plane of incidence (i.e. horizontal) will generate a reflected wave having electric field components in both the vertical and horizontal directions. It is, therefore, customary to write the reflection and transmission coefficients in matrix form

$$\boldsymbol{R} = \begin{vmatrix} R_{11} & R_{12} \\ R_{21} & R_{22} \end{vmatrix} \qquad (10.21a)$$

and

$$\boldsymbol{T} = \begin{vmatrix} T_{11} & T_{12} \\ T_{21} & T_{22} \end{vmatrix} \qquad (10.21b)$$

The subscript 12 implies conversion from an incident wave with its electric field in the vertical plane to a reflected wave with a horizontal electric field, and 21 implies conversion from an incident wave with its electric field horizontal to a reflected wave with a vertical electric field. The coefficients R_{11} and R_{22} are often called the reflection coefficients; R_{12} and R_{21} are called the conversion coefficients and were previously denoted by ${}_{\parallel}R_{\parallel}$, ${}_{\perp}R_{\perp}$, ${}_{\parallel}R_{\perp}$, and ${}_{\perp}R_{\parallel}$ respectively. When a wave is produced by several reflections the resulting reflection coefficient matrix is the product of the matrices for the individual reflections. Its value depends on the order in which the reflections occur (Budden 1985, Sec. 11.4). The reflected wave having its electric vector in the vertical plane is called the normal component, and the wave having its electric vector horizontal is called the abnormal component.

In the presence of the magnetic field the refractive index can be represented approximately by the quasi-longitudinal approximation

$$\mu_{o,e}^2 \approx 1 - i(\omega_r/\omega) \exp(\pm i\tau), \tag{10.22a}$$

where

$$\omega_r/\omega = \omega_N^2/\omega(\nu^2 + \omega_L^2)^{1/2} \tag{10.22b}$$

and

$$\tan\tau = \omega_L/\nu \tag{10.22c}$$

10.3.4 Effects of Earth–ionosphere curvature

The most important effect of curvature is that it restricts the angle of incidence to values less than $\sin^{-1}[a/(a + h)]$, where a is the radius of the Earth and h is the height of reflection, for which there is grazing incidence at the ground. Detailed theory has been given by Wait (1970). The net effect is to lower the strength of the received signal by as much as a factor of 3 below that calculated on the basis of a plane-Earth plane-ionosphere model.

Earth–ionosphere curvature reduces the speed of propagation v below the free-space speed. The reduction is greatest on the lowest order mode; this may be expected because the higher order modes are incident at smaller angles, and the differences between the angles of incidence for the flat and curved cases become smaller as the mode order increases. Earth–ionosphere curvature affects the efficiency with which a given mode can be launched into the waveguide (e.g. Pappert 1973). The excitation factor Λ_n of the nth mode is given approximately by

$$\Lambda_n \approx \frac{h}{aC_n^2}\left[1 - \left(1 - \frac{2h}{aC_n^2}\right)^{1/2}\right]^{-1} \tag{10.23}$$

With a flat Earth–ionosphere guide, $a \to \infty$ and $\Lambda_n \to 1$. With finite a, Λ_n is less than unity. The excitation factor departs considerably from unity as the height increases from 70 km to 100 km and as wave frequency increases on frequencies above about 16 kHz (Wait 1970).

10.3.5 Nonreciprocity of VLF propagation

An interesting feature of VLF propagation is the difference between east-to-west propagation and west-to-east propagation. Figures 10.7 and 10.8 illustrate the dependence of attenuation and phase speeds, respectively, on azimuth under conditions typical of the daytime ionosphere. With north–south (180°) and south–north (0°) propagation there is little difference whereas for azimuths of 90° and 270° there are marked differences. These nonreciprocal features result from the differences in electron orbits in the plane of propagation (Barber and Crombie 1959; Davies 1969). Snyder and Pappert (1969) reported more complexity in modal behavior due to the geomagnetic field for conditions typical to nighttime. Among other things, they reported a much enhanced polarization mixing, both in middle and in equatorial latitudes, and a rapid variation in

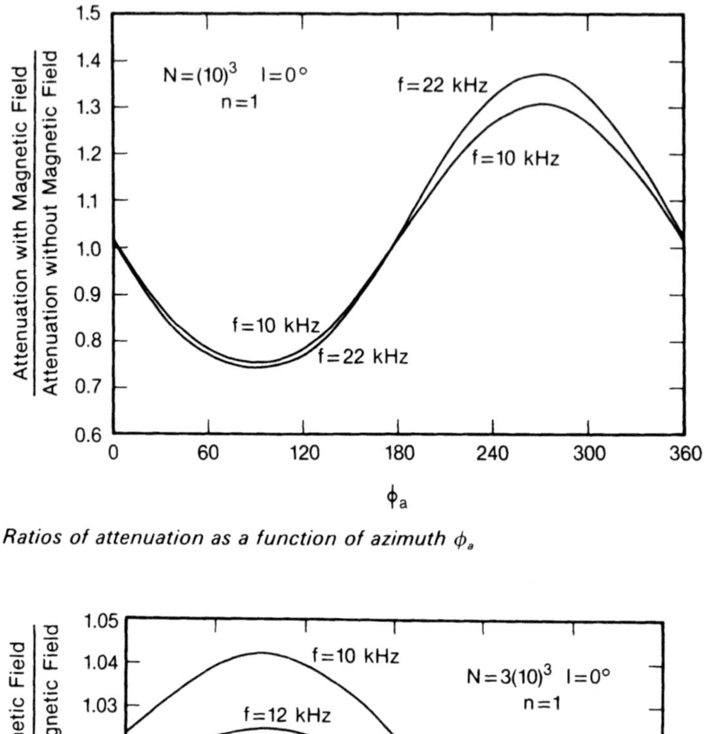

Fig. 10.7 *Ratios of attenuation as a function of azimuth* ϕ_a

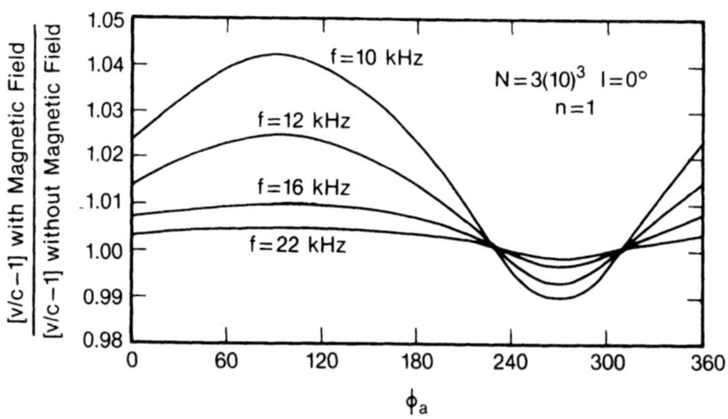

Fig. 10.8 *Variations of phase velocities with azimuth* ϕ_a

modal constants in the neighborhood of north–south and south–north propagation. Further, they showed that the maximum total signal attenuation occurs for north–south propagation rather than for east–west propagation. This fact was demonstrated by the experimental results of Bickel *et al.* (1970).

10.3.6 Effects of inhomogeneities – mode interference

The D region is not sharply bounded, and it has horizontal gradients, for example, near sunrise and sunset and near the auroral zones during disturbed periods. An important consequence of a horizontal gradient is the creation of new modes at the discontinuity (e.g. where the night ionosphere meets the day ionosphere). This can lead to 'cycle slippage' in the phase of VLF navigation systems (Kikuchi and Ohtani 1984). Computer programs have been developed that calculate phase and amplitude, taking into account mode conversion in the Earth–ionosphere waveguide (e.g. Pappert and Ferguson 1986).

The effects of mode conversion and mode interference are particularly evident at sunrise and on frequencies near and above 20 kHz (Reder and Westerlund 1970; Frisius 1970). Mode conversion increases with increasing angle between the path and the terminator. When modes are comparable in amplitude, cycle slippage can occur. Slippage can occur also when the phase varies very rapidly compared with the time constant of the receiver (Reder and Westerlund 1970). Another significant source of mode conversion is the sea–ice boundary at the coast of Greenland.

10.4 Propagation of ELF waves

Extremely low frequencies (< 3 kHz) have very limited practical value because of the low bandwidth and the difficulty of launching a usable power (large antenna). The main natural sources are atmospherics from lightning. As is the case with VLF, the propagation of ELF waves is best represented by waveguide theory, restricted to use of only one mode (Wait 1970). An important feature of ELF propagation is that the distance between source and receiver may be comparable with the wavelength. For example, on 300 Hz the wavelength is 1000 km, so interpretation of observations may require that the near-field effects of the source be taken into account. On extremely low frequencies the ionosphere behaves more like a conductor than a conducting dielectric, and the displacement current ($\dot{D}$) is small. Because of the large skin depth in the D layer for ELF waves, the effective reflection heights ($h \approx 90$ km) are larger than is the case for VLF waves.

An important source of ELF signals is the US Navy's Wisconsin Test Facility (WTF). The frequencies lie in the 40–50 Hz and 70–80 Hz ranges. This facility, part of Project Sanguine (IEEE 1974), was designed for communication with submarines. ELF propagation is discussed extensively in IEEE (1974). The WTF has two 22·5 km quasi-orthogonal antennas. In middle latitudes, the attenuation rate on 75 Hz is about 1·2 dB/Mm by day and 0·8 dB/Mm at night. At night, a multilayer resonance occurs because the extremely long wavelength allows energy to penetrate the D and E layers (Davis 1974). In the auroral zone, sporadic E can have an effect on ELF propagation as do polar cap absorption events and magnetic storms. Sporadic E has marked effects on ELF propagation (Barr 1977; Pappert 1985).

The effect of an external vertical magnetic field is to increase the attenuation by a factor F, where

$$F = \frac{\text{attenuation with magnetic field}}{\text{attenuation without magnetic field}} \approx \frac{\cos(\tau/2)}{(\cos\tau)^{1/2}} \tag{10.24}$$

and where τ is given by (10.22*c*). When $\tau \ll 1$ (i.e., $\omega_L \ll \nu$) $F \approx 1$. At the level where ELF waves are reflected, $\nu \approx \omega_L \approx 10^6$, $\tau \approx 45°$, and $F \approx 1{\cdot}85$. As in the case of VLF, the E–W and W–E propagations of ELF waves are non-reciprocal. However, in general, Wait (1970) concluded that only the vertical component of the Earth's magnetic field is really significant in ELF propagation and, furthermore, the ionosphere is effectively an isotropic conductor even in the presence of the Earth's magnetic field. Wait (1970) successfully applied the waveguide propagation of ELF pulses to the interpretation of data on the 'slow tails' of atmospherics.

The abnormal or horizontal component of the electric field is generated by 'return stroke' lightning discharges (i.e. between thunderclouds and ground) (Chapman and Jones 1964*a*). The abnormal component can be produced by two processes: (i) The source discharge may have a significant horizontal dipole moment, and (ii) the ionosphere can alter the polarization, thus generating an abnormal component (Jones 1974). It appears that on frequencies below 1 kHz source effects dominate, and above 1 kHz the horizontally polarized component is mainly caused by propagation.

The electromagnetic spectrum between about 5 Hz and 50 Hz is characterized by the Schumann resonance modes in the Earth–ionosphere resonant cavity. The fundamental resonance f_1 is near 8 Hz, and the harmonics are given by

$$f_n = f_1\,[n(n+1)/2]^{1/2} \tag{10.25}$$

that is, at approximately 14 Hz, 20 Hz, and 25 Hz (Balser and Wagner 1964; Volland, 1982). The diurnal variation of the strengths of these resonances corresponds closely to the mean worldwide thunderstorm activity; high levels of resonance occur during geomagnetic disturbances (Chapman and Jones 1964*b*; Polk 1983).

10.5 Phase and amplitude variations on VLF

10.5.1 Quiet-day variations of phase

10.5.1.1 Short distances

It is convenient to describe the diurnal phase change $\Delta\phi$ in terms of a change Δh of height h of a fictitious mirror reflector. For a ground range of $2d$ and wavelength λ we have

$$\Delta h = \Delta\phi\,\frac{\lambda}{2\pi}\,\frac{(d^2 + h^2)^{1/2}}{2h} \tag{10.26}$$

where $\Delta\phi$ is in radians.

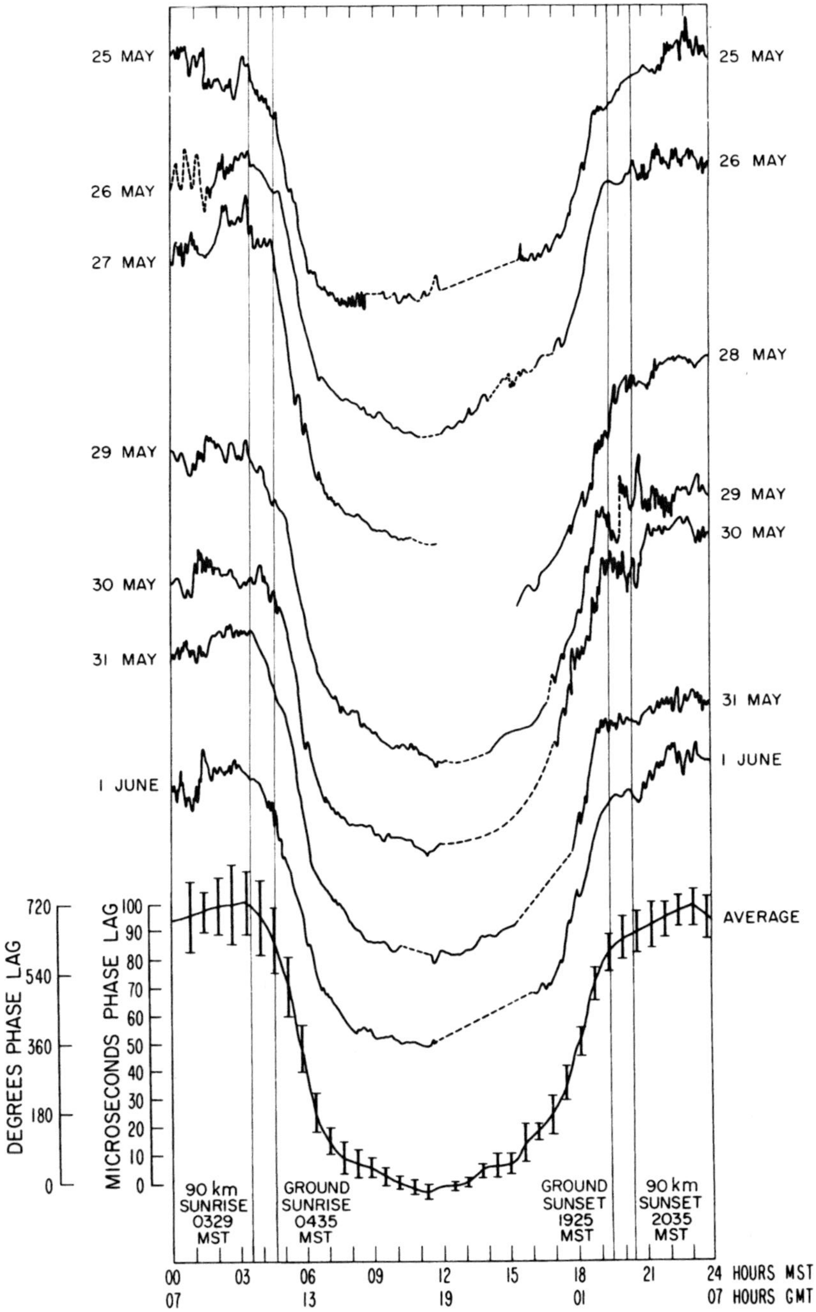

Fig. 10.9 *Diurnal variation of phase of WWVL 20 kHz over the 113 km path from Fort Collins to Wiggins, CO (From Chilton et al. 1964)*

The diurnal variation in the phase of WWVL (on 20 kHz) over a 113 km path (Fort Collins to Wiggins) in Colorado is shown in Fig. 10.9. This dish-shaped variation with minimum near noon is similar to that observed on 16 kHz over 90 km in England (Rugby to Cambridge) by Straker (1955). The diurnal phase shift of $\approx 720°$ corresponds to a height change of about 18 km. These results show that the effective height of reflection $h(\chi)$ varies in a Chapman-like manner:

$$h(\chi) = h_0 + H \sec \chi \tag{10.27}$$

for $\chi < 85°$. At Cambridge h_0 is about 69 km and the scale height H varies from 4 km in January and June to about 7 km in October and April, with a mean of about 5·5 km.

Observations of the abnormal component over a distance of 540 km (Bain *et al.* 1952) showed that the diurnal change starts about 1 h before ground sunrise at the path midpoint, whereas at 90 km the change commences about 10 min after ground sunrise. Over this distance (540 km), the phase change reaches a maximum value near ground sunrise and then decreases for about 2 h thereafter. This phase reversal, together with a much smaller one occurring during the main phase change (neither being observed on the 90 km path), may be due to reflections from different D layers. It is possible, however, that these effects may be due to interference between multiple rays or modes.

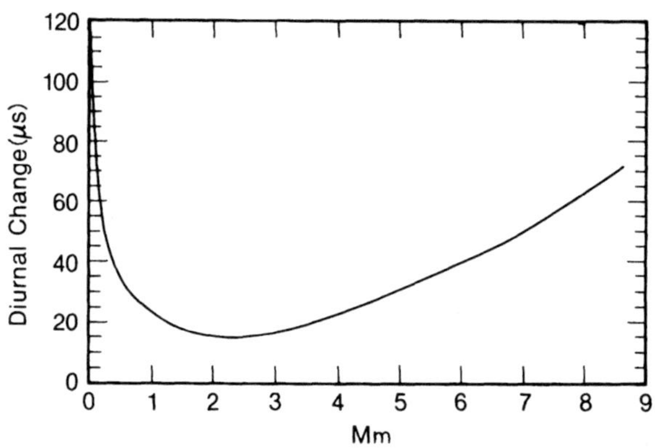

Fig. 10.10 *Diurnal variation of the time of arrival of total sky wave versus distance (1 Mm = 10⁶ m) (After Chilton et al. 1964)*

10.5.1.2 Long distances

The magnitude of the diurnal phase change is large for small distances, decreases to a minimum at about 2000 km and thereafter increases as the length of the daylit path increases (Fig. 10.10). For east-west distances of about 5000 km the diurnal phase change begins about half an hour before sunrise at the eastern terminal. The phase then remains roughly constant until a half hour before sunset at the eastern terminal. This gives the trapezoidal variation shown in Fig. 10.11. It must be remembered that this trapezoidal representation is highly idealized and, in practice, the diurnal pattern undergoes considerable variation from day to day and from season to season.

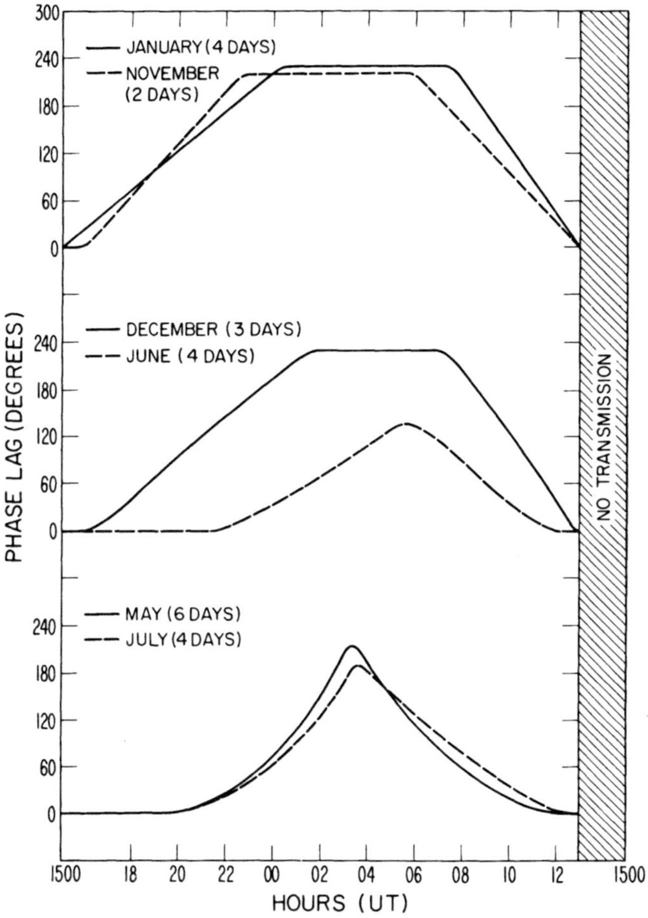

Fig. 10.11 *Diurnal variation of the phase of GBR (Rugby, England) VLF signal (16 kHz) observed at Boulder, CO (From Chilton et al. 1964)*

Observations of the phase of the GBR (Rugby, England) signal in New Zealand (Crombie *et al.* 1958) can be explained in terms of the times at which the sunrise or sunset line crossed the nearly north–south path, and the angle between these lines and the path. It is necessary to assume that the height changes suddenly from its daytime to its nighttime value at the time of sunset, and vice versa at sunrise. The magnitude of the phase change can be explained in terms of a waveguide mode theory.

If the height h changes by an amount Δh, the corresponding phase change $\Delta\phi$ is given (in radians) by

$$\Delta\phi = -\frac{2\pi d}{\lambda}\left(\frac{h}{2a} + \frac{\lambda^2}{16h^2}\right)\frac{\Delta h}{h} \qquad (10.28)$$

in which d is the great circle distance of the receiver from the transmitter and h is the mean height of reflection. The day-to-night height changes calculated from the observed phase changes for various circuits are given in Table 10.1.

Table 10.1 *Diurnal phase shifts for 10–20 kHz radio waves over long path distances**

Frequency (kHz)	Path and distances (km)	Path composition (percent land)	Diurnal phase shift (μs)	Δh	$\frac{c\Delta t}{d}$
16	Rugby–New Zealand 11 000	–	70	16·5	1·91
16	Rugby–Boston 5200	0	34	16·9	1·96
10·2	Haiku–San Diego 4180	0	53	21·0	3·8
11·2	Haiku–San Diego 4180	0	46	20·5	3·3
19·8	Haiku–San Diego 4180	0	26	16·0	1·87
18	Panama–San Diego 4670	90	29	17·0	1·93
10·2	Haiku–Boulder 5360	25	61	18·0	3·4
10·2	Haiku–Forestport, N.Y. 7833	50	87	17·5	3·33
18	Panama–Haiku 8417	0	50	15·5	1·79
18	Panama–Forestport, N.Y. 3851	0	28·5	18·5	2·22
18	Panama–Boulder 4300	30	27·8	16	1·94
16	Rugby–Boulder 7500	40	42	14	1·66

*From Wait (1970)

Phase shift is sometimes expressed as a time delay. The phase shift Δt (in seconds) is related to the phase shift $\Delta\phi$ (in radians) by

$$\Delta t = \Delta\phi/2\pi f \qquad (10.29)$$

A word of caution is necessary regarding the use of time delay as a measure of phase change. This is misleading since delay time refers to group propagation rather than phase propagation.

Curves showing the phase velocity as a function of reflection height and wave frequency are shown in Fig. 10.5. As a particular path is crossed by a day/night boundary, the relative phase seen at the receiver will depend on the amounts of the path that are in daylight and in darkness at the instant concerned. If d_0 and d_n are the distances in daylight and darkness, and β_0 and β_n are the respective phase changes per unit length, the total phase lag ϕ_t between transmitter and receiver is given by

$$\phi_t = d_0\beta_0 + d_n\beta_n = d_0(\beta_0 - \beta_n) + \beta_n d \tag{10.30a}$$

where

$$d = d_0 + d_n \tag{10.30b}$$

For an east–west path this gives the trapezoid variation of phase noted above; for a north–south propagation path, the entire path tends to pass quickly from daylight to night, giving a rectangular variation (Fig. 10.12).

10.5.1.3 Influence of geographical latitude

The diurnal curves in Fig. 10.11 and 10.12 illustrate an effect of latitude. The GBR–Boulder path lies in much higher latitudes than the NBA (Panama–Boulder) path. If ground sunset and sunrise are considered, part of the GBR–Boulder path becomes completely dark during June, but this is not reflected in the June curve. If, however, sunset and sunrise at the level of the D region are considered, then the path never really becomes dark, and thus we should not expect the full diurnal variation to develop. Examination of the diurnal phase changes during the months for which the path does become completely dark suggests that there is a seasonal change in the diurnal phase shift. In December and January, the mean phase change is 210°, whereas in March it is 260°.

Observations over paths in very high latitudes (e.g. GBR to Point Barrow, Alaska, 6000 km) show that the diurnal variation of phase is only about 30° even when two-thirds of the path is illuminated. The observed phase change represents a change of effective height of 13 km for that part of the path that has a day-to-night change. One of the most conspicuous latitudinal effects on VLF propagation is *not* caused by the ionosphere. This is the large propagation loss in transit across the Greenland icecap (Westerlund *et al.* 1969).

10.5.1.4 Phase speed

The observations discussed above give only the change of phase, but measurements of phase at two stations separated by a known distance have been used to measure the phase speed. Using waveguide theory, Wait (1970) showed that on 10 kHz the phase speed at night is approximately the free space value c and by day it is about $1{\cdot}03c$. On 14 kHz the corresponding speeds are $0{\cdot}98c$ and c, and on 16 kHz the night and day speeds are about $0{\cdot}97c$ and $0{\cdot}99c$ respectively.

The phase speed v_p of a vertically polarized wave for the lowest order mode is

$$v_p = c/[1 - (\lambda^2/16h^2)]^{1/2} \tag{10.31}$$

The change Δv_p of speed introduced by a change Δh of height is

$$\frac{\Delta v_p}{v_p} \approx \frac{\lambda^2}{16h^3}\Delta h \tag{10.32}$$

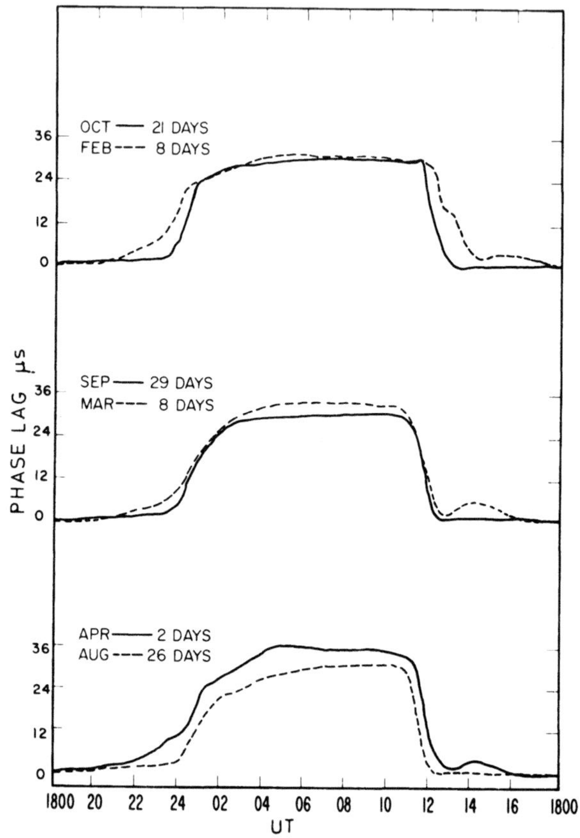

Fig. 10.12 *Diurnal variation of the phase of NBA VLF signal (18 kHz) from Panama observed at Boulder, CO (From Chilton et al. 1964)*

10.5.2 Quiet-day variations of amplitude

10.5.2.1 Reflection coefficient

The reflection coefficient on 16 kHz over a short distance (GBR–Cambridge) varies from day to night and with season. In general the reflection coefficient is higher by night than by day and higher on a winter day than on a summer day. Typical values lie in the range 0·1–0·5.

Over short distances (< 300 km) the skywave is essentially circularly polarized, so the ratio R_{12}/R_{11} is nearly unity with a 90° phase shift. Over longer distances (500–800 km) the polarization is nearly linear, R_{11} being much larger than R_{12}. Over distances greater than 1000 km the polarization is approximately vertical during daytime (i.e. $R_{11} \gg R_{12}$) and becomes elliptical at night ($R_{12}/R_{11} \approx 10/1$).

10.5.2.2 Amplitude

The amplitude of a VLF signal propagating over short distances is highly variable because of interference between the various components and particularly between the ground wave and the once-reflected signal. For a given distance, the amplitude is very sensitive to the magnitudes and phases of the reflection coefficients, and to the height of reflection.

This sensitivity is seen in the diurnal variation where the night-to-day ratios are about 6 dB in CW signal strength and as much as 20 dB in the level of broadband sferics. In addition, rapid changes of amplitude occur near sunrise and sunset along the path. The amplitude usually goes rapidly through a minimum or maximum before reaching its steady daytime or nighttime level (Fig. 10.13).

Changes in the usual pattern of the field behavior are introduced by such factors as a land–sea boundary or changes in ground characteristics. At these boundaries, standing wave patterns are set up which modify the field distribution over distances of several wavelengths.

10.5.2.3 Attenuation rate

The variation with distance of the vertical field E of the far field in a waveguide mode is given approximately by

$$E = \frac{300}{h}\left[\frac{p\lambda}{a \sin(d/a)}\right]^{1/2} \exp(-\alpha d) \tag{10.33}$$

where p is the radiated power (kW), α is the attenuation rate, and the other symbols have their usual meanings. Expressing (10.33) in decibels we have

$$20 \log E + 10 \log \sin(d/a) = \text{a constant} - \alpha d \tag{10.34}$$

where α is in decibels per unit distance (usually 1 Mm) and all the other factors are lumped together in the constant. Therefore, if the left-hand side of (10.34) is plotted against d, the slope of the resulting straight line will give the value of α. Heritage *et al.* (1947) observed signals from VLF stations on frequencies of 16·6 kHz and 18·6 kHz. An example is shown in Fig. 10.14. These data give an attenuation rate of 3·0 dB/Mm.

The day value of α over the North Atlantic averages about 3 dB/Mm on frequencies near 20 kHz. The night values are consistently lower than the day values. Paths over the Arabian Sea and the Indian Ocean show rates of about 2 dB/Mm at great distances. Daytime measurements in the Pacific give average rates of 1·7 dB/Mm.

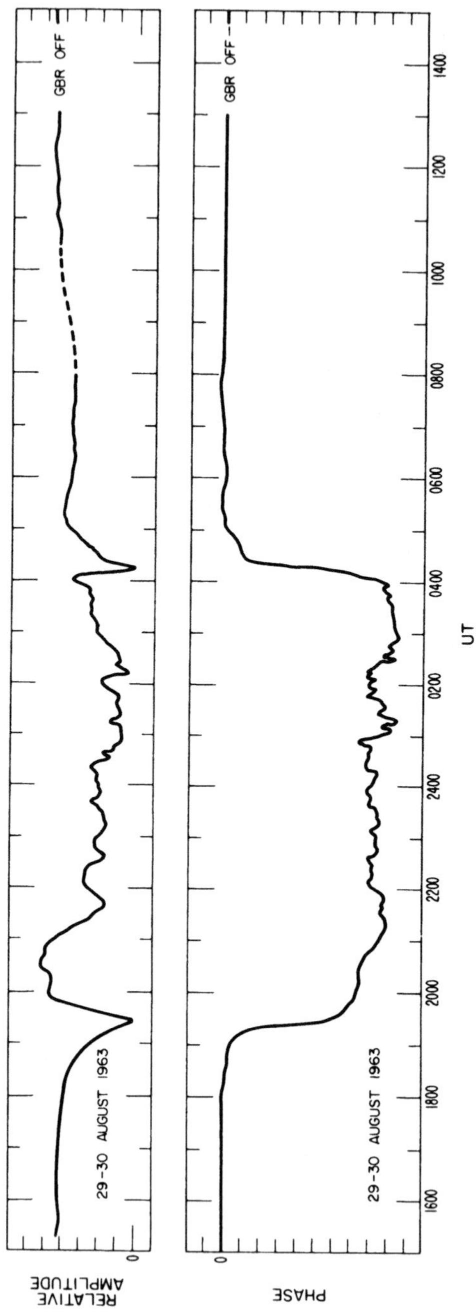

Fig. 10.13 *Diurnal variation of amplitudes and phases of GBR VLF signal (16 kHz) observed at Frankfurt (840 km)*

Observations on sferics show that the best conditions for a signal of 15–20 kHz occur for west-to-east propagation over sea at night. The rate under these conditions is a minimum and is about 1·0 dB/Mm. The maximum attenuation for this frequency range occurs for a land path from east to west during daylight. In assessing the propagation loss suffered by a VLF signal the losses due to the following must be added to the nighttime sea-water value: (i) propagation over land, (ii) daytime propagation, (iii) nonreciprocity, and (iv) propagation over icecaps (e.g. Greenland or Antarctica).

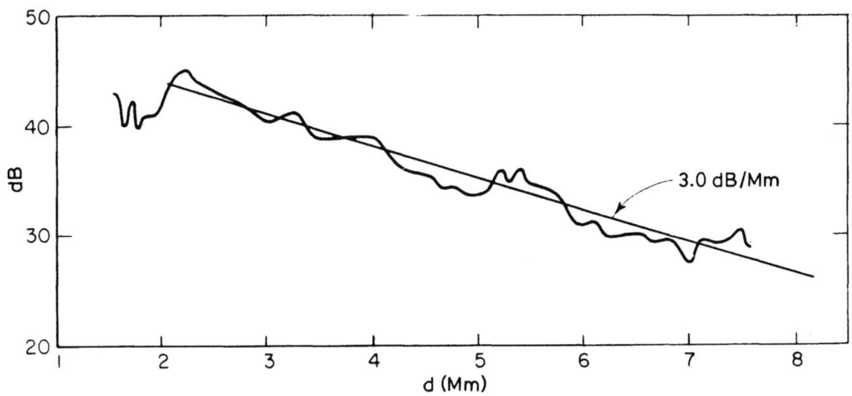

Fig. 10.14 *Variation of amplitude with distance over the Pacific (Courtesy S. Weisbrod)*

10.5.3 VLF disturbances

10.5.3.1 Sudden ionospheric disturbances

Aspects of solar X-ray flares (discussed in Sec. 9.3) include sudden phase anomaly (SPA), sudden enhancement of signal strength (SES), and sudden decrease (SDA) or enhancement (SEA) of atmospherics on a frequency near 27 kHz (*Solar-Geophysical Data*, No. 499, March 1986). Of these the SPA is the most important from a practical point of view because the additional ionization lowers the effective height of reflection, which introduces a shift in the position contours of navigation systems.

On ELF ($<$ 1 kHz), signal enhancement occurs whereas between 1 kHz and 10 kHz there is a signal decrease (SDA) (Leroy 1979; Davis 1976). Data on a number of SPAs over several circuits have been given by Chilton *et al.* (1965), which show phase advances up to 250° that correspond to a change of effective reflection height of $\approx$ 10 km. The magnitude of an SPA depends upon the intensity of the solar X-ray flare and on the solar zenith angle. SPAs can last up to about 2 h. Some effects of ionospheric disturbances on the phase speeds and attenuation rates are given in Table 10.2. It has been found that an average duration is 40 min for an SPA. The rise time to maximum is approximately 6 min (IEEE 1965).

10.5.3.2 Sudden commencements

Phase advances have been observed during magnetic sudden commencements on paths that traverse geomagnetic latitudes near the auroral zone ($\approx 65°$) as illustrated in Fig. 10.15.

Table 10.2 *Phase velocity and attenuation rates for models of perturbed and unperturbed ionosphere; f = 13600 Hz**

	c/v_p				
	θ(W–E)		θ(Polar)	θ(E–W)	
	0°	45°	90°	45°	0°
Winter night	1·00203	1·00193	1·00172	1·00246	0·997
Summer day	0·99965	0·99959	0·99947	0·99935	0·99923
Quiet day	0·99956	0·99951	0·99942	0·99935	0·99931
SID	0·99942	0·99938	0·9993	0·99926	0·99923
SID	0·99968	0·99962	0·9995	0·99939	0·99929
Auroral abs.	0·9997	0·99964	0·99949	0·99936	0·99922
PCA day	0·99627	0·99626	0·99625	0·99624	0·99624
PCA night	0·99685	0·99684	0·99682	0·9968	0·9968
	α in dB/1000 km				
	θ(W–E)		θ(Polar)	θ(E–W)	
	0°	45°	90°	45°	0°
Winter night	0·15	0·44	1·06	2·7	2·5
Summer day	2·07	2·15	2·55	3·37	3·84
Quiet day	1·65	1·77	2·23	3·04	3·49
SID	1·55	1·67	2·12	2·85	3·26
SID	1·95	2·04	2·46	3·34	3·84
Auroral abs.	2·21	2·30	2·74	3·63	4·11
PCA day	3·73	3·77	3·86	3·96	4·07
PCA night	3·73	3·79	3·89	4·01	4·06

*From Galejs (1970)

10.5.3.3 PCA disturbances

Figure 10.16 shows that the phase delays of 10·2 kHz and 13·6 kHz between Norway and Barbados can be disturbed for several days during a PCA and may result in errors of about 10 km in position lines. Large polar cap disturbances cover the entire polar cap above 60° magnetic latitude and last up to 12 days. However, the later stages of these events are largely caused by electron precipitation along the auroral zone rather than by proton precipitation over the polar cap. The time variations of the OMEGA 10·2 kHz Hawaii-to-Norway phase are

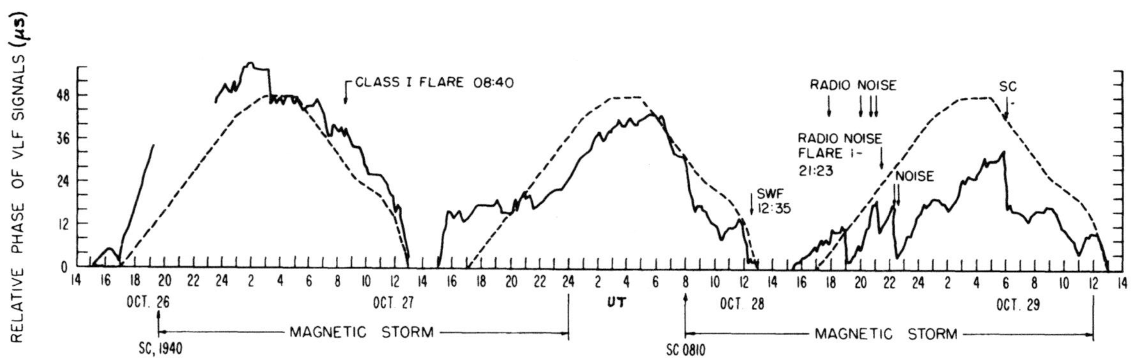

Fig. 10.15 *Phase of the 16 kHz signal from GBR, received in Boulder, illustrating the effects of magnetic disturbance on long-distance VLF propagation (From Chilton et al. 1964)*

closely related to the time profiles of the logarithm of the fluxes of energetic protons precipitating into the polar cap, measured by instruments on polar-orbiting and geostationary satellites (Sauer *et al.* 1987). The phase advance A (in centicycles) is related to the 6 MeV GOES-2 differential proton flux J as follows:

$$A = 8{\cdot}24 \ln (8{\cdot}7J) \tag{10.35}$$

The agreement between observed phase advances and the phase advances calculated using (10.35) is shown in Fig. 10.17.

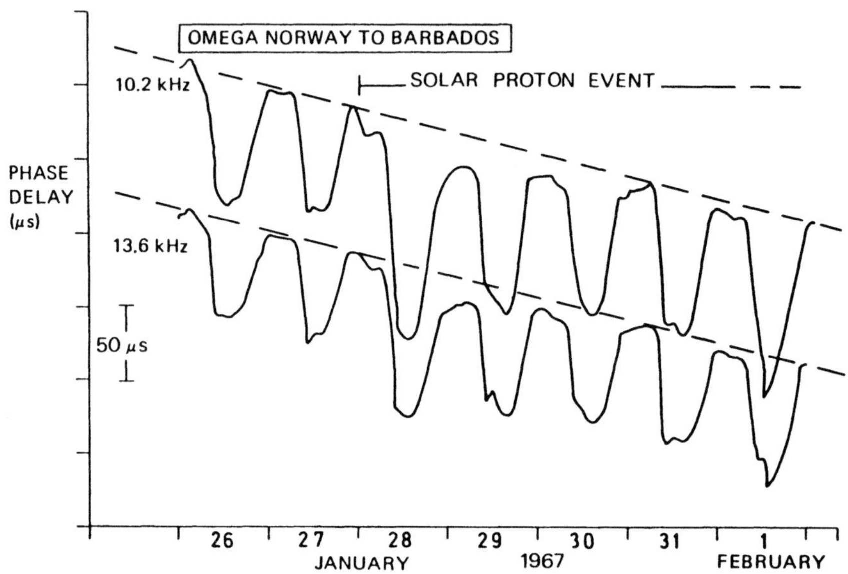

Fig. 10.16 *Phase delay of 10·2 and 13·6 kHz from OMEGA–Norway to Barbados (From Burgess 1970, with permission)*

10.5.3.4 Auroral disturbances

VLF signals are sensitive to nighttime auroral disturbances, and this is particularly true for the received wave polarization as illustrated in Fig. 10.18. The north loop is in the geographic meridian, and the west loop is perpendicular to the meridian. During quiet periods the two signals behave alike. Near the onset of auroral disturbances, VLF signals often respond before other detectors such as riometers, photometers, and magnetometers (Reder and Westerlund 1970). By day, auroral effects are small, suggesting that any additional ionization occurs above the level of reflection (70 km).

Storm aftereffects are illustrated in Figure 10.19. The change of phase at sunrise occurs later, (and the phase change at sunset occurs earlier) than during quiet days, and when it occurs it takes place suddenly. On May 19 an aftereffect appeared that lasted for several days.

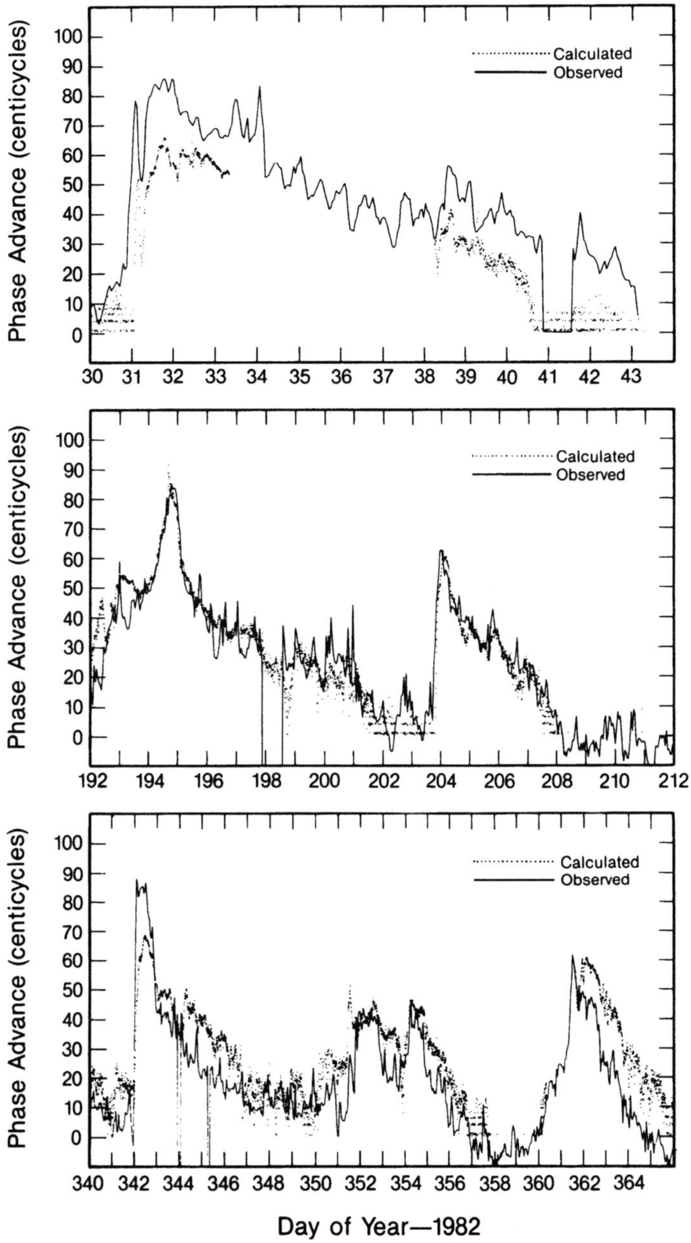

Fig. 10.17 *Comparison of observed phase of OMEGA 10·2 kHz (Hawaii–Norway) with values calculated using (10.35) (From Sauer et al. 1987)*
top January 30–February 12, 1982
middle July 11–31, 1982
bottom December 6–31, 1982

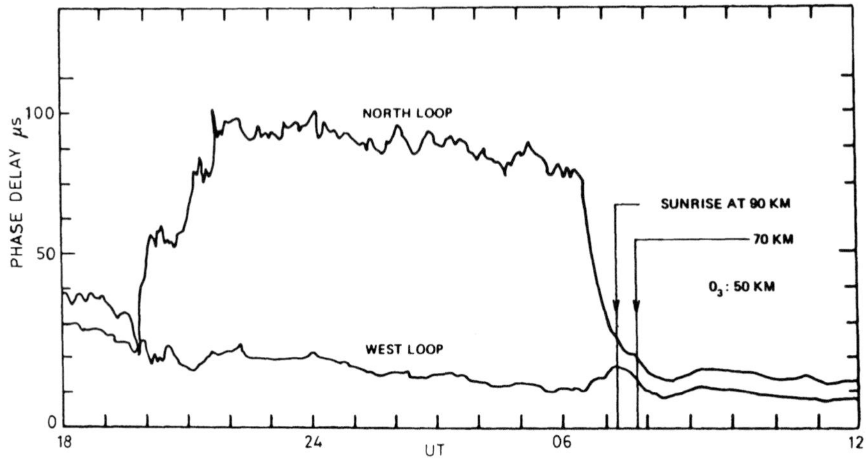

Fig. 10.18 *Auroral phase anomalies for the path GBR to Kiruna, Sweden, January 18–19, 1964 (16 kHz) when GBR was recorded with two crossed loops (From Reder and Westerlund 1970, with permission)*

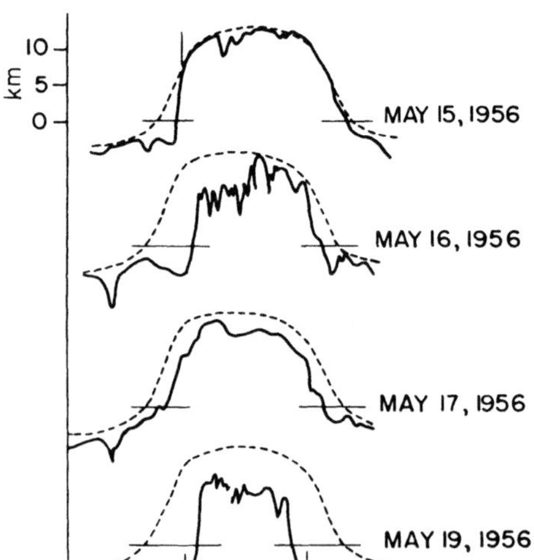

Fig. 10.19 *Observations of the phase height of 16 kHz waves (GBR to Cambridge, England) during an ionospheric storm (From Belrose 1964, with permission)*
May 15, 1956, is a normal day; May 16 and 17 have the primary storm effects, and May 19 has the aftereffect. The dashed line is the expected quiet variation

10.5.4 Propagation effects on OMEGA

10.5.4.1 The OMEGA system

OMEGA is a VLF navigation system operating on frequencies between 10 and 14 kHz from a number of (10 kW) transmitters situated around the world and having essentially 24 h operation. It is estimated that OMEGA is the most widely deployed ground-based navigation aid (Swanson 1977, 1982). Its principle is based on the fact that a receiver, at which there is a given phase difference between two signals from fixed sources, lies on a given hyperbola (line of position or LOP). Thus the use of two pairs of transmitters locates the receiver. The frequencies are stabilized by cesium frequency oscillators and are time-shared in intervals of 10 s. As in all phase comparisons, the system has a navigational lane ambiguity of one cycle ($\approx$ 15 km) but this can be corrected by the use of additional frequencies (e.g. 11·33 kHz and 13·6 kHz). Thus OMEGA has reliability, redundancy, and range.

10.5.4.2 Propagation characteristics

Figure 10.20 shows the diurnal variation on 19·8 kHz which is more complicated than the normal trapezoid. The steps during transitions are thought to be caused by modal conversion at the terminator (Crombie 1966). The field strength repeats from day to day with a standard deviation of 1 dB on 10·2 kHz. The stabilities of signals over paths between about 4 Mm and 10 Mm are given in Table 10·3 (Swanson 1982). Note that, on the higher frequency, the phase is more stable near midday than at night. Navigation position errors can be partially corrected by transforming solar radiation (e.g. proton) fluxes into phase error corrections (Argo 1975) (Fig. 10.21). The model corrects the phase deviations to $\pm$ 1 km. Propagation blackouts of the type experienced on high frequencies are virtually unknown on VLF.

10.5.5 VLF noise

In discussion of the day-to-day repeatability of phase measurements it is important to distinguish between two types of noise:

(i) the atmospheric noise from thunderstorms, which is superimposed on the received signal and is referred to as additive noise
(ii) fluctuations in phase imposed on the signal by the ionosphere, before it reaches the receiver, called multiplicative noise

Normally, effects of additive noise can be overcome by increasing the transmitter power level, for instance, but this is not the case with multiplicative noise.

Atmospheric noise is highly impulsive, and proper receiver techniques may yield improvements of about 15 dB relative to Gaussian noise. Although in middle and low latitudes the noise level is high, in the Arctic the ambient noise is low and reception of weak signals is common.

When atmospheric noise dominates and the receiver bandwidth W is suffi-

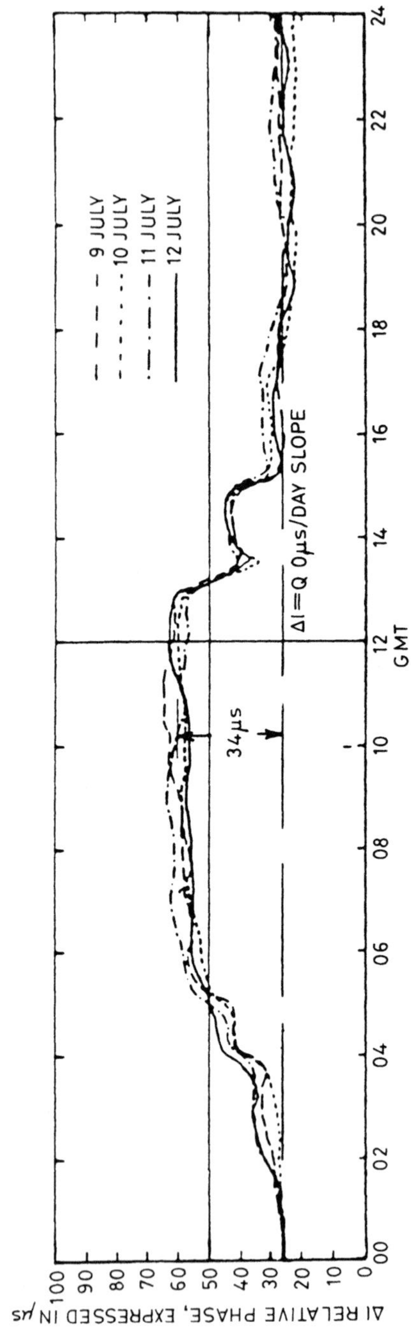

Fig. 10.20 *Diurnal variation of the phase of signal from station NPM, Hawaii (19·8 kHz) recorded in San Diego, July 9–12, 1965 (From Swanson 1977, with permission)*

ciently small that the atmospheric noise behaves as thermal noise, the standard deviation of phase σ_ϕ is

$$\sigma_\phi = n/2\pi s \tag{10.36}$$

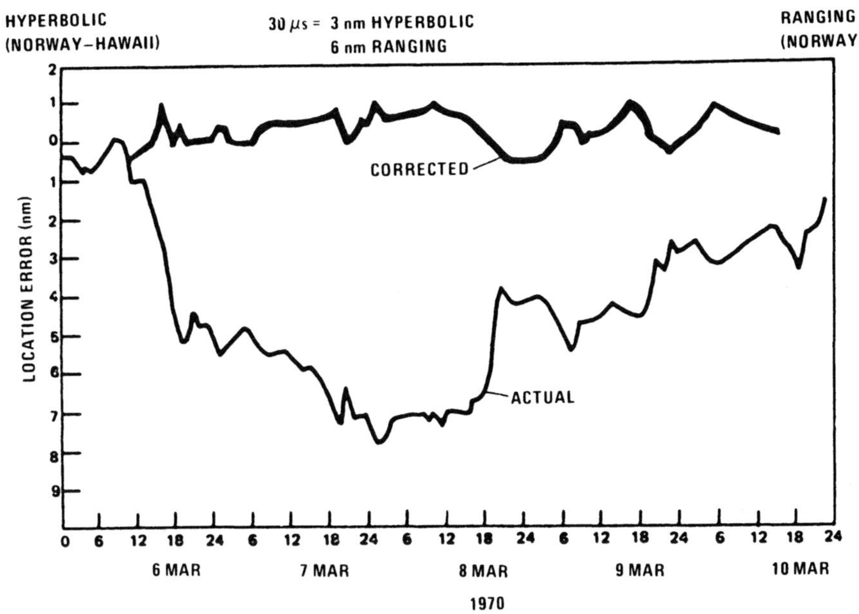

Fig. 10.21 *OMEGA location errors (in nautical miles) during a solar proton event (From Larsen 1979)*
Actual = errors in measured location
Corrected = error when ionosphere effect is taken into account

Table 10.3 *Instabilities of VLF signals expressed as standard deviation**

Period	SD on 10·2 kHz (μs)	SD on 13·6 kHz (μs)
Day	3	2
Night	5	4
Transition	4	4

*From Swanson (1977)

where n and s are, respectively, the (additive) RMS noise and signal levels in the bandwidth W. The noise is proportional to $\sqrt{W}$ down to a value of $W \approx 0{\cdot}01$ Hz. For smaller bandwidths multiplicative (propagation) noise sets the limit. This noise is produced by irregularities in the medium that change with time and thus impose random phase variations about the mean phase delay.

In the measurement of group delay, t_g, by phase measurements ϕ_1 and ϕ_2 (in cycles) on frequencies f_1 and f_2, let the standard deviation be σ_ϕ and the correlation coefficient be ϱ. Then the standard deviation σ_g of the group delay is

$$\sigma_g = \sigma_\phi[2(1 - \varrho)]^{1/2}/(f_2 - f_1) \tag{10.37}$$

When the phase fluctuations are highly correlated, $\varrho = 1$ and $\sigma_g = 0$. When the fluctuations are uncorrelated, $\varrho = 0$ and

$$\sigma_g = 2^{1/2}\sigma_\phi/(f_2 - f_1) \tag{10.38}$$

Thus to determine suitable frequency spacings it is important to know how ϱ varies with frequency.

The RMS error E in the received frequency f measured over a time interval T is

$$E = n/2\pi sft \tag{10.39}$$

where s/n is the signal/noise ratio. Thus, for a given frequency and noise level, the RMS error can be decreased by (i) increasing the received signal, and/or (ii) increasing the integration period T. On a frequency of 10 kHz, s/n of 5 and an observation period of 1 h give a precision of about 1 part in 10^9.

Crombie (1970) considered the case of a set of irregularities concentrated near the path midpoint and moving horizontally with uniform speed. The motion imposes Doppler shifts Δf on the reflected signals, and the resulting spectrum $W(f)$ is Gaussian:

$$W(f) \propto \frac{1}{\sigma_a} \exp\left[-(f - f_0)^2/2\sigma_a^2\right] \tag{10.40a}$$

where

$$\sigma_a \approx \Delta f/2{\cdot}5 \tag{10.40b}$$

and f_0 is the center frequency. The autocorrelation in the time domain is

$$\varrho_a(\tau) = \exp\left[-(2\pi\tau\sigma_a)^2\right] \tag{10.41}$$

where τ is the time between observations. Thus for two frequencies f_2 and f_1 the correlation coefficient between the envelopes of the received signals is also Gaussian.

In the case of spatial correlation we are concerned with the way in which fluctuations in the signal decrease as the distance d between receivers increases. By analogy with (10.41) the correlation coefficient for receivers in the plane of propagation is

$$\varrho_L(d) = \exp\left[-(2\pi\sigma_L d)^2\right] \tag{10.42a}$$

where

$$\sigma_L \approx \lambda/320L^2 \tag{10.42b}$$

L is the scale size of the irregularities.

When separation is perpendicular to the plane of propagation

$$\varrho_T(d) \approx \exp[-(2\pi\sigma_T d)^2] \tag{10.43a}$$

where

$$\sigma_T \approx 1/20L \tag{10.43b}$$

Observations indicate that the transverse correlation falls to $1/\varepsilon$ for $d \approx 210\,\text{km}$, which corresponds to irregularity sizes of about 65 km. Similarly ϱ_a falls to $1/\varepsilon$ when $\tau \approx 3000\,\text{s}$, which corresponds to a horizontal drift of some $35\,\text{m}\,\text{s}^{-1}$. There is a high correlation over the frequency range from 11·7 to 13·8 kHz; the correlation bandwidth is about 10 kHz.

10.6 Phase and amplitude on low frequencies

10.6.1 Characteristics of LF propagation

Low frequencies (30–300 kHz) have many of the properties of very low frequencies as far as phase stability and relative immunity to blackout are concerned. LF waves penetrate higher into the ionosphere than do VLF waves and therefore are affected more by irregularities and disturbances than are VLF waves. The higher frequencies allow for pulse amplitude modulation systems such as Loran-C on 100 kHz.

10.6.2 Coupling on low frequencies

We have assumed that, having been excited, the ordinary and extraordinary waves propagated independently in the ionosphere. This is not true on low frequencies when energy interchange, or coupling, takes place between the two characteristic waves as well as between forward- and backward-propagating waves. These effects have been observed extensively at night on 150 kHz; the reflected signal includes a main echo and a 'coupling echo' (Ratcliffe 1959, Sec. 13.10). A rapid change in polarization of the main echo (from left-handed to right-handed and vice versa) is called 'flip-over'. No daytime coupling effects have been observed (Budden 1985, Sec. 16.6).

10.6.3 Navigation systems that use LF

10.6.3.1 The Decca system

The Decca system operates on frequencies in the range 70–100 kHz. It is a hyperbolic CW system designed for high accuracy over medium ranges. About 50 Decca chains have been installed throughout the world, each chain consisting of one master and two or three slave stations.

The master and slave stations transmit on different frequencies, which are converted in the receiver to a single frequency for phase comparison. Navigation depends upon the groundwaves and, because of the phase stability of the groundwave, very accurate positioning can be obtained. At steep incidence during the day the wave reflected from the ionosphere is heavily absorbed on these frequencies. However, beyond a few hundred kilometers the ionospheric reflection becomes stronger and, because the groundwave on 100 kHz is fairly rapidly attenuated, interference between groundwaves and skywaves may occur and reduce the accuracy of the system. The range of the Decca system is therefore only about 300–400 km, and will vary with ionospheric conditions. Accuracies of about 100 m may be obtained for pure groundwave propagation. For ionospheric propagation to greater distances the difference in the reflecting properties of the ionosphere within the frequency range may be a problem.

10.6.3.2 The Loran-C system

Loran-C is an accurate navigation system using pulsed radio signals with a carrier frequency of 100 kHz (Doherty and Johler 1982). A chain consists of a master and two or more slave stations and may cover a geographical region with a radius of up to 2000 km. An example is the existing chain in the North Atlantic, which has a master station at the Faroe Islands and slaves in Iceland at Jan Mayen, and in northern Norway. The pulses transmitted from stations in the chain are synchronized with great precision, and the user determines his position by measuring the time delay between signals received from different stations. The great advantage of the Loran-C is that the groundwave pulse, because it arrives first at the receiver, can normally be separated from the skywave. Accuracies of position determination of 150 m up to distances of 1800 km are quoted as typical in the literature (Potts 1972). Each station transmits pulses in a coded pattern that ensures correct identification and prevents interference.

Loran-C has been used for time disemmination. The pulse duration is 150–120 μs. Measurements obtained at Aleutian slave stations (Attu and Sitkinak) show that the propagation is nonreciprocal. West-to-east propagation is better than east-to-west propagation (Doherty 1964). For paths in the USA ($\approx$ 2500 km) the diurnal variation of phase is approximately rectangular, changing by $+20\,\mu$s at sunset and $-20\,\mu$s at sunrise. During day the phase variations are less than $\pm 1\,\mu$s and at night the variations are less than $\pm 4\,\mu$s (Hefley 1972). Typical diurnal amplitude changes are about 20 dB; however, amplitude variations are seldom accompanied by corresponding phase shifts except those associated with solar X-ray bursts.

10.6.4 Regular phase variations on low frequencies

10.6.4.1 Diurnal variations over short distances

The curves in Fig. 10.22 represent the diurnal variations of the 'phase' height of reflection of LF waves for two distance ranges. On the shorter range the variation is dish-shaped, and the daytime heights follow a variation of the type expressed by (10.27). Note that the height starts to change about 10–15 min before ground sunrise at the path midpoint and that afternoon changes are slower than morning changes. The reflection height changes from about 72 km near noon to about 90 km at night. Over distances less than about 300 km, LF waves are reflected near 80–85 km near noon and 90–100 km at night.

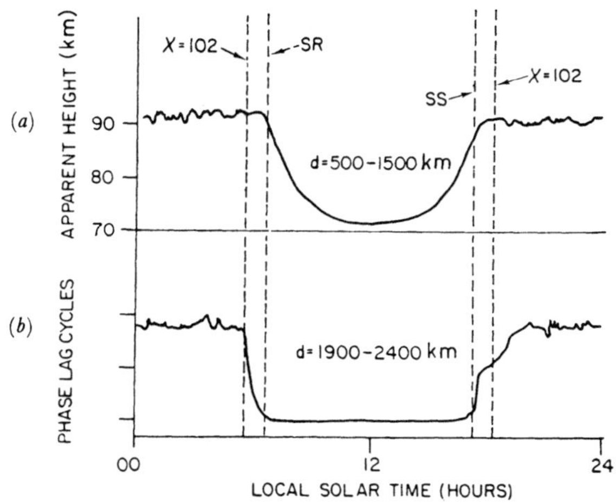

Fig. 10.22 *Diurnal variations of 'phase height' of reflection of LF waves (From Belrose 1964, with permission)*

10.6.4.2 Diurnal variations over long distances

Over distances of about 2000 km, the diurnal variation of height is nearly trapezoidal (Figs 10.22*b* and 10.23). The phase height begins to decrease about 1 h before ground sunrise at the path midpoint (when $\chi \approx 101°$) and decreases abruptly shortly after sunrise. The height is roughly constant during daytime. The evening increase in phase height begins when $\chi \approx 94°$ and is slower than the morning decrease. Nocturnal fading begins about 2–3 h after sunset in the D region ($\chi \approx 101°$).

10.6.5 Irregular variations of phase

10.6.5.1 Solar X-ray effects

Large solar X-ray flares produce phase advances in middle latitudes, as seen in Fig. 10.23 (near 1700 GMT), along with an amplitude enhancement of 6–20 dB. Smaller solar flares have caused phase retardations (Doherty 1964) accompanied by an initial amplitude decrease and a subsequent amplitude increase. Some X-ray bursts produce phase retardations together with amplitude decreases only. Thus, with phase retardation, the amplitude may increase, decrease, or remain the same. The response of the ionosphere to an X-ray flare depends upon the latitude (Doherty 1964).

10.6.5.2 Polar cap disturbances

Low-frequency waves are sensitive to solar proton events (SPEs) such that weak events (undetected by riometers) produce marked LF phase changes. Major polar cap disturbances (PCDs) affect LF propagation for up to 15 days. During a PCD the nighttime phase height is lower than normal. When the disturbance is intense the waves may be reflected from heights as low as 50 km, and there is little or no diurnal change. The response of a circuit to a solar proton event depends, not only on the strength of the event, but also on the geomagnetic latitude because the protons penetrate more easily above the cutoff latitude.

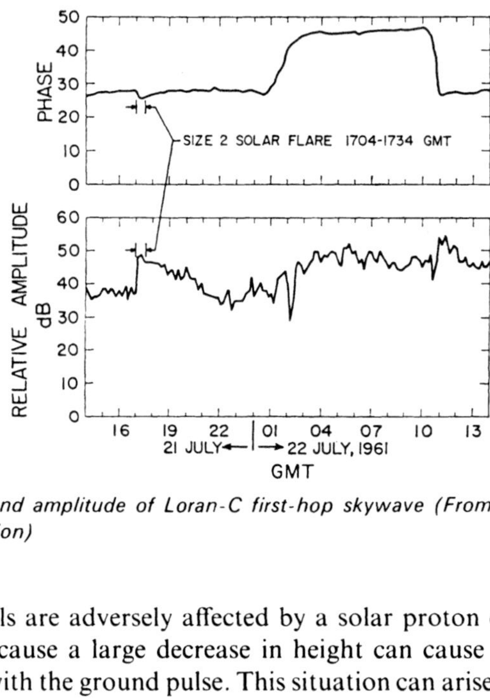

Fig. 10.23 *Phase and amplitude of Loran-C first-hop skywave (From Doherty 1964, with permission)*

Loran-C signals are adversely affected by a solar proton event (Larsen and Thrane 1977) because a large decrease in height can cause the once-reflected echo to overlap with the ground pulse. This situation can arise when the groundwave suffers a large delay across poorly conducting or irregular ground.

10.6.6 Amplitudes on low frequencies

10.6.6.1 Variations over short distances

Over distances less than 100 km the variation is dominated by interference between the groundwave and the skywave. If the amplitude of the vertical electric field is measured on the ground, or in an airplane flying at a constant height, and allowance is made for the inverse distance attenuation, a pattern of the type shown in Fig. 10.24 is obtained. This pattern is called the Hollingworth pattern. From the positions and signal strengths of the turning points it is possible to deduce the characteristics of the downcoming wave.

Few experimental data are available at long distances but numerical calculations by Johler and Berry (1964) for 100 kHz waves in a model ionosphere show periodic fluctuations out to at least 1500 km.

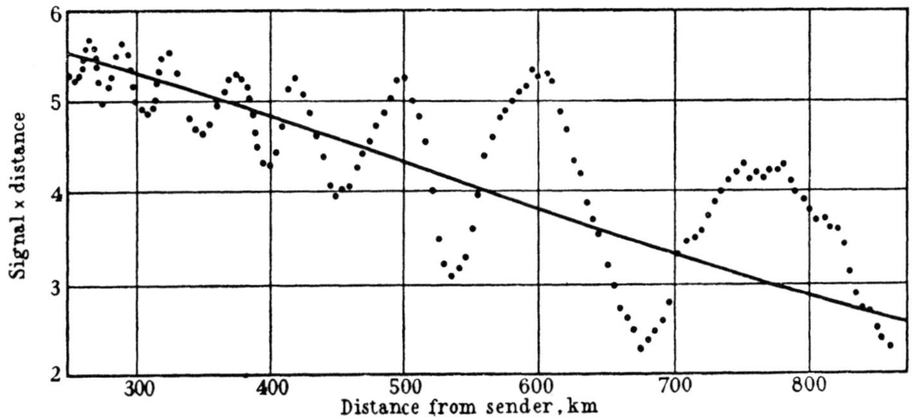

Fig. 10.24 *The Hollingworth interference pattern observed on November 3, 1949, during the midday hours on a frequency of 85 kHz (From Bracewell et al. 1951 © Inst. Elect. Engr.; reprinted with permission)*
Full line is the calculated groundwave signal

10.6.6.2 Temporal variations

Steeply incident radio waves on 16 kHz appear to be reflected more strongly at sunspot minimum than at sunspot maximum, whereas on 245 kHz the reverse is true. The VLF phenomenon may be attributed to a sharpening of the electron-density gradient at reflection, whereas the decrease of amplitude on 245 kHz may result from increase of absorption below reflection. Monthly median values of $\log \varrho$ at constant solar zenith angle are linearly related to sunspot number R on 245 kHz up to $R \approx 150$ when a saturation occurs.

The diurnal variations of amplitudes on 100 kHz show that on some days there are dips near sunrise and sunset (Doherty 1964). Some of the more interesting features are the oscillatory variations of amplitudes (in the frequency range 70–100 kHz) near dawn and dusk. The skywave amplitude begins to decrease (near dawn) when the solar zenith angle is approximately 100°, and reaches a minimum when the angle is 97°. The signal then passes through a marked maximum for χ between 96° and 95°.

The variation of the reflection coefficient $R_{\parallel}$ with the equivalent vertical frequency f_v ($= f\cos\phi$ where ϕ is the angle of incidence) is shown in Fig. 10.25*a*. Over short distances, the amplitude of the skywave decreases rapidly with increase of frequency, and the minimum amplitudes occur on equivalent vertical frequencies in the range 100–300 kHz. The amplitude differences between sunspot maximum and sunspot minimum years is shown in Fig. 10.25*b*. On frequencies lower than about 100 kHz the amplitudes are stronger for sunspot maximum than for sunspot minimum, whereas on frequencies higher than about 100 kHz the reverse holds (Belrose 1982). It should be noted that, for short distances, since the downcoming wave is essentially circularly polarized, the reflection coefficient R_{11} and the conversion coefficient R_{12} are nearly equal.

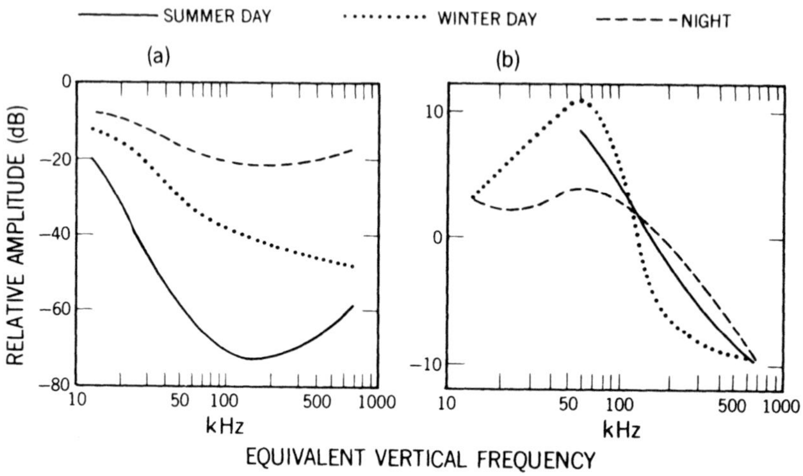

Fig. 10.25 *(a) Variation of skywave amplitude with equivalent vertical frequency at sunspot minimum*
(b) Ratio of amplitudes at sunspot maximum and minimum as function of equivalent vertical frequency
(After Belrose 1982)

10.7 Field strengths on frequencies below about 30 kHz

Methods of calculating the amplitude and phase of VLF/ELF signals have been given by CCIR (1986, Report 895–1). The older method gives the field strength E, at a distance D km on the ground, of a wave of length λ km reflected at a height h km, as

$$E = E_0 \exp[-i(kd + \pi/4)] \sum_n \Lambda_n \exp(-ikS_n d) \qquad (10.44a)$$

where

$$E_0 = \frac{300}{h} \frac{p^{1/2}\lambda^{1/2}}{[a \sin(d/a)]^{1/2}} \qquad (10.44b)$$

Here p = radiated power (kW)
$k = 2\pi/\lambda$
Λ_n is the excitation factor for the nth mode (10.23)
S_n is defined in (10.7).

In general, the terms Λ_n and S_n are complex. The amplitude E can, therefore, be expressed as the sum of its real and imaginary parts:

$$E = E_r + iE_i \tag{10.45}$$

This gives the total amplitude

$$|E| = (E_r^2 + E_i^2)^{1/2} \tag{10.46a}$$

and the relative phase ϕ is given by

$$\phi = \tan^{-1}(E_i/E_r) \tag{10.46b}$$

The field strength due to a given mode n depends on Λ_n and S_n. These are functions of wavelength, ionospheric profile, ground conductivity, collision frequency profile, direction and magnitude of the Earth's magnetic field, and the angle of incidence. Over long distances the use of full-wave or WKB solutions is recommended (CCIR 1986, Report 895-1).

10.8 References

AL'PERT, YA. L. (1983): 'The near earth and interplanetary plasma. Vols. I and II' (Cambridge University Press, Cambridge)

ARGO, P. E. (1975): 'Modeling Omega PCA phase advances'. Tech. Rep. TR 1950, Naval Electronics Lab. Center, San Diego, CA

BAIN, W. C., BRACEWELL, R. N., STRAKER, T. W., and WESTCOTT, C. H. (1952): 'The ionospheric propagation of radio waves of frequency 16 kc/s over distances of about 540 km'. Proc. IEE, Monograph 37R

BALSER, M. and WAGNER, C. A. (1964): 'Thunderstorm excitation of the earth–ionosphere cavity' *in* BLACKBAND, W. T. (Ed.): 'Propagation of radio waves at frequencies below 300 kc/s'. AGARDograph 74 (Macmillan Co., New York) p. 403

BARBER, N. F., and CROMBIE, D. D. (1959): 'VLF reflections from the ionosphere in the presence of a transverse mangetic field', *J. Atmos. Terr. Phys.*, **16**, p. 37

BARR, R. (1977): 'The effect of sporadic-E on the nocturnal propagation of ELF radio waves', *J. Atmos. Terr. Phys.*, **39**, p. 1379

BELROSE, J. S. (1964): 'Present knowledge of the lowest ionosphere' *in* BLACKBAND, W. T. (Ed.): 'Propagation of radio waves at frequencies below 300 kc/s'. AGARDograph 74, (Macmillan Co., New York) p. 3

BELROSE, J. S. (1982): 'LF propagation: An overview' *in* BELROSE, J. S. (Ed.): 'Medium, long and very long wave propagation'. AGARD Conf. Proc. No. 305, Advisory Group on Aerospace Research and Development, North Atlantic Treaty Organisation, p. 22–1

BICKEL, J. E., FERGUSON, J. A., and STANLEY, G. V. (1970): 'Experimental observations of magnetic field effects on VLF propagation at night', *Radio Sci.*, **5**, p. 19

BRACEWELL, R. N., BUDDEN, K. G., RATCLIFFE, J. A., STRAKER, T. W., and WEEKS, K. (1951): 'The ionospheric propagation of low frequency radio waves over distances less than 1000 km', *Proc. IEE*, **98**, Pt. III, p. 221

BUDDEN, K. G. (1961): 'The wave-guide mode theory of wave propagation' (LOGOS Press, London)

BUDDEN, K. G. (1985): 'The propagation of radio waves' (Cambridge University Press, Cambridge)

BURGESS, B. (1970): 'VLF phase delay variability and the design of long range navigation aids' *in* DAVIES, K. (Ed.): 'Phase and frequency instabilities'. AGARD Conf. Proc. No. 33, Advisory Group for Aerospace Research and Development, North Atlantic Treaty Organisation, p. 19

CCIR (1986): 'Propagation in ionized media'. Recomendations and Reports of the CCIR VI, 1986, Int. Telecom. Union, Geneva

CHAPMAN, F. W. and JONES, D. G. C. (1964*a*): 'The influence of the ionosphere on the polarization of VLF and ELF radio waves' *in* BLACKBAND, W. T. (Ed.): 'Propagation of radio waves at frequencies below 300 kc/s'. AGARDograph 74 (Macmillan Co., New York) p. 361

CHAPMAN, F. W. and JONES, D. L. (1964*b*): 'Earth–ionosphere cavity resonances and effective ionosphere parameters' *in* BLACKBAND, W. T. (Ed.): 'Propagation of radio waves at frequencies below 300 kc/s'. AGARDograph 74 (Macmillan Co., New York) p. 413

CHILTON, C. J., CROMBIE, D. D., and JEAN, A. G. (1964): 'Phase variations in VLF propagation' *in* BLACKBAND, W. T. (Ed.): 'Propagation of radio waves at frequencies below 300 kc/s'. AGARDograph 74 (Macmillan Co., New York) p. 257

CHILTON, C. J., CONNER, J. P., and STEELE, F. K. (1965): 'A comparison between solar X-ray emission and VLF sudden phase anomalies', *Proc. IEEE*, **53**, p. 2018

CROMBIE, D. D. (1966): 'Further observations of sunrise and sunset fading of very-low-frequency signals', *Radio Sci.*, **1**, p. 47

CROMBIE, D. D. (1970): 'Effects of normal ionospheric fluctuations on propagation of VLF signals to great distances' *in* DAVIES, K. (Ed.): 'Phase and frequency instabilities'. AGARD Conf. Proc. No. 33, Advisory Group for Aerospace Research and Development, North Atlantic Treaty Organisation, p. 35

CROMBIE, D. D., ALLAN, A. H., and NEWMAN, M. (1958): 'Phase variations of 16 kc/s transmissions from Rugby as received in New Zealand', *Proc. IEE*, **105B**, p. 301

DAVIES, K. (1969): 'Ionospheric radio waves' (Blaisdell Publishing Co., Waltham, MA)

DAVIS, J. R. (1974): 'The influence of certain ionospheric phenomena on extremely low frequency (ELF) propagation', *IEEE Trans.*, **Comm–22**, p. 484

DAVIS, J. R. (1976): 'Localized nighttime D-region disturbances and ELF propagation', *J. Atmos. Terr. Phys.*, **38**, p. 1309

DOHERTY, R. H. (1964): 'Oblique incidence pulse measurements at 100 kc/s' *in* BLACKBAND, W. T. (Ed.): 'Propagation of radio waves at frequencies below 300 kc/s'. AGARDograph 74 (Macmillan Co., New York) p. 133

DOHERTY, R. H. and JOHLER, J. R. (1982): 'LORAN-C: An overview of operational and propagation characteristics' *in* BELROSE, J. S. (Ed.): 'Medium, long and very long wave propagation'. AGARD Conf. Proc. No. 305, Advisory Group for Aerospace Research and Development, North Atlantic Treaty Organisation, p. 39–1

DUNCKEL, N. and HELLIWELL, R. A. (1977): 'Spacecraft observations of man-made whistler-mode signals near the electron gyrofrequency', *Radio Sci.*, **12**, p. 821

FERGUSON, J. A., MORFITT, D. G., and HANSEN, P. M. (1985): 'Statistical model for low-frequency propagation', *Radio Sci.*, **20**, p. 528

FRISIUS, J. (1970): 'Observations of diurnal amplitude and phase variations on 16 kHz transmission path and interpretation by a simple propagation model' *in* DAVIES, K. (Ed.): 'Phase and frequency instabilities'. AGARD Conf. Proc. No. 33, Advisory Group for Aerospace Research and Development, North Atlantic Treaty Organisation, p. 76

GALEJS, J. (1970): 'Phase velocity variations of VLF waves' *in* DAVIES, K. (Ed.): 'Phase and frequency instabilities'. AGARD Conf. Proc. No. 33, Advisory Group for Aerospace Research and Development, North Atlantic Treaty Organisation, p. 45

HEFLEY, G. (1972): 'The development of Loran-C navigation and timing'. NBS Monograph 129, US GPO, Washington, DC 20402

HELLIWELL, R. A. (1965): 'Whistlers and related ionospheric phenomena' (Stanford University Press, Stanford, CA)

HERITAGE, J. L., WEISBROD, S., and BICKEL, J. E. (1947): 'A study of signal versus distance data at VLF'. VLF Symposium Record **4**, p. 77

IEEE (1965): 'Nuclear test detection issue', *Proc. IEEE*, **53**, (12), December

IEEE (1974): 'Special issue on Project Sanguine', *IEEE Trans*, **Comm-22** (4), April

JOHLER, J. R. and BERRY, L. A. (1964): 'On the propagation of long wavelength terrestrial radio waves – Two theoretical techniques' *in* BLACKBAND, W. T. (Ed.): 'Propagation of radio waves at frequencies below 300 kc/s'. AGARDograph 74 (Macmillan Co., New York) p. 101

JOHLER, J. R. and HARPER, J. D. (1962) 'Reflection and transmission of radio waves at a continuously stratified plasma with arbitrary magnetic inclination', *J. Res. NBS*, **66**, p. 81

JONES, D. L. (1974): 'Extremely low frequency (ELF) ionospheric radio propagation studies using natural sources', *IEEE Trans.*, **Comm–22**, p. 477

KIKUCHI, T. and OHTANI, A. (1984): 'Anomalous interference in OMEGA VLF wave propagation on east-to-west equatorial paths', *J. Atmos. Terr. Phys.*, **46**, p. 697

LARSEN, T. R. (1979): 'Effects of D-region ionization on radio wave propagation' *in* DONNELLY, R. F. (Ed.): 'Solar-terrestrial predictions proceedings II', US GPO, Washington DC 20402, p. 617

LARSEN, T. R. and THRANE, E. V. (1977): 'Ionospheric effects on Loran-C in polar regions' *in* 'Propagation limitations of navigation and positioning systems'. AGARD Conf. Proc. No. 209, Advisory Group for Aerospace Research and Development, North Atlantic Treaty Organisation, p. 14–1

LEROY, B. (1979): 'Relations between solar X-ray bursts, sudden enhancements of atmospherics and regions of solar activity' *in* DONNELLY, R. F. (Ed.): 'Solar–terrestrial predictions proceedings III', US GPO, Washington, DC 20402, p. C-152

PAPPERT, R. A. (1973): 'Excitation of the Earth–ionosphere waveguide by point dipoles at satellite heights', *Radio Sci.*, **8**, p. 535

PAPPERT, R. A. (1985): 'Calculated effects of traveling sporadic E on nocturnal ELF propagation: Comparison with measurement', *Radio Sci.*, **20**, p. 229

PAPPERT, R. A. and FERGUSON, J. A. (1986): 'VLF/LF mode conversion model calculations for air to air transmissions in the Earth–ionosphere waveguide', *Radio Sci.*, **21**, p. 551

POLK, C. (1983): 'Natural and man-made noise in the earth-ionosphere cavity at extremely low frequencies', *Space Sci. Rev.*, **35**, p. 83

POTTS, C. E. (1972): 'Precise time and frequency dissemination via the Loran-C system', *Proc. IEEE*, **60**, p. 530

RATCLIFFE, J. A. (1959): 'The magneto-ionic theory' (Cambridge University Press, Cambridge)

REDER, F. and WESTERLUND, S. (1970): 'VLF signal phase instabilities produced by propagation medium: Experimental results' *in* DAVIES, K. (Ed.): 'Phase and frequency instabilities', AGARD Conf. Proc. No. 33, Advisory Group for Aerospace Research and Development, North Atlantic Treaty Organisation, p. 103

RORDEN, L. H., SMITH, R. L., and HELLIWELL, R. A. (1964): 'The LOFTI experiment' *in* BLACKBAND, W. T. (Ed.): 'Propagation of radio waves at frequencies below 300 kc/s'. AGARDograph 74 (Macmillan Co., New York) p. 75

ROSENBERG, T. J. and DUDENEY, J. R. (1986): 'The local time, substorm and seasonal dependence of electron precipitation at $L = 4$ inferred from riometer measurements', *J. Geophys. Res.*, **91**, p. 12032

SAUER, H. H., SPJELDVIK, W. N., and STEELE, F. K. (1987): 'Relationship between long-term phase advances in high-latitude VLF wave propagation and solar energetic particle fluxes', *Radio Sci.*, **22**, p. 405

SNYDER, F. P. and PAPPERT, R. A. (1969): 'A parametric study of VLF modes below anisotropic ionospheres', *Radio Sci.*, **4**, p. 213

STRAKER, T. W. (1955): 'The ionospheric reflection of radio waves of frequency 16 kc/s over short distances', *Proc. IEE*, **102C**, p. 396

SWANSON, E. R. (1977): 'Propagation effects on OMEGA' *in* 'Propagation limitations of navigation and positioning systems'. AGARD Conf. Proc. No. 209, Advisory Group for Aerospace Research and Development, North Atlantic Treaty Organisation, p. 15–1

SWANSON, E. R. (1982): 'OMEGA' *in* BELROSE, J. S. (Ed.): 'Medium, long and very long wave propagation', AGARD Conf. Proc. No. 305, Advisory Group for Aerospace Research and Development, North Atlantic Treaty Organisation, p. 36–1

VOLLAND, H. (Ed.) (1982): 'CRC handbook of atmospherics: Vols. I and II' (CRC Press, Boca Raton, Florida)

WAIT, J. R. (1970): 'Electromagnetic waves in stratified media' (Pergamon Press, Oxford)

WAIT, J. R. (1986): 'Introduction to antennas and propagation' (Peter Peregrinus Ltd., London)

WATT, A. D. (1967): 'VLF radio engineering'. Int. Monographs in Electromagnetic Waves **14** (Pergamon Press, Oxford, England)

WESTERLUND, S., REDER, F. H., and ÅBOM, C. (1969): 'Effects of polar cap absorption events on VLF transmissions', *Planet. Space Sci.*, **17**, p. 1329

YABROFF, I. (1961): 'Computation of whistler ray paths', *J. Res. NBS*, **65D**, p. 485

Chapter 11

Medium frequencies

11.1 Uses of medium frequencies

The medium-frequency band (300–3000 kHz) is dominated by the amplitude modulation (AM) broadcast band ($\approx$ 500–1700 kHz). This frequency band has an enormous economic, social, and political impact on everyday life and it is intensively used. Although the MF band is an important part of the radio spectrum, our knowledge of propagation in the band is rather poor because of (i) spectrum congestion, (ii) high absorption, (iii) the complicated role of the Earth's magnetic field, (iv) difficulty in separating deviative from nondeviative effects, (v) collisions, and (vi) the fact that, on the lower frequencies, the medium is not slowly varying. For efficient use of the MF band it is essential for users to share the same channel with minimum interference. Thus the estimation of the signal strength in distant regions is an important consideration in channel sharing with potentially interfering transmitters (Wang 1982). Atmospheric radio noise in the MF band has been modeled by Herman and DeAngelis (1987) and researchers cited by them.

11.2 Some characteristics of medium waves

11.2.1 Signal strength

Medium-frequency propagation is characterized by a marked diurnal variation of signal strength brought about by the diurnal variation of absorption in the D region. During sunlight the skywave is weak and groundwave propagation over short distances ensures interference-free broadcasting. During darkness, the absorption in the D region is small and signals propagate to long distances; the resulting interference between channels creates problems. Ionospheric absorption has a marked magnetic dependence because the electron gyrofrequency (0·8–1·6 MHz) lies in the medium-frequency band. Thus the absorption of the extraordinary wave, approximately proportional to $(f - f_H)^{-2}$, is large and the polarization of the emitted wave is important in the determination of the fraction of the incident power that goes into the extraordinary wave. This is especially true near the dip equator where the magnetic field is nearly horizontal and the emitted electric field is usually vertical.

Because of the high variability of the D region, from hour to hour and from night to night, the signal strengths of MF waves exhibit high temporal variability during night and from one night to the next (Knight 1982). Near sunrise and sunset, and near the limit of the ground range, marked interference occurs between skywave and groundwave. MF skywaves are sensitive to D-region disturbances in high latitudes (Hunsucker *et al.* 1987).

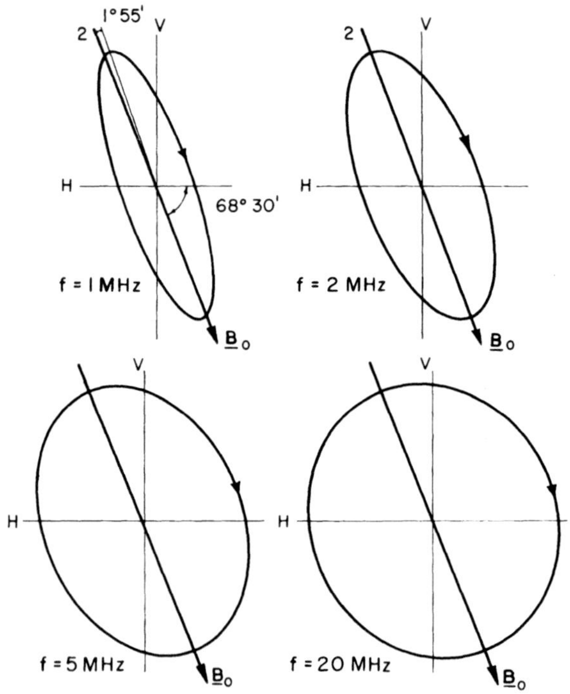

Fig. 11.1 *Polarization ellipses for different frequencies for geographical west–east propagation of the ordinary wave in the USA*
As the wave frequency increases the ellipse becomes more nearly circular and the major axis approaches the direction of the geomagnetic field

11.2.2 Polarization

The polarization of the characteristic waves becomes quite elliptical (Fig. 11.1) for geographical east–west propagation of the ordinary wave in the United States. On higher frequencies the polarization is nearly circular, and the major axis is essentially in the direction of the magnetic field. On 1 MHz, however, the polarization is highly elliptical and collisions produce a rotation of the axes (Sec. 3.4.5).

11.2.3 Ray paths with vertical propagation

Ray paths on medium frequencies are sensitive to the geomagnetic field. This is illustrated in Fig. 11.2 for a case of vertical propagation. The nighttime profile for Boulder has two valleys in the height range between 100 km and 200 km. Because $Y > 1$, the extraordinary wave is reflected at much higher altitude ($X = 1 + Y$) than is the ordinary wave. The sign of $\tan \alpha_x$, where α_x is the angle between the vertical and the extraordinary ray direction (Sec. 3.6), is determined by the sign of $(\mu^2 - 1)$, so the ray path will veer to the north and south as μ^2 goes above and below unity.

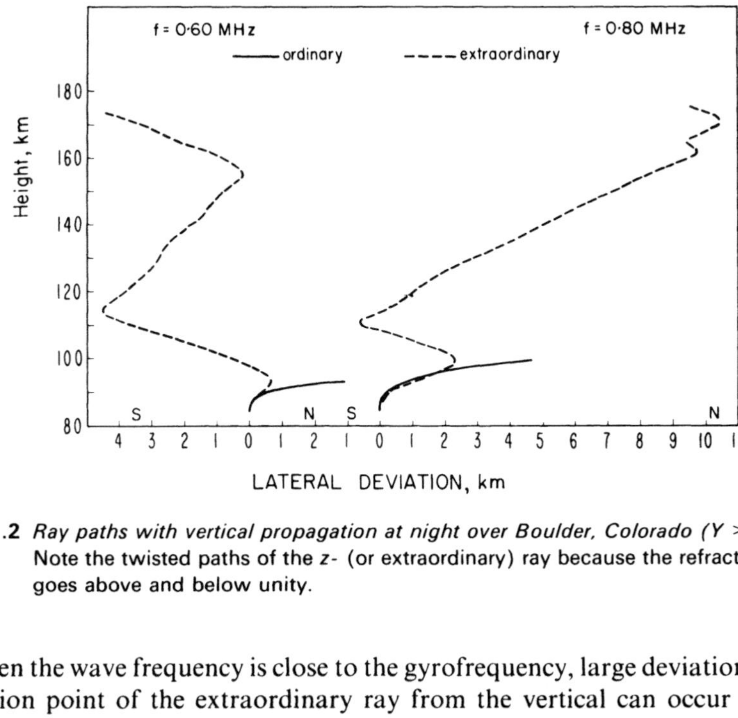

Fig. 11.2 *Ray paths with vertical propagation at night over Boulder, Colorado ($Y > 1$)* Note the twisted paths of the *z*- (or extraordinary) ray because the refractive index goes above and below unity.

When the wave frequency is close to the gyrofrequency, large deviations of the reflection point of the extraordinary ray from the vertical can occur (Davies 1969, Fig. 7.8). Indeed the reflection point of the extraordinary wave may lie north or south of the sounding station. Lateral deviations of the reflection points of the ordinary and extraordinary waves for various frequencies in the MF band are shown in Fig. 11.3. On frequencies less than the gyrofrequency the deviation of the ordinary wave is small when reflection occurs from the E layer, but increases rapidly as the wave penetrates the E layer. The deviation decreases with frequency for reflection from the F layer as the effect of underlying retardation decreases. On lower frequencies the displacement of the point of reflection of the extraordinary wave lies to the south of the entry point, and near the gyrofrequency ($\approx 1{\cdot}48$ MHz) the reflection point lies to the north of the point of entry.

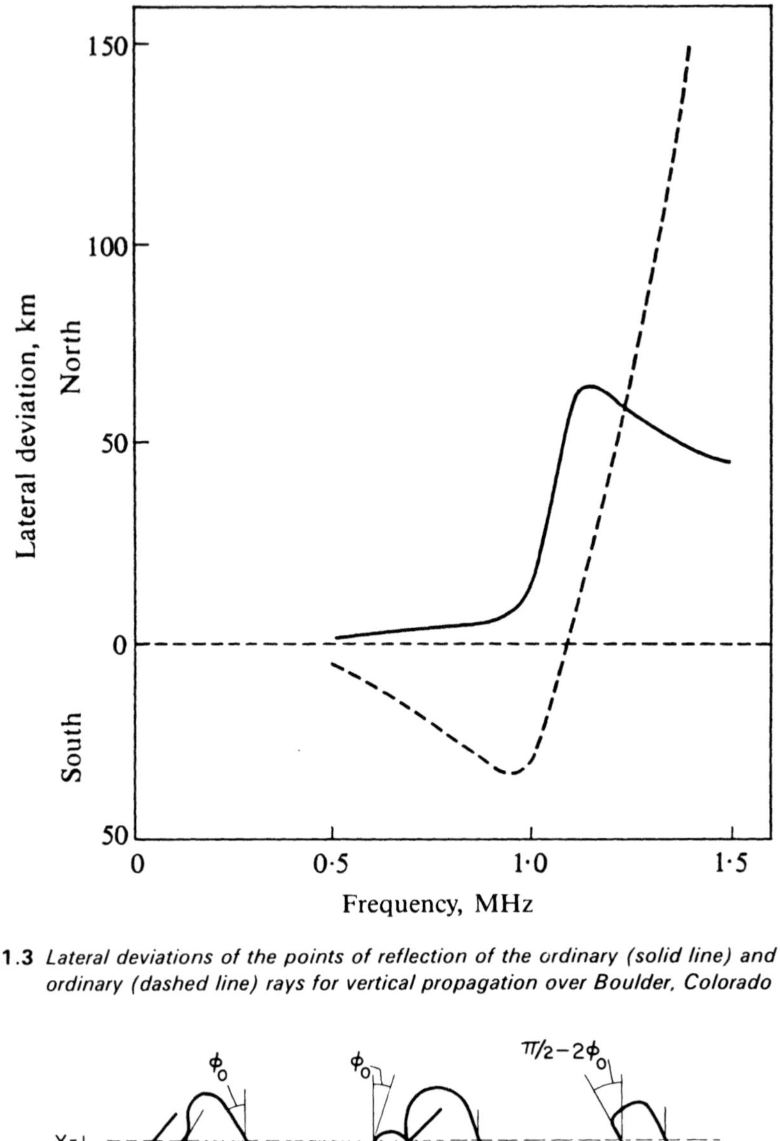

Fig. 11.3 *Lateral deviations of the points of reflection of the ordinary (solid line) and extraordinary (dashed line) rays for vertical propagation over Boulder, Colorado*

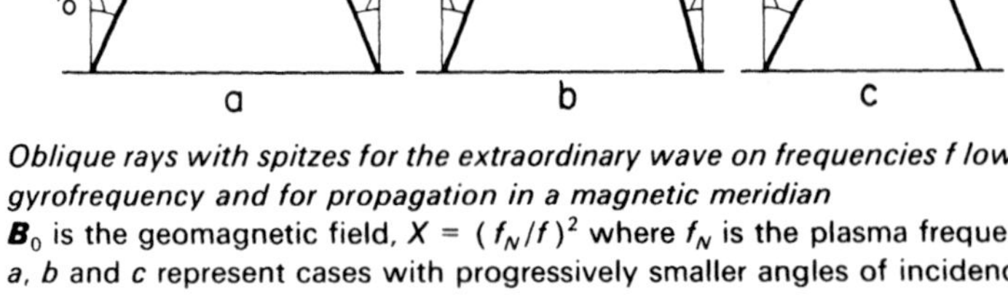

Fig. 11.4 *Oblique rays with spitzes for the extraordinary wave on frequencies f lower than the gyrofrequency and for propagation in a magnetic meridian*

$\boldsymbol{B}_0$ is the geomagnetic field, $X = (f_N/f)^2$ where f_N is the plasma frequency

a, *b* and *c* represent cases with progressively smaller angles of incidence ϕ_o

11.2.4 Ray paths with oblique propagation

On frequencies below the gyrofrequency, the refractive index of the extraordinary (or *z*-) wave is greater than unity for $0 < X < 1$ (see Sec. 3.2) and has a discontinuity at $X = 1$ (Budden 1985, Sec. 10.11), as seen in Fig. 6.42*b*. With propagation in a magnetic meridian, a spitze occurs below the level of reflection (Sec. 6.13.3.1) (Fig. 11.4). This spitze occurs for all angles of incidence less than

$$\sin^{-1}\left[\left(\frac{Y}{Y-1}\right)^{1/2}\cos I\right]$$

that is, for dip angles greater than

$$\cos^{-1}\left[\left(\frac{Y-1}{Y}\right)^{1/2}\right]$$

11.3 Medium-frequency soundings

11.3.1 LF–MF ionograms

The ionosonde technique (Hunsucker 1990, Ch. 3), which has been used for sounding with high-frequency waves, has not been used extensively on medium frequencies for the following reasons:

(i) The excessive interference caused by broadcast stations
(ii) The greater power required because of the large ionospheric absorption on the lower frequencies
(iii) The larger antennas needed for efficient radiation of power.

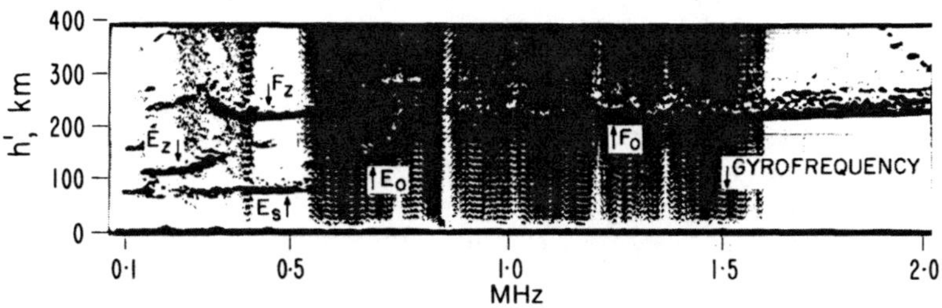

Fig. 11.5 *Low- and medium-frequency ionogram at Boulder, Colorado, December 2, 1959, 2250 105° WMT*
Below 1 MHz, *o*- and *z*-echoes are obtained from the F layer and from the E layer. Low-altitude E_s echoes are also seen. The *z*-echo refers to the extraordinary wave on frequencies lower than the gyrofrequency

The first pulse sounder specifically designed to sound on medium frequencies (Watts 1958) covered the frequency range from 50 kHz to 2000 kHz and was operated first at Sterling, Virginia, and later at Boulder, Colorado. An ionogram from this sounder (Fig. 11.5) shows that the extraordinary trace reflected from the F layer is continuous down to approximately 130 kHz. In this case the ordinary-wave (o) trace is continuous above 470 kHz. There is a discontinuous jump of the o-trace from the 100 km level due to penetration of the E layer. The traces marked E_z and F_z are reflected from the level at which $X = 1 + Y$ since there is no reflection at $X = 1 - Y$ because $Y > 1$.

Several ionosondes have been modified or built to cover the MF band. An ionogram from one of these sounders (Fig. 11.6) has several features of interest:

(i) The E echo at about 120 km is transparent, which suggests that it has a patchy structure.
(ii) There is an intermediate layer located near the 200 km level that has ordinary and extraordinary traces.
(iii) The F-layer z-trace below the gyrofrequency, marked F_z, exhibits marked retardation near the gyrofrequency.
(iv) The ordinary-wave trace shows no retardation near the gyrofrequency.

Medium-frequency soundings, on a set of fixed frequencies, have been made at Tsumeb in southern Africa by Elling (1960).

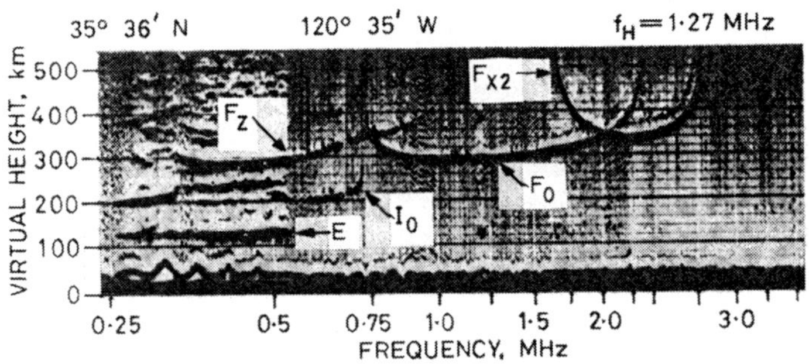

Fig. 11.6 *Medium- and high-frequency ionogram, Point Arguello, California, February 7, 1965, 0506 120° WMT*

Note the large number of echoes present on frequencies below about 0·6 MHz and the retardation of the F_x trace at frequencies above the gyrofrequency ($\approx$ 1·5 MHz). I_o denotes an ordinary-wave echo from the intermediate layer

11.3.2 Wave interaction

Several indirect radio techniques have been used to obtain electron density profiles and collision frequency profiles in the D region, including wave interaction and partial reflection techniques.

The wave interaction or cross-modulation technique is a development of the Luxembourg effect (Sec. 14.4.1) in which the modulation of a powerful (unwanted) wave is imposed on a weak (wanted) wave by nonlinear interaction in the D region (*Radio Science* 1965; Ratcliffe and Shaw 1948; and Hunsucker 1990, Ch. 6).

Consider a powerful disturbing wave of frequency f_D in which the power P is modulated by a fraction M ($M \ll 1$) with a modulation frequency ω. Note that M is essentially twice the amplitude modulation. The modulation of the disturbing wave produces a modulation of the electron collision frequency of magnitude A, where

$$A = \frac{2\kappa_D P_0 M}{N Q_0 \nu_0 G \left[1 + \left(\frac{\omega}{\nu_0 G}\right)^2\right]^{1/2}} \tag{11.1}$$

Here P_0 is the power of the disturbing wave in the absence of modulation, κ_D (see eqn. 3.42) is the amplitude absorption coefficient of the disturbing wave, N is the mean electron density in the absorbing region, $Q_0 = 3KT_0/2$ is the electron energy in the absence of the disturbing wave, G is the electron cooling rate, and $\nu_0 = 1{\cdot}5\,\nu_m$ is the collision frequency corresponding to an electron energy Q_0. Therefore, the absorption of the wanted wave (the cross modulation) will vary cosinusoidally with magnitude:

$$\Delta L_W = A L_{WO} \tag{11.2}$$

L_{WO} is the absorption of the wanted wave in the absence of the disturbing wave. In the pulsed wave interaction technique, the powerful radio pulse heats the electrons; this increases their thermal speeds and collision frequencies and, hence, increases the absorption of another probing radio wave traversing the heated region. The technique (Fig. 11.7) involves the emission of wanted pulses from a transmitter at a repetition rate of $2\,f_r$ per second, and the reflected signal is received nearby. Powerful disturbing pulses are emitted from a nearby transmitter at a repetition rate of f_r per second. Thus successive wanted pulses are disturbed and undisturbed for easy comparison. The relative timing of the wanted and the disturbing pulses is adjusted so that the downcoming wanted pulse meets the upgoing disturbing pulse at a predetermined height in the D region below the height of reflection of the wanted wave.

The changes in refractive index ($\Delta\mu$) and absorption coefficient ($\Delta\kappa$) are given by

$$\Delta\mu = \left(\frac{\partial\mu}{\partial\nu}\right)\Delta\nu \tag{11.3a}$$

and

$$\Delta\kappa = \left(\frac{\partial\kappa}{\partial\nu}\right)\Delta\nu \tag{11.3b}$$

$\Delta\nu$ is the change of collision frequency, and the gradients $(\partial\mu/\partial\nu)$ and $(\partial\kappa/\partial\nu)$ are obtained from (3.41) and (3.42), for example. The heating of the electrons, and hence the increase in collision frequency, can be calculated from the effective radiated power of the disturbing transmitter and an assumed collision frequency profile.

The amplitude (A) interaction coefficient T_A and the phase interaction coefficient T_ϕ are defined as follows:

$$T_A = \frac{\Delta A}{A} = \int_0^{h_i} \frac{\partial \kappa}{\partial \nu} \Delta \nu dh \qquad (11.4a)$$

and

$$T_\phi = \Delta\phi = \frac{\omega}{c} \int_0^{h_i} \frac{\partial \mu}{\partial \nu} \Delta \nu dh \qquad (11.4b)$$

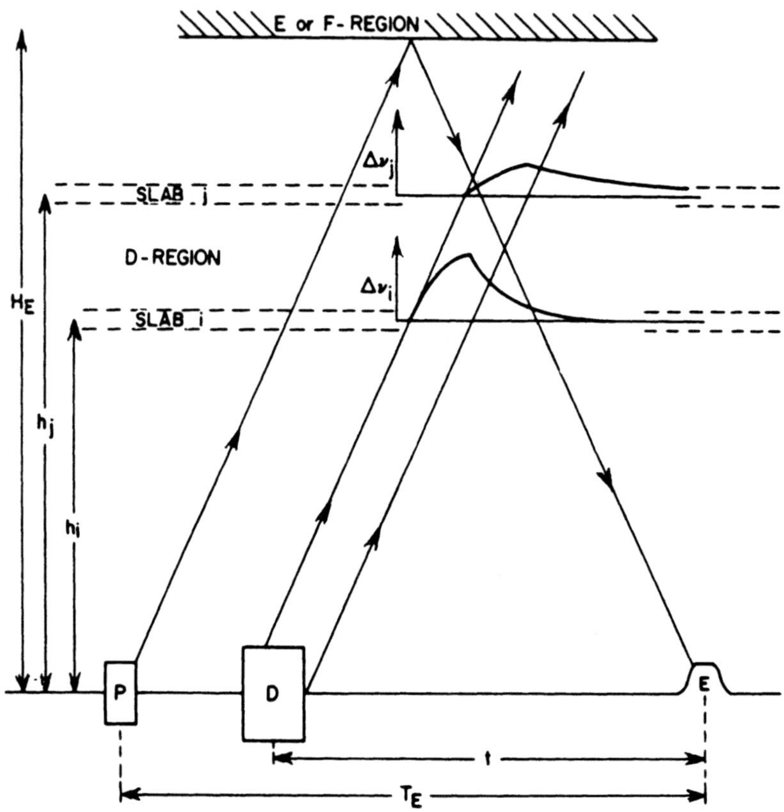

Fig. 11.7 *Time–height configuration of a wave-interaction experiment (From Coyne 1973 © American Geophysical Union)*
P is the probing pulse, D the disturbing pulse. The time variation of collision frequency is indicated for two heights

ΔA is the change in amplitude A, $\Delta\phi$ is the change of phase of the wanted (or probing) wave caused by the absorption of the disturbing wave, and h_i is the height of the interaction slab.

In order to derive electron-density profiles from the interaction measurements assumptions must be made about the height variations of the electron collision frequency ν_m and about the cooling rate of the heated electrons (Thrane 1974; Fejer 1970). A suitable collision frequency profile (Thrane *et al.* 1968) is

$$\nu_m = 8{\cdot}10^{15}\, p\ \mathrm{s}^{-1} \tag{11.5}$$

where ν_m is the collision frequency of electrons with energy KT (see eqn. 3.38) and p is the total atmospheric pressure in newtons m^{-2}. For small temperature increases, the cooling is exponential with a time constant

$$\tau = 1/(G\ \nu_m) \tag{11.6}$$

where G is the electron cooling per collision. An acceptable value is

$$G = 14{\cdot}1\ T_g^{-3/2} \tag{11.7}$$

where T_g is the gas temperature. At 85 km and 70 km typical values of τ are 640 μs and 60 μs respectively.

By using these data on collision frequencies and cooling, the electron-density profile can be obtained through a trial and error procedure in which a height profile is chosen and the calculated wave interaction – or cross-modulation – is compared with observations. The profile is adjusted for a best fit.

A typical system may have a wanted frequency of 2·3 MHz, a pulse repetition frequency of 50 Hz, a disturbing frequency of about 2·0 MHz with pulse repetition frequency of 25 Hz, and a power of 50 kW. The meeting height h_i may be changed from 60 km to 100 km, in 10 km steps, by adjusting the relative timing of wanted and disturbing pulses. Some sample $N(h)$ profiles obtained using this technique are given in Fig. 11.8.

Because a powerful radio wave heats the ionosphere and causes increased absorption, the heating at the crest of the modulation cycle is greater than at the trough. This leads to self-demodulation or self-distortion of the wave (Maslin 1974; Cutulo *et al.* 1982). The amplitude of the wave is decreased and the phase is advanced. When absorption takes place in a region where $\nu^2 \gg \omega^2$ the absorption decreases (see eqn. 7.19) and the modulation is enhanced. The percent demodulation depends on several factors such as (*a*) the electron-density profile, (*b*) the collision-frequency profile in the D region, (*c*) the wave frequency, and (*d*) the modulation frequency. The demodulation can vary from about -9% to $+27\%$, and the phase shift can be as great as 27°. Thus, increasing the power beyond a certain level is inefficient (Megill 1965; Maslin 1976).

To maximize the cross-modulation (or demodulation) the disturbing wave should be an extraordinary wave with a frequency near the gyrofrequency on which the absorption is high. The frequency of the disturbing wave should not be too close to the gyrofrequency because there the interpretation becomes very difficult (Maslin 1976). It is generally assumed that the heating of the electrons affects the collision frequency only and that the electron density is not affected. This is not true in general because the electron loss process may be temperature dependent.

11.3.3 Partial reflection technique

The electron-density distribution in the D region can be derived by measuring the differential absorption of ordinary and extraordinary waves scattered from plasma irregularities in the D region (Hunsucker 1990, Ch. 6). The amplitudes

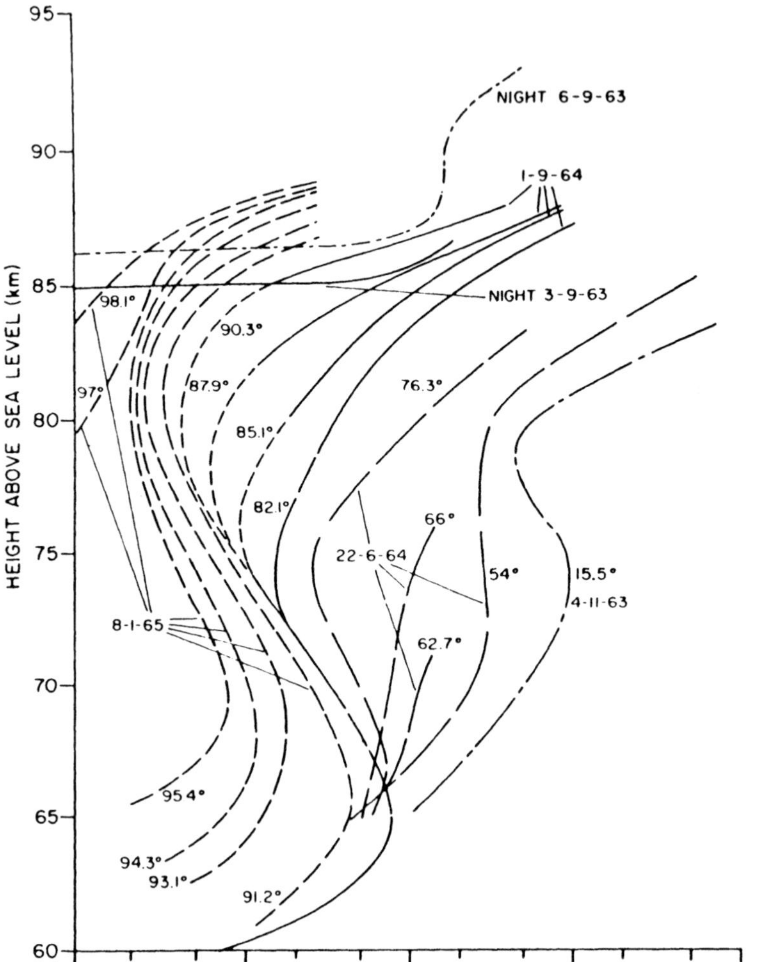

Fig. 11.8 *D-region profiles obtained by wave-interaction technique at Armidale, Australia, for night, dawn, and daytime in 1963–65 (From Smith et al. 1967)*

of waves scattered from such irregularities are determined partly by the scattering coefficient (ρ) and partly by absorption below the scattering level. The received amplitude ratio

$$\frac{A_x}{A_o} = \frac{\rho_x}{\rho_o} \exp\left[-2\int_0^h (\kappa_x - \kappa_o)dh\right] \tag{11.8}$$

where $\rho_{o,x}$ are the scattering coefficients and $\kappa_{o,x}$ are the absorption coefficients. The scattering is usually treated as being of the Fresnel type which, under nondeviating conditions and longitudinal propagation, gives (see eqn. 3.41)

$$\frac{\rho_x}{\rho_o} \approx \frac{\delta n_x}{\delta n_o} = \frac{1 - Y}{1 + Y} \frac{\mathscr{C}_{3/2}\left(\dfrac{1 - Y}{Z_m}\right)}{\mathscr{C}_{3/2}\left(\dfrac{1 + Y}{Z_m}\right)} \tag{11.9}$$

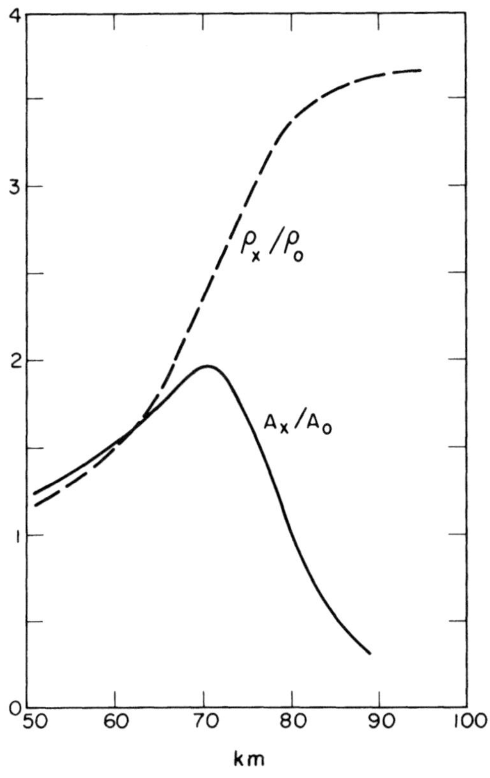

Fig. 11.9 *Variation with height of the calculated ratios ρ_x/ρ_o and A_x/A_o measured on 2·66 MHz (From Belrose 1970 © Pergamon Press, Inc.; reprinted with permission)*

The difference between the absorption coefficients is (3.42)

$$\kappa_x - \kappa_o = \frac{5}{4} \frac{e^2 N}{\varepsilon_0 m c \nu_m} \left[\mathscr{C}_{5/2}\left(\frac{1 - Y}{Z_m}\right) - \mathscr{C}_{5/2}\left(\frac{1 + Y}{Z_m}\right) \right] \tag{11.10}$$

The ratio ρ_x/ρ_o can be calculated from an assumed collision-frequency profile $\nu_m(h)$. The electron-density profile is obtained by inversion of integral (eqn. 11.8).

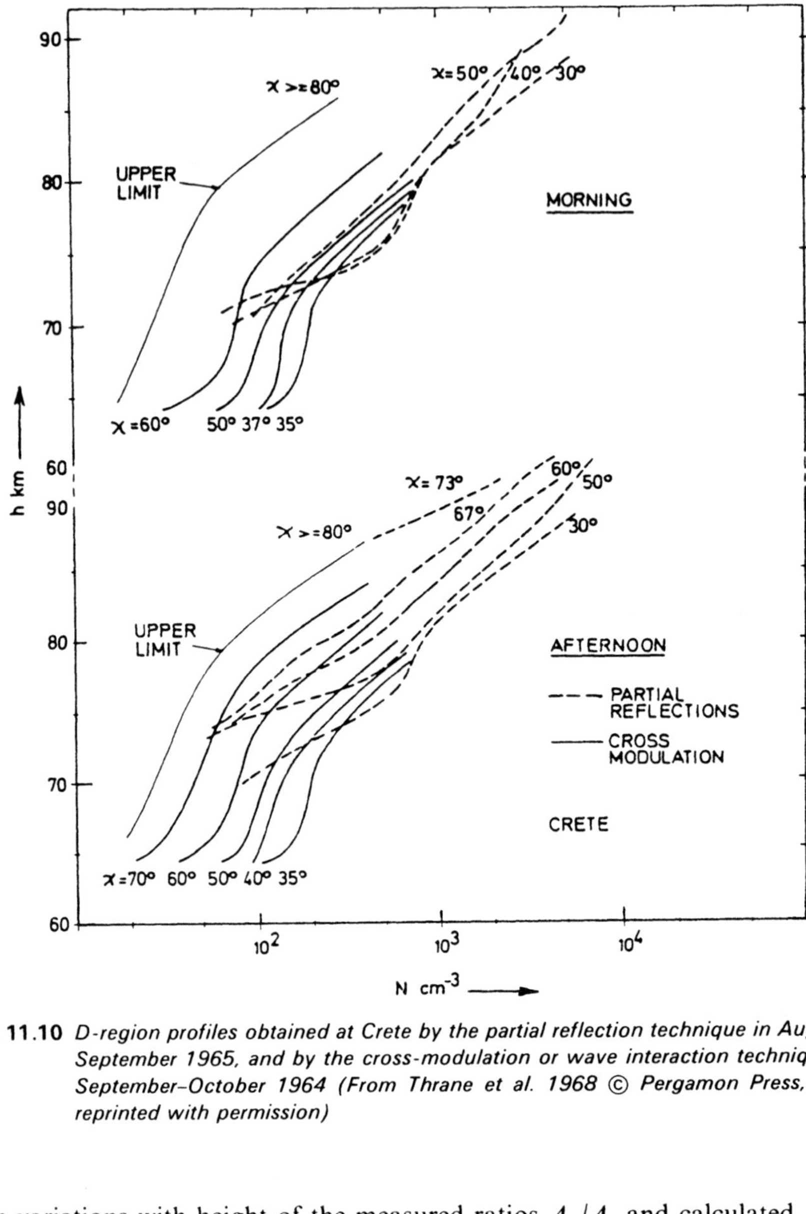

Fig. 11.10 *D-region profiles obtained at Crete by the partial reflection technique in August–September 1965, and by the cross-modulation or wave interaction technique in September–October 1964 (From Thrane et al. 1968 © Pergamon Press, Inc.; reprinted with permission)*

The variations with height of the measured ratios A_x/A_o and calculated ρ_x/ρ_o are shown in Figure 11.9. Note that at low altitudes, where $\kappa_x - \kappa_o \approx 0$, $\rho_x/\rho_o \approx A_x/A_o$ (Belrose 1970) and the measured A_x/A_o yields information on the electron-collision frequency (Thrane 1974). For further details on the validity of the assumptions see Belrose (1970) and Thrane (1974). Sample profiles obtained using the partial reflection technique and the wave interaction technique are presented in Fig. 11.10.

In principle, the electron-density profile can be determined from measurements of the differential phase

$$\begin{aligned}\Delta\phi &= \frac{2\omega}{c}\int(\delta n_o - \delta n_x)dh \\ &= -\frac{e^2\delta N}{\omega mc\varepsilon_0 Z_m^2}\left[(1+Y)\mathscr{C}_{3/2}\left(\frac{1+Y}{Z_m}\right) - (1-Y)\mathscr{C}_{3/2}\left(\frac{1-Y}{Z_m}\right)\right]\end{aligned} \tag{11.11}$$

However, this approach has had limited use (von Biel *et al.* 1970).

The partial reflection technique has been used extensively, with spaced receivers and directional (interferometer) antennas, to study motions in the D region (Vincent 1984; Hocking 1985; Adams *et al.* 1986; Manson *et al.* 1985).

11.3.4 Rocket techniques

Some of the more accurate D-region profiles have been obtained by radio signals transmitted from the ground to rockets traveling in the lower ionosphere. The usual method is to measure the Faraday rotation (Sec. 8.3.2) and absorption changes as the position of the rocket varies. When both Faraday rotation and differential absorption measurements are available, both electron density and collision frequency can be derived (Thrane 1974). A D-region profile (Hall 1973) can be obtained from rocket measurements of the strength of an LF wave transmitted from the ground (Hunsucker 1990, Ch. 9).

11.4 The E layer at night

11.4.1 Critical frequency of the E layer at night

The variation of the critical frequency of the E layer at night is important for the propagation of medium waves. If the maximum frequency reflected from the layer with oblique propagation falls below the operating frequency, then propagation is possible only by reflection from the F region. The signal strength of F echoes as well as their fading characteristics may differ appreciably from those of E echoes.

From ionograms of the type discussed in Sec. 11.3.1, the penetration frequencies of the nighttime E layer in the United States at sunspot minimum and sunspot maximum have been studied (Wakai 1971). The layer near 150 km is that referred to above as the *intermediate* layer. It has been found that, after the time when the solar zenith angle χ reaches 99°, the penetration frequency falls by approximately 0·2 MHz during the night, reaching a minimum of about 0·7 MHz a few hours before sunrise. Although seasonal variations seem to be lacking, a clear sunspot cycle effect has been found. On the average the penetration frequency is about 0·25 MHz lower at sunspot minimum than at sunspot maximum. From the practical viewpoint of MF propagation, the ever present sporadic E near 100 km may be the decisive factor. Until oblique ionograms are obtained for the MF band, this question will remain unsolved.

The observations described above are substantially in agreement with the observations of Elling (1960) at Tsumeb in South-West Africa. Elling's results on $f_o E$ and $f_o I$ (intermediate) are shown in Fig. 11.11. The $f_o I$ data agree with the above findings; the $f_o E$ values are lower but not as low as those given by Phillips (1953) for London, which show values of $f_o E$ below 0·5 MHz.

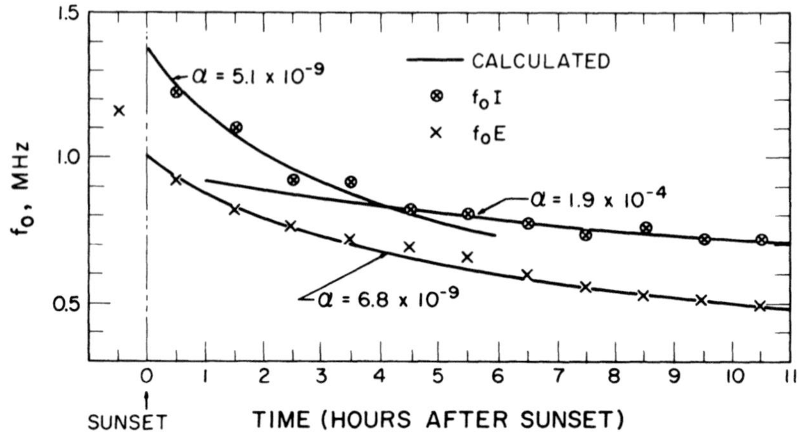

Fig. 11.11 *Post-sunset decay of the critical frequencies of the intermediate (I) layer and the E layer at Tsumeb (After Elling 1960)*
The full lines represent the decay expected for different values of the effective recombination coefficient

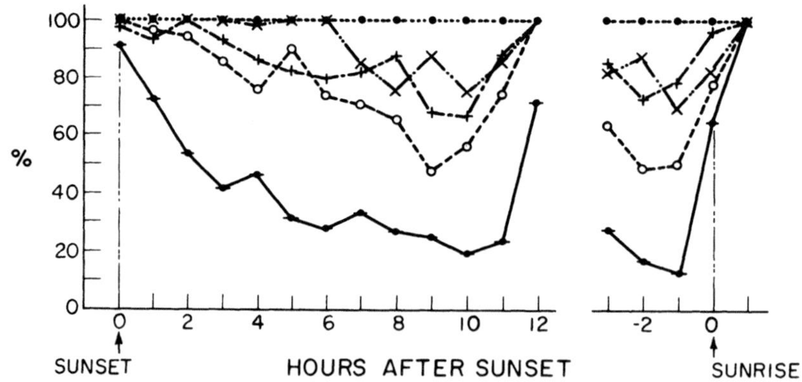

Fig. 11.12 *Frequency of occurrence of E-layer echoes for five medium frequencies (After Elling 1960)*

The frequency of occurrence of echoes originating from the E layer for different wave frequencies has been plotted by Elling (Fig. 11.12). On 365 kHz, reflections are observed during the entire night. The frequency of occurrence of

E echoes decreases with increasing wave frequency and with time after sunset. The frequency of occurrence of F echoes increases with increasing wave frequency and with time after sunset. Echoes from the intermediate layer are observed most often late at night and on frequencies between 600 kHz and 850 kHz particularly.

11.4.2 The night-E layer

The night-E layer is quite distinct from the nighttime E layer discussed above. It is essentially a disturbance phenomenon and occurs at any time during the night (Wakai 1971; French 1969). The layer is thick and produces cusps on both E- and F-ionogram traces. The critical frequency often varies rapidly, e.g. 1 MHz or more in a few minutes. Values of the critical frequency vary from 1·0 MHz to >4·5 MHz (Wakai 1971). The latitudinal variation of f_oE (Fig. 11.13) clearly indicates a peak in the auroral zone.

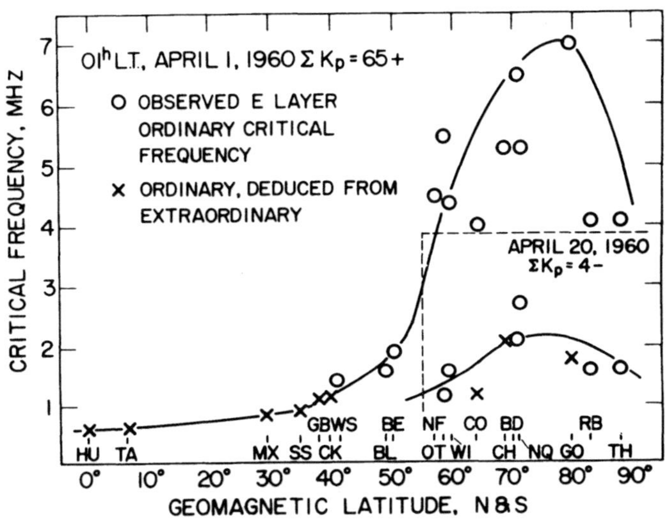

Fig. 11.13 *Latitudinal variation of E-layer critical frequencies at night during disturbed (April 1, 1960) and quiet (April 20, 1960) conditions (From Wakai 1971 © Communications Research Laboratory, Tokyo, reprinted with permission)*

The night-E layer can be found in all latitudes except perhaps in the equatorial belt (King 1962) but the higher values of f_oE are confined to auroral latitudes. At a low-latitude station like Raratonga, an f_oE of 1·5 MHz will be exceeded only a few times a year, whereas in the auroral zone, values of 1·5 MHz occur every day and 4 MHz is common. The diurnal variation shows a peak between about midnight and about 0600 local time. The presence of night-E during the day raises the values of f_oE above those expected for the normal E layer and its presence also affects the dependence of f_oE on solar zenith angle.

11.4.3 Importance of the E layer
Echoes from the E region are observed all night on frequencies below about 500 kHz. For propagation in middle latitudes the important parameter is the equivalent vertical frequency $f \cos \phi_o$. For a critical frequency of 500 kHz and a distance of 600 km all signals on frequencies below about 1500 kHz will have an E-layer echo. It is, of course, possible to have additional F-layer (and/or I) reflections. Only on the higher frequencies in the broadcast band and for short distances (less than say 400 km) will the E layer be unable to reflect medium-frequency waves. To determine whether F-layer echoes are possible it is necessary to determine whether the radio frequency is greater than the E-layer cutoff frequency.

11.5 Skywave signal strength

11.5.1 Factors affecting signal strength
It has already been pointed out that the intensity of a medium-frequency signal shows a marked increase from day to night because of the diurnal variation in ionospheric absorption. This absorption is one of several factors that affect the loss of power in transmission between a transmitter and a receiver. The main sources of power loss are as follows:

(i) Losses in and near the terminal equipments (e.g. transmission line losses, Earth-induced currents, line mismatch, antenna losses)
(ii) The spatial or distance loss due to the spreading of the wave
(iii) Polarization matching losses
(iv) Ionospheric absorption and scattering
(v) Losses due to defocusing of the beam by the curvature of the ionosphere (this may be a negative loss, i.e. a gain).

Because this work is concerned with the propagation aspects we do not consider losses of type (i).

Certain of the losses mentioned above are true losses in the sense that power is always lost to the beam; these include (ii) and (iv). On the other hand, (iii) and (v) do not necessarily involve a net loss of power but rather a redistribution. For example, concerning polarization losses, a loss to the ordinary wave is compensated by a gain to the extraordinary wave, which may be lost through absorption. Similarly, focusing involves a redistribution of power flux over the ground so that a decrease in one area is balanced by a greater concentration elsewhere.

11.5.2 Temporal variations

11.5.2.1 Solar cycle dependence
Measurements in Europe (Ebert 1962) over the frequency range 164–1466 kHz and over paths from 500 km to about 2700 km show that the relationship between the 11-month running-mean signal (F_R) and the corresponding sunspot

number R_{11} is given by

$$F_R = F_0 - 0{\cdot}02\ R_{11} \tag{11.12}$$

Now F_0 is about 60 dB, so

$$F_R \approx 60(1 - 0{\cdot}0003R_{11}) \tag{11.13}$$

This dependence on sunspot number is about one-twentieth that for the high-frequency case (7.38). It should be remembered that the MF signal is measured near local midnight whereas the HF results are for noon. The long-term variation (Fig. 11.14) reveals dependence both on sunspot number and on magnetic disturbance. The sensitivity to sunspot number depends critically on the hour of observation. In Europe the reference hour is normally 6 h after sunset. In North America the customary reference is sunset + 2 h at which the difference between sunspot high and sunspot low, during 1944–47, was 14 dB compared with only 8 dB at sunset + 6 h.

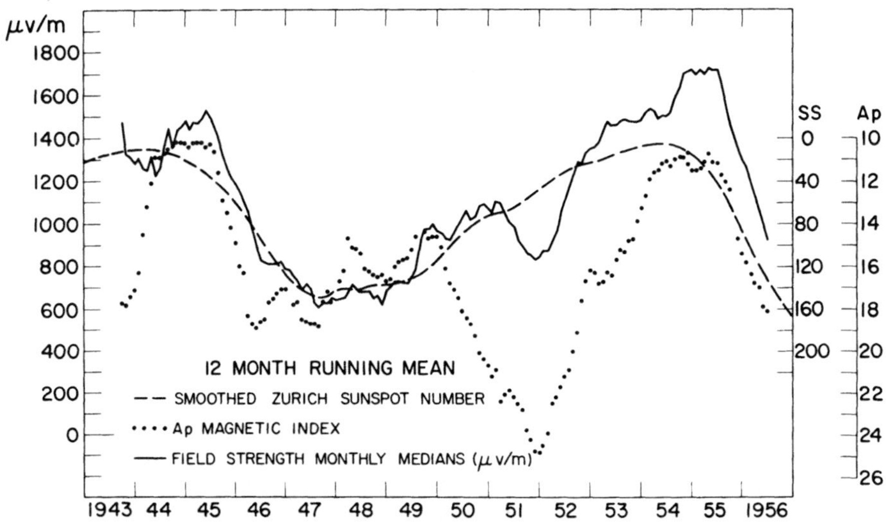

Fig. 11.14 *Long-term variations of sunspot numbers and magnetic A_p, and the strength of the (700 kHz, 50 kW) MF skywave from Cincinnati to Atlanta (600 km)*

11.5.2.2 Seasonal changes

Contours of the temporal variations of 845 kHz signals in Europe (Fig. 11.15) show maxima in March and in September (CCIR 1986, Report 431–4) and minima in summer and winter, the summer minimum being the more pronounced. The overall variation may be as much as 15 dB at the lower end of the MF band, decreasing to about 3 dB at the upper end. In low latitudes over the Western Hemisphere there is little seasonal variation whereas in middle latitudes (USA) there is a slight minimum of signal strength in the summer months.

Table 11.1 shows the seasonal variation in the average daily number of hours of reception at Fairbanks, Alaska, of signals from five MF stations in the continental United States and Canada. The data show that in summer, when the paths are sunlit for most of the day, the hours of reception are drastically reduced (Hunsucker et al. 1988).

Table 11.1 *Reception at Fairbanks, Alaska, of MF signals from Continental USA and Canada (five stations)**

KTWO, Casper, WY (1030 kHz); KFAX, San Francisco, CA (1100 kHz); KOTZ, Kotzebue, AK (720 kHz); KGA, Spokane, WA (1510 kHz); CFRN, Edmonton, Alberta (1260 kHz)

Season	No. of hours per day
Winter	13·3
Spring	13·3
Summer	4·0
Autumn	11·0

*Data obtained from Hunsucker and DeLana (1986)

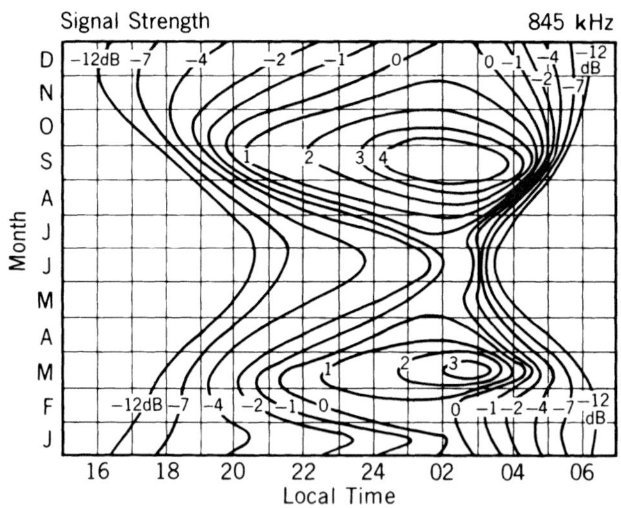

Fig. 11.15 *Diurnal and seasonal contours of MF (845 kHz) signals in Europe (After CCIR 1986, VI, p. 376)*

11.5.2.3 Diurnal variations

Fig. 11.15 shows that, in Europe, the signal strength maximizes near 0200 LT. The quiet-day variation shows that the signal starts increasing before sunset and increases relatively slowly until about 0200 LT and then decreases slowly near

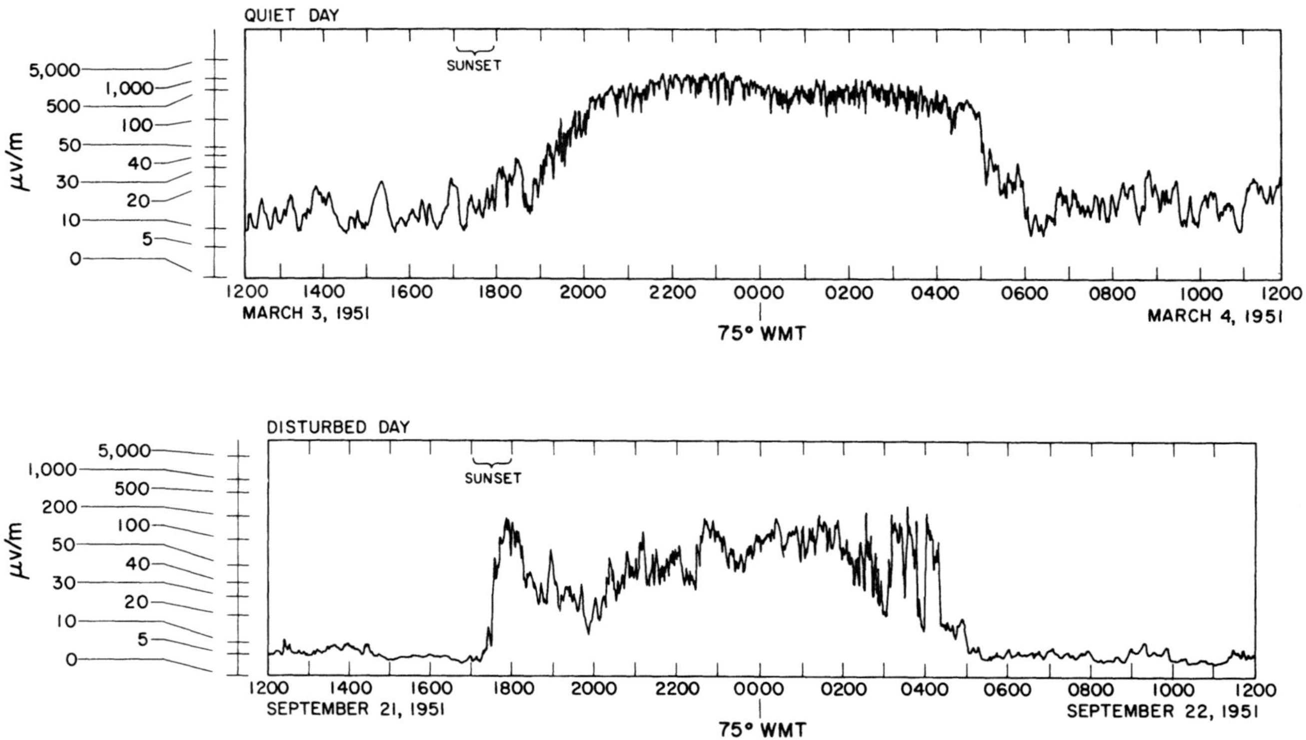

Fig. 11.16 *Diurnal variations of the signal strength of 700 kHz, 50 kW signals from Cincinnati to Baltimore (665 km) on a quiet day and on a disturbed day*

sunrise. In Europe the median daytime signal in winter is about 25 dB less than the nighttime value before sunrise.

The diurnal variations over North America on a quiet day and on a magnetically disturbed day are shown in Fig. 11.16. It is seen that, on a quiet day, the skywave is present during daytime and there is a gradual transition of signal strength from daytime to nighttime. On a disturbed day the day signal strength is lower than on a quiet day and the transitions from night-to-day and day-to-night are sharper. The effect of another large ionospheric storm on MF signals in middle latitudes is seen in Fig. 11.17. August 18 and 19 were quiet days on which the diurnal variation was normal. The effect of the disturbance on August 20 was to depress the signals during the night and, to some extent, enhance them during the day.

When comparing results in different geographical regions it is important to remember that the reference times may differ. For example, over the Americas (Region 2) the reference time is 2 h after sunset, in Europe (Region 1) it is 6 h after sunset, and in Australia it is midnight. The field strength at 6 h after sunset is typically 2·5 dB stronger than that at 2 h after sunset. Values higher than 2·5 dB can occur for long east–west paths (Wang 1979).

On 1655 kHz the median signal strength F (in decibels), for ± 2 h around sunrise (F_R) and sunset (F_S), is given by

$$F_R = -6{\cdot}616 - 10{\cdot}112t - 2{\cdot}034t^2 + 0{\cdot}423t^3 \qquad (11.14a)$$

$$F_S = -7{\cdot}938 + 10{\cdot}050t - 2{\cdot}310t^2 - 0{\cdot}577t^3 \qquad (11.14b)$$

in which t is measured in hours from sunrise and sunset at the midpoint of the path. Note that t is negative before sunrise and sunset at the midpoint of the path. For discussion of signal strengths in the frequency range 1605–1705 kHz, see Wang (1985).

11.5.3 Frequency dependence

Because the D region largely disappears at night, an appreciable part of the absorption of MF signals occurs near the level of reflection in the E region. Hence the inverse square law (7.18) should not be expected to hold. The path loss in Europe is given by

$$L = 10 \log d + 20 \log f + 0{\cdot}00176 f^{0{\cdot}26} d - 7{\cdot}75 \qquad (11.15)$$

where L is in decibels, and d is distance in kilometers (Radio Spectrum Utilization 1964). Note that for distances less than, say, 300 km it is more accurate to use the slant path distances. Equation (11.15) shows that the signal strength decreases with frequency. On the other hand, in the USA, measurements indicate that the combined absorption loss L_a (in dB) and polarization loss L_p (Norton 1959) is given by

$$L_p + L_a(\phi_o) = 26 f^{-0{\cdot}4} \cos^{0{\cdot}6} \phi_o \qquad (11.16)$$

where ϕ_o is the angle of incidence.

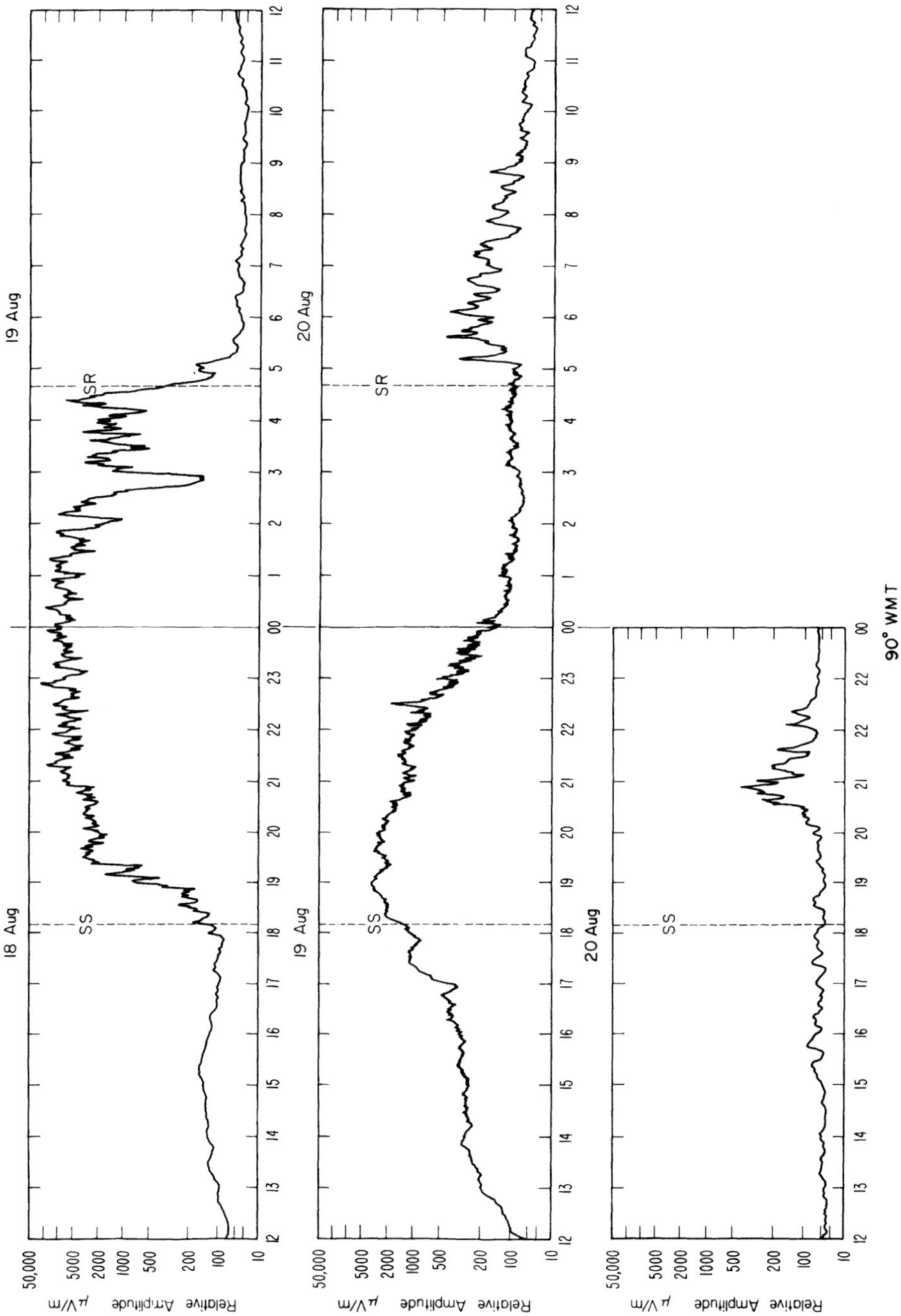

Fig. 11.17 *Diurnal variations of signal strength on 700 kHz between Cincinnati and Baltimore (665 km)*
August 18 and 19 were quiet days; August 20 was a disturbed day

The USSR method of predicting field strength F uses

$$F = 105{\cdot}3 - 20\log d - 0{\cdot}0019 f^{0{\cdot}16} d - 0{\cdot}0024 f^{0{\cdot}4} d(\tan^2\Phi - 0{\cdot}57) \tag{11.17}$$

where Φ is magnetic latitude (CCIR 1986, Report 575–3).

Because the frequency dependence of the signal strength is weak over the entire range from 500 kHz to 1600 kHz, it is often ignored and the field strength on 1000 kHz is taken as representative of the entire band.

11.5.4 Dependence on magnetic latitude

The basic loss factor k (CCIR 1986) is given by

$$k = 3{\cdot}2 + 0{\cdot}19 f^{0{\cdot}4} \tan^2(\Phi + 3) \tag{11.18}$$

where f is in kilohertz and Φ is magnetic latitude. Note that for $|\Phi|$ greater than 60° the value of k at 60° is used. For distances less than 3000 km, Φ may be taken as the mean of the magnetic latitudes at the transmitter and receiver. Paths longer than 3000 km are divided into two equal sections that are considered separately. The magnetic latitude is taken as the mean of the latitudes of the two half-paths. To a sufficient accuracy the geomagnetic latitude can be obtained by finding the magnetic dip I, at a given geographical location, and substituting in (2.8).

The geomagnetic control of MF signal strengths has been known for many years. For example, Dellinger and Cosentino (1940) found that the MF signals received in Washington from South America were 15 to 55 times stronger than those received from Europe. They found that the night-to-night variation in signal strength from South America was about 6 : 1. This ratio was very much less than that for signals from Europe (90 : 1). These differences are caused by auroral zone (ionospheric) absorption and/or winter anomaly absorption (Hagg 1982).

The profound influence of the auroral zone on MF propagation has been shown by measurements in Fairbanks, Alaska, of broadcast stations in Alaska, Canada, and the continental USA. During quiet conditions the MF skywaves from the USA are not influenced by the auroral oval. When the planetary K_p index equals or exceeds 5, all stations monitored in Fairbanks show effects of the intensification and equatorward expansion of the auroral oval (i.e. enhanced D-region absorption and blanketing sporadic-E). Also, for these paths the behavior of MF skywaves during quiet conditions is determined primarily by the spatial and temporal behavior of the main ionospheric trough (Hunsucker *et al.* 1985). In general, paths that are tangent to the maximum of the auroral oval show marked absorption whereas circuits that lie mostly equatorward of the oval are affected less (Hunsucker *et al.* 1987). Fig. 11.18 shows the effects of auroral absorption (associated with the great geomagnetic storm of February 8–9, 1986) on MF skywave signals from San Francisco received at Fairbanks, Alaska. In the top panel, the diurnal variation of the MF skywave is typical for

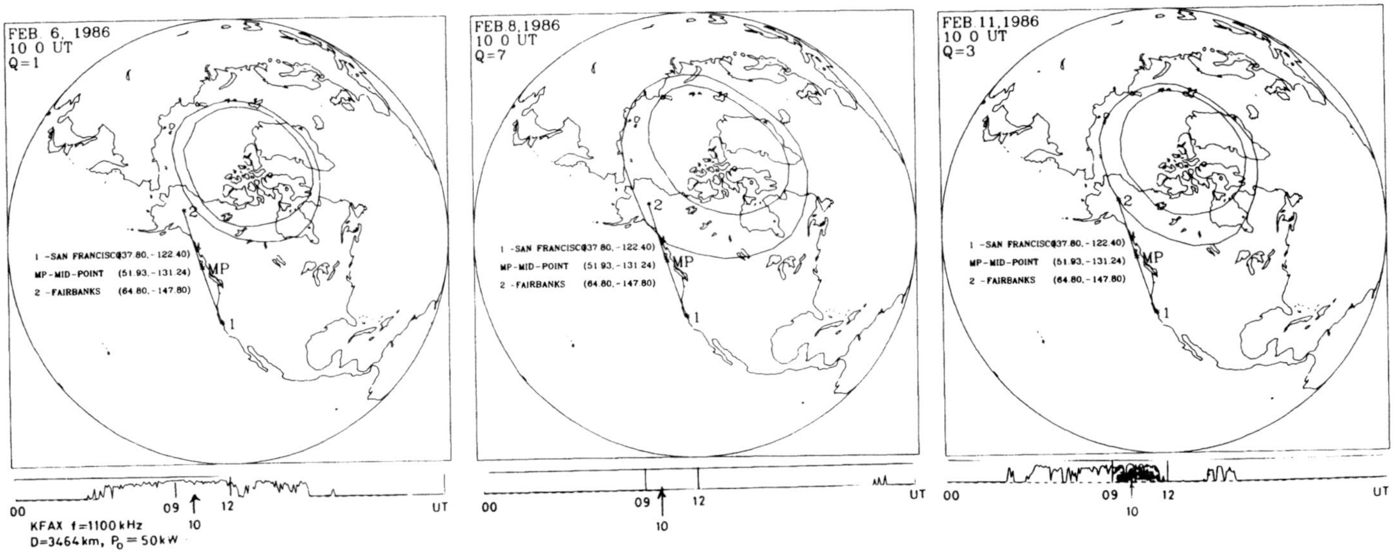

Fig. 11.18 *Effects of auroral absorption on the 1100 kHz signal from San Francisco to Fairbanks, Alaska (From Hunsucker et al. 1987)*
Signals are displayed beneath maps of the auroral oval

a magnetically quiet day at low sunspot numbers; nearly all the path lies outside the auroral oval. The middle panel indicates complete blackout of the signal near storm maximum at which time about 40% of the path lay within the auroral oval because of the equatorward expansion of the oval. The bottom panel shows that, even over this sub-auroral circuit, the signal did not regain its prestorm character although the auroral oval had essentially returned to its prestorm position. Indeed the signals did not return to quiet-day levels until February 14, 1986, some 5 to 6 days after the storm maximum (Hunsucker *et al.* 1987).

11.5.5 Distance variation

The curve in Fig. 11.19 shows the variation of the annual median value of skywave field strength with distance; it was adopted by the Regional Administrative LF/MF Broadcasting Conference held in Geneva in 1975 and is called the Cairo north–south curve. It is applicable only for distances greater than 300 km (CCIR 1986, Report 575-3). In using this curve one must subtract the coupling loss (Sec. 7.6) to allow for the power lost to the extraordinary wave. This is usually neglected in latitudes greater than about 45° because there is a good match between the vertical electric field and the Earth's magnetic field, so most of the power goes into the ordinary wave.

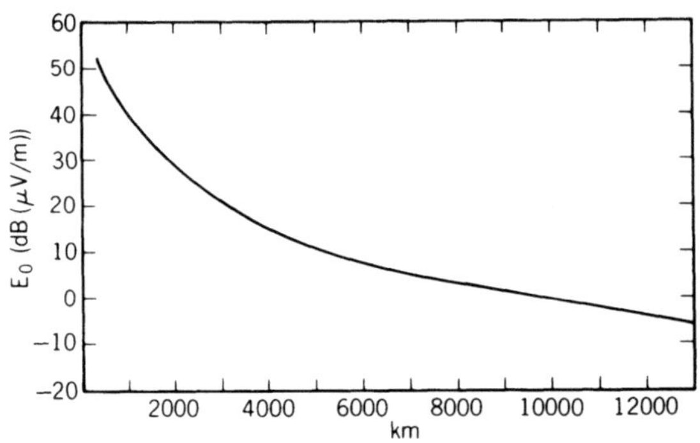

Fig. 11.19 *Annual midnight median value of skywave field strength of Cairo north–south curve*
(After CCIR 1986, Report 575-3)

11.5.6 Fading on medium frequencies

Over very short distances the groundwave dominates and there is little or no fading. Over long distances (> 1000 km) the signal is entirely skywave and the fading is similar to that discussed in Sec. 7.7. Over distances of about 100–500 km the skywave and groundwave are roughly comparable and deep severe fading can occur.

The fading speed n, (i.e. the number of maxima per hour) depends roughly on the equivalent vertical frequency (King 1958). For example,

$$n = 0{\cdot}3 f \cos \phi_o \qquad (11.19)$$

where f is in kilohertz. Thus for an angle of incidence ϕ_o of 60° the number of fades will vary from about 150/h on 500 kHz to about 450/h on 1500 kHz. There appear to be two dominant periods, one near 10 s and the other between 10 and 20 min.

At night in low geomagnetic latitudes ($\leqslant 40°$) the fading has a Rayleigh distribution such that the instantaneous field E exceeds the median E_m for p% of the time where

$$E = 2{\cdot}1\, E_m \left(\log \frac{100}{p}\right)^{1/2} \qquad (11.20)$$

This shows that the field strength value exceeded for 1% of the time is about 9·5 dB greater than the median, and the value exceeded 10% of the time is about 6 dB greater than the median. In high geomagnetic latitudes ($>60°$) the corresponding values are about 15 dB and 10 dB.

11.5.7 Nonreciprocity on medium frequencies

Measurements indicate that, in the USA during high sunspot numbers, the transmission loss for east-to-west propagation is greater than that for west-to-east propagation (Crombie 1979) for paths between 300 km and 2800 km. The difference amounts to about 9 dB for the whole MF band and increases to about 13 dB for frequencies greater than about 830 kHz. This nonreciprocity may be the result of polarization changes at ground reflection on multi-hop paths (Knight 1982). Thus part or all of the reflected wave can generate an extraordinary wave which, on second entry, is entirely absorbed. This is pronounced when waves are reflected from land at angles near the Brewster angle. Waves propagating from east to west suffer greater loss than waves traveling from west to east.

11.6 Skywave propagation curves

For purposes of allocating MF broadcasting channels, it is necessary to have some means of estimating the expected signal strengths in order to avoid undue interference between transmitters on the same frequency. At the same time it is important that the maximum use be made of each channel.

To aid in the allocation of broadcast channels, propagation curves have been published that can be used to calculate nighttime signal strength for various parts of the world. Methods for predicting skywave field strengths on frequencies between 150 kHz and 1600 kHz are given in CCIR (1986, pp. 311–390) and Pokempner (1980). As an example we consider briefly the method used by the CCIR (1986), which uses some of the foregoing data.

For a transmitter output power P dB (above 1 kW) the annual median field strength E, at a given time t relative to sunset or sunrise, is given by

$$E = V + E_0 - L_t \tag{11.21}$$

where

$$V = P + G \tag{11.22}$$

V is called the cymomotive force (field strength × distance) of the transmitter in decibels, above a reference cymomotive force of 300 V. This reference corresponds to a radiated power of 1 kW. G includes the antenna gains in the horizontal and vertical directions. L_t is the time loss factor (Fig. 11.20). The times of sunrise and sunset for various months and geographical latitudes are obtained by inserting $\chi = 90{\cdot}83°$ in (2.48*a*) and solving for h with a correction for the equation of time (2·48*c*). E_0 is the field for a transmitter cymomotive force of 300 V, and it is made up of several terms:

$$E_0 = G_s - L_p + A - 20\log p - 10^{-3}k_R p \tag{11.23}$$

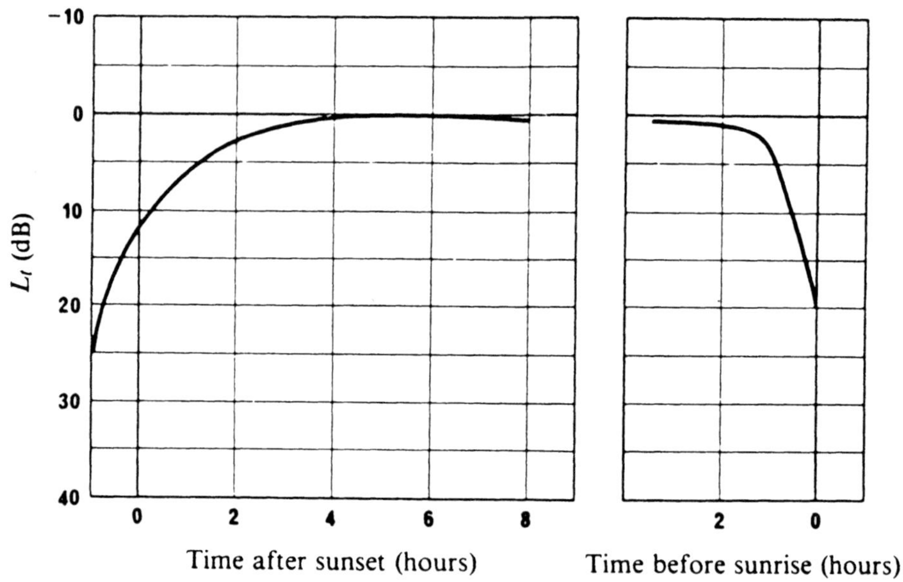

Fig. 11.20 *Variation of time loss factor L_t near sunset and sunrise (After CCIR 1986)*

G_s is called the sea gain and is important only when the terminals are near seawater.

$$A = 106{\cdot}6 - 2\sin\Phi \tag{11.24}$$

where Φ is the geomagnetic latitude.

$$p = (d^2 + 40\,000)^{1/2} \tag{11.25}$$

is the slant path length.

The polarization coupling loss is given roughly by

$$L_p = 180\,(36 + \theta^2 + I^2)^{-1/2} - 2 \tag{11.26}$$

where θ is the angle between the great-circle-path midpoint and magnetic west–east; that is,

$$\theta = 90 - \alpha + D. \tag{11.27}$$

α is the geographic bearing of the path, east of geographic north, D is the magnetic declination (Fig. 11.21), and

$$k_R = k + 0{\cdot}01\,bR_{12} \tag{11.28}$$

where R_{12} is the 12-month running sunspot number

b = 4 for North America
b = 1 for Europe and Australia
b = 0 elsewhere

and k is given by (11.18).

11.7 Calculation of signal strength

Determine the signal strength received at 40°N, 90°W from a 50 kW transmitter near Denver (40°N, 105°W) having a vertical antenna with gain of 2 dB over an isotropic radiator, under the following conditions:

Sunspot number	50
Month	September
Local time	2000
Frequency	850 kHz

From spherical geometry or from a map (see Sec. 12.3.1):

Distance	$d = 1276$ km	
Midpoint	40·26°N, 97·5°W	
Dip	$I = 69°$	(Fig. 2.11)
Geomagnetic latitude	$\Phi \approx 52{\cdot}5°$	(eqn. 2.8)
Declination	$D = +10°$	(Fig. 11.21)
Azimuth at MP	$\alpha = 90°$	(Geometry)
	$\theta = 10°$	(eqn. 11.27)

$$A = 105\,\text{dB}$$

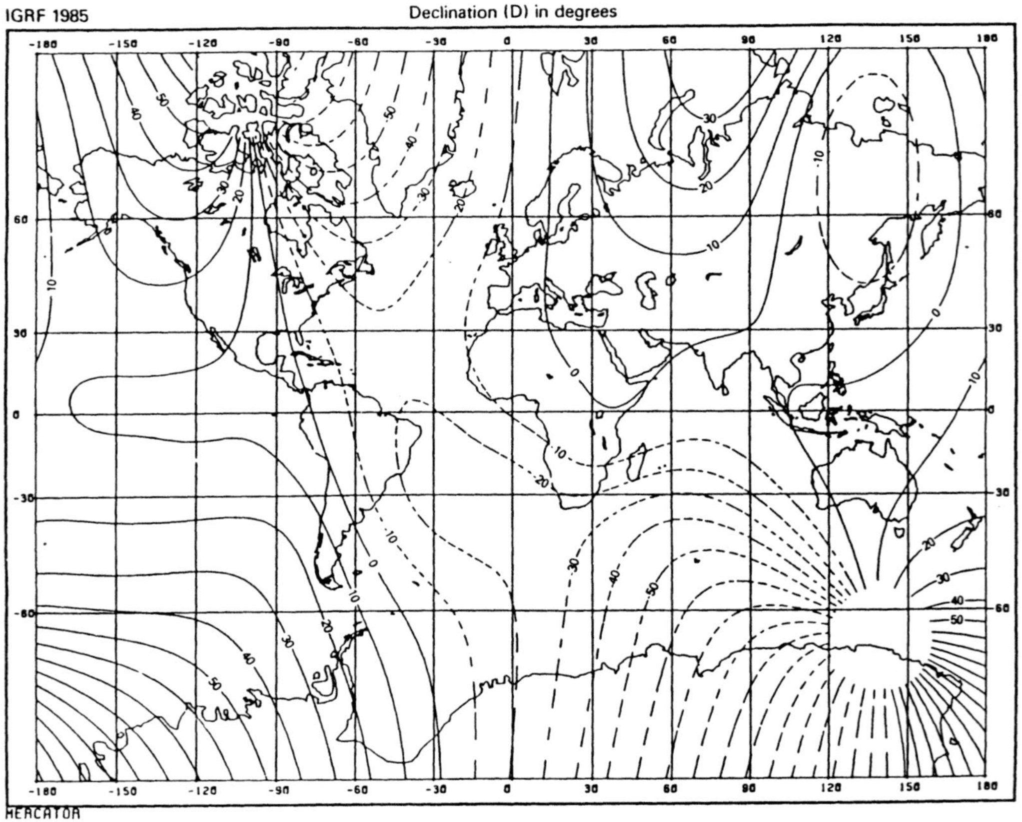

Fig. 11.21 *Magnetic declination D for the International Geomagnetic Reference Field 1985 (From Barraclough 1987 © Elsevier Science Publishers; reprinted with permission)*
Full lines denote positive declination (in degrees) east of north
Broken lines denote negative declination west of north

$$
\begin{aligned}
p &= 1292\,\text{km} \\
L_p &= 0{\cdot}6\,\text{dB} \\
k &= 9{\cdot}2\,\text{dB} \\
b &= 4 \\
k_R &= 9{\cdot}2 + 2 = 11{\cdot}2\,\text{dB} \\
\text{Sunset} &\approx 1800 \text{ (from eqn. 2.48, with } \chi = 90{\cdot}83^\circ) \\
t &= 2\,h \\
L_t &\approx 2{\cdot}5\,\text{dB (from Fig. 11.20)} \\
\text{Seasonal correction} &\approx 0\,\text{dB (from Fig. 11.15)} \\
E_0 &= -0.6 + 105 - 20\log 1292 - 10^{-3} \\
&\quad \times 11{\cdot}2 \times 1292 \\
&= 27{\cdot}70\,\text{dB} \\
V &= 10\log 50 + 2 = 17 + 2 = 19 \\
&\quad \text{(from eqn. 11.22)} \\
E &= 19 + 27{\cdot}70 = 46{\cdot}7\,\text{dB}\ (1\,\mu\text{V/m}) \\
&\approx 216\,\mu\text{V/m}
\end{aligned}
$$

11.8 Medium-frequency broadcasting

Most people are familiar with the standard broadcast band used for amplitude-modulated radio broadcasting. In the Americas this band is made up of 107 channels, each 10 kHz wide, occupying the frequency range 535–1605 kHz. There are roughly 4000 commercial standard broadcast stations in the United States alone with transmitter powers ranging from a few watts to 50 kW. Certain classes of stations in the standard broadcast band are protected for skywave service at night. Technical data relating to the use of this band can be found in FCC (1984). The standard broadcast band will soon be extended to 1705 kHz in the Americas. Technical standards were adopted by the International Telecommunications Union in 1988.

The adoption of this band for standard broadcasting seems to have been purely fortuitous. When the era of broadcasting began in 1922, it was still thought that the longer the wavelength the greater the strength of the wave diffracted around the Earth. However, the band from 300 kHz to 550 kHz had already been allocated to the ship-to-shore service in 1898. In the absence of any planned use of the radio spectrum, the US broadcasters settled on what appeared to be the next best frequency band, namely that between about 500 kHz and about 1500 kHz. This selection was partly based on economics, namely that the cost of equipment and antenna installations was less for this band than for frequencies below 300 kHz.

To maximize the number of channels available and minimize the co-channel interference, some stations are allowed to broadcast with relatively high power during the day but must either decrease the power and/or change antenna directivity, or go off the air, at about sunset to avoid interference with other stations that share the same channels. Other stations, called clear-channel stations, are allowed to broadcast continuously on higher power (e.g. 50 kW). From a propagation point of view it would be a more efficient use of the radio spectrum to move the local broadcasting stations to the VHF/UHF bands and utilize the long-distance propagation of the MF band for more appropriate uses. In practice, some rural communities (e.g. in Alaska) depend on AM broadcast as the sole source of emergency information.

11.9 References

ADAMS, G. W., BROSNAHAN, J. W., WALDEN, D. C., and NERNEY, S. F. (1986): 'Mesospheric observations using a 2.66-MHz radar as an imaging interferometer: Description and first results', *J. Geophys. Res.*, **91**, p. 1671

BARRACLOUGH, D. R. (1987): 'International Geomagnetic Reference Field: The fourth generation', *Phys. Earth Planet. Inter.*, **48**, p. 279

BELROSE, J. S. (1970): 'Radio wave probing of the ionosphere by the partial reflection of radio waves (from heights below 100 km)', *J. Atmos. Terr. Phys.*, **32**, p. 567

BUDDEN, K. G. (1985): 'The propagation of radio waves' (Cambridge University Press, Cambridge)

CCIR (1986): 'Recommendations and reports of the CCIR 1986, VI, Propagation in ionized media' (Int. Telecom. Union, Geneva)

COYNE, T. N. R. (1973): 'Analysis of the pulsed wave interaction experiment', *J. Geophys. Res.*, **78**, p. 206

CROMBIE, D. D. (1979): 'Comparison of measured and predicted signal strengths of nighttime medium frequency signals in the USA', *IEEE Trans.*, **BC-25**, p. 86

CUTULO, M., FELIZIANI, S., MORICONI, E., GAFFURI, G., and CARLOMUSTO, L. (1982): 'Papers on the self-modulation and distortion of MF waves in the ionosphere' *in* 'Medium, long and very long wave propagation'. AGARD Conf. Proc. No. 305, Advisory Group for Aerospace Research and Development, North Atlantic Treaty Organisation, p. 30–1

DAVIES, K. (1969): 'Ionospheric radio waves' (Blaisdell Publ. Co. Waltham, MA)

DELLINGER, J. H. and COSENTINO, A. T. (1940): 'A radio transmission anomaly', *Proc. IRE*, **28**, p. 431

EBERT, W. (1962): 'Ionospheric propagation on long and medium waves'. Tech. Doc. 3081, European Broadcasting Union, Brussels

ELLING, W. (1960): 'Reflecting properties of the ionosphere between 350 and 1500 kc/s at Tsumeb' *in* BEYNON, W. J. G. (Ed.): 'Some ionospheric results – International Geophysical Year' (Elsevier Publ. Co., Amsterdam) p. 252

FCC (1984): 'Federal Communications Commission: Rules and Regulations'. US Govt. Printing Office, Washington, DC 20402

FEJER, J. A. (1970): 'Radio wave probing of the lower ionosphere by cross-modulation techniques', *J. Atmos. Terr. Phys.*, **32**, p. 597

FRENCH, A. G. (1969): 'An intermediate ionospheric layer at high latitudes', *J. Atmos. Terr. Phys.*, **31**, p. 1435

HAGG, E. L. (1982): 'Reduction in MF skywave field strength at night due to magnetic-storm and winter-anomaly-related absorption' *in* BELROSE, J. S. (Ed.): 'Medium, long and very long wave propagation'. AGARD Conf. Proc. No. 305, Advisory Group for Aerospace Research and Development, North Atlantic Treaty Organisation, p. 29–1

HALL, J. E. (1973): 'Lower ionosphere electron densities from rocket measurements employing LF radio propagation and DC probe techniques', *Planet. Space Sci.*, **21**, p. 119

HERMAN, J. R. and DEANGELIS, X. A. (1987): 'Bandwidth expansion effects on the voltage deviation parameter (V_d) of MF and HF atmospheric radio noise', *Radio Sci.*, **22**, p. 26

HOCKING, W. K. (1985): 'Measurement of turbulent energy dissipation rates in the middle atmosphere by radar techniques: A review', *Radio Sci.*, **20**, p. 1403

HUNSUCKER, R. D. (1990): 'Radio techniques for probing the terrestrial ionosphere' (Springer-Verlag, New York)

HUNSUCKER, R. D. and DELANA, B. S. (1986): 'High-latitude field-strength measurements of standard broadcast skywave transmission monitored at Fairbanks, Alaska'. Geophys. Institute, University of Alaska, Fairbanks, AK, 99775-0800

HUNSUCKER, R. D., DELANA, B. S., and WANG, J. C. H. (1985): 'Propagation characteristics of medium frequency skywave signals from the continental U.S. and Canada received in Fairbanks, Alaska' *in* 'Propagation effects on military systems in the high latitude region', AGARD Conf. Proc. No. 382, Advisory Group for Aerospace Research and Development, North Atlantic Treaty Organisation, p. 3–4

HUNSUCKER, R. D., DELANA, B. S., and WANG, J. C. H. (1987): 'Effects of the February 1986 magnetic storm on medium frequency skywave signals recorded at Fairbanks, Alaska' *in* GOODMAN, J. M. (Ed.): 'Ionospheric effects symposium 1987, Proceedings'. Naval Research Laboratory, Washington, DC, p. 197

HUNSUCKER, R. D., DELANA, B. S., and WANG, J. C. H. (1988): 'Medium frequency skywave propagation at high latitudes: Results of an FCC sponsored study'. Proc. 42nd Broadcast Engineering Meeting, Nat. Assoc. Broadcasters

KING, G. A. M. (1962): 'The night-E layer' *in* SMITH, E. K. and MATSUSHITA, S. (Eds.): 'Ionospheric sporadic E' (Macmillan Co., New York) p. 219

KING, J. W. (1958): 'The fading of radio waves reflected at oblique incidence', *J. Atmos. Terr. Phys.*, **12**, p. 26

KNIGHT, P. (1982): 'Medium frequency propagation: A survey' *in* BELROSE, J. S. (Ed.): 'Medium, long and very long wave propagation'. AGARD Conf. Proc. No. 305, Advisory Group for Aerospace Research and Development, North Atlantic Treaty Organisation, p. 28–1

MANSON, A. H., MEEK, C. E., SMITH, M. J., and FRASER, G. J. (1985): 'Direct comparisons of prevailing winds and tidal wind fields (24-, 12-h) in the upper middle atmosphere (60–105 km) during 1978–1980 at Saskatoon (52°N, 107°W) and Christchurch (44°S, 173°E)', *J. Atmos. Terr. Phys.*, **47**, p. 463

MASLIN, N. M. (1974): 'Theory of energy flux and polarization changes of a radio wave with two magnetoionic components undergoing self demodulation in the ionosphere', *Proc. Roy. Soc.*, **341A**, p. 361

MASLIN, N. M. (1976): 'Estimating ionospheric cross-modulation', *Proc. Roy. Soc.*, **351A**, p. 277

MEGILL, L. R. (1965): 'Self-distortion of radio signals in the D-region', *Radio Sci.*, **69D**, p. 369

NORTON, K. A. (1959): 'Transmission loss in radio propagation — II'. NBS Tech. Note 12, National Bureau of Standards, Boulder, CO

PHILLIPS, G. J. (1953): 'The influence of the ionosphere on medium wave broadcasting', *B.B.C. Quarterly*, **VIII**, p. 1

POKEMPNER, M. (1980): 'Comparison of available methods for predicting medium frequency sky-wave field strengths'. NTIA Report 80–42 National Telecommunications and Information Administration, Boulder, CO 80303

Radio Science (1965): 'Special issue on non-linear processes in the ionosphere'. **69D**, (1), pp. 1–145

'Radio spectrum utilization' (1964) (IEEE, New York, N.Y.)

RATCLIFE, J. A. and SHAW, I. J. (1948): 'A study of the interaction of radio waves', *Proc. Roy. Soc.*, **193A**, p. 311

SMITH, R. A., COYNE, T. N. R., LOCH, R. G., and BOURNE, I. A. (1967): 'Small perturbation wave interaction in the lower ionosphere. Part 3; Measurements of electron densities' *in* BELROSE, J. S., BOURNE, I. A., and HEWITT, L. W. (Eds.): 'Ground-based radio wave propagation studies of the lower ionosphere I'. Defence Research Board, Ottawa, Canada, p. 335

THRANE, E. V. (1974): 'Ionospheric profiles up to 160 km — A review of techniques and profiles' *in* RAWER, K. (Ed.): 'Methods of measurements and results of lower ionosphere structure'. Akademie – Verlag, Berlin, p. 3

THRANE, E. V., HAUG, A., BJELLAND, B., ANASTASSIADES, M., and TSAGAKIS, E. (1968): 'Measurements of D-region electron densities during the International Quiet Sun Years', *J. Atmos. Terr. Phys.*, **30**, p. 135

VINCENT, R. A. (1984): 'MF/HF radar measurements of the dynamics of the mesopause region — A review', *J. Atmos. Terr. Phys.*, **46**, p. 961

VON BIEL, H. A., FLOOD, W. A., and CAMNITZ, H. G. (1970): 'Differential-phase partial-reflection technique for the determination of D-region ionization', *J. Geophys. Res.*, **75**, p. 4863

WAKAI, N. (1971): 'Study on the nighttime E region and its effects on the radio wave propagation', *J. Radio Res. Labs.*, **18**, p. 245

WANG, J. C. H. (1979): 'Medium-frequency sky-wave field strength in North America', *IEEE Trans.*, **BC-25**, p. 79

WANG, J. C. H. (1982): 'Inteference and sharing at medium frequency: sky-wave propagation considerations' *in* 'Propagation aspects of frequency sharing, interference and systems diversity'. AGARD Conf. Proc. No. 332, Advisory Group for Aerospace Research and Development, North Atlantic Treaty Organisation, p. 9-1

WANG, J. C. H. (1985): 'A skywave propagation study in preparation for the 1605–1705 kHz broadcasting conference', *IEEE Trans.* **BC-31**, p. 10

WATTS, J. M. (1958): 'The interpretation of night-time low-frequency ionograms', *J. Geophys. Res.*, **63**, p. 717

Chapter 12

High-frequency propagation predictions

12.1 Aim of the Chapter

This Chapter will give the reader a brief idea of the ways in which ionospheric data are applied to high-frequency communication problems. Several computer programs, available for both mainframe and personal computers, make it possible to determine optimum frequencies, signal strengths, broadcast coverages, and so forth. For the determination of required transmitter power, for a specified grade of service, maps of worldwide distribution of radio noise and its variability are available in CCIR (1988). Ionospheric predictions are useful in planning systems, selecting frequencies, and assessing interference (intentional and otherwise) between systems. They are valuable in the overall frequency management of the high-frequency band. Ionospheric data are of value to (i) short-term operators who have to react to circuit interruption because of MUF failure, high absorption, excessive fading, and (ii) long-term system planners and frequency managers who are involved with the installation of terminal equipment, antenna design, compatibility and cost.

This Chapter is concerned, primarily, with the second category since the first has been dealt with, to some extent, in Secs. 9.7 and 9.8. We also consider real-time channel assessment in which probing signals (e.g. oblique soundings) are used to evaluate the channel capacity. Such evaluations are of particular use to traffic controllers and circuit operators (e.g. Radio Spectrum Ultilization 1964). We consider predictable characteristics, their temporal and spatial variabilities, some specific prediction systems, and applications to communications.

12.2 Predictable characteristics

12.2.1 Long-term indices

The various solar indices for calculating long-term variations of ionospheric characteristics (e.g. R_{12}, $F_{10 \cdot 7}$) have been discussed in Chapter 2. Liu *et al.* (1983) derived an ionospheric index that is independent of the sunspot number and is designated as *IG*. It is an ionospherically based index for use with the existing data in CCIR Report 340–5 (CCIR 1986*a*). Noontime monthly median values of f_oF_2 observed at 13 middle-latitude stations throughout the world are used to determine what value of the sunspot number in Report 340 is needed to produce the observed values. The value of the sunspot number thus determined is designated as the value of *IG* for the month in question. The 12-month smoothed value of *IG* is denoted by IG_{12}.

The CCIR has recommended (CCIR 1986*a*, Recommendation 371–5) that the 12-month smoothed sunspot number R_{12}, or the 12-month smoothed value of the ionospheric index IG_{12}, or the index $F_{10\cdot7}$ be used for predicting median values of f_oF_2 and $M(3000)F_2$ up to 12 months ahead of the last observed values. Further, the CCIR has recommended that the index $F_{10\cdot7}$ be used to predict monthly median values of f_oE and f_oF_1 up to 12 months ahead of the last observed values. In Australia, the T-index is used for predicting all the median values of all the ionospheric parameters (Wilkinson 1982).

Observed and predicted sunspot numbers are published every month in *Solar-Geophysical Data* prompt reports. Such predictions are given in Table 12.1 for cycle 22.

The CCIR recommends the use of R_{12}, IG_{12}, or the index $F_{10\cdot7}$ for less-than-12-month predictions of monthly median f_oF_2 and $M(3000)F_2$. It also recommends that $F_{10\cdot7}$ be used for the prediction of f_oE and f_oF_1. The 12-month running means are closely correlated, and any two can be determined if the third is known (CCIR 1986*a*, Recommendation 371–5).

$$F_{10\cdot7} = 63\cdot7 + 0\cdot728R_{12} + 8\cdot9 \times 10^{-4}R_{12}^2 \tag{12.1}$$

$$R_{12} = 11\cdot44 + 0\cdot478\ IF_2 + 0\cdot00278\ IF_2^2 \tag{12.2}$$

$$IF_2 = (2\cdot05 + 0\cdot001\ R_{12})R_{12}^{(0\cdot946 - 0\cdot00047\ R_{12})} - 20 \tag{12.3}$$

IF_2 is defined in Sec. 5.2.3.

12.2.2 Prediction of ionospheric characteristics

12.2.2.1 Critical frequencies of the E and F_1 layers

The prediction of monthly median values of the critical frequencies of the E and F_1 layers, for a given index, is based on the formulas given in Secs. 5.2.2 and 5.2.3 or, alternatively, on the numerical maps of Leftin (1976) and Rosich and Jones (1973). For many practical applications (5.1) and (5.4) are sufficient. The critical frequency of sporadic E is highly variable and is best obtained from the numerical maps of f_oE_s (Leftin *et al.* 1968) and f_bE_s (Leftin and Ostrow 1969).

12.2.2.2 Critical frequencies of the F_2 layer

Models of the monthly median values of the critical frequencies of the F_2 layer are available in the form of numerical coefficients (CCIR 1986*a*, Report 340–5)

Table 12.1 *Smoothed (observed and predicted) sunspot numbers: cycles 21 and 22*

Year	Jan.	Feb.	Mar.	Apr.	May	Jun.	Jul.	Aug.	Sep.	Oct.	Nov.	Dec.
1980	164	163	161	159	156	155	153	150	150	150	148	143
1981	140	142	143	143	143	142	140	141	143	142	139	138
1982	137	133	129	124	120	117	115	109	101	96	95	95
1983	93	90	86	82	77	70	66	66	68	68	67	64
1984	60	56	53	50	48	46	44	40	34	29	25	22
1985	20	20	19	18	18	18	17	17	17	17	17	15
1986	14	13	13	14	14	14	14	13	12*	13	15	16
1987	18	20	22	24	26	28	31	35	39	44	47	51
1988	58	65	71	78	84	91	99	108	115	122	128	132
						(3)	(6)	(9)	(12)	(15)	(19)	(22)
1989	136	139	148	155	161	166	169	172	179	184	186	187
	(22)	(21)	(19)	(19)	(20)	(24)	(28)	(31)	(34)	(36)	(38)	(41)
1990	187	186	184	179	173	169	167	165	158	149	142	138
	(43)	(46)	(47)	(47)	(44)	(40)	(38)	(39)	(38)	(37)	(34)	(31)

* September 1986 marks the onset of Sunspot Cycle 22.

The numbers in parentheses are uncertainties in the predicted values (+ and −).

From *Solar-Geophysical Data*, December 1988.

which are defined in Sec. 5.3.3. Recently new sets of coefficients have been developed (Rush *et al.* 1984, 1989). M(3000)F_2 and the minimum virtual height of the F_2 layer are available also in the form of numerical coefficients (CCIR 1986*a*, Report 340–5). The degree of detail in a worldwide map of a given characteristic will depend on two things: (i) the spacing of the stations in the observation network, and (ii) the highest order of the harmonic coefficients used in the representation. The spacing of stations has never been uniform, and a number of stations have been operated for a limited time only. For reasons of economy the harmonic series has to be truncated, which results in smoothing of the maps and prevents representation of some features (e.g. troughs, auroral features) with strong horizontal gradients. Each hourly monthly median CCIR map of f_oF_2 is defined by 988 coefficients.

12.2.3 Prediction of system parameters

12.2.3.1 MUF and FOT

When the monthly median critical frequencies and heights [or $M(3000)F_2$] have been determined the maximum usable frequency (MUF) can be obtained by graphical methods (nomograms) or by parabolic layer theory (Sec. 6.6). The behaviors of the E and F_1 layers are such that the smaller M factor roughly compensates for the higher critical frequency of the F_1 layer, so the product (the MUF) is roughly the same. The E layer is usually the controlling layer for circuit distances up to about 2000 km, and the F layer for distances greater than 2000 km.

The F_2 layer determines the propagation over distances greater than 3000 km, and single-hop propagation of the low-angle ray is limited to about 4000 km. It has been found (empirically) that, as the distance of propagation is increased beyond the one-hop F_2 limit (about 4000 km), the propagation is not suddenly cut off. Some physical reasons for this (discussed in Secs 6.7 and 6.8) are ionospheric and ground scatter, the high-angle ray, and ionospheric tilts. Empirically, it has been found that propagation does not appear to fail until the ionosphere fails to support the propagation at one of two 'control points' on the great circle path at 2000 km from each end. This is true, within limits, over wide ranges of path lengths, geographical locations, and time. In the practical solution of determining the MUF for F_2-layer propagation, the control points 2000 km from each end are chosen. The MUF for the path is the lower of the two MUFs, for a distance of 4000 km, for each control point as the midpoint of a 4000 km path.

When it is desired to apply the procedure to the E, E_s, and F_1 layers, an additional pair of control points 1000 km from each end may be used. However, there is less justification for this course than is the case with F_2 propagation.

The f_oF_2 maps are usually given for 12-month smoothed sunspot numbers 0 and 100; the value for any given index is obtained by linear interpolation, or extrapolation except that, for sunspot numbers greater than 150, the f_oF_2 is put equal to the value for 150. For prediction purposes the zero distance MUF ($= f_x F_2 \cong f_o F_2 + 0{\cdot}5 f_H$) is used. The other quantity given in the prediction systems is the $M(3000)F_2$ together with suitable procedures for interpolation, for distances less than 3000 km, and extrapolation for distances between 3000 km and 4000 km. This can be done graphically by means of the nomogram in Fig. 12.1. From the F_2(ZERO)MUF and the $M(3000)F_2$ one can obtain F_2(3000)MUF = [F_2(ZERO)MUF − $0{\cdot}5 f_H$]* $M(3000)F_2$. Mark this value on the 3000 km line and draw a straight line between the zero distance MUF (left-hand side) and extrapolate to the 4000 km line. The $F_2(d)$MUF for any distance d less than 4000 km can be read off from the intersection of the constructed line with the distance d line and the right-hand scale. The M factors for E- and F_1- layer propagation can be obtained by use of parabolic layer theory.

The optimum working frequency (FOT from the French initials) is usually taken as 0·85 of the monthly median MUF. This ensures that the FOT is below the monthly median maximum frequency for 90% of the days of the month at the particular hour specified. Another frequency of interest is the HPF (highest probable frequency) that exceeds the MUF 90% of the number of days in the month. Originally, the term 'optimum frequency' was interpreted to mean that frequency on which the signal was the strongest. It was based on the fact that the signal strength is greater the closer the frequency is to the MUF. This has to be weighed against spectrum congestion. Thus use of a frequency lower than the FOT may be optimum if the lower signal strength is more than offset by the lower interference. However, the frequency on which the signal/noise is a maximum may not be optimum from the point of view of multipath (time spread) which can give rise to intersymbol interference.

12.2.3.2 Radiation angle

Radiation angle depends on the effective (specular) height of reflection and, if this is known or assumed, the elevation angles can be obtained from Figs. 6.36 and/or 12.2. The effective layer heights are included in the various computer prediction programs that are now used in place of the older graphical methods. For an effective reflection height h, the elevation angle Δ for a hop length d is

$$\Delta = \tan^{-1}[\cot(d/2a) - a \operatorname{cosec}(d/2a)/(a+h)] \tag{12.4}$$

where a is the Earth's radius.

12.2.4 Radio noise

In the absence of interference, noise is the limiting factor that determines whether the signal is usable for the transmission of information. In principle, if all noise could be eliminated from a system, only an infinitesimal transmitted power would be required and the amplification in the receiver could be unlimited. In the presence of noise, there is a limit to the amount of useful amplification because signal and noise are amplified to the same extent. Thus, in choosing a receiving site, it is important to have a knowledge of the ambient noise with

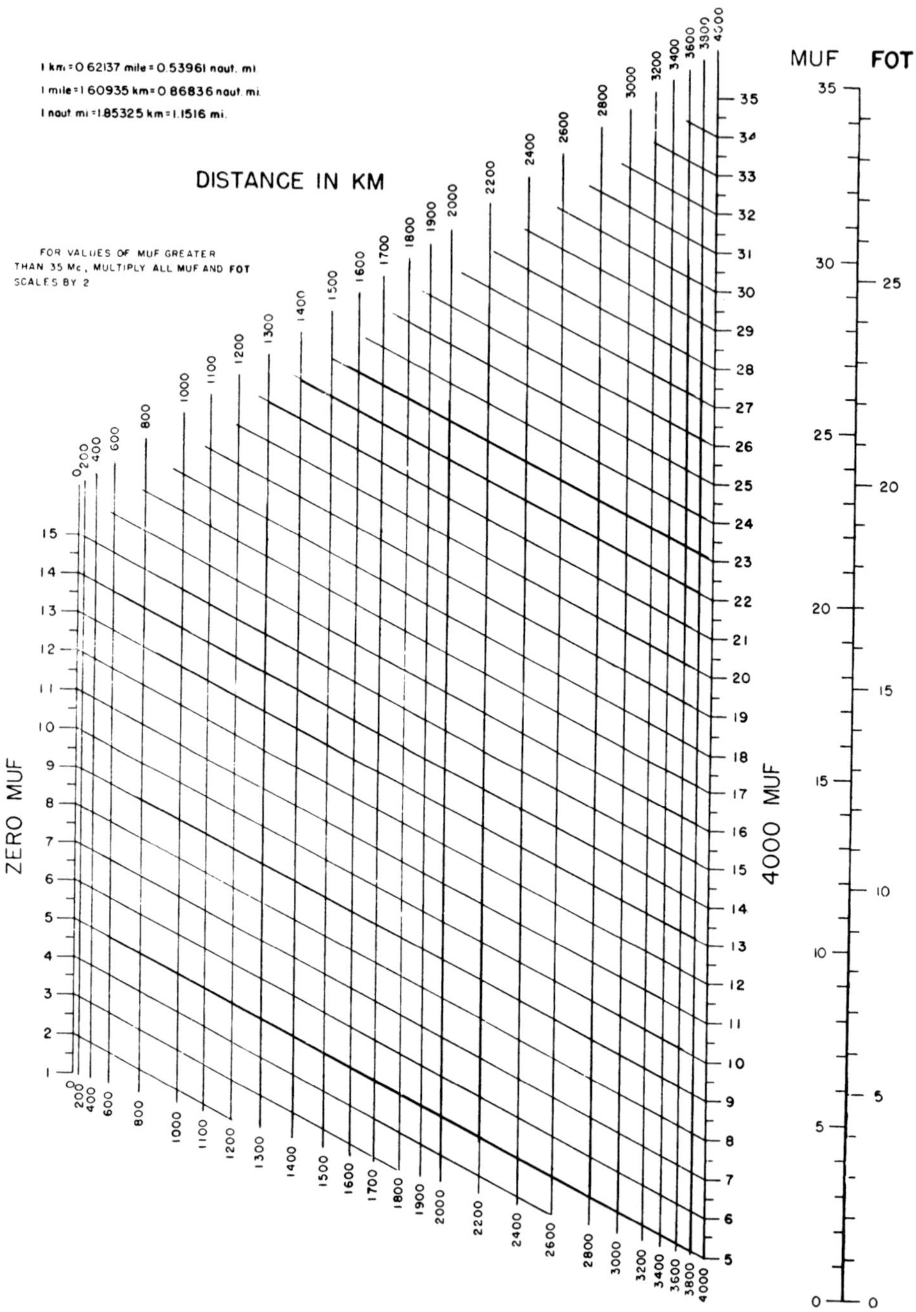

Fig. 12.1 *Nomogram for transforming F_2(ZERO)MUF and F_2(4000)MUF to MUF for intermediate distances*

which the desired signal must compete so that power sufficient to override the noise may be transmitted.

In general, system performance is dependent on the detailed statistical characteristics of both the signal and the noise as well as on the signal-to-noise ratio. There are several noise sources: (*a*) the noise internally generated by the receiving system, (*b*) natural noise, i.e. atmospheric and galactic, (*c*) unintentionally radiated man-made noise, and (*d*) intentionally radiated noise, e.g. jamming. Depending on frequency, time, and location, one of these noise sources may dominate and all need to be considered. As in the case of ionospheric characteristics the noise is expressed by means of numerical maps. The existing noise models are contained in CCIR (1988). See also Spaulding and Washburn (1985) on atmospheric noise and CCIR (1986*a*, Report 258–4) and Spaulding and Disney (1974) on man-made noise.

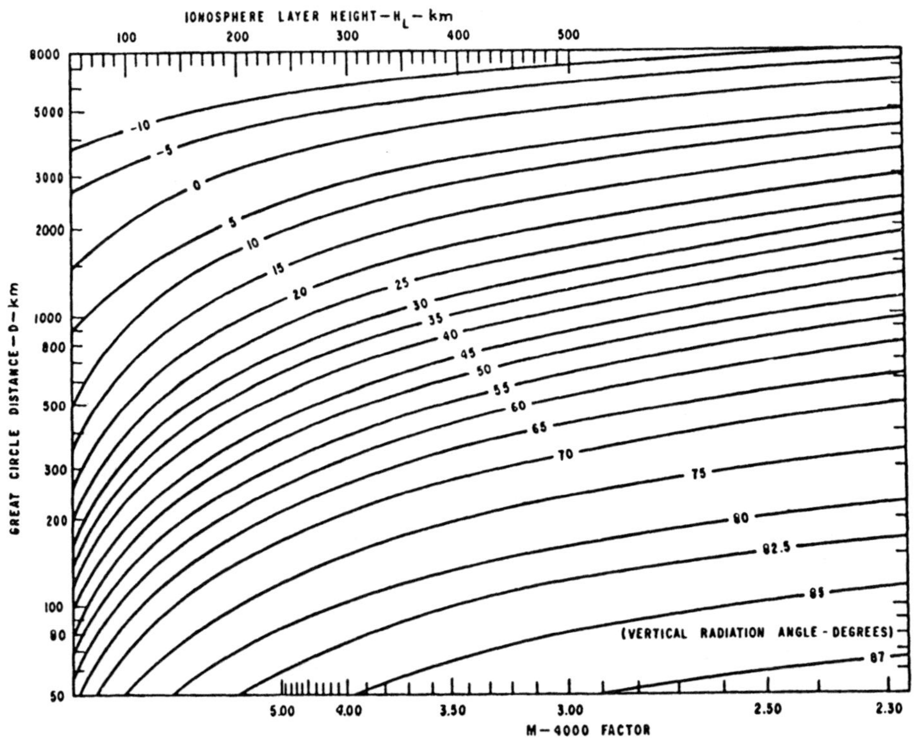

Fig. 12.2 *Radiation angles as a function of great-circle distance and ionospheric layer height*

The noise power is defined as the external noise power available from a lossless antenna. Let s represent the signal power and n the average noise power in watts. The signal-to-noise ratio (s/n) can be expressed in decibels:

$$10 \log(s/n) = 10 \log s - 10 \log n = S - N \quad (12.5)$$

We can define a receiving system operating noise factor f such that

$$n = fKT_0b \tag{12.6}$$

K is Boltzmann's constant, T_0 is a reference temperature taken as 288 K, and b is the noise power bandwidth in hertz. The operating noise figure F is defined as

$$F = 10 \log f \tag{12.7}$$

The total power P_n (in decibels above 1 W) available at the terminals of a lossless antenna is given by

$$P_n = F_a + B - 204. \tag{12.8}$$

$B = 10 \log b$, $10 \log KT_0 = -204$ (for further details see Spaulding and Stewart 1987) and

$$F_a = 10 \log(p_n/KT_0b) = 10 \log(T_a/T_0) \tag{12.9}$$

where T_a is an effective antenna temperature due to external noise.

The current noise maps give worldwide values of the median F_{am}, in each of the following 4 h local time blocks: 00–04, 04–08, 08–12, 12–16, 16–20, 20–24, as shown in Fig. 12.3. The maps give F_{am} for a frequency of 1 MHz; the value of F_{am} on another frequency f is given by 'frequency-law' curves such as those shown in Fig. 12.4*a*. The statistical distribution about the median is given by the upper, D_u, and lower, D_l, decile values, that is, the values exceeded 10% and 90% of the time. In addition, the corresponding season-time block values are given for the four seasons: winter (December, January, February), spring (March, April, May), summer (June, July, August) and autumn (September, October, November) and the reverse in the Southern Hemisphere. The sigmas represent a local variability and year-to-year variability. Sailors and Brown (1982; 1983) developed a simplified numerical noise map suitable for use on minicomputers. Fig. 12.5 shows man-made noise levels (Spaulding and Washburn 1985; CCIR 1988).

To illustrate the use of these noise data consider the case of F_a and its variation for the winter season, 1200–1600 time block, for Boulder, Colorado (40°N, 105°W) on 3 MHz. From Fig. 12.3, F_{am} on 1 MHz is 42 dB. From Fig. 12.4*a* this translates to 27 dB on 3 MHz. Hence the *probable* noise power on 3 MHz in an effective bandwidth of 1 Hz is 27 dB above the reference value of KT_0. On the assumption that the noise power is proportional to bandwidth we find that the noise power in a 6 kHz channel is $27 + 10 \log 6000 = 64.8$ dB above KT_0. Substitution in (12.8) gives $P_n = -139{\cdot}2$ dB above 1 W.

Note that the total uncertainty σ_T is the root-mean-square value of the individual uncertainties. For example, for the upper decile,

$$\sigma_T = (\sigma^2_{F_{am}} + \sigma^2_{D_u})^{1/2} \tag{12.10}$$

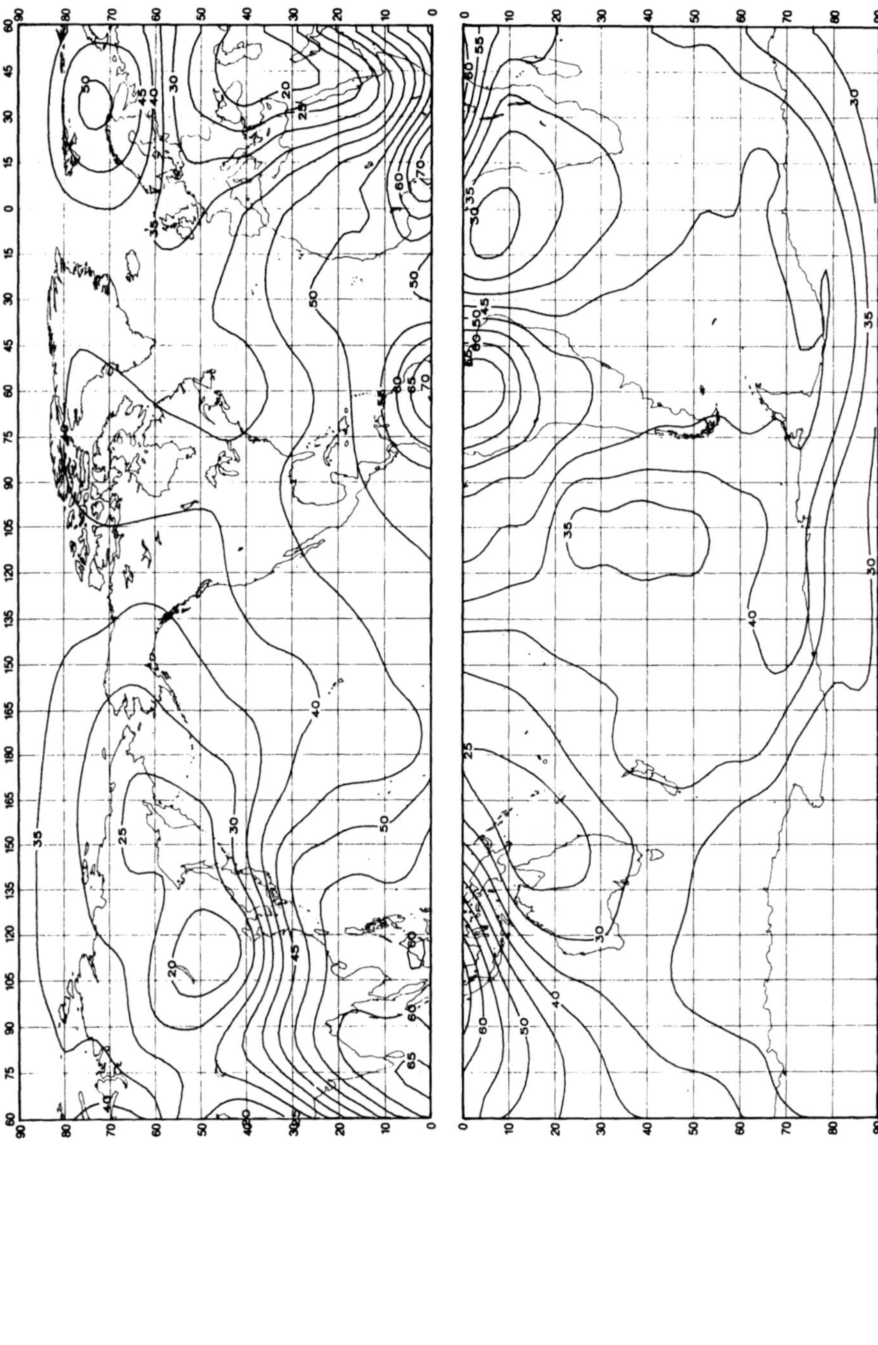

Fig. 12.3 *World map of atmospheric radio noise F_{am} (dB above KT_0b at 1 MHz) (From CCIR 1988)* Winter (December, January, February), 1200-1600 LT

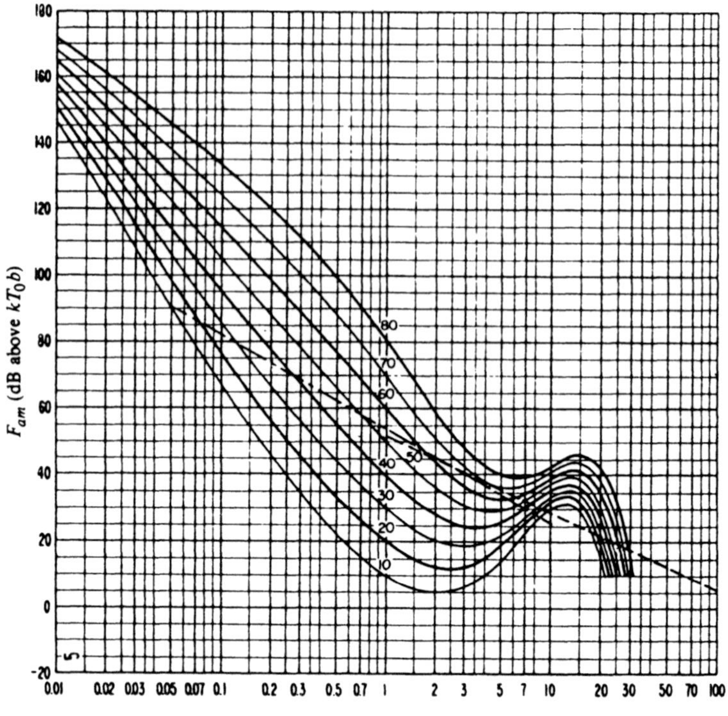

F_{am} (dB above kT_0b)
Frequency (MHz)
80
70
60
50
40
30
20
10
5

a

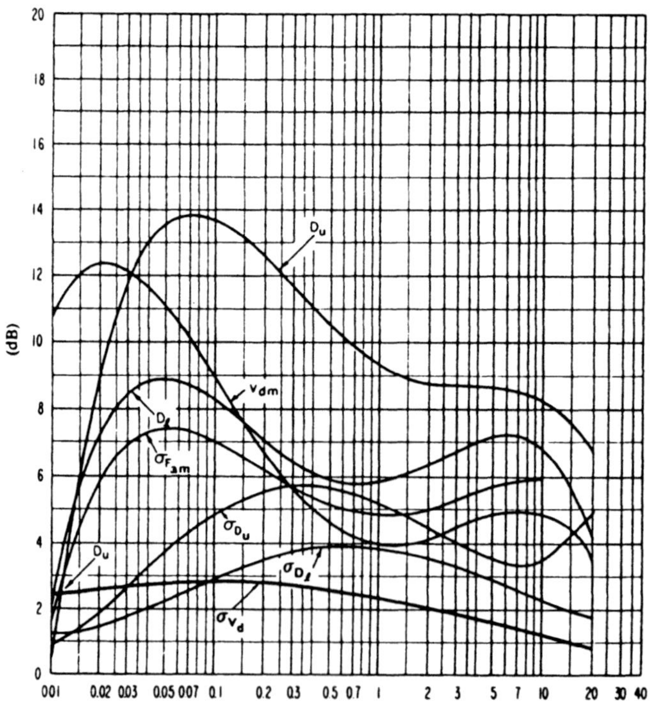

(dB)
Frequency (MHz)
D_u
V_{dm}
D_l
$\sigma_{F_{am}}$
σ_{D_u}
σ_{D_l}
σ_{V_d}
D_u

b

12.2.5 System performance

To measure the performance of a given system the following definitions are used: The skywave availability Q is defined as the fraction of the number of days that radio signals can propagate at a given hour over a given path (Maslin 1979). For example, on the median MUF, $Q = 0{\cdot}50$; on the FOT, $Q = 0{\cdot}90$; and on the HPF, $Q = 0{\cdot}10$. Circuit reliability ρ is defined as the fraction of days that successful communication may be expected at a given hour within the month on a specified operating frequency

$$\rho = QP \tag{12.11}$$

P is the probability that the mean received signal/noise ratio exceeds a specified level R. Service probability is the fraction of time that a specified grade of service (e.g. a specified signal/noise) or better is achieved.

The grade of service defines the quality of communication desired as, for example, the percentage of error-free messages in teletype transmitted or the percentage of satisfied customers of a given service. The required signal/noise ratio for a minimum acceptable grade of service is given in Barghausen *et al.* (1969) and also in CCIR (1986*b*, Report 525–2; 1986*c*, Recommendation 240–4).

The magnitude of ρ depends on two opposing effects: (i) P increases with increasing frequency – higher signal, lower noise; (ii) Q decreases with increasing frequency. The higher the frequency the lower the chance of ionosphere reflection.

12.2.6 System loss

To determine the total signal at the receiver terminals we need to include the following losses:

(*a*) Ground loss L_g (which occurs on reflection in multi-hop propagation)
(*b*) Distance loss due to spatial spreading of the wave, L_{sp}
(*c*) Ionospheric absorption L_a
(*d*) Fading and polarization losses: L_f, L_p
(*e*) Antenna gain

Fig. 12.4 *(a) Variation of radio noise with frequency; Winter, 1200–1600 LT*
—— Atmospheric noise;
–·–·– Man-made noise;
– – – Galactic noise
The numbers on the curves (10 to 80) are the values of F_{am} on 1 MHz
(b) Data on noise variability
(CCIR 1988)
$\sigma_{F_{am}}$ = standard deviation of values of F_{am}
D_u = ratio of upper decile to median value, F_{am}
σ_{D_u} = standard deviation of values of D_u
D_l = ratio of median value, F_{am}, to lower decile
σ_{D_l} = standard deviation of values of D_l
V_{d_m} = expected value of median deviation of average voltage
The values shown are for a bandwidth of 200 Hz.
σ_{V_d} = standard deviation of V_d.

(*a*) *Ground loss* depends on the conductivity of the ground and on the angle of elevation of the ray. The losses for a good conducting surface (e.g. sea water) are less than for a poorly conducting surface such as bare rock or desert. For two types of surfaces the loss L_g can be obtained, with a knowledge of the wave frequency and elevation angles (Fig. 12.6).

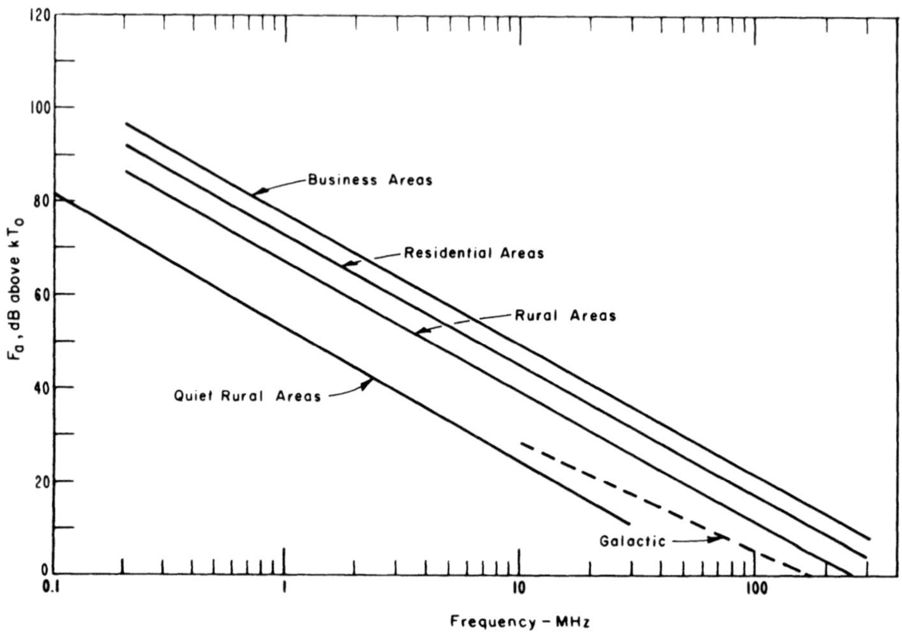

Fig. 12.5 *Median values of man-made noise power for a short vertical lossless grounded monopole antenna (From Spaulding and Disney 1974)*

(*b*) *The distance loss* is usually taken into account by equating the effective distance d_e (Sec. 7.3.1) with the geometrical distance because it is too complicated to take into account the detailed effects of focusing. For long-distance transmission ($\geqslant$ 3000 km) the effective distance is, essentially, the ground range.

(*c*) *Ionospheric absorption*: During the day nondeviative absorption in the D region is usually the dominant loss mechanism (see Sec. 7.4) except possibly for the high-angle modes. Some prediction techniques consider the detailed structure of the received signal in computing the total signal but others lump all paths and express the loss in a single expression (e.g. see Barghausen *et al.* 1969; Schultz and Gallet 1970):

$$L_a = \frac{286(1 + 0{\cdot}0087x_n)(1 + 0{\cdot}005R_{12})W\cos^m(\chi_{12})}{\nu^2/4\pi^2 + (f + f_H)^2} \cdot \frac{\cos(0{\cdot}893\chi)}{\cos(0{\cdot}893\chi_{12})} N\sec\phi \qquad (12.12a)$$

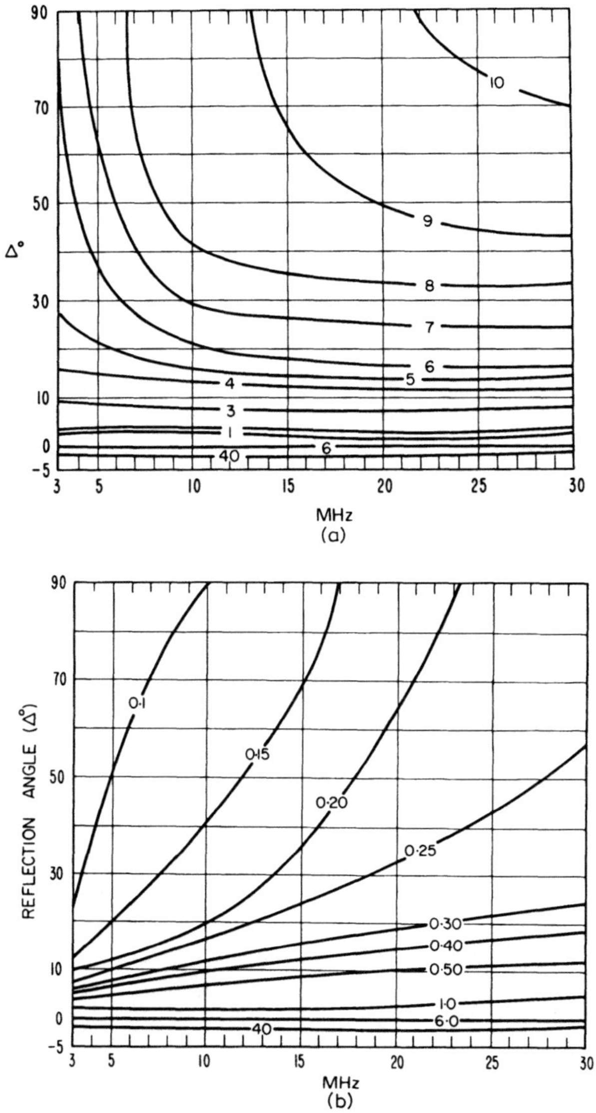

Fig. 12.6 *(a) Power loss (dB) at poorly conducting ground reflection with random polarization*
(b) Power loss (dB) at sea reflection with random polarization

where

$$m = 2{\cdot}25 - 0{\cdot}032\, x_n \quad (m \geqslant 1{\cdot}3) \qquad (12.12b)$$

x_n = geographic latitude measured positive from the equator
R_{12} = 12-month smoothed sunspot number
χ = solar zenith angle

χ_{12} = solar zenith angle at local noon
ϕ = angle of incidence at 100 km
ν = collision frequency
W = winter anomaly factor (Sec. 7.5.2)
N = number of hops

The quantity 0·893 accounts for the fact that the Sun rises earlier at 100 km than at the ground. At night the quantity

$$(1 + 0{\cdot}005\, R_{12}) \cos^{m}(\chi_{12}) \cos(0{\cdot}893\chi)/\cos(0{\cdot}893\,\chi_{12})$$

is set equal to 0·01 to agree with observations. Values of W are given in Barghausen *et al.* (1969, their Table 7.1).
(*d*) *Fading and polarization loss*: To allow for the day-to-day variations Barghausen *et al.* (1969) added an excess system loss (Y_p) that can exceed 20 dB. This excess loss takes into account effects of fading and polarization.
(*e*) *Antenna gain*: In order to determine the system loss, it is essential to know the gains of the transmitting and receiving antennas for various angles of elevation and, where appropriate, azimuth. This type of information can be obtained from the geometry of the antenna and its surroundings. However, in some of the existing prediction programs there are 'built-in' antennas that can be selected.

12.3 Propagation calculations

12.3.1 Distances of less than 4000 km

The average height of the F_2 layer (330 km) corresponds to a one-hop distance of about 4000 km. Hence, this distance provides a convenient dividing line between long-distance and intermediate-distance transmission. In this Section we consider an example to illustrate the various steps that are followed in the determination of optimum frequencies, required transmitter powers, etc. Since different prediction systems use slightly different procedures and different formulas, the following will not agree with detailed calculations from any of the various computer programs.

Problem 1
Find the optimum working frequency for a path between Philadelphia (40°N, 75°W) and Boulder, Colorado (40°N, 105°W) at 1800 UT in December 1986.
Step 1: From spherical geometry the midpoint is 41°31′N, 90°W.

The path distance d between a transmitter at latitude ϕ_T and longitude λ_T and a receiver at latitude ϕ_R and longitude λ_R and the midpoint coordinates are obtained from the spherical geometry of Fig. 12.7:

$$\cos(d/a) = \sin\phi_T \sin\phi_R + \cos\phi_T \cos\phi_R \cos(\lambda_R - \lambda_T) \qquad (12.13)$$

The bearing β_R of the receiver from the transmitter and the bearing β_T of the transmitter from the receiver are obtained from

$$\frac{\sin(PT/a)}{\sin\beta_T} = \frac{\sin(PR/a)}{\sin\beta_R} = \frac{\sin(d/a)}{\sin(\lambda_R - \lambda_T)} \tag{12.14}$$

The geographical coordinates (ϕ_M, λ_M) of the midpoint M are obtained from spherical triangle PTM:

$$\sin\phi_M = \sin\phi_T\cos(d/2a) + \cos\phi_T\sin(d/2a)\cos\beta_R \tag{12.15}$$

$$\lambda_M = \lambda_T + \sin^{-1}\left[\frac{\sin(d/2a)\sin\beta_R}{\cos\phi_M}\right] \tag{12.16}$$

and the bearing β_M of the path TR at M is

$$\beta_M = \sin^{-1}\left(\frac{\cos\phi_T\sin\beta_R}{\cos\phi_M}\right) \tag{12.17}$$

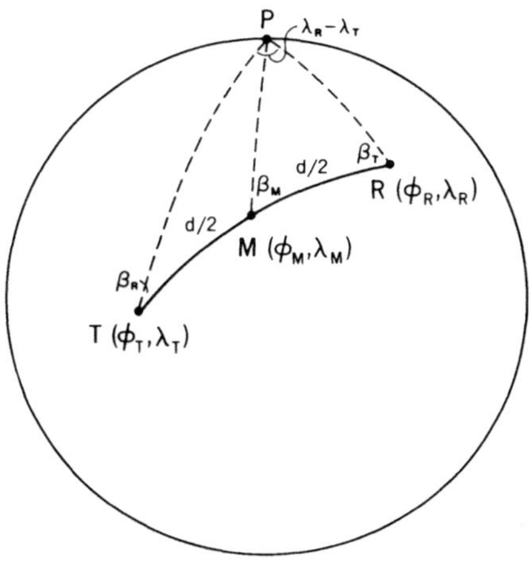

Fig. 12.7 *Spherical geometry for determination of path parameters*

Note that it is customary to define bearing as east of north. Thus the bearing of T from R is $(360 - \beta_T)$.
Step 2: Circuit length = 2540 km, elevation = 4°
Step 3: From *Solar-Geophysical Data* (1987) we have $R_{12} = 10$
Step 4: From numerical maps $f_x F_2$ = 6·8 MHz, MUF(4000)F_2 = 22 MHz
Step 5: By interpolation (Fig. 12.1) MUF (2540)F_2 = 18·0 MHz
Step 6: Solar zenith angle χ at the midpoint is obtained from (2.48), $\chi = 68°$
Step 7: From (5.1) f_oE = 2·9 MHz

Step 8: Angle of elevation for 1-hop $E < 0$
Step 9: For two-hop E, MUF(1270) $\approx$ 14 MHz
Step 10: E cutoff frequency $\approx$ 14 MHz
Step 11: FOT(F_2) = 0·85 × 18·0 = 15·3 MHz (From Step 5); FOT(E) = 14 MHz
Step 12: Check sporadic E, f_oE_s = 2·5 MHz from numerical map.
Therefore, $f_oE_s < f_oE$ and regular E is the more important, and the one-hop F_2 path will dominate, so FOT = 15·3 MHz.

Problem 2
Find the transmitter power required to ensure performance of an A3 telephony double-sideband system with a 6 kHz bandwidth and a carrier on the above FOT (15·3 MHz). The required grade of service is marginally commercial for 90% of the hour with half-wave horizontal dipoles (receiver and transmitter) at a height of 2 m (required carrier/noise = 27 dB).

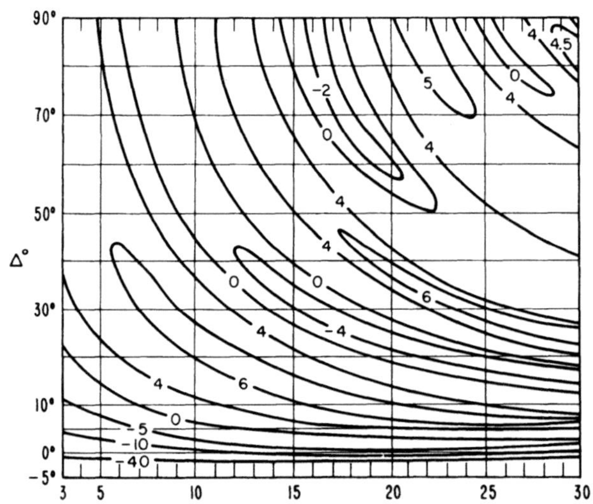

Fig. 12.8 *Space-wave radiation of a half-wave horizontal antenna 20 m above poorly conducting Earth*

To determine the required power it is necessary to find the system loss (path loss − antenna gains) and the ambient noise.
Step 13: From a numerical map (e.g. Fig. 2.10) $f_H \approx$ 1·5 MHz
Step 14: Determine ionospheric absorption (12.12)

$\phi(100)$ = 80° (from geometry)
W = 1·28 (Barghausen *et al.* 1969)
x_n = 41°, $\chi_{12} = \chi$ = 68°, N = 1
L_a = 4·3 dB

Step 15: Distance loss (geometry) $\approx$ 126 dB

Step 16: Total path loss = 130·3 dB
Step 17: One antenna gain (see Fig. 12.8) = −5 dB
Two antenna gain = −10 dB
Step 18: Excess system loss (Barghausen *et al.* 1969) = 9 dB
Step 19: Net system loss L_s = 130·3 + 10 + 9 = 149·3 dB
Step 20: Noise power (from maps and graphs such as Figs. 12.3 and 12.4)
On 1 MHz $F_{am} \approx$ 42 dB (CCIR 1988)
On 15·3 MHz $F_{am} \approx$ 37 dB (CCIR 1988)
For 90% operation we add an upper decile D_u = 9 dB and an allowance for variability of D_u, σ_{D_u} = 4 dB (Fig. 12.4*b*)
F_{au} = 37 + 9 + 4 = 50 dB
From (12.8) for a 6 kHz band
P_n = 50 + 37·8 − 204 = −116·2 dB
Step 21: Required receiver power:
P_r = 27 − 116·2 = −89·2 dB
Step 22: Required transmitter power
$P_t = P_r + L_s$ = −89·2 + 149·3 = 60·1 dB
That is, P_t = 1·0 MW

12.3.2 Distances between 4000 km and 9000 km

The procedures for the 4000–9000 km range are similar to those given above, but there are a few differences. For example:

(i) The MUF is determined at control points (or areas) rather than at the path midpoint.
(ii) Allowance has to be made for reflection losses at the Earth's surface.
(iii) Higher order paths have to be considered in detail.

For example, consider the circuit London (51·5°N, 0°W) to Washington, DC (38·9°N, 77°W). The relevant path data are from Sec. 12.3.1:

Path length	5900 km
Midpoint	52°N, 42°W
Midpoint gyrofrequency	1·4 MHz
Control point *A*:	55°N, 30°W
Control point *B*:	50°N, 57°W.

At 1200 UT in December 1986 at control points *A* and *B*, the zero distance MUF (for R_{12} = 10) is obtained by interpolation between values for R_{12} = 0 and R_{12} = 100. The zero-distance MUFs are, at *A*, 4·6 MHz; at *B*, 4·2 MHz. We choose the lower value at *B* (i.e. 4·2 MHz). The MUF(4000)F_2 at R_{12} = 10 is obtained from numerical maps by interpolation, which gives 16·5 MHz. Hence

$$\text{FOT} = 0{\cdot}85 \times 16{\cdot}5 = 14\,\text{MHz}$$

In calculating the absorption from (12.12*a*) it is sufficiently accurate to determine the absorption corresponding to the values of χ, χ_{12}, x_n, at the midpoint of the path (χ = 75°).

The next step is to determine the dominant propagation paths (e.g. 2F, 3F, 3E). We neglect high-angle rays and scatter propagation because these paths tend to be unreliable. We now find the maximum elevation angle for reflection from the E layer and the minimum F-layer angle Δ that can penetrate the E layer. These considerations show that

(*a*) 2F and 3F are possible
(*b*) 3E and 4E are impossible because signals penetrate E
(*c*) 4F is impossible because signal penetrates F_2.

The system loss and the noise power (at Washington) are found as in the previous example. A surface reflection loss must be included which is of the order of 1 dB for both 2F and 3F paths.

Effective frequency = 14 + 1·4 = 15·4 MHz

	2F	3F	
Radiation angle	2°	8°	
Angle with vertical at 100 km	79°·7	77°·15	
Absorption (dB)	4·5	5·4	
Ground reflection loss (dB)	1·0	1·0	(Fig. 12.6*b*)
Distance loss (dB)	130	130	
Path loss (dB)	135·5	136·4	
Excess system loss	9	9	
Net antenna loss ($R_x + T_x$)	10	−8	
Total system loss	154·5	137·4	
Noise power at Washington, DC, LT = 0700 ($F_{am} \approx 30$ dB, $D_u = 9$, $\sigma = 6{\cdot}7$)	−118·5	−118·5	(CCIR 1988).

Because of the antenna performance, the total system loss is much greater on the 2F hop than on the 3F hop. We can, therefore, ignore the 2F path. The 3F hop is called the most reliable mode (MRM).

Required signal/noise	=	30 dB
Required receiver power	=	30 − 118·5 = −88·5 dB
Required transmitter power	=	−88·5 + 137·4 = 48·9 dB above 1 W
	=	78 kW

Alternatively, we could start with a given transmitter power, operating frequency, and angle of elevation and calculate the path length and, knowing the antenna characteristics at the transmitter and receiver, determine the power delivered at the terminals of the receiver. From the diurnal variation of the MUF, FOT, LUF, and noise level, the hours during which an acceptable signal is available can be calculated. Fig. 12·9 shows contours of signal strengths for a given transmitter location and effective radiated power. The signal does not, in general, drop to zero as the frequency f increases from just below to just above the MUF. Relative to the loss at the MUF, the loss L_m (in dB) on a higher frequency f is (Bradley 1979):

$$L_m = \left(130 \frac{f}{\mathrm{MUF}} - 1\right)^2 \tag{12.18}$$

Hence, it may be possible to use a frequency greater than the MUF.

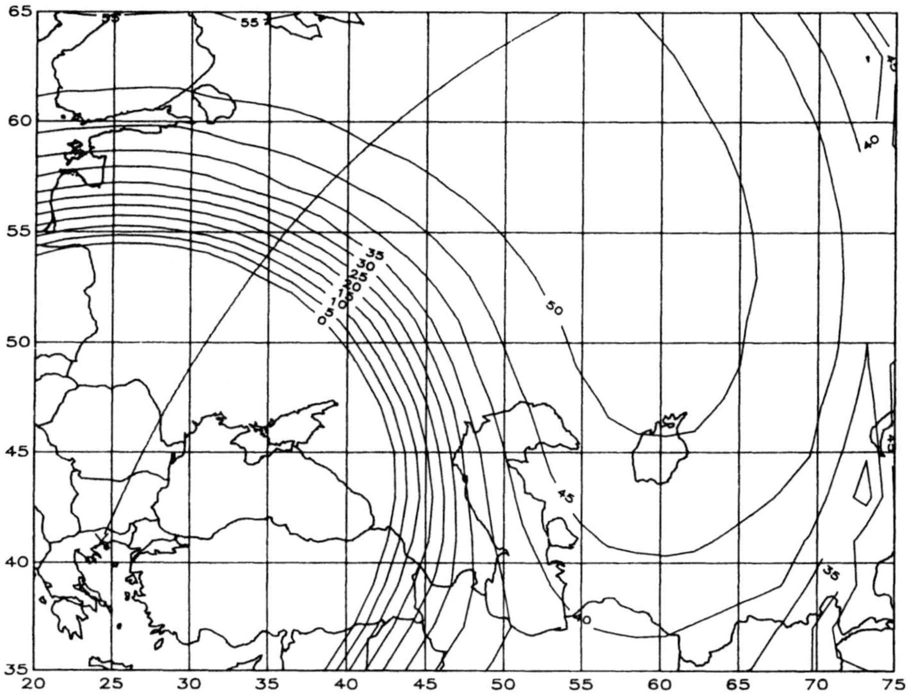

Fig. 12.9 *Calculated contours of the field strength (in dBu, i.e., dB above 1 μV) for a 250 kW transmitter on 11·855 MHz at Kavalla, Greece, at 2000 UT for March 1985 (From Rush 1986)*

12.3.3 Distances greater than 9000 km

Over distances greater than 9000 km the echo structure becomes very complicated and such paths as chordal paths, M paths, and scatter paths become relatively important. Various methods have been developed to treat the case of propagation paths at these distances (Barghausen *et al.* 1969; Teters *et al.* 1983; CCIR 1986*d*). Most of the methods involve calculating the signal strength by dividing the path into a minimum number of hops, none of which exceeds 4000 km. Some methods, however, treat the propagation calculation in a purely empirical fashion (CCIR 1986*a*, Report 894–1, for example).

12.4 Choice of operating frequency

12.4.1 Optimum working frequency (FOT)

The optimum working frequency (0·85 MUF) can be calculated for each hour of the day by the methods described in Sec. 12.2 and Sec. 12.3. In principle, it would be desirable to change the frequency slowly throughout the day in order to make use of the optimum conditions. However, in practice this is both inconvenient and, in most cases, impossible because of international methods of allocating frequencies. In general, a user is allocated only two or three frequencies for use on a given circuit. A knowledge of the diurnal variation of the FOT will, however, aid the circuit operator in anticipating circuit failure due to penetration of the ionosphere by the signal. Such practice ensures that the operating frequency is less than the FOT, and therefore the frequency will rarely be 'optimum'.

12.4.2 Lowest useful frequency

For a given transmitter output power, as the operating frequency is decreased, the available power at the receiver normally decreases, owing to increased ionospheric absorption. Furthermore, the noise power increases, so the signal-to-noise ratio deteriorates and the circuit reliability decreases. The minimum frequency, below which the reliability is unacceptable, is called the lowest useful (high) frequency (LUF). The LUF depends upon transmitter power, the factors that determine the path loss (e.g. frequency, season, geographical location), and the noise level (mainly frequency and receiver location). One of the main factors is ionospheric absorption, and hence, since this normally varies with solar zenith angle, the LUF peaks around noon. Consequently, it is necessary to ascertain if a selected frequency exceeds the LUF.

12.4.3 Operating frequency

The diurnal variations of the FOT and the LUF for a circuit between the Canary Islands and London in a December, when the average sunspot number was 29, are shown in Fig. 12.10. During the night, it is necessary to use a frequency of about 6 MHz to ensure that the operating frequency does not exceed the FOT by an appreciable amount for too long a time. This frequency band cannot be used during the day because it lies below the LUF. During the day it would be better to use a frequency of about 15 MHz.

Occasionally the MUF may drop to an exceptionally low value, as during a magnetic disturbance, or increased D-region absorption will raise the LUF, as during a shortwave fadeout. Under these circumstances even the FOT may be unusable. Also, near dawn it is rather difficult to communicate, especially on two-hop paths where one of the paths is in darkness (low MUF) and the other in daylight (high LUF). The best propagation is usually shortly after sunset when the D-region absorption drops quickly but the F_2 MUFs drop slowly.

12.5 Ionospheric prediction programs

The principles discussed in the preceding Section have been consolidated into comprehensive computer programs, which have replaced the older graphical methods. One of the most widely used programs is the Ionospheric Communications Analysis and Prediction (IONCAP) program developed at the US National Telecommunications and Information Administration (Teters *et al.* 1983). This program consists of a set of subroutines for the prediction of HF skywave system performance intended for the planning and operation of HF communication systems, and it is based on previous systems (e.g. Barghausen *et al.* 1969; Lucas *et al.* 1972; Headrick and Skolnik 1974). The propagation

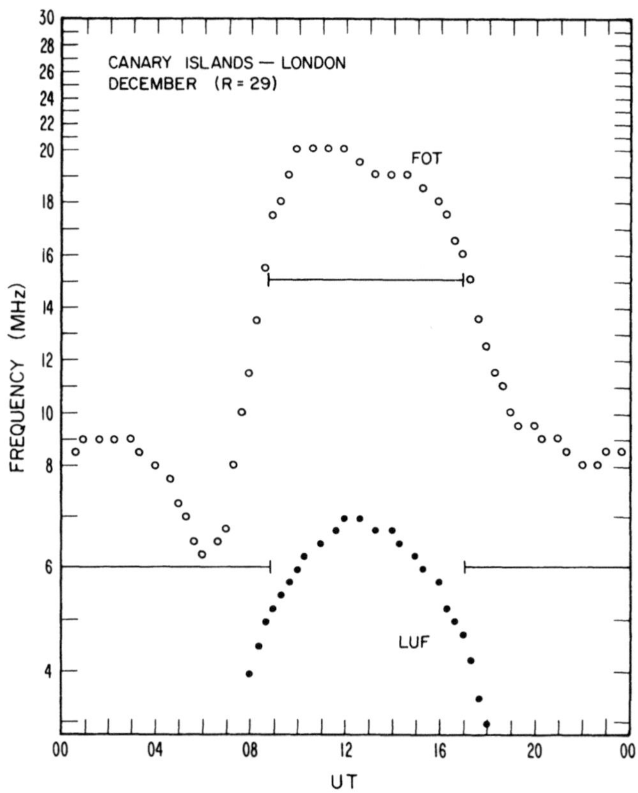

Fig. 12.10 *Sample diurnal variation of the optimum working frequency (FOT) and the lowest useful (high) frequency LUF*

includes refractive bending, scattering on frequencies above the MUF, and sporadic-E propagation. The predicted signal strength and noise environment can be used to select optimum frequencies, proper antennas, required transmitter powers, optimum times of operation, and broadcast coverages, as a function of time and/or frequency.

For a given circuit the following inputs may be given: time of day, month, year, sunspot number, antenna types, geographical locations of transmitter and receiver. The hour of the day can be either universal time or local time at the transmitter. The user can choose either the shorter or longer of the great circle paths – normally the short path is required (default value). Additional information may include man-made noise level, minimum takeoff angle, required circuit reliabilities, required signal-to-noise ratio, the maximum time spread for multipath (default value = 0·85 ms). For system performance calculations the allocated frequencies (up to 11) may be given.

Normally, the ionospheric parameters are obtained from the permanent data base, but for specific cases external data bases can be used for parameters such as the critical frequencies, semi-thicknesses, heights of maximum ionization, and/or true height profiles.

The antenna package in IONCAP includes, for example, (i) horizontal rhombic, (ii) vertical monopole, (iii) horizontal dipole, (iv) horizontal yagi, (v) vertical log periodic, (vi) curtain, (vii) sloping vee, (viii) inverted L, and (ix) sloping rhombic.

Required signal/noise ratios for certain radiotelephone and radioteletype services are given by Teters *et al.* (1983).

The sequence of calculations is essentially as follows:

(i) Number of hops
(ii) Angle of takeoff at transmitter and angle of elevation at the receiver
(iii) Time delay of the most reliable mode (MRM) (ms)
(iv) The virtual height of the MRM (km)
(v) Probability that the operating frequency exceeds the predicted MUF
(vi) Median system loss of the MRM (dB)
(vii) Signal strength, median (dB above 1 μV/m)
(viii) Signal power, median (dBW)
(ix) Noise power, median (dBW)
(x) Median signal/noise ratio (dB)
(xi) Lowest useful frequency (LUF).

The IONCAP program is now available for use with personal computers. Several other prediction programs are available for special purposes as, for example, that for HF broadcasting services (Rush 1986). For distances greater than 9000 km the field strength model of the Deutche Bundespost (FTZ) is used (Dambolt 1976). This program uses an empirical method for overall signal strength and avoids the calculations for individual paths. Another such program is the HFBC 84 method developed specifically for planning the use of the HF spectrum (CCIR 1986*a*, Recommendation 621). Programs such as MINIMUF (Sailors *et al.* 1986) have also been developed specifically for personal computers, but these have limited accuracy (Cole 1985). Prediction techniques can be applied to calculate the coverage of an HF/VHF broadcasting satellite service (Rush *et al.* 1986).

Although programs such as IONCAP are global, greater accuracy can be expected by concentrating on specific regions.

12.6 Short-term predictions

12.6.1 Short-term updating

Attempts have been made, with limited success, to update the prediction programs so as to predict a few days in advance. For example, in a method described by Bradley (1976) only standard vertical data (f_oF_2; f_oE; $M(3000)F_2$; $h'F_2$) are used, and therefore the method is independent of monthly median values. Other techniques use running means for a few (e.g. 5) days to extrapolate into the next few days.

In the program by Barghausen *et al.* (1969), short-term modification is achieved by means of a predicted magnetic K index. The percentage of dependence of the MUF on K is given by the equation

$$\mathrm{MUF}(K) = P \cdot \mathrm{MUF}(0)/100 \tag{12.19a}$$

where

$$P = P_0 + bK \tag{12.19b}$$

Here MUF(K) is the MUF for a magnetic index K, and P_0 and b depend on season, sunspot number, latitude, and local time, and are tabulated. The intercept P_0 is about 100, and b varies between about -13 and $+20$, with positive values in the evening and in low latitudes, and negative values during the morning and in high latitudes.

A difficulty with the short-term updating is that it is based on the assumption that the modification is superimposed on the monthly average value, whereas in general the daily map may bear little resemblance to the monthly median map. Thus caution must be exercised in updating the long-term predictions. Also, the spatial correlation coefficient of the day-to-day departures of f_oF_2 from the median values falls below 0·5 for distances beyond about 3000 km (e.g. Rush *et al.* 1974) so it may be inaccurate to predict the state of the ionosphere at one point from data for another location.

12.6.2 Short-term variability

The prediction of the existence of a particular path (or mode) on the basis of median values does not guarantee the existence of that mode on a particular day. For example, on an operating frequency equal to the median, for a single-hop path, the channel will be available for 50% of the time only. The fraction of the time (on a given hour, say) that a path is available is called the mode availability Q. As a percentage of time it is given by the following:

For $f \leqq f_m$

$$Q = 130 - \frac{80}{1 + \dfrac{1 - f/f_m}{1 - F_l}} \tag{12.20}$$

or 100, whichever is smaller.

For $f > f_m$

$$Q = \frac{80}{1 + \dfrac{\dfrac{f}{f_m} - 1}{F_u - 1}} - 30 \tag{12.21}$$

or 0, whichever is larger. In (12.20) and (12.21)

$$F_u = f_{D_u}/f_m, \quad \text{and} \quad F_l = f_{D_l}/f_m$$

where

f_{D_u} is the upper decile (frequency exceeded 10% of time)
f_{D_l} is the lower decile (frequency exceeded 90% of time).

Although both height variations and critical frequency variations contribute to the mode variability, the critical frequency changes are dominant. For E-layer modes the following values are useful: $F_u = 1{\cdot}05$ and $F_l = 0{\cdot}95$. For the F_2 layer, F_u and F_l are functions of sunspot number, latitude, and local time. Some values of F_u and F_l in middle latitudes (35° to 55°) are given in Table 12.2.

12.7 Real-time channel evaluation

12.7.1 Operational requirements

Long-term prediction programs are particularly useful in planning circuit requirements (transmitters, antennas, etc.) and allocating frequencies. On the other hand, for communication operations they have limited value, primarily because of the high day-to-day variations in ionospheric characteristics. For operational uses real-time evaluation systems have been developed (CCIR 1986*a*, Report 889–1).

12.7.2. Broadband sounding

The earliest of the real-time evaluation systems employed oblique sweep frequency sounders in parallel with a communication circuit. Oblique ionograms (see Sec. 6.7) for the particular path are inspected for the maximum observed frequency (MOF), the lowest observed frequency (LOF), signal strength, time spreading (multipath) and, possibly, noise and interference. The optimum frequency can be found by inspection, but the specific frequency must be selected from among the allocated channels. An allocated channel may not be optimum.

Table 12.2 *MUF variability parameters F_u and F_l for latitudes between 35° and 55°, for two ranges of sunspot numbers**

LT	$R < 50$						$50 < R < 100$					
	Winter		Summer		Equinox		Winter		Summer		Equinox	
	F_u	F_l	F_u	F_l	F_u	F_l	F_u	F_l	F_u	F_l	F_u	F_l
22–02	1·24	0·80	1·17	0·77	1·24	0·76	1·28	0·84	1·17	0·80	1·25	0·74
02–06	1·25	0·79	1·23	0·85	1·20	0·81	1·18	0·82	1·21	0·80	1·16	0·78
06–10	1·15	0·84	1·16	0·81	1·20	0·81	1·13	0·84	1·22	0·76	1·17	0·82
10–14	1·12	0·86	1·17	0·83	1·17	0·77	1·15	0·84	1·18	0·78	1·20	0·76
14–18	1·12	0·87	1·15	0·84	1·16	0·81	1·16	0·86	1·16	0·80	1·13	0·76
18–22	1·24	0·77	1·16	0·85	1·26	0·77	1·30	0·79	1·17	0·80	1·19	0·76

* After CCIR (1980)

Referring to Fig. 12.11, note that the signal of minimum multipath may be a high-angle path, which may be unstable. Furthermore, there may be frequency ranges (e.g. near 27 MHz in Fig. 12.11) in which there are no echoes, and, in general, multipath increases with decrease of frequency before dropping just above the LOF. Thus, for frequency selection, amplitude (energy) information is required as illustrated in Fig. 12.12.

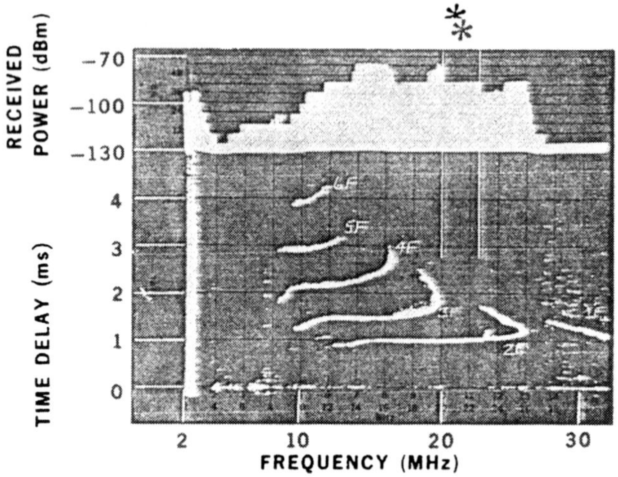

Fig. 12.11 *Information for frequency selection, obtained by using real-time oblique sweep frequency sounding (From Barry and Fenwick 1976, with permission)*

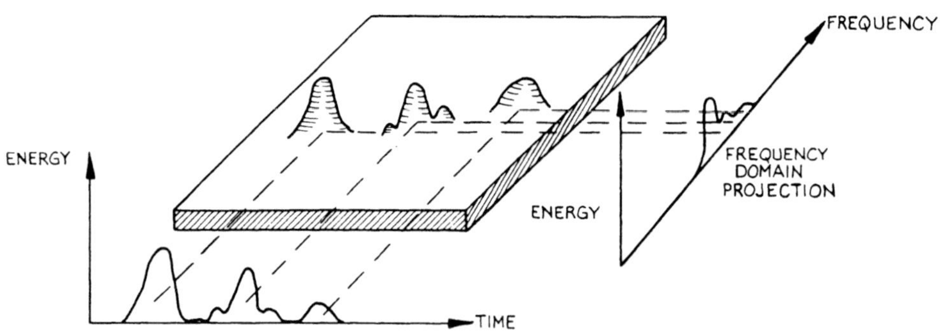

Fig. 12.12 *Time-delay, amplitude, and frequency information available from an oblique ionogram (From Darnell 1976, Fig. 1, with permission)*

The most critical aspect of such a test system is the interval between measurements. If the circuit characteristics are measured too infrequently, changing ionospheric conditions can catch the operator unaware. On the other hand, when the measurements are frequent, the system generates excessive interference. With large communication systems, networks of oblique sounders may be necessary (Probst 1968) for acceptable frequency management.

12.7.3 Channel sounding

An alternative to sounding over the entire HF band is to sound on the allocated frequencies only. This has the distinct advantage of limiting the interference to one's own channels. Although this technique does not provide the detailed information provided by the oblique ionosonde, the equipment is simpler, installation costs are lower, and the soundings are made on frequencies of interest (Stevens 1968). When the optimum channel is identified a message is dispatched to the transmitter site, which can commence transmission. This technique of channel evaluation assumes reciprocity; that is, the base-to-remote path is an adequate representation of the remote-to-base path. This is usually a good assumption, provided allowance is made for differences in antenna and noise characteristics at the two sites (Goodman 1987, Session 1).

12.7.4 Adaptive systems

In order to capitalize on a real-time sounding assessment it is essential to have a communication system that can respond sufficiently rapidly to changes in the state of the ionosphere. Such a system needs (*a*) fast frequency response, (*b*) rapid antenna changes, (*c*) ability to adjust rates of information and transmission (Fenwick and Woodhouse 1979).

12.7.5 Adaptive frequency management

With evaluation techniques and adaptive systems it is possible to maximize usage of the HF spectrum by the largest number of users. It is possible to replace some of the older approaches to ensuring communications. For example: (*a*) Increase transmitter power and, hence, interference, and (*b*) transmit on all possible frequencies simultaneously.

The term 'channel estimation' (Darnell 1976) is used to describe the process of monitoring and measuring the relevant parameters of a communication channel with the aim of describing quantitatively the absolute or relative states of a given set of channels in order to optimize the use of the dispersive, time-varying HF propagation medium. Real-time adaptive systems are of particular value when one of the terminals is mobile (e.g. a ship, aircraft, or vehicle) and the other is fixed. In this case high-power transmitters and directional antennas can be used at the fixed site.

One advantage of real-time channel evaluation (RTCE) over prediction is that a communicator can use with confidence that part of the spectrum above the monthly median MUF where the interference level is lower than it is on the 'optimum working frequency' (Clarke 1985).

12.8 References

BARGHAUSEN, A. F., FINNLEY, J. W., PROCTOR, L. L., and SCHULTZ, L. D. (1969): 'Predicting long-term operational parameters of high-frequency sky-wave telecommunications systems', ESSA Tech. Rep. ERL 110–ITS 78, US Dept. of Commerce, Boulder, CO

BARRY, G., and FENWICK, R. B. (1976): 'Techniques for real-time HF channel measurement and optimum data transmission' *in* BLACKBAND, W. T. (Ed.): 'Radio systems and the ionosphere'. AGARD Conf. Proc. No. 173, Advisory Group for Aerospace Research and Development, North Atlantic Treaty Organisation, p. 15-1

BRADLEY, P. A. (1976): 'A new computer-based method of HF sky-wave signal prediction using vertical-incidence ionosonde measurements' *in* BLACKBAND, W. T. (Ed.): 'Radio systems and the ionosphere'. AGARD Conf. Proc. No. 173, Advisory Group for Aerospace Research and Development, North Atlantic Treaty Organisation, p. 11-1

BRADLEY, P. A. (1979): 'Propagation at medium and high frequencies, 2: Long and short-term models' *in* 'Aerospace propagation media modelling and prediction schemes for modern communications, navigation, and surveillance systems'. AGARD Lecture Series No. 99, Advisory Group for Aerospace Research and Development, North Atlantic Treaty Organisation, p. 9–1

CCIR (1980): 'Second CCIR computer-based interim method for estimating sky-wave field strength and transmission loss at frequencies between 2 and 30 MHz'. Supplement to Report 252–2, Int. Telecom. Union, Geneva

CCIR (1986*a*): 'Propagation in ionized media. Recommendations and reports of the CCIR 1986, VI'. Int. Telecom. Union, Geneva

CCIR (1986*b*): 'Protection ratios required for spectrum utilization investigations. Report 525-2' *in* 'Recommendations and reports of the CCIR 1986: I. Spectrum utilization and monitoring'. Int. Telecom. Union, Geneva

CCIR (1986*c*): 'Signal-to-interference protection ratios. Recommendation 240-4' *in* 'Recommendations and reports of the CCIR 1986: III. Fixed service at frequencies below about 30 MHz'. Int. Telecom. Union, Geneva

CCIR (1986*d*): 'CCIR interim method for estimating skywave field strength and transmission loss at frequencies between the approximate limits of 2 and 30 MHz. Report 252-2' *in* 'Recommendations and reports of the CCIR 1986, VI'. Int. Telecom. Union, Geneva, p. 237

CCIR (1988): 'Characteristics and applications of atmospheric radio noise data, CCIR Report 322-3'. Int. Telecom. Union, Geneva

CLARKE, R. H. (1985): 'A real-time channel evaluation technique for medium-speed digital skywave communications' *in* COLE, D. C., and McNAMARA, L. F. (Eds.): 'Proc. URSI–IPS Conf. on the ionosphere and radio wave propagation'. IPS Radio and Space Services, Darlinghurst, NSW, Australia

COLE, D. G. (1985): 'Ionospheric forecasting for specific applications' *in* COLE, D. G., and McNAMARA, L. F. (Eds.): 'URSI–IPS Conf. on the ionosphere and radio wave propagation'. IPS Radio and Space Services, Darlinghurst, NSW, Australia

DAMBOLT, T. H. (1976): 'A comparison between the Deutche Bundespost ionospheric HF radio propagation predictions and measured field-strengths' *in* BLACKBAND, W. T. (Ed.): 'Radio systems and the ionosphere'. AGARD Conf. Proc. No. 173, Advisory Group for Aerospace Research and Development, North Atlantic Treaty Organisation, p. 12–1

DARNELL, M. (1976): 'Channel estimation techniques for HF communications' *in* BLACKBAND, W. T. (Ed.): 'Radio systems and the ionosphere'. AGARD Conf. Proc. No. 173, Advisory Group for Aerospace Research and Development, North Atlantic Treaty Organisation, p. 16-1

FENWICK, R. B., and WOODHOUSE, T. J., (1979): 'Real-time adaptive HF frequency management' *in* COYNE, V. J. (Ed.): 'Special topics in HF propagation'. AGARD Conf. Proc. No. 263, Advisory Group for Aerospace Research and Development, North Atlantic Treaty Organisation, p. 5–1

GOODMAN, J. M. (Ed.) (1987): 'The effect of the ionosphere on communication, navigation, and surveillance systems'. Naval Research Laboratory, Washington, DC

HEADRICK, J. M., and SKOLNIK, M. I. (1974): 'Over-the-horizon radar in the HF band', *Proc. IEEE*, **62**, p. 664

LEFTIN, M. (1976): 'Numerical representation of monthly median critical frequencies of the regular E-region (f_oE)'. OT Report 76–88, US Govt. Printing Office, Washington, DC 20402

LEFTIN, M., and OSTROW, S. M. (1969): 'Numerical maps of f_bE_s for solar cycle minimum'.

LEFTIN, M., OSTROW, S. M., and PRESTON, C. (1968): 'Numerical maps of f_oE_s for sunspot cycle minimum and maximum'. ESSA Tech. Rep. ERL 73-ITS 63, US Dept. of Commerce, Boulder, CO 80303

LIU, R. Y., SMITH, P. A., and KING, J. W. (1983): 'A new solar index which leads to improved f_oF_2 predictions using the CCIR Atlas', *Telecom. J.*, **50**, p. 408

LUCAS, D. L., LLOYD, J. L., HEADRICK, J. M., and THOMASON, J. F. (1972): 'Computer techniques for planning and management of OTH radars'. NRL MEMO Report 2500, Naval Res. Lab., Washington, DC

MASLIN, N. M. (1979): 'Assessment of HF communications reliability' *in* COYNE, V. J. (Ed.): 'Special topics in HF propagation'. AGARD Conf. Proc. No. 263, Advisory Group for Aerospace Research and Development, North Atlantic Treaty Organisation, p. 6

PROBST, S. E. (1968): 'The CURTS concept and current status of development' in FOLKESTAD, K. (Ed.): 'Ionospheric radio communications' (Plenum Press, New York) p. 370

'Radio Spectrum Utilization' (1964) IEEE, New York

ROSICH, R. K., and JONES, W. B. (1973): 'The numerical representation of the critical frequency of the F_1-region of the ionosphere'. OT Report 73-22, US Govt. Printing Office, Washington, DC 20402

RUSH, C. M. (1986): 'Ionospheric radio propagation models and predictions: A mini-review', *IEEE Trans.*, **AP-34**, p. 1163

RUSH, C. M., MILLER, D., and GIBBS, J. (1974): 'The relative daily variability of f_oF_2 and h_mF_2 and their implications for HF radio propagation', *Radio Sci.*, **9**, p. 749

RUSH, C. M., POKEMPNER, M., ANDERSON, D. N., PERRY, J., STEWART, F. G., and REASONER, R. (1984): 'Maps of f_oF_2 derived from observations and theoretical data'. *Radio Sci.*, **19**, p. 1083

RUSH, C. M., STEWART, F. G., POKEMPNER, M., and REASONER, R. (1986): 'Some ionospheric factors affecting the coverage of an HF/VHF direct broadcasting satellite service', *IEEE Trans.*, **AP-34**, p. 475.

RUSH, C. M., FOX, M., BILITZA, D., DAVIES, K., McNAMARA, L., STEWART, F., and POKEMPNER, M. (1989): 'Ionospheric mapping — An update of f_oF_2 coefficients', *Telecom. J.*, **56**, p. 179

SAILORS, D. B., and BROWN, R. P. (1982): 'Development of a minicomputer atmospheric radio noise model'. NOSC Tech. Rep. 778, Naval Ocean Systems Center, San Diego, CA

SAILORS, D. B., and BROWN, R. P. (1983): 'Development of a minicomputer atmospheric noise model', *Radio Sci.*, **18**, p. 625

SAILORS, D. B., SPRAGUE, R. A., and RIX, W. H. (1986): 'MINIMUF-85: An improved HF MUF prediction algorithm'. Tech. Rep. NOSC-TR1121, Naval Ocean Systems Center, San Diego, CA

SCHULTZ, L. D., and GALLET, R. M. (1970): 'Normal ionospheric absorption measurements'. ESSA Prof. Paper 4, US Dept. of Commerce, Washington, DC 20402

SPAULDING, A. D., and DISNEY, R. T. (1974): 'Man-made radio noise. I: Estimates for businesses, residential and rural areas'. NTIA Report 74-38, US Dept. of Commerce, Washington, DC 20402

SPAULDING, A. D., and STEWART, F. G. (1987): 'An updated noise model for use in IONCAP'. NTIA Report 87-212, US Dept. of Commerce, Boulder, CO 80303

SPAULDING, A. D., and WASHBURN, J. S. (1985): 'Atmospheric radio noise; worldwide levels and other characteristics'. NTIA Report 85-173, US Dept. of Commerce, Boulder, CO 80303

STEVENS, E. E. (1968): 'The CHEC sounding system' *in* FOLKESTAD, K. (Ed.): 'Ionospheric radio communications' (Plenum Press, New York) p. 359

TETERS, L. R., LLOYD, J. L., HAYDON, G. W., and LUCAS, D. L. (1983): 'Estimating the performance of telecommunications systems using the ionospheric transmission channel'. NTIA Report 83-127, US Dept. of Commerce, Boulder, CO 80303

WILKINSON, P. J. (1982): 'A comparison of monthly indices of the ionospheric F-region'. IPS Series R, Report IPS-R41, Austr. Dept. Sci., Darlinghurst, NSW, Australia

Chapter 13

Propagation on very high frequencies

13.1 Introduction

The very-high-frequency range is defined as between 30 MHz and 300 MHz. The ionosphere can reflect radio signals on frequencies of > 50 MHz but over most of the VHF range the main effect of the ionosphere is to scatter radio energy by irregularities. Such irregularities exist in the D region (turbulence), in the E region (meteors and sporadic E), and in the F region, especially near the dip equator and in high latitudes (plasma instabilities).

On lower frequencies (e.g. less than 50 MHz) propagation effects such as Faraday rotation (Sec. 8.3.2), angular refraction, and time delay can be important. The value of $Y(=f_H/f)$ is small ($<0{\cdot}05$), and under most conditions the propagation is essentially quasi-longitudinal (e.g. on 60 MHz for $\theta \lesssim 89{\cdot}28°$ with $f_H = 1{\cdot}5$ MHz) and the ray path is almost straight. The maximum electron densities in the ionosphere are rarely sufficient to reflect radio waves on frequencies much above 30 MHz except for relatively short periods (e.g. in the equatorial regions and in higher latitudes near the maximum of the sunspot cycle). For communications on the ground the main ionospheric mechanism is scattering by plasma irregularities.

The chief irregularities in the ionosphere are:

(i) Turbulence in the D region
(ii) Ionized trails due to the passage of meteors through the atmosphere
(iii) Spread-F (field aligned)
(iv) Sporadic E
(v) Artificial irregularities

13.2 Characteristics of VHF scatter

Ionospheric propagation in the 30–100 MHz range is characterized by the following features:

(i) Ionoscatter, which is regular but weak scatter from the D region, is useful for continuous single-frequency communication over paths of about 1000 km to 2000 km.
(ii) Meteorscatter, which consists of intermittent reflections from ionized meteor trails (from 30 MHz to about 110 MHz) over distances up to 2000 km.
(iii) Sporadic (and strong) reflections from field-aligned F irregularities over distances up to 4000 km.
(iv) Sporadic-E reflections can be important (especially as sources of interference) up to distances of about 2000 km and on frequencies as high as 80 MHz.

The chief disadvantages of VHF propagation, from the point of view of radio communications, are these:

(i) Low signal strengths, because of the relative inefficiency of the scattering mechanism
(ii) Rapid fading
(iii) Limited ground range (2000 km)
(iv) Mutual interference between channels normally independent.

Scatter systems become operational on wavelengths that are comparable with the sizes of the irregularities in the propagation medium. For the longer ranges ≈ 2000 km (for ionoscatter and meteorscatter) antennas with low angles of radiation (1° to 5°) are required, so the use of elevated sites is advantageous. There is a marked difference in the frequency dependencies of the signal/noise ratios: For ionoscatter systems the s/n varies as f^{-5} whereas for meteorscatter systems it varies approximately as $f^{-2.5}$ (Ince *et al.* 1978).

13.3 D-region scatter

13.3.1 Scatter from irregularities in the D region

Irregularities are produced by the action of turbulence, wind shear, etc., on the electron distribution in the height range between about 70 km and about 90 km. These irregularities in electron density result in corresponding fluctuations in refractive index. Since turbulence merely brings about a redistribution of the electrons, these types of irregularities will exhibit marked solar control as does the overall electron content of the D region. There is also evidence for stratification in the D region, which means that the refractive index changes abruptly with height (see Sec. 11.3.3).

The effect of these rapid changes, both in time and space, on very-high-frequency waves is to scatter the energy out of the direction of the incident beam. Since most of the energy is scattered in directions making small angles with the direction of the incident beam, the phenomenon is called 'forward scatter'.

The geometry of forward scatter is sketched in Fig. 13.1. The ratio of the received power p_r to transmitted power p_t is of the form (Bailey *et al.* 1955; Wheelon 1960)

$$p_r/p_t = K(f_N)^4/f^m l^2 [\sin(\gamma/2)]^n \tag{13.1}$$

Here f is the wave frequency, l is the distance from the transmitter (or receiver) to the scattering volume, γ is the scattering angle, and the proportionality factor K depends on antenna, incident polarization, size of scattering volume, and the spectrum of turbulence. Equation (13.1) applies for single scattering and isotropic irregularities. For isotropic turbulence $m = 4$ and $n = 5$ but experimental results indicate that usually $6 < n < 8$, n having extremes of 4 and 12 (Kirby 1968) between different circuits, seasons, etc. (Ince 1978). Equation (13.1) emphasizes the joint frequency–distance dependence of received power that is characteristic of scatter propagation. It will be seen that the received power decays as the inverse square of the distance rather than as the inverse fourth power as in the radar equation. The reason for this is that, for a given circuit, the scattering volume increases with increase of range l. The power decreases rapidly with both frequency and scattering angle.

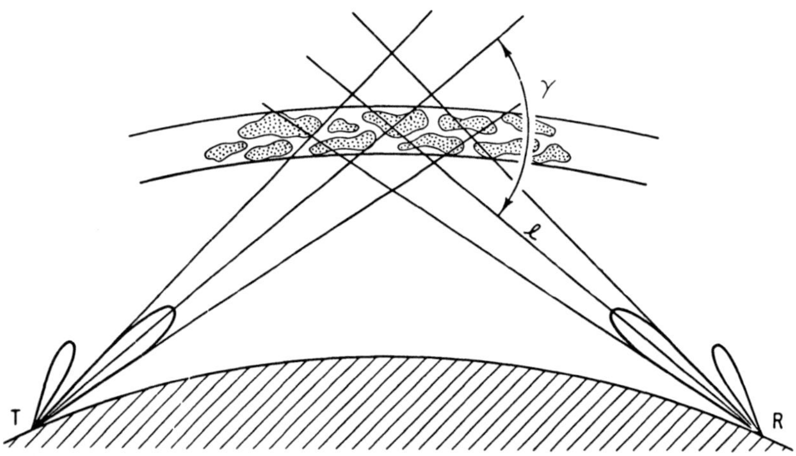

Fig. 13.1 *Geometry of forward scatter from the D region (From Davies 1965)*
γ is the angle through which the beam is scattered; ℓ is the distance between the scattering volume and the receiver *R*

13.3.2 Height of scattering

Experimental evidence suggests that the scattering of VHF waves is produced primarily from two layers. During the daytime the predominant heights are around 70 km; this scattering layer is subject to solar control and, therefore, is thought to be due to turbulence. During nighttime the scattering appears to come from the 85–90 km region; the ionization at these altitudes is thought to be produced by the influx of meteors. Some observations seem to be best explained on the basis of simultaneous scattering from both layers.

13.3.3 Signal strengths in middle and high latitudes

13.3.3.1 Long-term variations

From hourly median values of signal strengths it has been found that, in the frequency range of 25–108 MHz, over distances of 1000–2000 km, the system loss ranges from about 140 dB to about 210 dB. The loss depends on several factors such as wave frequency, type of antenna, geography, and time. Median values of system loss (Bailey *et al.* 1955) indicate that on a frequency of about 50 MHz the diurnal variations of signal strength are as follows:

Middle latitudes: A maximum occurs near noon and a minimum near 2000 h local time. The diurnal curve is broader in summer than in winter, as would be expected from the solar illumination.
High latitudes: The diurnal maximum is broader and tends to be centered earlier in the day, compared with middle latitudes. The system loss for a given hour varies considerably from day to day. This variation is greater in high latitudes than in middle latitudes.

In middle and high latitudes the seasonal variation is characterized by a strong maximum in summer for all hours of the day. No definite long-term (solar cycle) variation has been established.

The above results were obtained over paths about 1200 km long and with narrow-beam (6°) rhombic antennas aimed along the great circle.

13.3.3.2 Short-term variations

Observed amplitude distributions show that, during most of the day, the envelope is, essentially, distributed in a Rayleigh manner. The distributions are peaked somewhat at the high-signal end during the late night and early morning hours. Fading rates normally lie between 0·2 Hz and 3 Hz, the rate increasing with frequency. Most observations show that this rate depends on carrier frequency raised to a power between 0·75 and 1·25. One would expect a linear dependence (exponent = 1·0) if the drift speed of the scattering region were dominant compared with the changing structure of irregularities. This departure from linear dependence at times favors the idea of a turbulent D region as against meteor-produced ionization. Fading rates are higher for broad-beam systems, and for off-great-circle paths, than for great-circle transmissions with narrow beams.

13.3.3.3 Frequency dependence

The value of m in (13.1) varies with time between about 7 and 9·5 and has a median of about 8. Daytime values are 1 to 2 greater than nighttime values.

13.3.3.4 Geographical variations

Observations have shown that, except for paths near the equator, the transmission loss for paths in the Arctic, the United States, and Europe differ by less than ± 5 dB from the overall median value. Within these limits, the signal strengths on the Arctic paths are, statistically, higher than those in middle latitudes.

Near the magnetic equator the signal intensities are, on the whole, higher than in the United States. The background scatter signal is about the same, but this is obscured for a large portion of the time by strong enhancements (up to 50 dB) which appear to be reflected from field-aligned ionization in the E region. Rapid fading is often characteristic of these enhancements.

There is little correlation between VHF transmission loss in temperate latitudes and f_oE, or HF attenuation. There appears to be little variation of signal intensity with sunspot number but there is a statistical correlation (positive) with the magnetic K index.

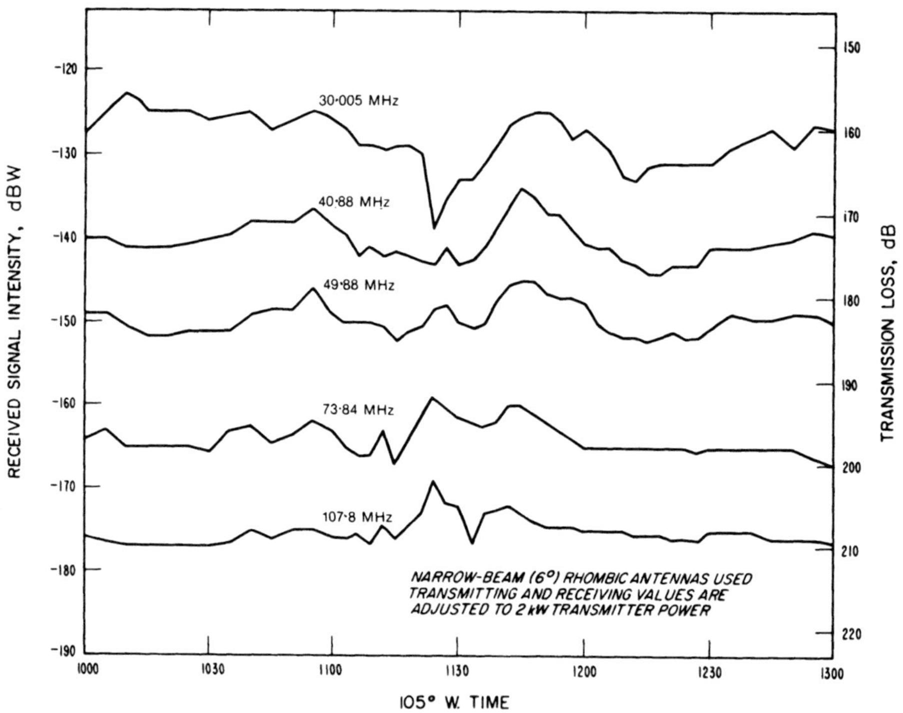

Fig. 13.2 *Signal behavior on five frequencies during a sudden ionospheric disturbance on March 29, 1958 (After Blair et al. 1961)*

13.3.4 Abnormal behavior

13.3.4.1 Sudden ionospheric disturbances

These have been discussed in Sec. 9.3, but a few more comments are appropriate because it appears that the behavior of the received VHF signal varies with frequency. The increased ionization created by the flare radiation appears to produce two effects: intensity enhancement (more efficient scattering) and increased D-region absorption. On the lower frequencies, where the absorption effects are greatest, the net result is a decrease in signal. On the higher frequencies the overall effect is an increase in signal. On intermediate frequencies little effect is seen in many instances, from which it may be concluded that the two processes just cancel. The effects on the received signals on five frequencies are shown in Fig. 13.2 for signals transmitted over a 1295 km path (United States) employing narrow-beam antennas. A definite frequency dependence is evident. On the other hand, with parallel broad-beam systems, there is increased attenuation on all frequencies up to 75 MHz (the highest frequency used).

13.3.4.2 Correlation with auroral phenomena

HF propagation in the auroral zone is characterized by periods of high absorption (Sec. 9.5.4). These periods are not, in general, periods of high absorption of VHF waves. On the contrary, high signal intensities, on 50 MHz, are usually observed during periods of enhanced HF absorption and strong geomagnetic activity. Marked increases in signal level often coincide with isolated intervals of HF blackout.

13.3.4.3 Sputter

An additional effect associated with aurorae is the occurrence of rapid fading called 'sputter', which is due to a multipath signal. There have been occasions when 'sputter' speeds of 200–300 Hz have been observed, that is, about 100 times the normal fading rate. The effect is minimized with narrow-beam antennas. It may be observed, to some degree or other, 100 h and more per year (Bailey *et al.* 1955).

13.3.4.4 Polar cap absorption effects

Forward-scatter signals are affected during PCA events (see Sec. 9.4) in high latitudes. When the onset of the proton flux (which gives rise to the PCA) occurs at night, the VHF signals are enhanced (fairly abruptly), and strong absorption follows during the ensuing daylight. The fact that the scattered signals are absorbed during the PCA and enhanced during auroral blackouts indicates that the height at which scattering occurs lies below the region of auroral absorption (90 km) but above the height at which polar cap absorption occurs (60 km).

Polar cap absorption events are generally confined to paths with midpoints in geomagnetic latitudes greater than about 60°. The onset of this absorption (in the lower latitudes) often follows the onset of a magnetic storm and often at the time of a strong (Forbush) decrease in the intensity of cosmic rays which is, in general, many hours after the commencement of the PCA itself in higher latitudes. A high correlation has been found between simultaneous riometer and forward-scatter signal strength (Collins and Maynard 1968).

13.3.5 Communication applications

The main advantage of ionoscatter is the continuity of the signal. With the development of satellite communication systems, interest in applications of ionoscatter has declined. However, there are military applications (Bartholomé 1968; Ince *et al.* 1978) employing 'automatic request for repetition' (ARQ) techniques in which an error-detecting code makes possible a request to repeat any incorrect character. A typical ionoscatter teletype system for a distance of 1000–2000 km on a frequency of 50 MHz might include a 1 kW transmitter, yagi antenna to provide at least 10 dB signal/noise ratio for a 100 Hz band during 99% of the hours in a year when the noise is of galactic origin. Meteor echoes and auroral signals can introduce multipath spread of a few milliseconds (Kirby 1968). There are two problems: (i) meteor echoes with large (kHz) Doppler shifts, and (ii) long-delayed F echoes. Communication systems using D-region scatter are rare.

13.4 Meteor scatter

13.4.1 Ionization by meteors

In recent years there has been renewed interest in meteor burst systems because of several technical developments such as microprocessors. Further, the nuclear survivability of meteor burst systems is superior to other systems, such as satellite and HF skywave systems, thus making them attractive to the military (Oetting 1980; Brown 1985; Weitzen 1988).

Meteorscatter systems depend on the fact that meteors entering the atmosphere from outer space produce ionized trails in the height range 80–120 km. Billions of these trails are produced daily; they diffuse rapidly and usually disappear within a few seconds. During their brief existence, however, these ionized columns will reflect VHF waves, particularly in the frequency range 30–110 MHz. The lower frequency limit is set by the need to be above the regular maximum usable frequency, and the upper limit is set by receiver sensitivity limitations, since the signal/noise ratios on the higher frequencies are relatively low.

The ionized trail is the result of conversion of the kinetic energy of the meteor into potential energy of ionization as the meteor is slowed down by collisions with the neutral molecules. The ionization is distributed in the form of a long, thin paraboloid of revolution. The electron density per unit length of the trail is proportional to the mass of the particle.

The height distribution of trails varies with particle characteristics. The higher velocity particles (of a given mass) produce trails at the higher (120 km) heights; the mean trail height increases by about 10 km as the velocity increases from less than 15 km/s to more than 60 km/s. For a given velocity, particles of higher mass produce maximum trail ionization at lower heights. Over the mass range 10^{-7} to 10^3 g the height variation is about 44 km. There is also a height variation with the zenith angle (χ) of the trail, larger angles corresponding to greater heights. A variation of about 13 km corresponds to the range of χ from 0° to 90°.

A useful equation relates average height h (km) of reflection to wave frequency f, in MHz (Greenhow and Hall 1960):

$$h = 124 - 17 \log f \qquad (13.2)$$

The lengths of the trails depend primarily on the mass and zenith angle. Typical lengths range up to 50 km, the most probable value being 15 km, although this depends on the definition of 'length'.

Visual studies give initial radii of the trails in the range 0–1·2 m (with a mean value of 0·65 m); radio observations give a range of 0·55–4·35 m. The 0·55 m radius is for a height of 81 km.

Let us consider now the dissipation of the ionization. Once the trail is formed it expands by diffusion at a relatively low rate, thus producing a radial distribution of material that is approximately Gaussian. After a time t the radius r is related to the initial radius r_0 by the expression

$$r^2 = r_0^2 + 4Dt \qquad (13.3a)$$

$$\log D \approx 0{\cdot}067\, h - 5{\cdot}6 \qquad (13.3b)$$

where D is the diffusion coefficient of the electrons at a height h and varies from 1 m^2/s at 85 km to about 130 m^2/s at 115 km. Hence, after 1 s, a trail will have a radius in the range 2–20 m. An empirical relation between the initial trail radius and the meteor height (CCIR 1986, Report 251–4) is

$$\log r_0 = 0{\cdot}035\, h - 3{\cdot}25 \qquad (13.4)$$

The practical lifetime of a trail depends on the means of detection. Most trails, detected by radio methods, are those resulting from small particles and last for a fraction of a second only. Larger meteors produce more densely ionized trails, and trails with durations of about 1 min are observed several times per day. Trails with durations measured in hours are rare.

The dissipation of trails is complicated further by the action of winds in the 80–120 km region. At the time of formation, trails are quite straight but they are deformed rapidly by winds (see Sec. 13.4.8) of the order of 25 m/s and vertical wind shears with maxima of about 100 m/s per km. A wind shear of this magnitude can rotate part of a trail through an angle of 5° in 1 s.

13.4.2 Meteor data

The term 'meteor' is restricted here to those particles entering the Earth's atmosphere that are completely burned up by frictional heating.

Particles in the mass range of 10^{-7} to 10^3 g and dimensions in the range 40 μm to 8 cm are primarily responsible for radio reflections. Before being swept up by the Earth's atmosphere these meteors move in elliptical orbits around the Sun, and they are thought to be of cometary origin.

Meteors can be divided into two classes:

(i) Shower meteors, although the most spectacular, account for only a small fraction of the total incidence of meteors. A meteor shower is often named after the stellar constellation in which the radiant point lies (e.g. Leonid for Leo).
(ii) Sporadic meteors move in random orbits. Most meteors of interest to radio work are sporadic meteors.

The radiants and times of occurrence of sporadic meteors are random. Their radiant points are for the most part concentrated toward the ecliptic plane (the plane of the Earth's orbit) and move in the same direction around the Sun as the Earth moves. The orbits are concentrated so as to produce a maximum incidence of meteors at the Earth in July and a minimum in February, This variation in the space density of meteors is shown in Fig. 13.3.

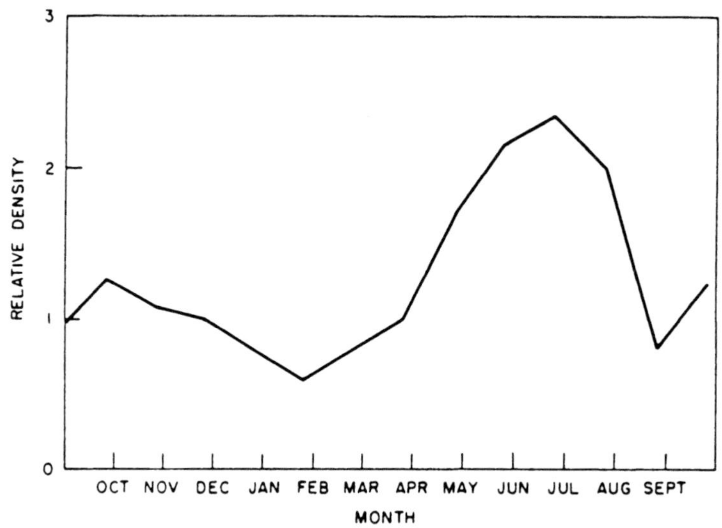

Fig. 13.3 *Variation of space density of meteors along the Earth's orbit (After Hawkins 1956)*

The rate of incidence of sporadic meteors at the Earth is further modified by two factors. The first of these, resulting in a regular diurnal variation in meteor rate, is illustrated in Fig. 13.4. On the morning side of the Earth, meteors are swept up by the forward motion of the Earth in its motion around the Sun. On the evening side the only meteors reaching the Earth are those that overtake it. This results in a maximum occurrence rate around 6 a.m. and a minimum rate around 6 p.m. The ratio of maximum to minimum depends on the latitude of the observer. A further minor seasonal variation is introduced because of the tilt of the Earth's axis relative to the ecliptic plane. This variation, also dependent on the latitude of the observer, can change the expected hourly rates by factors as large as 1·4.

The mass distribution of sporadic meteors is such that there are approximately equal total masses of each size of particle. There are, for example, 10 times as many particles of mass 10^{-4} g as there are particles of mass 10^{-3} g. This approximate relation between particle mass and number is given in Table 13.1. The mass distribution of shower meteors is somewhat similar to that for the sporadic meteors, the important difference being that shower meteors have more large particles relative to the number of smaller particles.

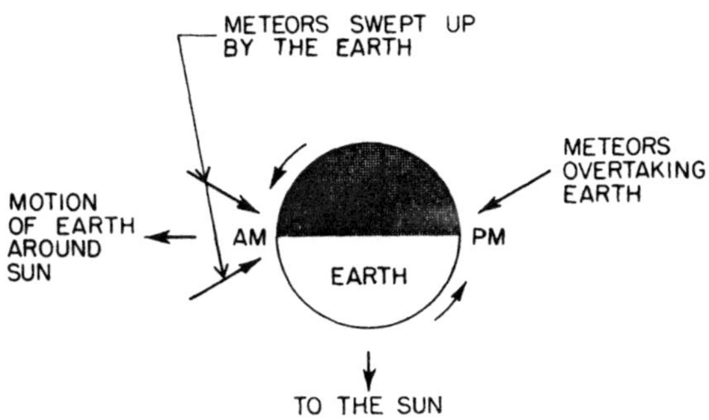

Fig. 13.4 *Diurnal variation of meteor rates (After Sugar 1964)*

The velocities of meteors approaching the Earth are in the range 11·3–72 km/s. The lower limit is the escape velocity for a particle leaving the Earth and is, therefore, the lowest velocity that a particle falling toward the Earth can have. The upper limit is the sum of two-components, a 30 km/s component associated with the velocity of the Earth in its orbit around the Sun, and a 42 km/s component associated with the meteor itself. This latter velocity is the escape velocity for a particle leaving the solar system. Observations indicate that meteor velocities are within the range 11–72 km/s, which suggests that meteors are members of the solar system.

13.4.3 Reflection from a meteor trail

13.4.3.1 Important parameters

The distribution of energy reflected by a meteor trail is a function of many variables. The ionization density distribution across and along the trail, the orientation of the trail, the radio wavelength, the polarization of the incident wave relative to the trail, motion of the trail either as part of the process of formation or due to ionospheric winds, and the straightness of the trail are all significant. In discussing the reflection properties it is convenient to divide the trails into two classes – underdense trails and overdense trails – and to examine the properties of each class independently. Underdense trails are those

Table 13.1 *Order-of-magnitude estimates of the properties of sporadic meteors*

	Mass (grams)	Visual magnitude	Radius	Number of this mass or greater swept up by the Earth each day	Electron line density (electrons per meter of trail length)
Particles pass through the atmosphere and fall to the ground	10^4	−12·5	8 cm	10	–
Particles totally disintegrated in the upper atmosphere	10^3	−10·0	4 cm	10^2	–
	10^2	−7·5	2 cm	10^3	–
	10	−5·0	0·8 cm	10^4	10^{18}
	1	−2·5	0·4 cm	10^5	10^{17}
	10^{-1}	0·0	0·2 cm	10^6	10^{16}
	10^{-2}	2·5	0·08 cm	10^7	10^{15}
	10^{-3}	5·0	0·04 cm	10^8	10^{14}
	10^{-4}	7·5	0·02 cm	10^9	10^{13}
	10^{-5}	10·0	80 μm	10^{10}	10^{12}
Approximate limit of	10^{-6}	12·5	40 μm	10^{11}	10^{11}
radar measurements →	10^{-7}	15·0	20 μm	10^{12}	10^{10}
	10^{-8}	17·5	8 μm	?	?
Micrometeorites (Particles float down unchanged by atmospheric collisions)	10^{-9}	20·0	4 μm	Total for this group estimated as high as 10^{20}	Practically none
	10^{-10}	22·5	2 μm		
	10^{-11}	25·0	0·8 μm		
	10^{-12}	27·5	0·4 μm		
Particles removed from the solar system by radiation pressure	10^{-13}	30	0·2 μm	–	–

wherein the electron density is low enough that the incident wave passes through the trail and the trail can be considered a partial reflector. Overdense trails are those wherein the electron density is high enough to prevent complete penetration of the incident wave and to cause reflection of waves in the same sense as ordinary ionospheric reflections. A rough sorting of trails into these two categories can be done on the basis of trail lifetime or duration. On long wavelengths (relative to the VHF band) the underdense trails have durations of less than about 1 s; the overdense trails tend to have longer durations. The analysis can be further divided in terms of the radio wavelength. On long wavelengths the effective duration of a trail is large compared with the time it takes the trail to form, and the trail may be considered to have a cylindrical shape. On short wavelengths, however, the duration is much shorter and the column must be considered to be of the paraboloidal form previously indicated.

The solutions to be considered here are at best useful approximations to the physical problems. They will apply quite well to some of the trails observed and rather poorly to others. A complete analysis of the reflections from even a relatively simple trail would be far too complex to be of any practical use or interest. In the following treatment the terms long wavelength and short wavelength are used in a relative manner to indicate the lower frequency end and the upper frequency end of the VHF band.

13.4.3.2 Long-wavelength radar reflections from low-density trails

We assume that the trail is an infinitely long, right-circular cylinder of electrons whose diameter is very small compared with the wavelength, so the trail electron density is low enough that the incident wave passes through the trail without major modification. Consider the case of radarlike reflections, for which transmitter and receiver are at the same location. Figure 13.5*a* illustrates the reflection geometry and the trail model. Let R be the length of the perpendicular from the radar set to the trail. The signal received can be computed by summing the energy backscattered by each electron in the trail while taking proper account of the phase relations of these contributions. When this is done it is found that the principal contribution of energy from a trail is from its first Fresnel zone, a region of length $(2\lambda R)^{1/2}$, centered at the point on the trail where the line from the radar is normal to the trail. For example, on 50 MHz a trail 200 km away will have a first Fresnel zone 1·6 km in length. As the trail expands by diffusion, the phase difference of the contributions from the electrons on opposite sides of the trail increases, so the received power decreases. The signal amplitude will have fallen to $1/\varepsilon$ of its initial value ($\varepsilon = 2{\cdot}718$) at the time the radius of the trail is $\lambda/2\pi$. On 50 MHz this radius is 1 m.

The transmission equation can be written in terms of the scattering cross section σ in the following way:

$$\frac{p_r}{p_t} = \frac{g_t g_r \lambda^2}{16\pi^2 R^4}\sigma A_1 A_2 \tag{13.5}$$

where p_t and p_r = the transmitted and received power respectively
g_t and g_r = power gains of the transmitting and receiving antennas, respectively, relative to an isotropic radiator in free space
λ = wavelength (m)
R = distance from transmitter to the trail (m)
A_1 and A_2 = attenuation factors, near unity, which depend on the initial radius and time
σ = scattering cross section defined as the ratio of the power scattered per unit solid angle to the power incident per unit area.
If it is assumed that the trail is initially in the form of a line of electrons (an initial trail radius of zero), the scattering cross section may be written as

$$\sigma = \left[r_e q \left(\frac{R\lambda}{2} \right)^{1/2} \right]^2 \exp\left(- \frac{32\pi^2 D}{\lambda^2} t \right) \tag{13.6}$$

where $r_e = \mu_0 e^2/4\pi m = 2{\cdot}8178 \times 10^{-15}$ m is the classical radius of the electron, and where
μ_0 = permeability of free space
e and m = charge and mass of the electron
q = electron line density of the trail (electrons/m)
D = diffusion coefficient (m^2/s)
t = time measured from the formation of the trail (s)
Values of σ are on the order of $5 \times 10^5\,m^2$.

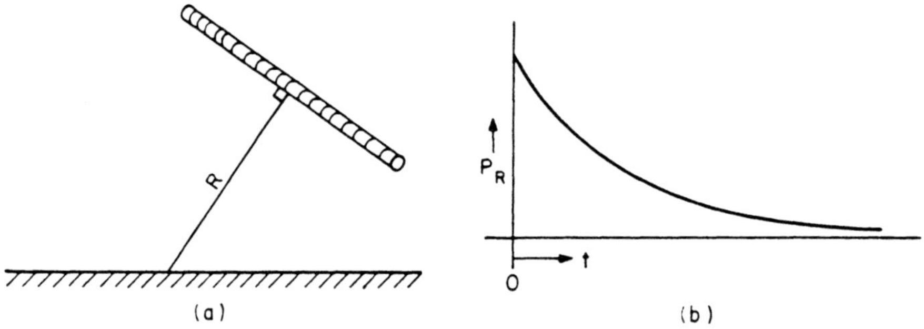

Fig. 13.5 *Low-density long-wavelength reflections*
(*a*) Geometry
(*b*) Time variation of received signal

The first factor in (13.6) represents the scattering cross section of the initial line distribution of electrons, and the second factor represents the attenuation with time as the trail expands and destructive interference begins. The first factor may be interpreted as follows: The effective cross section of the trail is the same as would be obtained for in-phase addition of scattered fields associated with each electron within one-half of the first Fresnel zone. Since the power scattered by an electron is proportional to r_e^2, the field is proportional to r_e. The number of electrons in half of the first Fresnel zone is $q\sqrt{R\lambda/2}$, and the field is, therefore, proportional to $r_e\sqrt{R\lambda/2}$. The total effective cross section is just the square of this product.

The second factor in (13.6) results from the integration across the trail. The distribution of electrons across the trail has a Gaussian shape and an effective radius of $(4Dt)^{1/2}$. The signal amplitude falls to $1/\varepsilon$ when the effective radius is $\lambda/2\pi$. When the initial radius is nonzero, the peak amplitude is reduced from that given by (13.4). The form of the time variation of received power is sketched in Fig. 13.5*b*.

In the above relation, since it was assumed that trails were of infinite length, every trail had a first Fresnel zone and, therefore, gave a reflection. In practice, however, since trails are of finite length, this zone may not lie on the trail but rather on an extension of it. In such a case the signal returned from the trail would be a relatively weak one, and in practice these returns are of little importance. Thus a radar can 'see' only a fraction of all the trails incident on the ionosphere within its range because most trails do not have the proper orientation for the line of sight to be perpendicular to the trail. The requirement for proper orientation can be restated as a requirement that the trail be tangent to a sphere centered at the radar.

13.4.3.3 Separate transmitter and receiver

When the transmitter and receiver are separated, as in the case of point-to-point propagation, the reflection requirement is that the incident ray and reflected ray make equal angles with the trail axis. Geometrically this is equivalent to a requirement that the trail be tangent to an ellipsoid of revolution with foci at the transmitter and receiver. For this case two important changes take place because of the oblique geometry. First, the length of the first Fresnel zone increases and, therefore, more electrons contribute to the received signal. Secondly, the maximum possible propagation path difference over the cross section of the trail decreases with increasing obliquity. This decrease is equivalent to an increase in wavelength and results in both a larger initial signal and a slower rate of signal decay as the trail expands. In this case it can be shown that the peak received power varies as $\lambda^3 q^2$, and the signal duration as λ^2 (Sugar 1964).

The geometry of this bistatic case is shown in Fig. 13.6, which shows three planes of interest: (i) the horizontal plane containing the transmitter and receiver, (ii) the plane containing the transmitter and receiver and the point P of closest approach of the meteor trail, and (iii) the plane that is tangent at P to the ellipsoid of revolution. The distance PP^1 $(=L)$ is half the Fresnel radius and is given by

$$L = \left[\frac{\lambda R_1 R_2}{(R_1 + R_2)(1 - \sin^2\phi \cos^2\beta)}\right]^{1/2} \tag{13.7}$$

where R_1, R_2 are the lengths of the paths from P to the transmitter and receiver respectively that make equal angles ϕ with the normal to the tangent plane. The angle β is the angle between the meteor trail and the intersection of planes (ii) and (iii). Note that in the radar case $R_1 = R_2 = R$, $\phi = 0$, and $2L = (2\lambda R)^{1/2}$ as in Sec. 13.4.3.2.

When T and R are separated, the value of L is a maximum when the trail is in the plane of propagation ($\beta = 0°$) and is a minimum when $\beta = 90°$. For example, on 50 MHz ($\lambda = 6$ m) with $R_1 = R_2 = 500$ km and a reflection height of 100 km, $\phi = 78°$, $L_{min} \approx 1{\cdot}2$ km, $L_{max} \approx 6{\cdot}1$ km.

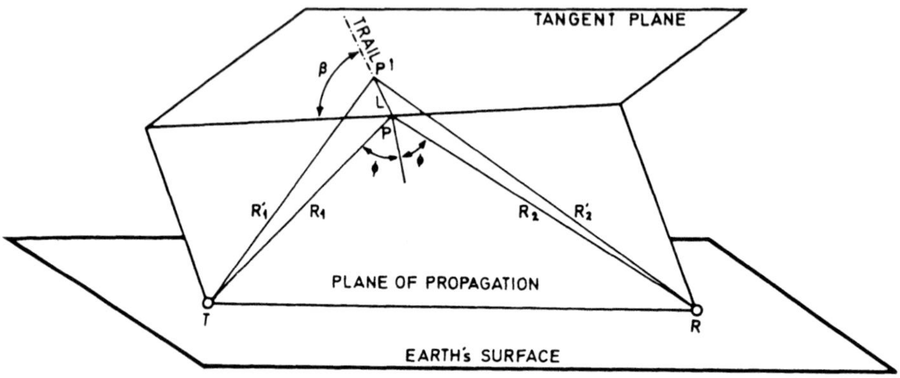

Fig. 13.6 *Geometry of meteor-trail reflection with oblique propagation*
L is half the Fresnel zone

13.4.3.4 Long-wavelength reflections from high-density trails

With high electron densities the assumption that the incident wave passes through the trail essentially unmodified is no longer valid. For this case it is assumed that the wave penetrates the trail until reaching a surface sufficiently high in electron density to reflect the wave. The model used is that of an expanding cylindrical reflector as shown in Fig. 13.7*a*. After a time the electron density everywhere within the trail falls below the critical value and the underdense model is once more applicable. However, by this time the radius is quite large and the signal contribution is small.

In the case of backscatter, the transmission equation is based on reflection from a cylinder and is given by

$$\frac{p_r}{p_t} = \frac{g_t g_r \lambda^2}{64\pi^2 R^3}\left[4\,Dt\ln\left(\frac{t'}{t}\right)\right]^{1/2} \tag{13.8}$$

This applies until approximately the time when the second factor becomes zero, which occurs at a time

$$t = t' = \frac{r_e q \lambda^2}{4\pi^2 D} \tag{13.9}$$

After this time the relation for underdense trails is applicable. The form of the time dependence of received power predicted by (13.8) is illustrated in Fig. 13.7*b*.

For the overdense case the maximum received power varies as $q^{1/2}$ in contrast to the q^2 variation for the underdense case. The λ^3 dependence is the same for both cases. The dividing line between low-density and high-density trails is commonly taken to be $q \approx 0{\cdot}75 \times 10^{14}$ electrons/m, which corresponds to ionization produced by meteors of mass $\approx 10^{-3}$ grams. Heavier meteors will produce overdense trails, and lighter meteors will produce underdense trails (Sugar 1964).

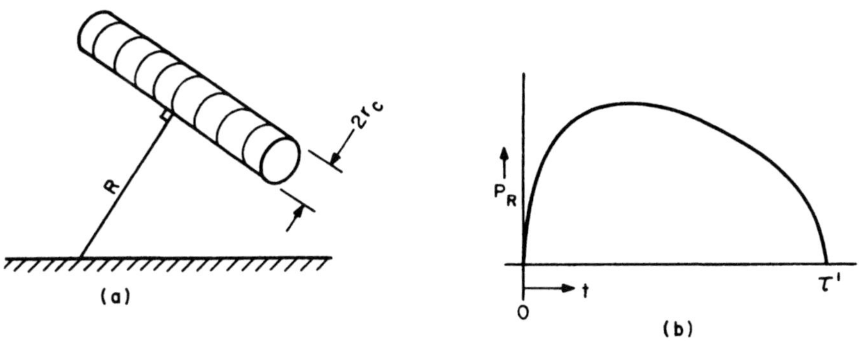

Fig. 13.7 *High-density long-wavelength reflections*
(*a*) Geometry
(*b*) Time variation of received signal

13.4.3.5 Short-wavelength reflections from low-density trails

In the preceding Sections it has been assumed that the time of formation of a meteor trail is quite small compared with the total time that radio waves are scattered by the trail. The formation of the trail was in effect considered to be a transient state preceding the desired 'steady state' trail, and this transient was neglected. On sufficiently short wavelengths the transient state is important.

Consider first the problem of estimating the shortest wavelength for which the long-wavelength results are applicable. The time formation of a trail is essentially the time required for the incoming particle to traverse one-half the first Fresnel zone as defined for the long-wavelength case. A rough estimate of the transition wavelength between longwave and shortwave cases may be obtained by equating the time of formation for a trail and the duration of a longwave echo, and solving for the wavelength. For example, consider backscatter reflections from an underdense trail. The time to traverse half the first Fresnel zone is $(R\lambda/2)^{1/2}/v$, where v is the meteor velocity, and the nominal echo duration (to the $1/\varepsilon$ amplitude) is $\lambda^2/16\pi^2 D$. Equating these and solving for λ_T, the transition wavelength, we obtain

$$\lambda_T = 8\pi^2 D \sqrt{2R}/v \qquad (13.10)$$

Using as typical values $R = 500$ km, $v = 40$ km/s, and $D = 3\,m^2/s$ gives a transition wavelength of 3·3 m.

Since in practice there can be wide variations in the values of the parameters in (13.10) it is necessary to evaluate λ_T for each specific case. When the practical extremes of all the variables for both the backscatter and forward scatter cases are investigated, it is found that the transition wavelength is in the range 0·1–100 m. For a random sample of meteors the transition wavelength is likely to be near 2·9 m for backscatter and in the range 0·5–1·6 m for forward scatter.

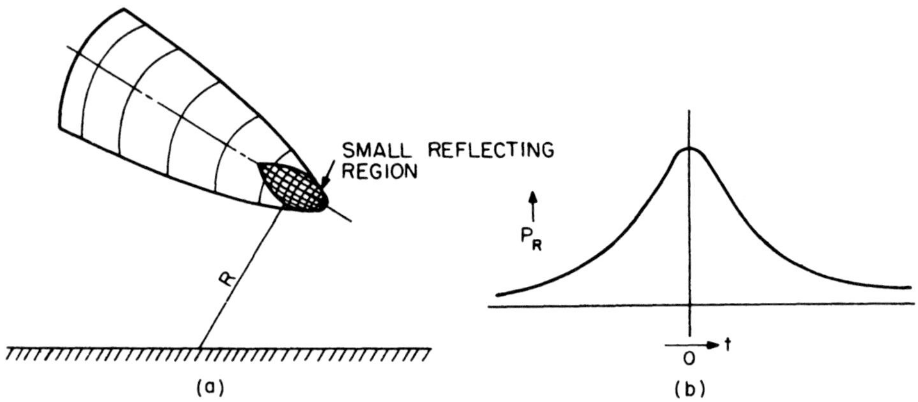

Fig. 13.8 *Low-density short-wavelength reflections*
(*a*) Geometry
(*b*) Time variation of received signal

The second (short-wavelength) case is treated by considering the effects of differential diffusion as the trail is being formed. Each part of the trail begins to diffuse as soon as it is formed, and thus the effective radius of the trail is smaller close to the meteor than it is farther back along the trail where diffusion has been going on for a longer time. The effective radius for any part of the trail is taken as $r = \sqrt{4Dt}$, the same value previously used for the whole trail. Now it can be shown that, except near the end of its path, the velocity of the meteor can be considered to be constant. Therefore, the position along the trail is proportional to t, and the relation $r = \sqrt{4Dt}$ indicates that the trail will have the form of a paraboloid of revolution. The geometry is shown in Fig. 13.8*a*, and the time variation in received power is given in Fig. 13.8*b*. The time $t = 0$ is the instant of closest approach of the meteor. In this case the peak power varies as q^2, as was true in the long-wavelength underdense trail case. With decreasing wavelength the power falls as λ^6 is compared with λ^3 for the long-wavelength underdense case. This rapid variation with wavelength is consistent with the fact that very few observations of meteor reflections have been made on frequencies above the VHF range. The case of short-wavelength reflections from high-density trails has been discussed by Sugar (1964). Statistical data on echo durations, line densities, and cross sections are given in Table 13.2.

13.4.4 Other aspects of reflections from meteor trails

We have considered reflections from meteor trails from a relatively idealized viewpoint, and have ignored many of the practical aspects. In this Section some of these are discussed. The discussion is directed toward the long-wavelength cases, since relatively few observations on short wavelengths have been made and analyzed.

Table 13.2 *Statistical characteristics of meteor data on 32 MHz**

	Echo duration (s)	Line density (electrons/m)	Cross section (m^2)
Lower decile	0·38	$1{\cdot}30 \times 10^{14}$	$1{\cdot}38 \times 10^{6}$
Lower quartile	0·48	$1{\cdot}75 \times 10^{14}$	$1{\cdot}63 \times 10^{6}$
Median	0·85	$4{\cdot}00 \times 10^{14}$	$2{\cdot}53 \times 10^{6}$
Upper quartile	1·20	$1{\cdot}03 \times 10^{15}$	$4{\cdot}70 \times 10^{6}$
Upper decile	7·50	$8{\cdot}60 \times 10^{15}$	$1{\cdot}30 \times 10^{7}$

* From Millman (1978)

13.4.4.1 Long-wavelength reflections during trail formation

The transient state associated with trail formation is of some interest since it accounts for some of the observed characteristics of trail reflections. As a trail is being formed, but before the meteor reaches the first Fresnel zone, a weak reflection is obtained from the incomplete trail. This comes primarily from the part of the trail corresponding to the shortest transmission path at the instant; when (as is usually the case) this is the head of the trail, the reflected signals are shifted in frequency because of the motion of the effective reflecting point. As the meteor approaches the first Fresnel zone for the trail this frequency shift approaches zero, and thus the received frequency decreases with time. The observed frequency will of course depend on trail orientation, meteor velocity, and observing wavelength. A maximum shift of about 5 kHz is possible on 50 MHz over a 1000 km path.

As the meteor crosses the first Fresnel zone, specular reflection occurs, accompanied by a large signal enhancement. Then as the particle continues on, the additional ionization from the lengthening trail creates a minor fluctuation of the signal (Fig. 13.9). Most trails do not have the proper orientation to produce a specular reflection, and in these cases only the weak Doppler-shifted component is observed.

13.4.4.2 Trail drift and distortion

The effects of ionospheric winds are appreciable for trails that last a second or more. A Doppler shift of the received frequency will be associated with the average wind velocity at the trail. For a velocity of 25 m/s this 'body Doppler' can be as large as 18 Hz for backscatter observations on 50 MHz and will be somewhat less for forward scatter observations.

To a first approximation the Doppler shifts produced by the winds (v) will have a Gaussian distribution (Millman 1978) given by

$$\Delta f(v) = \Delta f(\bar{v}) \exp\left[-\frac{(v-\bar{v})^2}{2\sigma^2}\right] \tag{3.11a}$$

where

$$\sigma = 2\left(\frac{2}{\pi}\right)^{1/2} \frac{\bar{v}}{\lambda}\left(\frac{a}{a+h}\right)\cos\xi \tag{3.11b}$$

and where $\bar{v}$ is the mean wind speed and ξ is the elevation angle to the meteor trail from the horizontal. Typical values of $\Delta f(\bar{v})$ on 40 MHz range from about 20 Hz, with $\xi = 0°$, to 0 Hz with $\xi = 90°$.

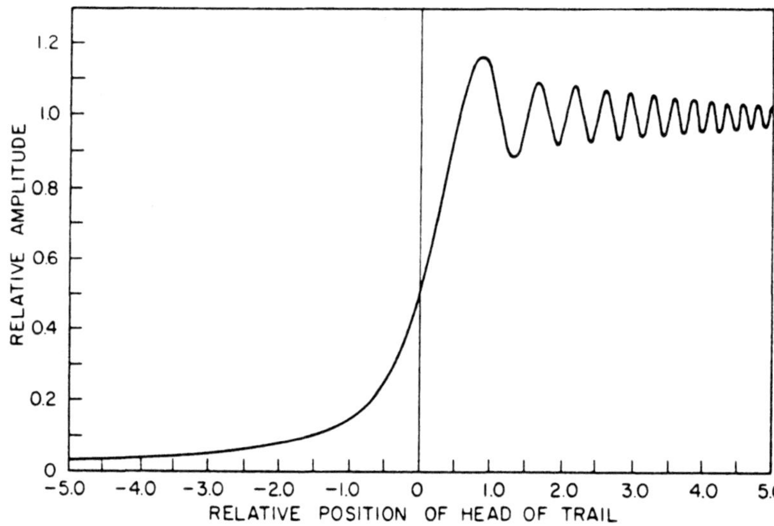

Fig. 13.9 *Amplitude variation of received signal as trail length increases*

The trail distortion resulting from wind shears can lead to the formation of several local first Fresnel zones for the trail since a distorted trail can have several points where the transmission path has a local minimum. These local minima, or 'glints', are strong scatterers, and the received signal is a composite of their contributions. Since they are moving at different velocities, the signals from each have different Doppler shifts and the composite signal fades in an irregular manner (see Herring and Forsyth 1980). On 50 MHz the fading rates observed for forward scatter are of the order of 1 to 10 Hz. In addition to producing the fading observed for long-enduring trails, the wind shears can rotate a trail sufficiently to produce reflections when the initial orientation was not suitable. Thus this type of trail loses its aspect sensitivity as time goes on, and if it lasts about 10 s it will scatter in all directions.

13.4.4.3 Diversity effects

As pointed out in Sec. 13.4.3, underdense trails act as small coherent sources and, therefore, CW signals scattered by them exhibit broad space and frequency correlation. Correspondingly, very little pulse broadening is observed when pulse signals are used. Over a forward scatter path about 1000 km long, the correlation coefficient observed (on 50 MHz) for single underdense trails falls to 0·5 for antenna spacings of about 150 km along the path, for antenna spacings of about 30 km across the path, and for frequency separations greater than 5 MHz. Three-microsecond pulses show no appreciable broadening under these conditions.

Overdense trails, in contrast to the underdense trails, tend to act like relatively large sources and, therefore, exhibit much poorer space and frequency correlation properties. The correlation for single overdense trails falls to 0·5 for antenna spacings of the order of 50 λ. Measurements obtained using short pulses over a 1000 km path indicate that the received composite signal can have a total time spread as large as 10 μs.

13.4.5 Short-term statistical characteristics

13.4.5.1 General remarks

From the viewpoint of radio propagation it is the collective properties of trail reflections rather than the individual reflections that are of primary interest. These statistical characteristics can be divided into two classes. First there are the short-term properties, those that would be observed during a single hour or less. Secondly are the long-term characteristics of the reflections, which include the diurnal, seasonal, and geographical variations. In the following Sections the initial radii are assumed to be zero.

13.4.5.2 Peak amplitude distribution

The peak amplitude of the reflection from an underdense trail is proportional to the electron line density q and, hence, proportional to the mass of the meteor. The number N of signals observed with peak amplitudes greater than A is given by

$$N \propto A^{-1} \quad \text{(for underdense trails)} \qquad (13.12)$$

$$N \propto A^{-4} \quad \text{(for overdense trails)} \qquad (13.13)$$

Between these two conditions there will be a transition region where contributions from both overdense and underdense trails will be important, and in this region the exponent is observed to be about $-1{\cdot}2$.

13.4.5.3 Duty cycle versus threshold

Duty cycle is defined as the fraction of time that a received signal exceeds a prescribed threshold. To a first approximation it is the fraction of time that the reception of messages will occur in a meteor-burst communication system and it is an important parameter in systems work.

The relationship between duty cycle T and the threshold A_1 for propagation by underdense trails is given by

$$T = \tau C A_1^{-1\cdot 2} \tag{13.14}$$

Here τ is the time taken for the amplitude to fall to (1/2·718. . .) of its initial value and C is a proportionality factor. The measured information duty cycle varies with wavelength as (Brown and Williams 1978)

$$T \propto \lambda^4 \tag{13.15}$$

13.4.5.4 Apparent location of trails

The geometrical conditions required for strong reflections from a trail combined with the radiant distribution for sporadic meteors favor trails occurring in some parts of the sky and discriminate against those in other parts of the sky. The result is that, for a given path, those meteors contributing to propagation occur in a relatively small region of the sky, the so-called 'hot spot'. For the longer transmission paths this region is near, but not at, the path midpoint (Brown 1985). The locations and intensities of the 'hot spots' vary with time of day, season, wave polarization, and path orientation, giving rise to the diurnal and seasonal variation in the arrival rate (Weitzen 1986; 1987).

13.4.6 Long-term statistical characteristics

13.4.6.1 Diurnal variations

The characteristics discussed in this Section include variations observed on an hour-by-hour or month-by-month basis and also variations associated with the characteristics of the observing system and its location. Nearly all data pertaining to these variations are relatively poor because of the large, seemingly random, hour-to-hour fluctuations observed in meteor activity. This random component of meteor activity is roughly comparable with a random variation in transmitter power output over a 10 dB range (Sec. 10.5.5).

The diurnal variation in meteor rate results in substantial diurnal variations in many of the characteristics of meteor propagation. Three basic diurnal factors are significant: (i) The meteor arrival rate varies; (ii) meteor velocities vary; and (iii) the effective radiants vary. The meteor arrival rate will have a roughly sinusoidal variation with a maximum at about 6 a.m. and a minimum at about 6 p.m. The average ratio of maximum to minimum is about 4.

The average meteor velocity also has a roughly sinusoidal variation; it reaches maximum at about 6 a.m. and minimum at about 6 p.m. The diurnal variation in velocity cannot exceed 60 km/s (twice the orbital velocity of the Earth) and, in practice, averages much less than this. A variation of 10–15 km/s above and below a daily mean velocity of 35–45 km/s is typical. The velocity variation is not in itself of much significance in radio work but its consequences are. The velocity variation leads to a diurnal variation in trail height; this, in turn, because of the variation of diffusion coefficient with height, leads to a diurnal variation in the duration of meteor bursts. Observations of the diurnal variation of the $1/\varepsilon$ duration of underdense bursts (Fig. 13.10) indicate that the duration varies over a 2 : 1 range through the day.

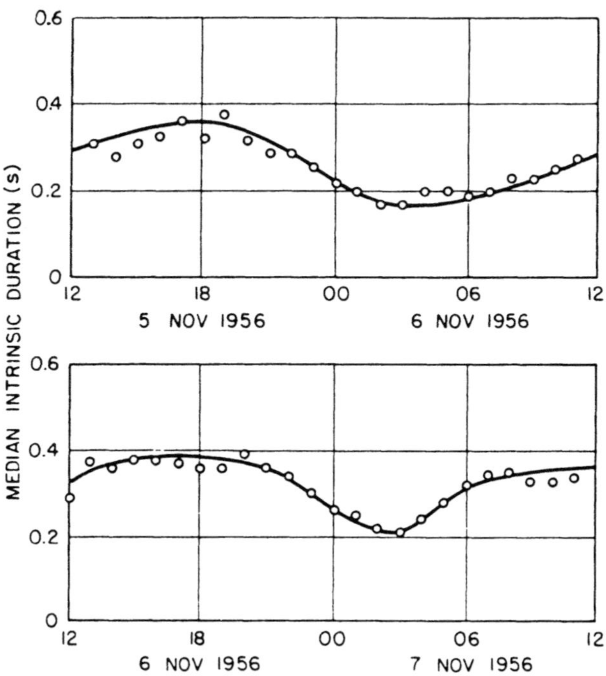

Fig. 13.10 *Diurnal variation of the $1/\varepsilon$ duration of underdense meteor bursts (From Keary and Wirth 1960 © Academic Press; reprinted with permission)*

Because of the orbital motion of the Earth, most meteors appear to have radiants concentrated in the hemisphere ahead of the Earth (the morning hemisphere); very few radiants appear to be behind the Earth. As the Earth rotates, the position of the dominant radiants changes relative to the transmission path, and the location of the 'hot spot' moves around. The effects of diurnal variation of radiants are indicated in Fig. 13.11 (Hines 1956), which depicts the predicted diurnal variation in meteor rates for each side of a north–south and an east–west transmission path. The differences in the shapes of the various curves indicate the effects of radiant changes relative to path

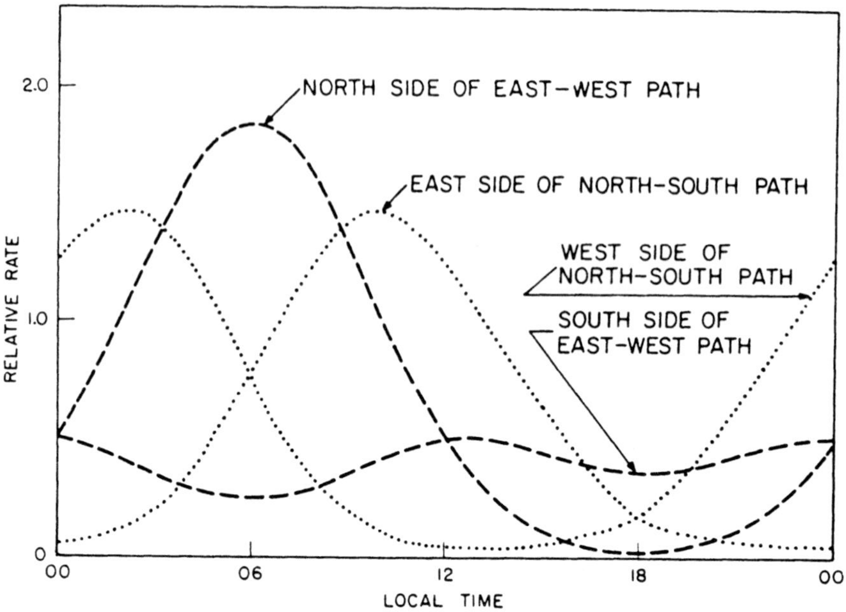

Fig. 13.11 *Predicted diurnal variation in meteor rates (From Hines 1956, © Pergamon Press; reprinted with permission)*

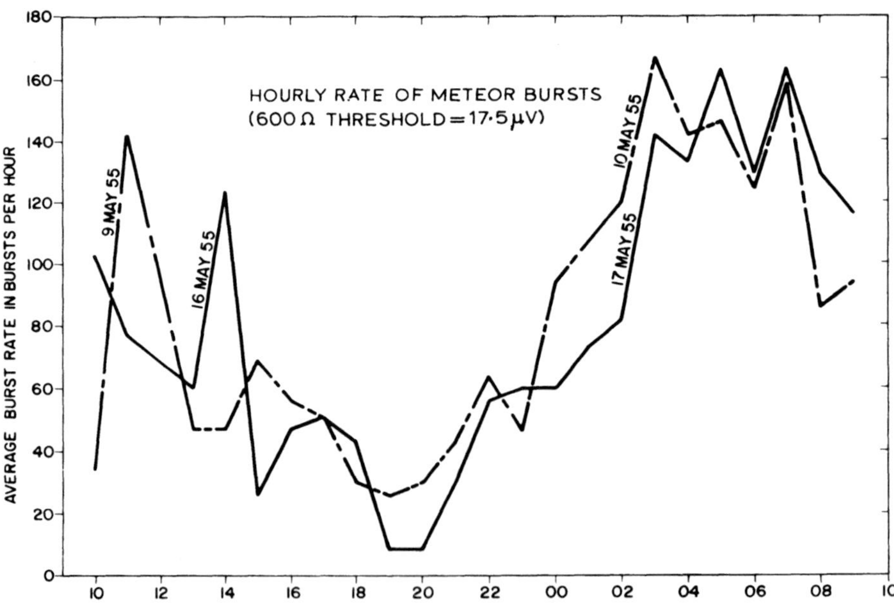

Fig. 13.12 *Observed diurnal variation of meteor burst rates (Cedar Rapids – Sterling) on 50 MHz (After Sugar 1964)*

orientation. The sum of all four curves would be an approximate indication of the diurnal variation in the arrival rate of meteors.

The combination of the three described variations with the random components of arrival rates results in rather irregular daily variations of meteor rates. Fig. 13.12 compares the rates observed on 50 MHz over a 1245 km path on two days, a week apart. Although the data for each day show the same general characteristics, the detailed variations on the two days are quite different.

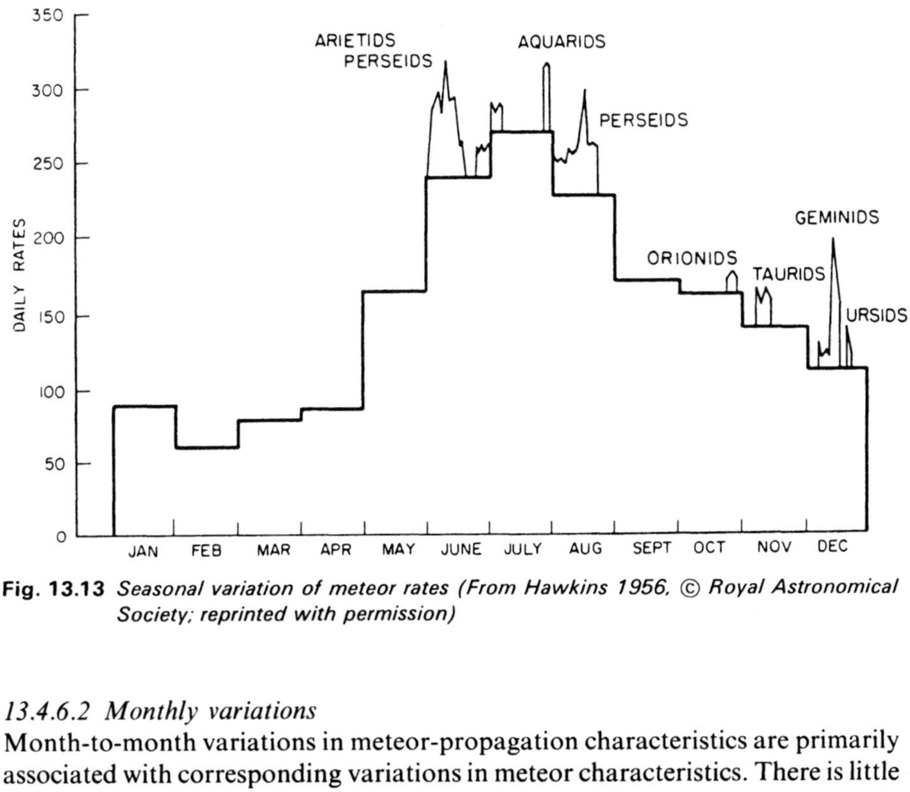

Fig. 13.13 *Seasonal variation of meteor rates (From Hawkins 1956, © Royal Astronomical Society; reprinted with permission)*

13.4.6.2 Monthly variations

Month-to-month variations in meteor-propagation characteristics are primarily associated with corresponding variations in meteor characteristics. There is little evidence that any of the normal month-to-month variations in ionospheric characteristics are reflected in meteor propagation characteristics, with the possible exception of D-region absorption. The seasonal variation of meteor rates is illustrated in Fig. 13.13 (Hawkins 1956). The echo rates shown are for meteor trails with an electron line density of $2{\cdot}75 \times 10^{14}$ electrons/m or greater. The contributions of shower meteors are shown above the histogram for the sporadic meteors. The relative contributions of shower meteors will decrease as the system sensitivity is raised, since the two types of meteors have different mass distributions.

13.4.6.3 Geographical variations

Information on geographical variations is available for Greenland (Ostergaard *et al.* 1985), India (Rao *et al.* 1975), and other countries (JATP 1978).

13.4.6.4 Abnormal absorption

Since most meteor trails are produced at heights above the absorbing region, the meteor signals are absorbed in much the same way as HF signals. On frequencies much greater than 30 MHz, the normal D-region absorption is negligible. Enhanced absorption in high latitudes can affect meteor propagation on frequencies as high as 100 MHz and higher (Ostergaard *et al.* 1985). For example, during a PCA, the 104 MHz duty cycle deviated by a ratio of 1000 : 1 from normal over the Goose Bay–Ottawa path (Maynard 1968). Polar cap absorption caused an extended series of blackouts on the Edmonton–Yellowknife circuit in July 1958. When there is sufficient ionization in the D region there can be sufficient Faraday rotation of the electric field to affect the operation of systems employing co-planar transmitting and receiving antennas (Cannon 1986; Cannon *et al.* 1985) and affect the apparent absorption (Baggaley 1979).

13.4.7 Use of meteor reflections for intermittent radio communications

13.4.7.1 General characteristics

Meteor-burst communication systems are basically weak-signal systems because the signal loss associated with the meteor-trail reflection is relatively high. For example, a typical system operating on 50 MHz over a 1300 km path with a transmitter power of 2 kW was commonly set to transmit messages whenever the signal at the receiver exceeded 2×10^{-14} W (2 μV open-circuit voltage for a 50 Ω source). The system loss is 170 dB; under similar circumstances ionospheric scatter propagation would exhibit a system loss of about 180 dB. Of the 170 dB total about 90 dB represents the attenuation associated with the length of the transmission path and 80 dB the scattering loss. Some parameters of a typical meteor-burst communication system are given in Table 13.3.

As is implied in the title of this Section, meteor comunication systems operate in an intermittent mode. They transmit messages only during the brief intervals when meteor propagation is present. Control of message transmission is achieved by providing for the two-way transmission over the propagation path, on adjacent frequencies, and transmitting only a control signal when a meteor trail is not present. When a meteor burst occurs, its presence is indicated by reception of the control signal, and message transmission can begin. These principles are incorporated in the JANET system (Maynard 1968), and in the COMET system (Bartholomé and Vogt 1968, Brown and Williams 1978), which uses automatic request for repetition (ARQ). ARQ is particularly suited to intermittent operation. In the interval between suitable meteor signals a system accepts messages and stores them for subsequent transmission. The system is, in essence, one that adapts its operation to the conditions of the transmission medium; it transmits only under favorable conditions. The aim is, of course, to obtain better overall transmission by avoiding the times when transmission is poor. Whether or not the intermittent mode of operation results in an overall improvement over continuous operation depends on the details of the equipment and on the characteristics of the propagation medium (Goodman 1987, Session 3A; Harper 1985).

Table 13.3 *Parameters of a typical meteor-burst communication system**

Carrier frequency	40–110 MHz
Transmitter power	200–2000 W
Bandwidth	100 kHz
Doppler spread	5 Hz
Time spread	1 μs
Information duty cycle	2·5–5%
Average message delay	10–80 s
Worst message delay	1–5 min

* Milstein *et al.* (1987)

The efficiency, or capacity, of a communication system can be measured by its throughput, which is the amount of correct data that can be passed per second. With meteor-burst systems the throughput is low because of the relatively long waiting time required to pass a message, that is, the time it takes to find a properly oriented meteor trail off which to reflect the signal, plus the propagation time (Weitzen *et al.* 1984; Weitzen 1988). Increasing the transmitter power reduces the waiting time by making possible the use of lower-density trails. Thus meteor-burst systems have an important role for uses in which waiting time is not critical, such as data collection from remote sites (Herman *et al.* 1987). With ground-to-ground links the background noise is cosmic noise.

13.4.7.2 Channel capacity

For an intermittent system operated at a fixed transmitter output and with band-limited Gaussian noise interference, the relation between channel capacity M and system bandwidth B is given by (Sugar 1964)

$$M \approx kB^{0\cdot 4} \tag{13.16}$$

where k is a constant of proportionality.

13.4.7.3 Models of meteor-burst systems

Several computer models have been constructed for the determination of meteor-burst channel characteristics (Brown 1985; Milstein *et al.* 1987; Weitzen 1986; Haakinson 1983). These models predict, for example, the occurrence rate of trails exceeding some threshold, the locations and intensities of the 'hot spots', the reliability of a system, and the waiting period required to transmit a message as functions of local time and season. In recent years artificial intelligence techniques have been introduced into the handling of meteor-burst data (Weitzen 1987).

13.4.8 Meteor wind measurements

Meteor radars have been used globally to measure winds in the meteor region (JATP 1978). One technique involves the measurement of Doppler shifts on two beams with known elevations and azimuths. A typical system (Avery *et al.* 1983) has one beam pointing northeast and the other pointing northwest (Fig. 13.14). From measurements of the Doppler shifts, Δf_{NE} and Δf_{NW}, on the two antennas, the wind speed W and the azimuth ψ can be obtained from the sum $S\,(=\Delta f_{NE} + \Delta f_{NW})$, and difference $D\,(=\Delta f_{NE} - \Delta f_{NW})$, of the Doppler shifts:

$$\psi = \tan^{-1}(S/D) \tag{13.17}$$

$$W = \frac{1}{\sqrt{2}}\frac{\lambda}{\sin\phi}(S^2 + D^2)^{1/2} \tag{13.18}$$

where λ is the wavelength and ϕ is the angle between the beams and the vertical.

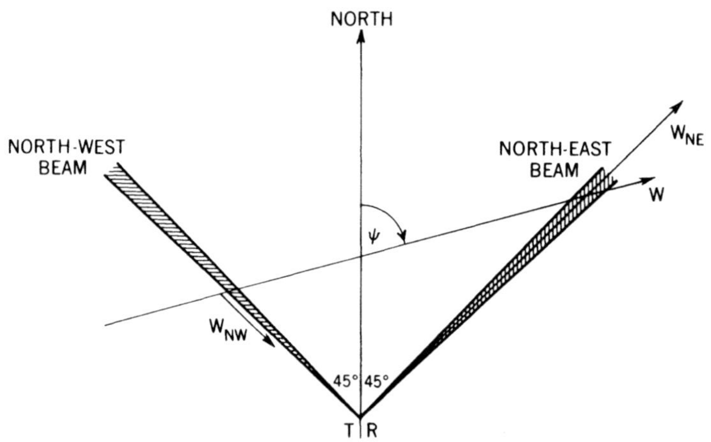

Fig. 13.14 *Geometry of Doppler radar system for wind measurement*

Typical wind speeds are in the range 0–25 m s^{-1}, and the direction rotates with diurnal, semi-diurnal, etc., components (Aso *et al.* 1979; Clark 1975; JATP 1978). Observations of range and elevation are used to measure height profiles (Fig. 13.15). Ellyett (1977) showed that there is an inverse relationship between meteor rates and sunspot numbers, which implies changes in neutral atmospheric densities.

13.5 Auroral scatter

13.5.1 Radio aurora

Radio aurora is ionization associated with auroral phenomena that produces radio echoes. Since ionization and visible emissions are not the same we should not expect one-to-one correlation between the spatial and temporal occurrences of these two phenomena. For example, radio aurora can occur during the daytime and it consists of magnetic field-aligned columns or sheets. Thus the reflections from radio aurora depend on both the ionization and the geometry (Chesnut 1972; Hanuise 1983). These aspect-sensitive reflections will be observed only when the transmitter and receiver are correctly positioned with respect to the Earth's magnetic field, but close positioning is not always required (Koehler *et al.* 1985).

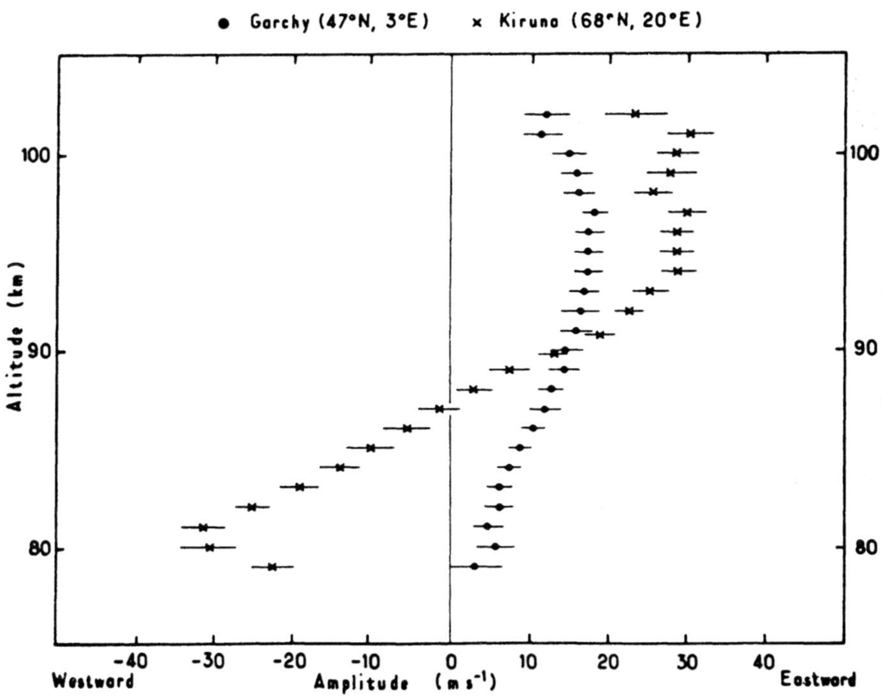

Fig. 13.15 *Prevailing winds at Garchy (France) and Kiruna (Sweden), 11–14 August 1974 (From Glass et al. 1978 © Pergamon Press; reprinted with permission)*

13.5.2 Bistatic observations

The term bistatic is used to denote the case in which the transmitter and receiver are geographically separated. Auroral observations usually involve continuous-wave transmissions. Canadian workers (Forsyth 1968) have identified the following types of signal enhancements: (i) E-events–probably from sporadic E, (ii) S-events that are not accompanied by magnetic disturbances, (iii) A_1 events with rapid (flutter) fading ($\approx$ 100 Hz) and strong aspect sensitivity, (iv) A_2 events with slow fading (1–0·01 Hz) and slight aspect sensitivity, and (v) A_3 events with slow fading and no aspect sensitivity (i.e. the signals arrive in the great circle and are caused by isotropic irregularities). Radio aurora has been observed on frequencies up to 3000 MHz.

13.5.3 Monostatic (radar) observations

In the case where receiver and transmitter essentially coincide, the ray path is perpendicular to the reflecting irregularity; thus for a given radar location the reflection height, elevation, and range are fixed by geometry. For example, taking the dip in the auroral zone as 78° and a ground range of 1000 km, the angle of elevation is near 3°, the reflection height is 133 km, and the echo range is 1018 km. For a given height of reflection there is a maximum range corresponding to zero elevation. Usually field-aligned echoes cannot be observed when the radar beam deviates more than a few (e.g. 10°) degrees from perpendicularity; this angle decreases with increase of wave frequency (e.g. 6° on 780 MHz). Field-aligned scattering is, in general, quite sporadic and is associated with aurora. Often, auroral echoes arise in the E region at heights from about 80 km to about 120 km. Sometimes they are reflected from heights in the range of about 250–600 km. Auroral radar echoes are of practical importance since they give rise to interference in radar (over-the-horizon) and communication systems (AGARD 1972). They are responsible also for certain types of spread-F on ionograms.

The scattering cross section σ diminishes exponentially with increasing frequency. The power reflection coefficient is about 10^{-4} on 106 MHz (Chesnut 1972). In treating backscatter it is therefore customary to invoke the Born approximation that signals scattered more than once are negligible. This type of scattering is sometimes called coherent scattering to distinguish it from incoherent scatter. Multiple scattering refers to the case where signals scattered more than once are not negligible.

In the case having a number of field-aligned (parallel) irregularities the received signal will be strongest when the individual echoes interfere constructively, that is, when the spacing is a multiple of $\lambda/2$. Thus a reflected wave tends to pick out structures that are related to the wave frequency. This must be borne in mind when echo features are interpreted in terms of ionospheric structures.

Auroral radars can be used to find the motions of radio auroras from the time rate of change of the range and/or from the Doppler shift (Schlegel and Nielsen 1985). The auroral drift motions are of the order of 300–1000 m s^{-1}. The motion tends to be from west to east before midnight and from east to west after midnight. This changeover in velocity occurs at the Harąng discontinuity (Schlegel and Nielsen 1985) where backscatter is weak. Some examples of spectra of auroral echoes are shown in Fig. 13.16.

The diurnal variation of 1 m radio aurora irregularities shows two maxima: one in the early morning and another in the afternoon. The seasonal variation of occurrence shows maxima at the equinoxes. The long-term occurrence has a maximum about 2 years after the peak of the sunspot cycle, as is the case with visual aurora.

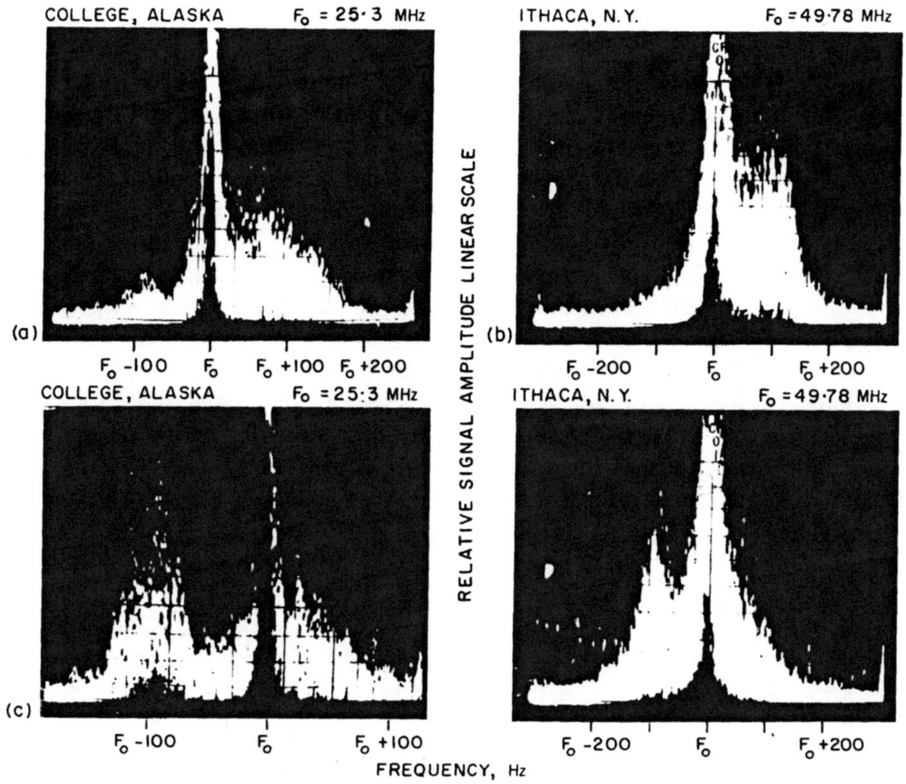

Fig. 13.16 *Spectra of auroral echoes reflected from radio aurora*

During quiet geomagnetic conditions the aspect-sensitive echoes are essentially perpendicular to the undisturbed magnetic field. During strong disturbances, the geomagnetic field can be distorted so that several new regions may exhibit the perpendicular condition and this, together with the increased strength of the irregularity itself, will produce stronger echoes. Refraction plays an important role in aspect sensitivity by changing the direction of the radar beam in the scattering region; therefore the geometry is not determined simply by the transmitter, the receiver, and the Earth's magnetic field.

Measurements indicate that the amplitude fading of auroral radar echoes on 93 MHz has a Rayleigh distribution that supports a model of a large number of independent scattering centers.

Auroral ionization has adverse effects on certain radar signals that pass through the ionosphere, as for example on radar signals that track space vehicles.

13.6 Equatorial scatter

13.6.1 E-layer scatter

E-region echoes rather similar to auroral radar echoes are observed near the magnetic equator. There the Earth's magnetic field is horizontal and the radar beam is perpendicular to the magnetic field lines for all angles in the east-west plane. The Doppler spectra are somewhat similar to those for auroral echoes shown in Fig. 13.16. At the equator, positive Doppler shifts are observed toward the east and negative shifts toward the west. No average frequency shift is observed overhead, so one would conclude that the irregularities are moving from east to west. However, observations show that the frequency shifts are independent of the elevation angle; these observations are inconsistent with a model of field-aligned columns moving from east to west. An alternative explanation is that the irregularities are plane waves. The radar reflections are obtained from those waves whose fronts are perpendicular to the beam. The waves travel essentially at the sound speed (on 50 MHz $\Delta f \approx$ 100 Hz) in the E region and are believed to be ion–acoustic waves produced by the equatorial electrojet (Farley 1985).

13.6.2 F-layer scatter

Equatorial F-scatter is most prevalent around the equinoxes and between 1800 and 2400 local time. Incoherent scatter records (Fig. 8.31) show that the small-scale (3 m) irregularities are in the form of plumes that arise from the lower boundary of the F_2 layer and rise to heights of about 1000 km. These irregularities give rise to spread F on vertical ionograms, to large increases in maximum observed frequencies on oblique ionograms (see Fig. 6.19), and intense scintillation on satellite-to-ground signals (see Sec. 8.5). The belt of equatorial irregularities lies between $+20°$ and $-20°$ of dip latitude. There is a broad spectrum of sizes of irregularities extending from about 1 m to more than 1 km. The irregularities are thought to be produced by plasma instabilities (Sudan and Keskinen 1984).

13.6.3 Communication effects

From the point of view of radio communications the most disastrous aspect of F-scatter is the occurrence of flutter fading during the prime listening hours (1800–2400 h). During this period voice modulation may be unrecognizable. The increase in maximum frequency can lead also to harmful interference. The signal strength (in dB) falls off rapidly with wave frequency, e.g. as f^{-11} (Kuriki *et al.* 1972).

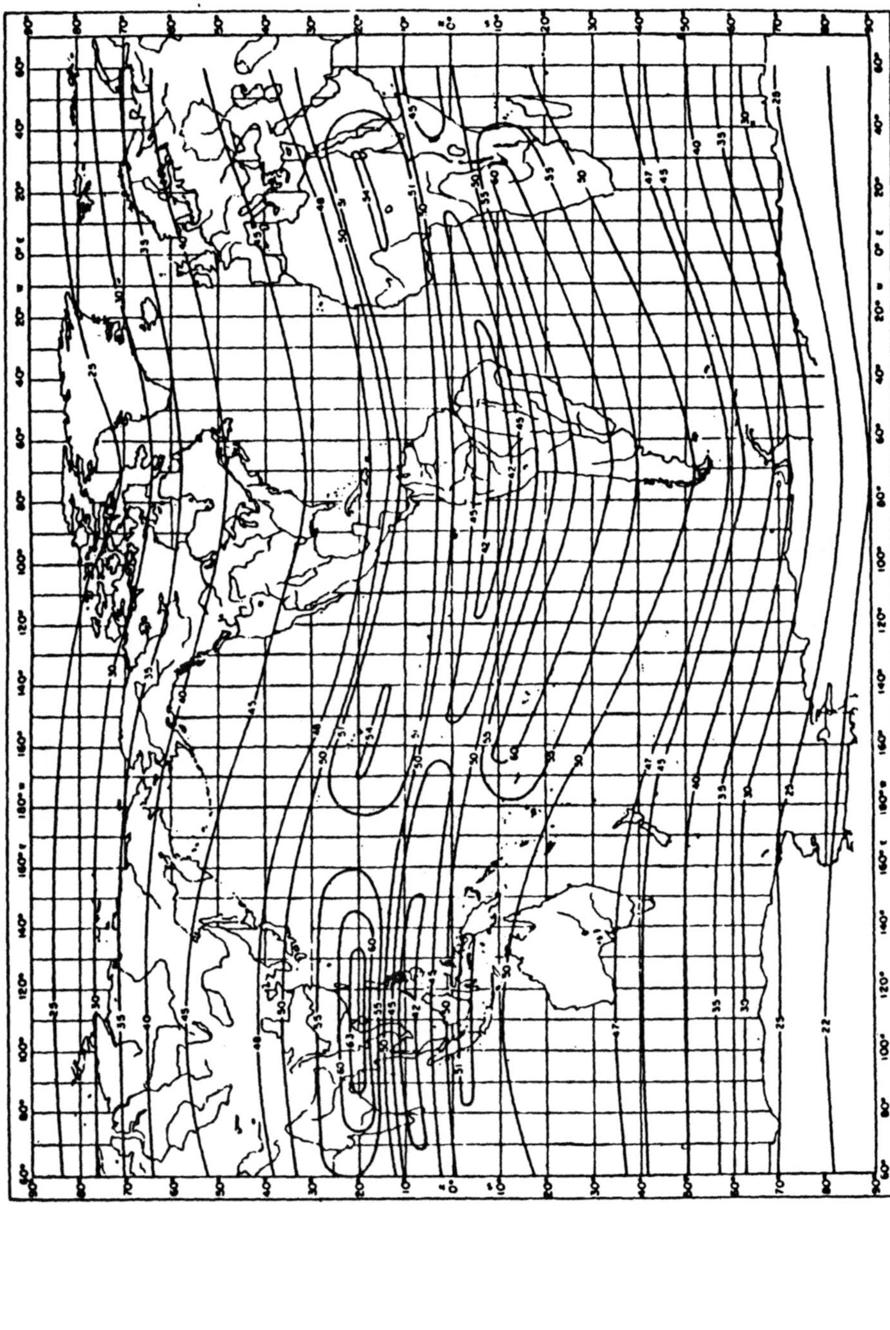

Fig. 13.17 *Areas where specified values of MUF(4000)F_2 are exceeded 1% of the time (Courtesy E. K. Smith)*

13.7 VHF propagation by ionospheric layers

13.7.1 F-layer propagation

Near the peaks of some large sunspot cycles long-distance propagation of VHF is possible via the F_2 layer in middle latitudes on frequencies of 50 MHz and higher. In low latitudes regular transmission occurs on frequencies of about 30 and 40 MHz, and it can occur on > 60 MHz by means of F–F paths (see Sec. 6.7.2.6). Such signals can be important because they cause interference. This is illustrated in Fig. 13.17, which shows areas where specified values of the MUF(4000)F_2 are exceeded during 1% of the time. Equatorial spread-F has a complicated response to magnetic disturbance (Watanabe and Oya 1986).

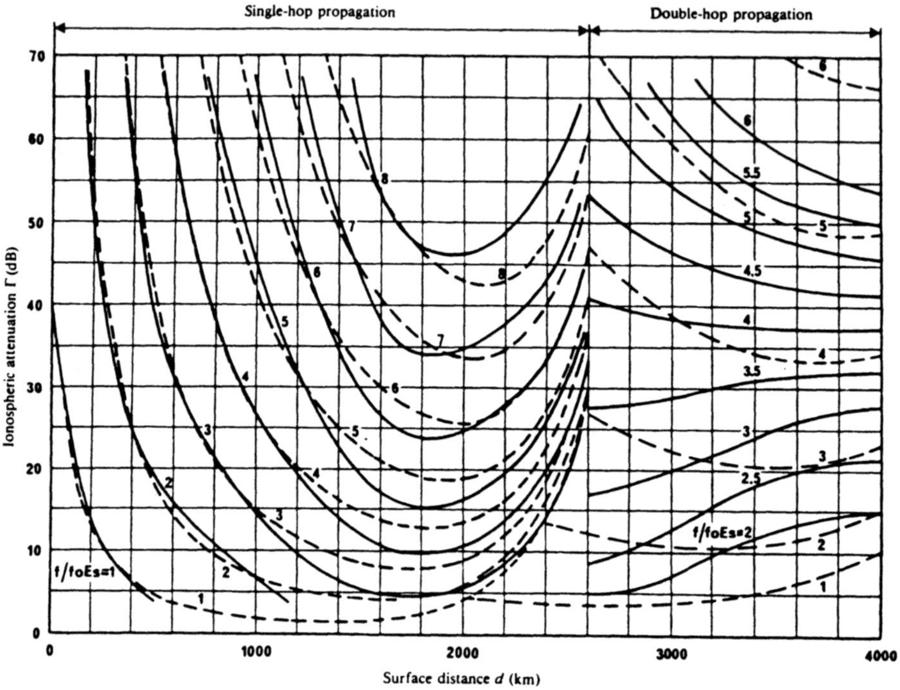

Fig. 13.18 *Signal attenuation for E_s propagation (From Miya et al. 1978, © American Geophysical Union)*
--- From observed values
—— From empirical formulas

13.7.2 Propagation by means of sporadic E

The occurrence of sporadic-E propagation decreases with increases of frequency. It is usually negligible on frequencies above about 90 MHz except as a source of interference to circuits with high reliability requirements. Sporadic-E propagation, on frequencies higher than 30 MHz, is rare over distances of less than 500 km.

There are three major geographical types of E_s:

(*a*) Equatorial (q-type) E_s is a regular daytime phenomenon associated with the equatorial electrojet.
(*b*) Temperate latitude E_s is a summer daytime feature (Smith 1978; Wagner 1987).
(*c*) Auroral sporadic E is most prevalent at night.

A statistical study of E_s in Europe (Edwards *et al.* 1984) on frequencies between 59 and 77 MHz shows an absence of echoes in the winter months, a marked increase in May, and a cessation in September. The echoes are present only during daylight, and there is no dependence on sunspots or wave polarization.

Miya *et al.* (1978) gave a method of calculating the strengths of signals transmitted by means of sporadic E as functions of distance and frequency (f/f_oE_s). The ionospheric attenuation for one-hop and two-hop paths is shown in Fig. 13.18. For one-hop paths the attenuation has minima for distances slightly less than 2000 km (CCIR 1986, Recommendation 534–2).

13.8 References

AGARD (1972): 'Radar propagation in the Arctic' *in* MILLMAN, G. H. (Ed.): AGARD Conf. Proc. No. 97, Advisory Group for Aerospace Research and Development, North Atlantic Treaty Organisation

ASO, T., TSUDA, T., and KATO, S. (1979): 'Meteor radar observations at Kyoto University', *J. Atmos. Terr. Phys.*, **41**, p. 517

AVERY, S. K., RIDDLE, A. R., and BALSLEY, B. B. (1983): 'The Poker Flat, Alaska MST radar as a meteor radar', *Radio Sci.*, **18**, p. 1021

BAGGALEY, W. J. (1979): 'D-region absorption and forward-scatter radio-meteor data', *J. Atmos. Terr. Phys.*, **41**, p. 671

BAILEY, D. K., BATEMAN, R., and KIRBY, R. C. (1955): 'Radio transmission at VHF by scattering and other processes in the lower ionosphere', *Proc. IRE*, **43**, p. 1181

BARTHOLOMÉ, P. J. (1968): 'Survey of ionosphere and meteor scatter communications' *in* FOLKESTAD, K. (Ed.): 'Ionospheric radio communications' (Plenum Press, New York) p. 145

BARTHOLOMÉ, P. J. and VOGT, I. M. (1968): 'COMET – A new meteor burst system incorporating ARQ and diversity reception', *IEEE Trans.*, **COM-16**, 2, p. 268

BLAIR, J., DAVIS, R. M., and KIRBY, R. C. (1961): 'Frequency dependence of D-region scattering at VHF', *J. Res. NBS*, **65D**, p. 417

BROWN, D. W. (1985): 'A physical meteor burst propagation model and some significant results for communications systems design', *IEEE J. Selec. Areas Comm.*, **SAC-3**, p. 745

BROWN, D. W. and WILLIAMS, H. P. (1978): 'The performance of meteor-burst communications at different frequencies' *in* INCE, A. N. (Ed.): 'Aspects of electromagnetic wave scattering in radio communications'. AGARD Conf. Proc. No. 244, Advisory Group for Aerospace Research and Development, North Atlantic Treaty Organisation, p. 24–1

CANNON, P. S. (1986): 'Polarization rotation in meteor burst communications systems', *Radio Sci.*, **21**, p. 501

CANNON, P. S., DICKSON, A. H., and ARMSTRONG, M. H. (1985): 'Meteor scatter radio communication at high latitudes' *in* SOICHER, H. (Ed.): 'Propagation effects on military systems in the high latitude region'. AGARD Conf. Proc. No. 382, Advisory Group for Aerospace Research and Development, North Atlantic Treaty Organisation, p. 9–1

CCIR (1986): 'Propagation in ionized media. Recommendations and reports of the CCIR, 1986 VI', Int. Telecom. Union, Geneva

CHESNUT, W. G. (1972): 'Polar propagation effects on VHF-UHF radars' *in* MILLMAN, G. H. (Ed.): 'Radar propagation in the Arctic'. AGARD Conf. Proc. No. 97, Advisory Group for Aerospace Research and Development, North Atlantic Treaty Organisation, p. 5-1

CLARK, R. R. (1975): 'Meteor wind measurements at Durham, New Hampshire (43°N, 71°W)', *J. Atmos. Sci.*, **32**, p. 1689

COLLINS, C. and MAYNARD, L. A. (1968): 'Simultaneous VHF riometer and forward-scatter observations of the disturbed lower ionosphere' *in* FOLKESTAD, K. (Ed.): 'Ionospheric radio communications' (Plenum Press, New York) p. 155

DAVIES, K. (1965): 'Ionospheric radio propagation'. NBS Monograph 80

EDWARDS, K. J., KERSLEY, L., and SHRUBSOLE, L. F. (1984): 'Sporadic-E propagation at frequencies around 70 MHz', *Radio Electr. Engr.*, **54**, p. 231

ELLYETT, C. D. (1977): 'Solar influence on meteor rates and atmospheric density variations at meteor heights', *J. Geophys. Res.*, **82**, p. 1455

FARLEY, D. T. (1985): 'Theory of equatorial electrojet plasma waves: New developments and current status', *J. Atmos. Terr. Phys.*, **47**, p. 729

FORSYTH, P. A. (1968): 'Auroral scatter' *in* THRANE, E. V. (Ed.): 'Scatter propagation of radio waves'. AGARD Conf. Proc. No. 37, Part 2, Advisory Group for Aerospace Research and Development, North Atlantic Treaty Organisation, p. 34–1

GLASS, M., BERNARD, R., FELLOWS, J. L., and MASSEHEUF, M. (1978): 'The French meteor radar facility', *J. Atmos. Terr. Phys.*, **40**, p. 923

GOODMAN, J. M. (Ed.) (1987): 'The effect of the ionosphere on communication, navigation, and surveillance systems'. Naval Research Laboratory, Washington, DC

GREENHOW, J. S. and HALL, J. E. (1960): 'The variation of meteor heights with velocity and magnitude', *Monthly Notices Roy. Astron. Soc.*, **121**, p. 174

HAAKINSON, E. J. (1983): 'Meteor burst communications model'. NTIA Report 83–116, US Dept. of Commerce, Boulder, CO 80303

HANUISE, C. (1983): 'High-latitude ionospheric irregularities: A review of recent radar results', *Radio Sci.*, **18**, p. 1093

HARPER, R. V. (1985): 'Performance analysis for adaptive meteor burst communications' *in* GOODMAN, J. M. (Ed.): 'Effect of the ionosphere on C^{3}I systems'. US Govt. Printing Office, Washington DC, 20402, p. 137

HAWKINS, G. S. (1956): 'A radio survey of sporadic meteor radiants', *Monthly Notices Roy. Astron. Soc.*, **116**, p. 92

HERMAN, J. R., DUGGAN, C. C., THOMPSON, W. I., COSTA, R. A., and DELUCA, D. M. (1987): 'Narrowband digital voice communications over a meteor burst channel', *Electron. Lett.*, **23**, p. 59

HERRING, R. W. and FORSYTH, P. A. (1980): 'The use of large aperture radio systems for meteor studies', *Can. J. Phys.*, **58**, p. 779

HINES, C. O. (1956): 'Diurnal variations in forward scattered meteor signals', *J. Atmos. Terr. Phys.*, **9**, p. 229

INCE, A. N. (Ed.) (1978): 'Aspects of electromagnetic wave scattering in radio communications'. AGARD Conf. Proc. No. 244, Advisory Group for Aerospace Research and Development, North Atlantic Treaty Organisation

INCE, A. N., VOGT, I. M., and WILLIAMS, H. P. (1978): 'A review of scatter communications' *in* INCE, A. N. (Ed.): 'Aspects of electromagnetic wave scattering in radio communications'. AGARD Conf. Proc. No. 244, Advisory Group for Aerospace Research and Development, North Atlantic Treaty Organisation, p. 21–1

JATP (1978): [10 companion papers]. *J. Atmos. Terr. Phys.*, **40**, 8, pp. 879–943

KEARY, T. J. and WIRTH, J. H. (1960): 'Statistical characteristics of forward scattered radio echoes from meteor trails' *in* DESIRANT, M. and MICHIELS, J. L. (Eds.): 'Electromagnetic wave propagation' (Academic Press) p. 277

KIRBY, R. C. (1968): 'Review of VHF forward scatter' *in* THRANE, E. V. (Ed.): 'Scatter propagation of radio waves; 2'. AGARD Conf. Proc. No. 37, Advisory Group for Aerospace Research and Development, North Atlantic Treaty Organisation, p. 38–1

KOEHLER, J. A., SOFKO, G. J., and MEHTA, V. (1985): 'A statistical study of magnetic aspect effects associated with VHF auroral backscatter', *Radio Sci.*, **20**, p. 689

KURIKI, I., TANOHATA, K., SAKAMOTO, T., and IGUCHI, M. (1972): 'Propagational mode deduced from signal strength in the VHF band on the trans-equatorial path', *J. Radio Res. Labs. (Japan)*, **19**, p. 175

MAYNARD, L. A. (1968): 'Meteor burst communications in the Arctic' *in* FOLKESTAD, K. (Ed.): 'Ionospheric radio communications' (Plenum Press, New York) p. 165

MILLMAN, G. H. (1978): 'HF scatter from overdense meteor trails' *in* INCE, A. N. (Ed.): 'Aspects of electromagnetic wave scattering in radio communications'. AGARD Conf. Proc. No. 244, Advisory Group for Aerospace Research and Development, North Atlantic Treaty Organisation, p. 6–1

MILSTEIN, L. S., SCHILLING, D. L., PICKHOLTZ, R. L., SELLMAN, J., DAVIDOVICI, S., PEVELCHEK, A., SCHNEIDER, A., and EICHMANN, G. (1987): 'Performance of meteor burst communication channels', *IEEE J. Selec. Areas Comm.*, **SAC–5**, p. 146

MIYA, K., SHIMIZU, K., and KOJIMA, T. (1978): 'Oblique incidence sporadic-E propagation and its ionospheric attenuation', *Radio Sci.*, **13**, p. 559

OETTING, J. A. (1980): 'An analysis of meteor burst communications for military applications', *IEEE Trans.*, **COM–28**, p. 1591

OSTERGAARD, J. C., RASMUSSEN, J. E., SOWA, M. J., QUINN, J. M., and KOSSEY, P. A. (1985): 'Characteristics of high latitude meteor scatter propagation parameters over the 45–104 MHz band' *in* SOICHER, H. (Ed.): 'Propagation effects on military systems in the high latitude region'. AGARD Conf. Proc. No. 382, Advisory Group for Aerospace Research and Development, North Atlantic Treaty Organisation, p. 9.2

RAO, B. R., RAO, M. S., RATNAM, S. R., KRISHNA RAO, D. A. V., and RAO, E. B. (1975): 'Study of meteors during December, 1974 by forward scatter of VHF radio signals between Dehradun and Waltair', *Indian J. radio Space Sci.*, **5**, p. 183

SCHLEGEL, K. and NIELSEN, E. (Eds.) (1985): 'Irregularities in the high-latitude ionosphere', Special Issue, *Radio Sci.*, **20**, 4, pp. 675–818

SMITH, E. K. (1978): 'Temperate zone sporadic-E maps (f_oE_s > 7 MHz)', *Radio Sci.*, **13**, p. 571

SUDAN, R. N. and KESKINEN, M. J. (1984): 'Unified theory of the power spectrum of intermediate wavelength ionospheric electron density fluctuations', *J. Geophys. Res.*, **89**, p. 940

SUGAR, G. R. (1964): 'Radio propagation by reflection from meteor trails', *Proc. IEEE*, **52**, p. 116

WAGNER, L. S. (1987): 'Characteristics of mid-latitude sporadic E observed with a wideband HF channel probe', *Radio Sci.*, **22**, p. 728

WATANABE, S. and OYA, H. (1986): 'Occurrence characteristics of low latitude ionospheric irregularities observed by impedance probe on board the Hinotori satellite', *J. Geomag. Geoelectr.*, **38**, p. 125

WEITZEN, J. A. (1986): 'Predicting the arrival of meteors useful for meteor burst communication', *Radio Sci.*, **21**, p. 1009

WEITZEN, J. A. (1987): 'A data base approach to analysis of meteor burst data', *Radio Sci.*, **22**, p. 133

WEITZEN, J. A. (1988): 'Meteor scatter propagation', *IEEE Trans.*, **AP–37**, p. 1813

WEITZEN, J. A. BIRKMEYER, W. P., and GROSSI, M. D. (1984): 'An estimate of the capacity of the meteor burst channel', *IEEE Trans.*, **COM–32**, p. 972

WHEELON, A. D. (1960): 'Relation of turbulence theory to ionospheric forward scatter propagation experiments', *J. Res. NBS*, **64D**, p. 301

Chapter 14

Ionospheric modification

14.1 Means of modification

The ionosphere can be affected (modified) by a variety of disturbances. Natural sources include, for example, severe weather, earthquakes, and volcanoes; these create mechanical disturbances (waves) in the neutral atmosphere which propagate to the ionosphere where they couple into the ionized gas. Anthropogenic disturbances include, for example, high explosives, nuclear explosions, heating by electromagnetic waves, and intentional and unintentional seeding of the ionosphere with chemicals.

14.2 Natural modifications

14.2.1 Effects of severe weather

There is considerable evidence that transient disturbances occur in the ionosphere as a result of weather in the troposphere (e.g. Davies and Jones 1972; JATP 1968). Although electrical coupling exists between the troposphere and the conducting ionosphere, the more noticeable effects are those transmitted mechanically in the form of acoustic–gravity waves that arise from hurricanes (e.g., Bauer 1958) and weather fronts (e.g. Bertin *et al.* 1975).

Using the HF Doppler technique (Sec. 7.8.3), several investigators (Hung *et al.* 1979; Okuzawa *et al.* 1986; Prasad *et al.* 1975) have found acoustic and gravity waves in the F region originating in severe thunderstorms like that represented in Fig. 14.1, for example. Such acoustic waves have two dominant periods (one near 3·5 min and the other near 5 min) as shown in the spectrum in Fig. 14.2. The following criteria have been established for detectable signals:

(i) Cloud tops must be greater than about 40 000 ft (12 km)
(ii) Thunderstorm cells must lie within 200–250 km from the point on the ground directly below the reflection point of the probing radio signal.

Thunderstorms meeting these criteria occur along severe weather fronts in the

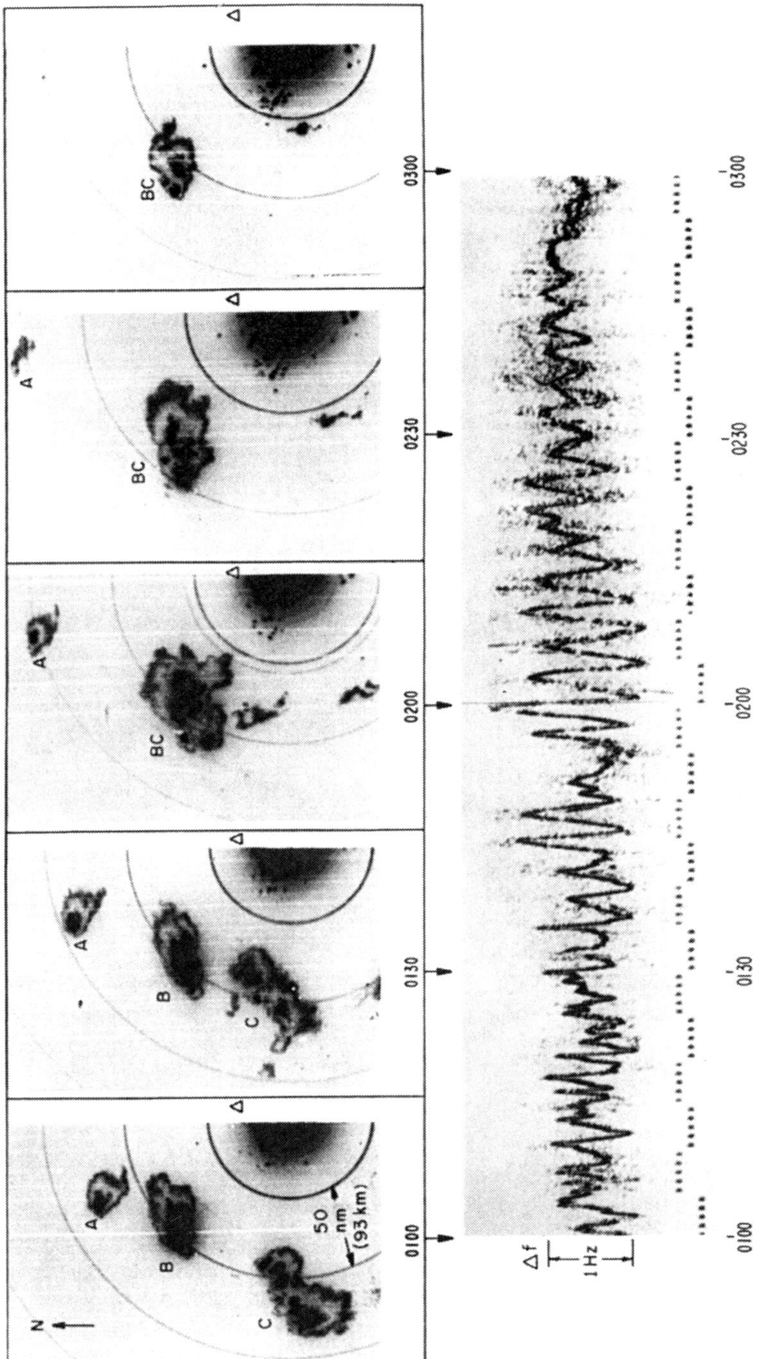

Fig. 14.1 *Above: Radar echoes of severe weather at Norman, Oklahoma*
Below: Related variations of ionospheric Doppler frequency
The ionospheric effects maximize when the clouds marked B and C merge

central USA (e.g. Texas, Oklahoma, Kansas – see Georges 1968; Davies and Jones 1973) mostly during April and May. The ionospheric disturbance can last for several hours as seen in Fig. 14.1. With measurements from spaced transmitters on several frequencies the velocity vector of the acoustic disturbance has been determined. Acoustic ray tracing in an atmospheric model can be used to determine the source location, which can be in agreement with a severe storm located by a weather radar (Fig. 14.3).

The 'peaky' spectrum of an acoustic wave is thought to result from atmospheric filtering of a broad-band noise source (Georges 1973). The power flux in an acoustic disturbance in the ionosphere is about $35 \times 10^{-6}\,\mathrm{W\,m^{-2}}$ which, neglecting acoustic wave attenuation, corresponds to a source power of about 2×10^6 W (Davies and Jones 1973).

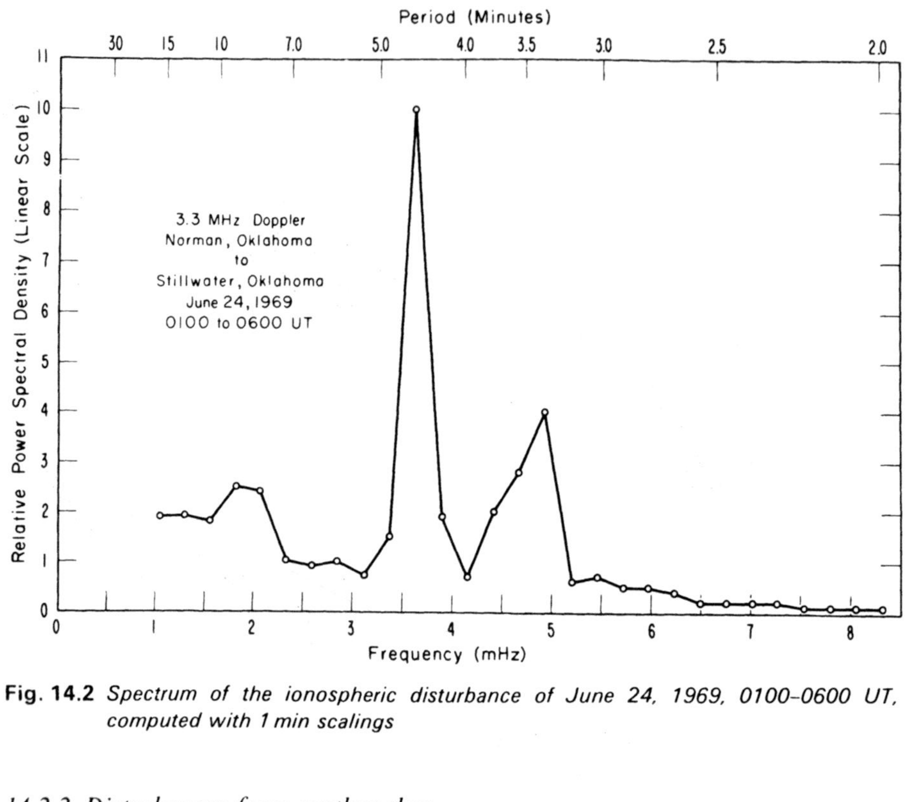

Fig. 14.2 *Spectrum of the ionospheric disturbance of June 24, 1969, 0100–0600 UT, computed with 1 min scalings*

14.2.2 Disturbances from earthquakes

Ionospheric disturbances have been detected which have been ascribed to earthquakes such as the Alaskan earthquake of 0336 UT, March 28, 1964 (Davies and Baker 1965; Leonard and Barnes 1965; Row 1966, 1967) and the Japan earthquake of 0048 UT on May 16, 1968 (Yuen *et al.* 1969). For the March 1964 event, short-period ($\approx$ 1 min) and long-period ($\approx$ 1 h) waves were detected at Boulder and elsewhere. The short-period waves are believed to have come from

Rayleigh (seismic) waves propagating near the Earth's surface with a speed of about 2·6 km s^{-1}. The long-period wave traveled in the atmosphere with a group speed of about 800 m s^{-1}, indicating a gravity wave ducted in the lower atmosphere (below 100 km).

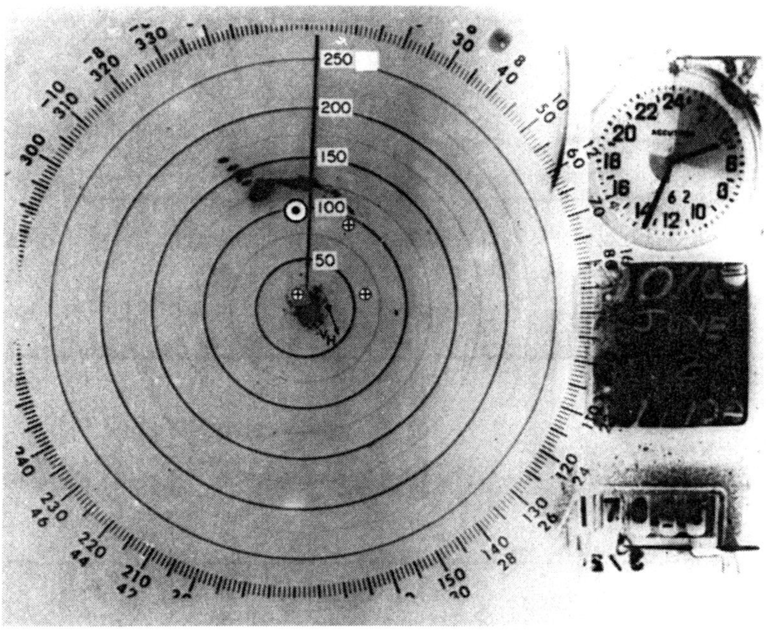

Fig. 14.3 *Oklahoma City weather radar photograph at 0434 UT (marked in nautical miles), with the horizontal trace velocity V_H for the ionospheric disturbance of June 11, 1970, 0300–0700 UT; $V_H \approx$ 850 m s^{-1}*
⊙ represents the point of origin calculated from ray tracing
⊕s represent locations of transmitters

Ionospheric waves with periods near 25 s have been observed in Hawaii from Rayleigh waves originating in the Hachinohe epicenter. The Rayleigh (ground) waves, pushing upward on the atmosphere, excited acoustic waves in the neutral atmosphere that propagated to the F_2 region. Fig. 14.4 shows a seismic record in Hawaii superposed on an HF Doppler oscillation record. The times are adjusted for the transit time from the ground to the ionosphere ($\approx$ 10 min). Good agreement between seismic records and HF Doppler records has been obtained for the Urakawa–Oki, Japan, earthquake of March 21, 1982 (Tanaka *et al.* 1984).

Disturbances resulting from the eruption of Mount St. Helens on May 18, 1980, produced pronounced atmospheric effects all over the world (Liu *et al.* 1982). These disturbances were detected at ground level as pressure fluctuations and in the F region, as oscillations of reflection height, by means of HF Doppler observations, and throughout the ionosphere by variations in Faraday content

(Roberts *et al.* 1982). Oscillations with periods ranging from 6 min to 15 min were recorded as far off as Europe and Taiwan (10 069 km); horizontal group speeds were near 300 m s^{-1}.

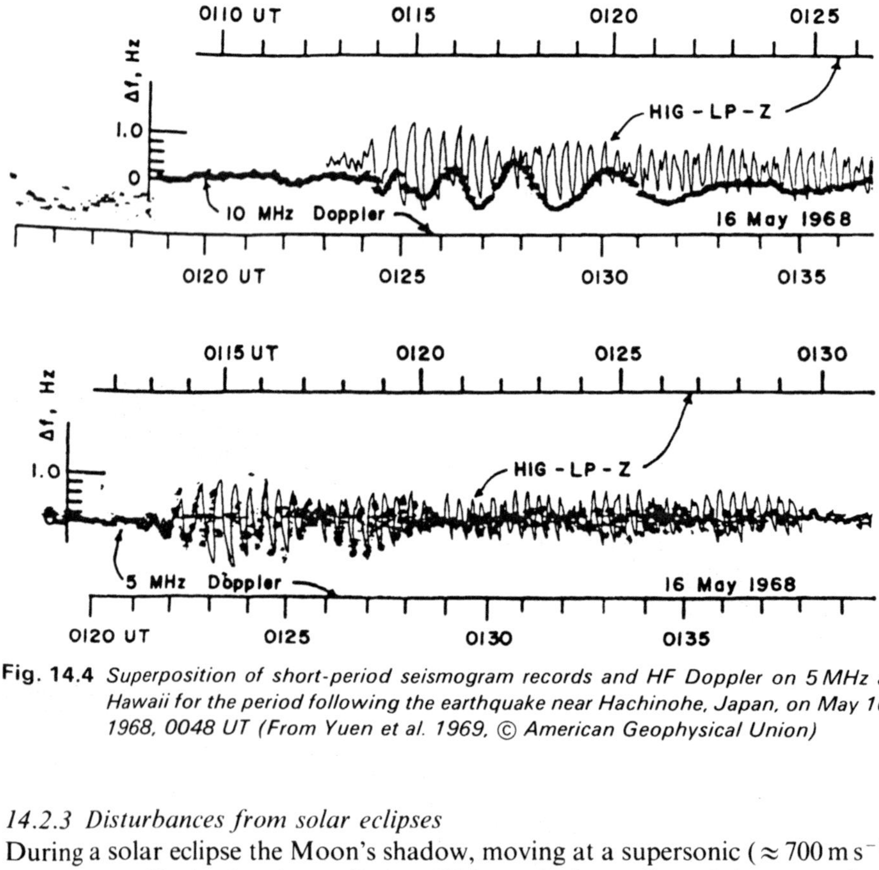

Fig. 14.4 *Superposition of short-period seismogram records and HF Doppler on 5 MHz at Hawaii for the period following the earthquake near Hachinohe, Japan, on May 16, 1968, 0048 UT (From Yuen et al. 1969, © American Geophysical Union)*

14.2.3 Disturbances from solar eclipses

During a solar eclipse the Moon's shadow, moving at a supersonic (≈ 700 m s^{-1}) speed, cuts off solar heating radiation. This results in cooling of the atmosphere particularly in the troposphere and the ozonosphere. Following its generation, a neutral atmosphere disturbance will propagate, and dispersion in the atmosphere will lead to an oscillatory signal at large distances (Chimonas 1970). Because of the curvature of the eclipse path, there will be a region where the energy will be focused.

Several attempts have been made to detect such waves with varying success. Following the eclipse over the eastern USA on March 7, 1970, evidence of 20 min waves in Faraday content over California was advanced (Davis and da Rosa 1970). Corroborative evidence was advanced by Sears (1972), Lerfald *et al.* (1972), and JATP (1972); negative results have also been reported (e.g. MacDougall 1970; Oetzel and Chang 1970). In connection with the eclipse of June 30, 1973, over North Africa, focusing should have occurred over South

Africa but observations there by Schödel *et al.* (1973) were negative as were observations by Hunter *et al.* (1975). Broche *et al.* (1976) reported short-period ($\approx$ 10 min) waves close to the eclipse path, but these were not those predicted by theory. On the whole the evidence for and against genuine eclipse effects is equally divided (e.g. Davies 1982; Cohen 1984).

14.3 Disturbances from explosions

14.3.1 Chemical explosions

Acoustic waves produced by a 500 ton TNT explosion at ground level have been observed in the F_2 region. Barry *et al.* (1966) saw clear explosion-induced disturbances about $9\frac{1}{2}$ minutes after detonation, and good agreement was found between observed changes in phase path and changes calculated by ray tracing. Jones and Spracklen (1974) reported an ionospheric disturbance from an explosion of a chemical plant.

14.3.2 Waves from nuclear explosions

Nuclear tests have provided powerful sources of atmospheric pressure waves, which have been detected at large distances. Both short-period and long-period waves have been detected.

14.3.2.1 Short-period waves

Figure 14.5 records an acoustic wave above Boulder, Colorado, following the United States megaton test near Johnston Island at 1602 UT October 30, 1962. The short-period ($\approx$ 1 min) waves arrived shortly after 2230 UT; that arrival time indicates a group speed of about 300 m s^{-1}. This suggests that the acoustic wave propagates in a ducted mode between about 50 km and the lower thermosphere ($\approx$ 110 km) with sufficient leakage on the upper side of the duct to produce detectable effects in the F-region. An expanded version of Fig. 14.5 shows that there is a time delay ($\approx$ 17 s) between the 5 MHz trace and the 4 MHz trace, indicating an upward trace speed of about 600 m s^{-1} (Baker 1968). During 1961 and 1962 wavelike disturbances were observed in the ionosphere above the USA following 75 nuclear explosions.

14.3.2.2 Long-period waves

Waves with periods of about an hour have been observed that are associated with large nuclear explosions (Kanellakos 1970). These waves have been detected by periodic changes in f_oF_2 (Fig. 14.6). In contrast to the speeds ($\approx$ 300 m s^{-1}) of the short-period waves, the speeds of the long-period disturbances are about 770 m s^{-1}. As seen in Fig. 14.6 the shapes and periods of the disturbances change with distance from the source. This is due, at least partly, to the dispersion of the signal as it propagates (Hines 1967). This dispersion depends on several factors such as the relative orientations of the source–observer path and the

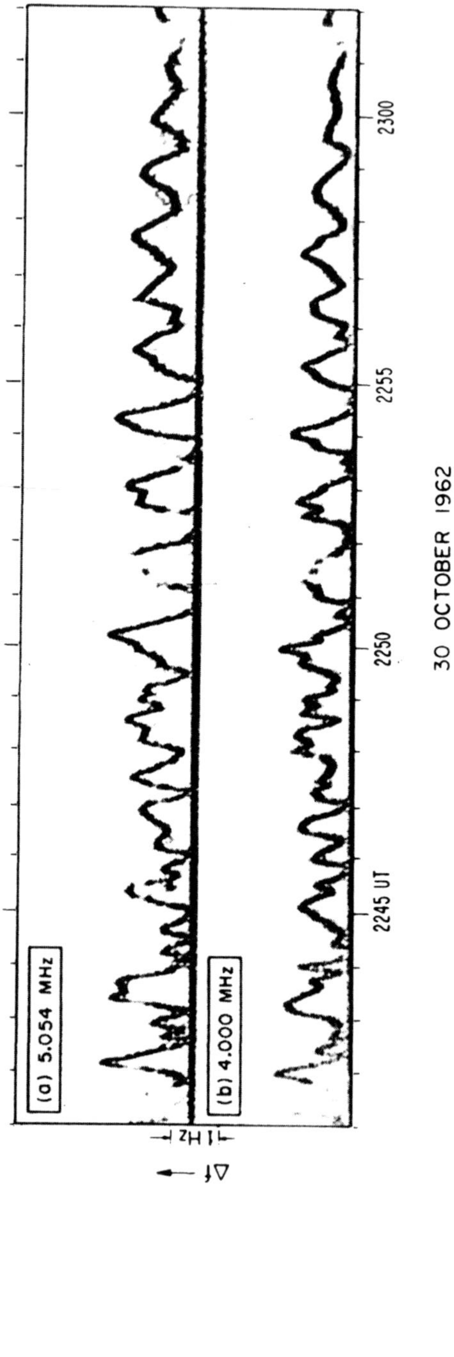

Fig. 14.5 *Acoustic-wave record of an ionospheric disturbance above Boulder following the nuclear explosion near Johnston Island on October 30, 1962 (1602 UT). Explosion yield was in the megaton range*

geomagnetic field because the electron motion induced in the F_2 region by the neutral wind is parallel to the magnetic field (Sec. 2.9.3).

14.4 D-region and E-region heating

14.4.1 The Luxembourg effect

The earliest indication that man could change the ionosphere occurred when Tellegen (1933) reported that the signals received in Holland from the Swiss radio station at Beromunster (650 kHz) appeared to be modulated with signals from the powerful station at Luxembourg (252 kHz, 200 kW). The large frequency is too complicated to be discussed here (Ginzburg 1970; Gurevich interaction was in the equipment. It was found that this phenomenon, called cross-modulation, was quite common when the unwanted, or disturbing, station was located near the transmission path from the wanted station to the receiver. The theory underlying the heating and cooling of electrons is essentially that for wave interaction given in Sec. 11.3.2.

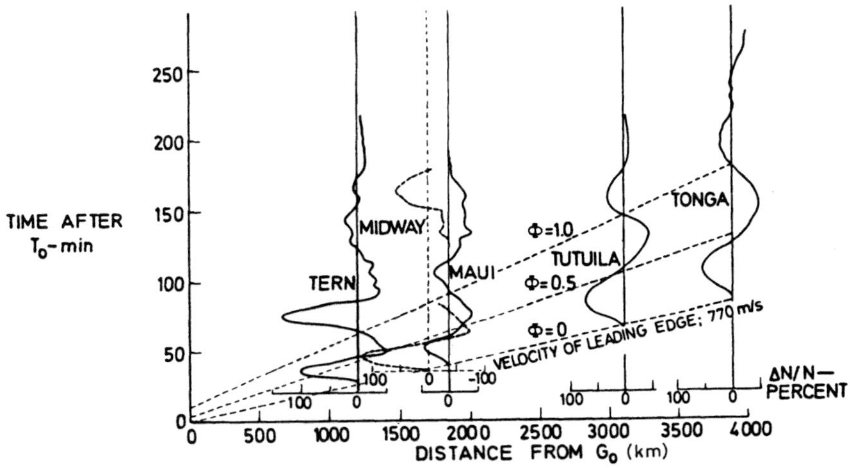

Fig. 14.6 *F_2-layer peak electron-density fluctuations from the Housatonic nuclear explosion as recorded at five stations (From Kanellakos 1970 © Advisory Group on Aerospace Research and Development, North Atlantic Treaty Organisation)*

Measurements of the cross-modulation imposed on signals with frequencies as low as 20 kHz, that traverse the heated D region, have been reported (Jones 1973). The magnitude of the cross-modulation, observed on WWVB (60 kHz), was much greater ($\approx$ 30%) with extraordinary-wave heating than with ordinary-wave heating ($\approx$ 10%).

14.4.2 Gyrointeraction

To obtain maximum wave interaction one requirement is that there be maximum absorption of the disturbing wave. One way of accomplishing this is to polarize the disturbing wave so that, on entry into the ionosphere, all the incident power goes into the extraordinary wave whose absorption coefficient is given by

$$\kappa \approx 4{\cdot}6 \times 10^{-2} \frac{N\nu(u)}{(\omega - \omega_{\mathrm{H}})^2 + \nu^2(u)} \ \mathrm{dB/km}$$

where the collision frequency $\nu(u)$ is a function of electron speed u. This expression shows that the absorption of the disturbing wave is large when the wave frequency ω approaches the gyrofrequency ω_H. The interaction process is called gyrointeraction. Actually the behavior of plasma waves around the gyrofrequency is too complicated to be discussed here (Ginzburg 1970; Gurevich 1978). A more detailed theory of gyrointeraction predicts that the absorption of the wanted wave will change with frequency of the disturbing wave with a double hump (Fig. 14.7) (Fejer 1974; Kissick and Ferraro 1975).

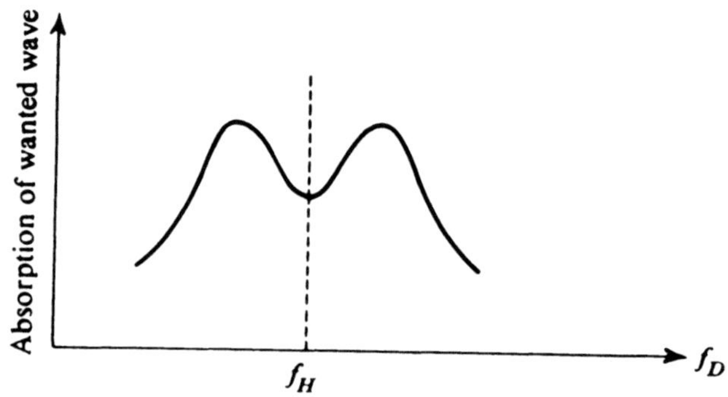

Fig. 14.7 *Theoretically predicted variation of absorption of the wanted wave as the frequency f_D of the heating wave is varied about the gyrofrequency f_H*

14.4.3 Self-distortion

Self-distortion refers to the effect on a radio signal of its own heating of the ionosphere. The theory of self-distortion (sometimes called self-demodulation) is similar to that of wave interaction. The best treatment is a numerical (computer) solution in which the energy absorbed from the wave is calculated and then used to determine the increase in electron energy and, hence, a modified electron collision frequency that is used to recalculate the absorption.

Numerical methods for calculating self-distortion, based on the above principle, have been given by Megill (1965) for the case of a nondeviating ionosphere in which the effects of the geomagnetic field are neglected. Let the power p in the incident wave be modulated according to

$$p(t) = p_0(1 - a \cos 2\pi f_a t)^2 \qquad (14.1)$$

in which f_a is the modulation frequency. The amplitude modulation depth M is

given by

$$M = \frac{A_{max} - A_{min}}{A_{max}} \tag{14.2}$$

where A is amplitude. In the absence of absorption,

$$M = \frac{2a}{1 + a} \tag{14.3}$$

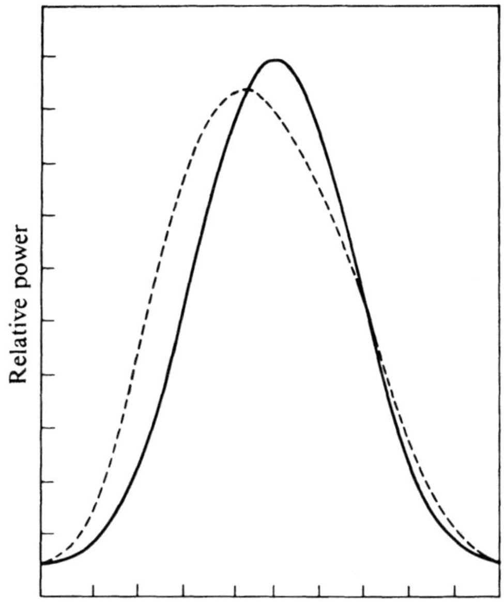

Fig. 14.8 *Form of the modulation before and after nonlinear absorption (From Megill 1965)*
Carrier frequency = 1·0 MHz
Modulation frequency = 300 Hz
Effective radiated power = 10^8W

In the presence of nonlinear absorption the following effects take place as shown in Fig. 14.8: (i) There is a change in amplitude modulation M; (ii) there is a phase shift in the modulation envelope; and (iii) a second harmonic is generated by the nonlinearity. When absorption occurs in a region in which the collision frequency is larger than the angular wave frequency, the demodulation will be negative and the modulation index will be increased. The amplitude change can be of about 5% (0·4 dB) and the phase advance about 20° (full modulation cycle = 360°). One consequence of self-demodulation is the adverse effect of increasing the effective power of radio transmitters. The higher the power, the greater the ionospheric absorption, so the net effect is self-defeating.

14.4.4 E-region heating

E-region heating by HF amplitude-modulated waves produces changes in the ionospheric (DC) conductivity, and the E layer radiates at the modulation frequency. This radiation has been observed at the heating facility near Tromsø, Norway, on frequencies ranging from ultra-low frequencies (1–30 Hz) through extremely low frequencies to very low frequencies (Stubbe *et al.* 1985). These signals are received on magnetic loop antennas. During the presence of the auroral electrojet (Sec. 2.4.4.2) typical field strengths in the range 2 to 6 kHz are about 10^{-3} nanoteslas. There is a rapid drop of signal strength on modulation frequencies below 2 kHz. Above 7 kHz the re-radiated signal strength decays as the inverse frequency. E-region irregularities have been produced by the Tromsø heater (Hibberd *et al.* 1983).

14.5 F-region heating

14.5.1 Resistive and resonance heating

In the heating discussed above, with the exception of gyrointeraction, the frequency of the disturbing wave is removed from the resonant frequencies of the plasma, and the ordered energy of the wave is converted into heat by electron–neutral collisions; this is called resistive heating. Heating of the F region, on the other hand, is of the resonance type in which the wave frequency is close to the plasma frequency of the layer, especially the critical frequency. For this resonance absorption the wave energy can be dissipated by means other than collisions (e.g. wave–wave interaction). Heating by high-power radio waves is normally classified as overdense or underdense respectively, according to whether the frequency of the heating wave is less than or greater than the critical frequency of the F region (Goodman 1987).

14.5.2 Heating facilities

Heating experiments have been conducted at several sites and, particularly, at Platteville, Colorado (Utlaut 1977); Arecibo, Puerto Rico (Gordon and Carlson 1977); and the Max-Planck facility near Tromsø, Norway (Stubbe *et al.* 1985). These installations are characterized by high powers and large antennas to provide high-power flux densities at F-region heights. In the USSR ionospheric modification has been achieved using a transmitter with a maximum equivalent power of 300 MW, and both ordinary and extraordinary polarizations (Migulin and Gurevich 1985).

14.5.3 Effects of heating

14.5.3.1 Sequence of processes

Several processes, with time scales ranging from abut 1 ms to tens of seconds, are involved in the modification of the ionospheric plasma as a result of HF radiowave heating. The time history of these processes is illustrated schematically in Fig. 14.9 (Jones *et al.* (1986). During the first few milliseconds the pondermotive forces dominate, and these result in enhanced ion and plasma lines in the backscatter spectrum. A few (e.g. 5) seconds after heater turn-on, nonlinear thermal effects lead to the growth of field-aligned striations in the plasma and to the onset of anomalous absorption. The energy extracted from the heating (or pump) wave causes an enhancement in the electron temperature and a volume expansion of the heated region. In addition, changes in the electron density occur with a decrease in the density high in the ionosphere where diffusion dominates, and an increase at the lower levels where the recombination losses are decreased as a consequence of temperature increase.

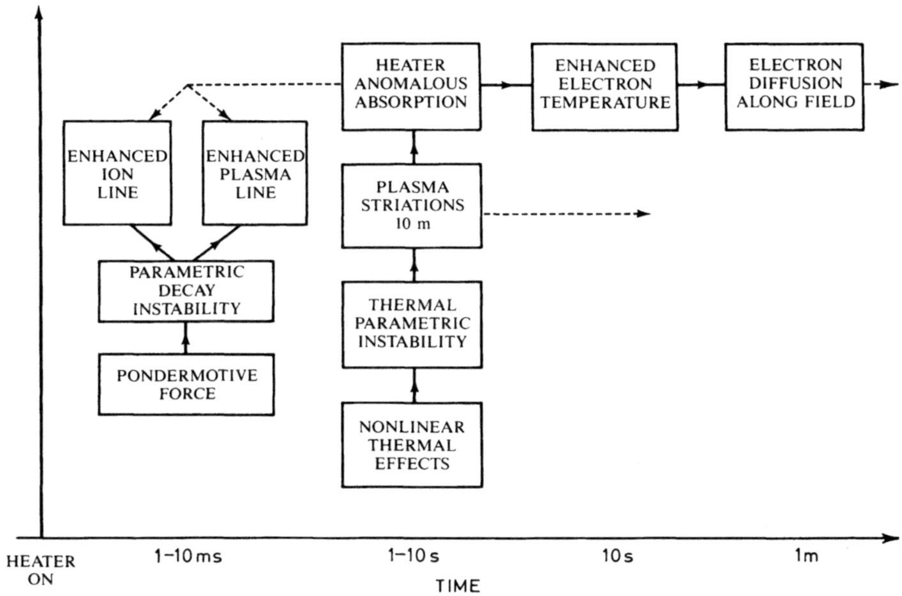

Fig. 14.9 *Schematic representation of the four stages of the heating process that account for observations (From Jones et al. 1986 © Pergamon Press, Inc.; reprinted with permission)*

Radio wave heating near the reflection level (with vertical propagation) is enhanced by the fact that as the refractive index goes to zero so does the magnetic field of the wave (see Sec. 3.4.3). Thus, at the (vertical) reflection height, the radio wave is an electric wave rather than an electromagnetic wave. Furthermore, the ordinary-wave electric field is parallel to the geomagnetic field. On the other hand, the electric field of the extraordinary wave at reflection is circularly polarized and is perpendicular to the geomagnetic field. The local electric field must be obtained from the wave equations by exact 'full-wave' solution (Hanson 1986). The most intense electric fields occur a fraction of a kilometer below the classical reflection level where solitons can be produced

(Payne *et al.* 1984); Birkmeyer and Hagfors 1986). Just below the reflection level, the incident and reflected heating waves form a standing electrostatic wave in which the (pondermotive) force produces depressions in the electron density in regions of maximum field strength. Thus there are quasi-periodic horizontal stratifications in electron density.

In the Soviet experiments, three different ionization effects have been identified: (i) Below 200 km there is an increase in ionization caused by a temperature dependence of the dissociative-recombination coefficient (see Sec. 2.9.2); (ii) at altitudes greater than 200 km, and with small pump (heating) wave power, the electron density decreases result from volume expansion; (iii) with large heater (pump) powers there is an increase of electron density produced by photoelectrons (with *o*-wave only). Plasma enhancements produce defocusing of both the probing signal and the heating wave.

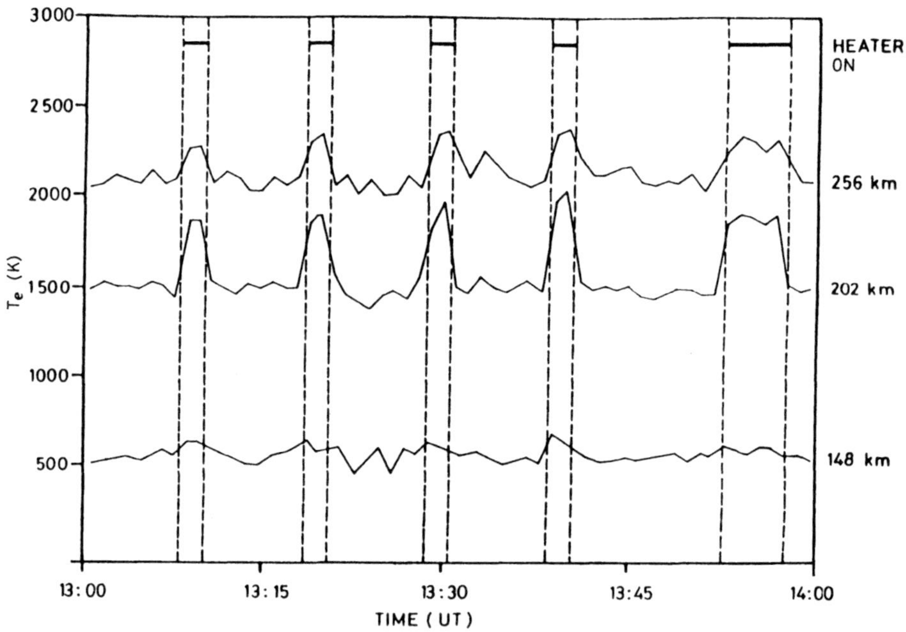

Fig. 14.10 *Electron temperature enhancement measured by the European Incoherent Scatter radar on March 17, 1985, during heating at 4·5 MHz (From Jones et al. 1986 © Pergamon Press, Inc.; reprinted with permission)*
The three height range gates are centered at 148, 202, and 256 km

14.5.3.2 Electron temperature

Electron temperature enhancements measured at the EISCAT facility (see Sec. 4.7) near Tromsø, Norway, are reproduced in Fig. 14.10. Note the time constants at heater turn-on and turn-off. The temperature enhancement results in expansion of the electron gas and a decrease of electron density which causes trapping (or ducting) of the heater wave and, hence, self-focusing of the heater wave.

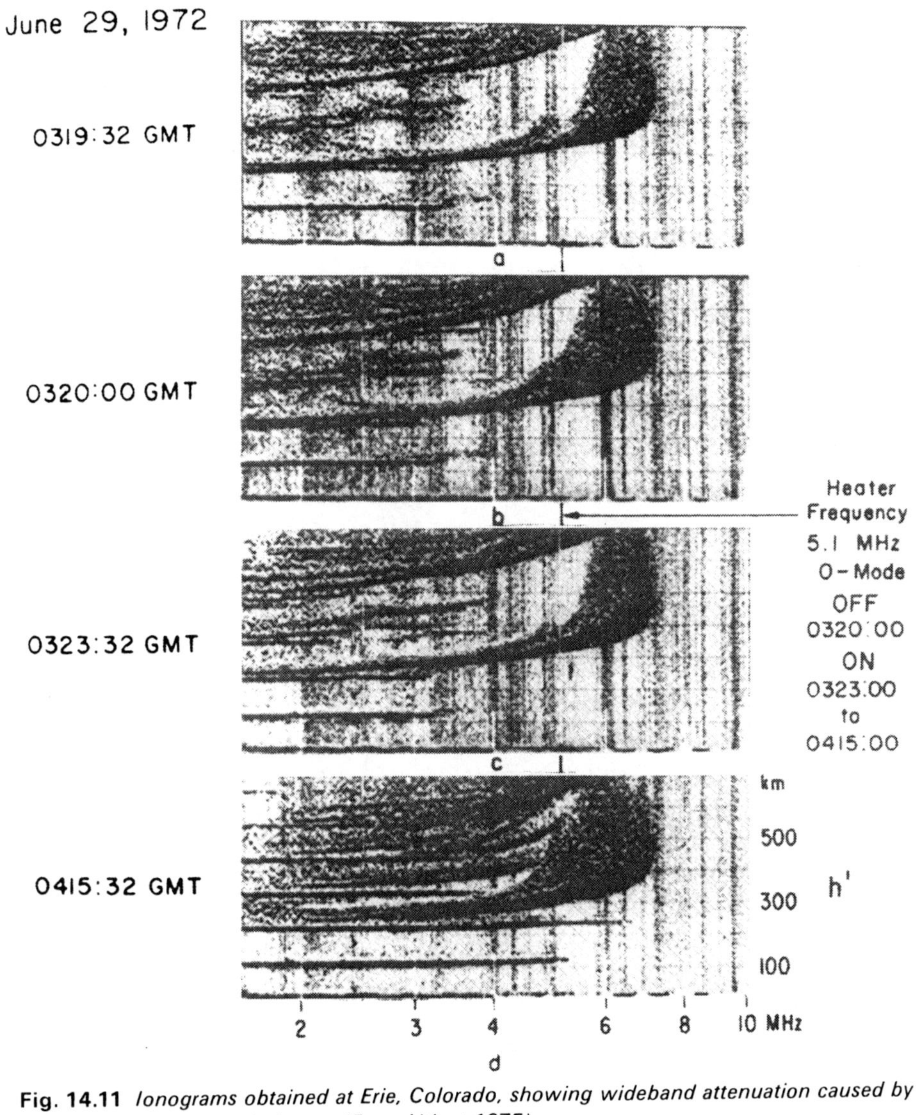

Fig. 14.11 *Ionograms obtained at Erie, Colorado, showing wideband attenuation caused by the Platteville heater (From Utlaut 1975)*

14.5.3.3 Anomalous wide-band attenuation (AWBA)

Many unanticipated phenomena have been observed with the heating experiments, and new understandings have been gained in the field of plasma physics because of the ability to carry out controlled, laboratory-type experiments on the ionosphere. Several of these phenomena have been described by Stubbe *et al.* (1985). One such effect, called wideband attenuation, was first discovered in the ionograms at Boulder (Fig. 14.11). It is observed only when ordinary-wave heating is used, as is predicted by theory. The AWBA results in the disappearance of the diagnostic *o*-wave at all frequencies above the heater frequency, and perhaps a few tens of kilohertz below. When AWBA occurs, its onset is within a few seconds of heater turn-on and it disappears within seconds of turn-off (Utlaut 1975). A more detailed study of AWBA at Tromsø is described by Jones *et al.* (1984).

Experiments in which the heater polarization was switched rapidly between ordinary and extraordinary showed that AWBA appeared and disappeared quickly as the heating wave was changed to and from the ordinary polarization. This indicates that the responsible mechanism must have a response time of about 1 min. Thermal parametric instabilities develop and quench rapidly (Weinstock 1974; Graham and Fejer 1976), and the plasma striations they produce are the principal cause of the anomalous absorption (Jones *et al.* 1984).

14.5.3.4 Airglow

Airglow from excited oxygen atoms at 6300 Å has been observed, particularly when the ordinary wave is used for heating (Duncan 1985). This enhancement of the normal airglow indicates the existence of collisional excitation of atomic oxygen by electrons. The process requires significant numbers of electrons with energies greater than 2 eV. For a Maxwellian distribution of energies this requires electron temperatures of about 2000 K. The likely explanation involves a nonthermal energy distribution having high velocity electrons produced by excitation of a plasma instability.

At Arecibo red-line enhancement and suppression have been observed. Suppression of roughly 1% is observed with extraordinary-wave heating, implying a temperature-dependent reduction in the recombination rate. Ordinary-wave heating results in enhancement (Gordon and Carlson 1974).

14.5.3.5 The plasma line

When a powerful heater on a frequency f ($<f_oF_2$) is used, an instability (enhancement) is created at the height at which the plasma frequency equals the heater frequency (e.g., Fejer *et al.* 1985). The line has been detected by use of incoherent scatter and coherent scatter. Other features seen include weak electromagnetic sidebands in the reflected heating wave. The heating wave is often called the pump wave. The spectrum of the received probing wave (e.g. incoherent scatter signal on 430 MHz) has a width of 20–60 Hz and is Doppler shifted (Duncan 1985). Observations of the plasma line at Tromsø have been discussed by Birkmeyer and Hagfors (1986).

14.5.4 Some propagation aspects of heating

14.5.4.1 Spread F

The ionosonde used to diagnose the Platteville modifications was located at Erie – about 26 km west of Platteville but well under the heated region. The first and most consistent phenomenon observed, after the Platteville transmitter was

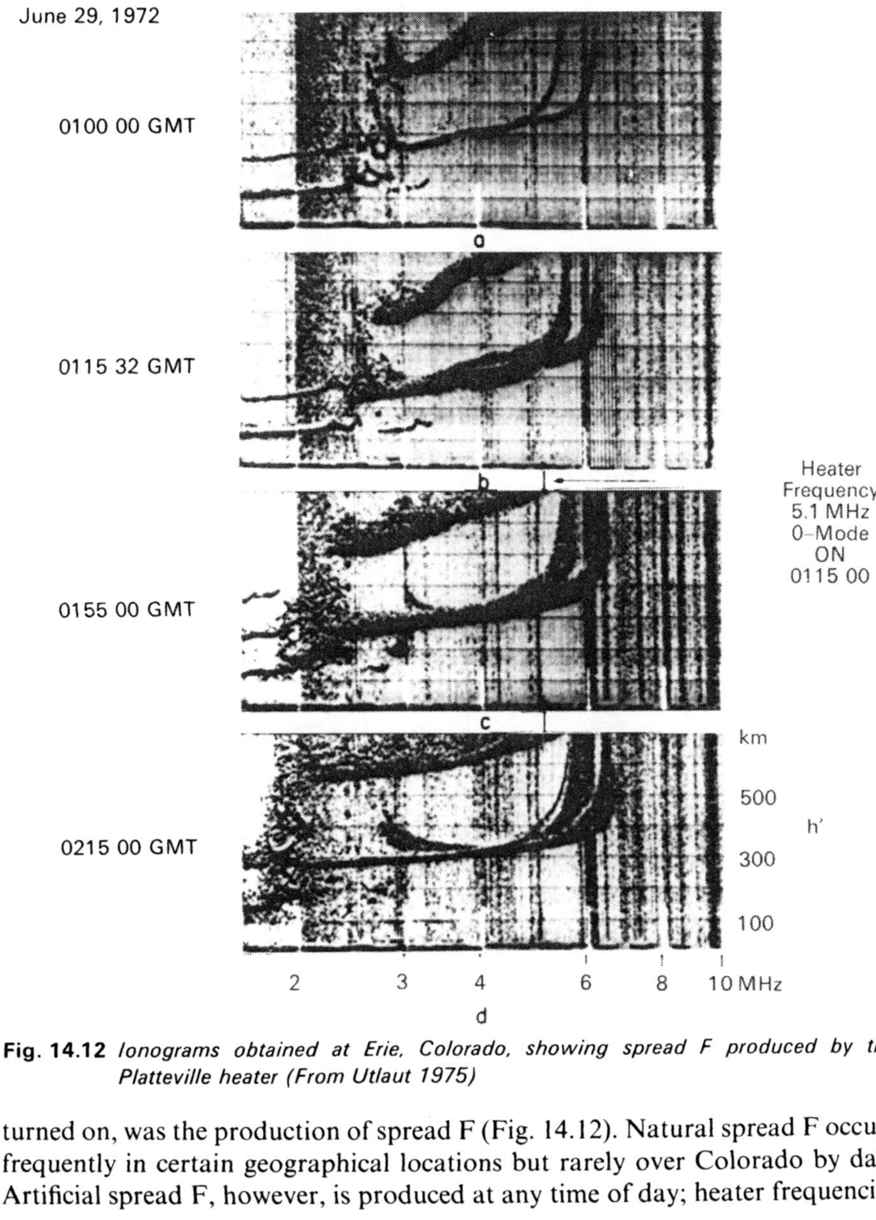

Fig. 14.12 *Ionograms obtained at Erie, Colorado, showing spread F produced by the Platteville heater (From Utlaut 1975)*

turned on, was the production of spread F (Fig. 14.12). Natural spread F occurs frequently in certain geographical locations but rarely over Colorado by day. Artificial spread F, however, is produced at any time of day; heater frequencies can range from at least 0·5 to 1·05 of the critical frequency and heating waves may be either ordinary or extraordinary. The diffuse echoes begin on ionograms within seconds of heater turn-on and they persist for varying durations after turn-off, depending on time of day. In the daytime, artificial spread F may persist for 5 to 10 min in the upper (F_2) region but only for about 1 min in the F_1 region. After sunset, it may persist for 10 to 20 min, and when generated after midnight, it has been observed for several hours until sunrise (Goodman 1982, 1985).

A second phenomenon observed, when the heater is on for 10 min or more, is a satellite trace paralleling the predisturbed echo trace on the ionogram (Fig. 14.12*d*). The trace tends to move, with time, away from the original trace as the heating continues and, after the heating ceases, to move back and eventually merge with the original trace. This sequence of events is evidence of the creation of a major bulge in the isoionic density contours as heated electrons are redistributed along the geomagnetic field (Utlaut 1974).

During heating the Boulder and Arecibo ionograms exhibit the same general characteristics, although details vary, possibly because of differences in field-line geometry and incident powers (Gordon and Carlson 1974). However, spread F has rarely, if ever, been observed at Tromsø.

14.5.4.2 Field-aligned scatter

The field-aligned plasma (FAS) striations produced during heating can scatter signals over a wide frequency band (3–500 MHz). The scattering process is of the Bragg type in which the spatial separation of the irregularities matches the half-wavelength of the probing signal.

Scattering of this type was first noted on ionograms at Boulder (Fig. 14.12) where marked spread-F types of reflections were observed during heating. Oblique backscatter was observed by Hedberg *et al.* (1983). The maximum backscatter power was obtained when the $\boldsymbol{k}$ vector of the radar signal was orthogonal to the direction of magnetic field-aligned striations. Backscatter from the heated E region on frequencies near 150 MHz has been reported by Hibberd *et al.* (1983) and on 435 MHz by Carpenter (1974).

The development of the field-aligned irregularities was observed by means of a large circular array about 11 km east of the Platteville heater (Allen *et al.* 1974). The observations showed that the angular spectra of the scattered signals have a systematic dependence on sounding frequency, range, Doppler shift, and heater power. The results of ray tracing simulation support the view that the spread F is caused by magnetic field-aligned ducts within the heated volume. A sequence of Doppler spectra and sky maps (Fig. 14.13) illustrates the dynamic behavior. The first map was obtained prior to heating and shows a single echo a few degrees from the zenith; this map shows the ionosphere to be smooth and slightly tilted. After 16 s of heating, several echoes appear along with a broadening spectrum. The broadening continues to develop with prolonged heating until a stable situation is attained after 3 min. The movement of the center of the echo pattern (toward the south and west) is caused by a traveling ionospheric disturbance.

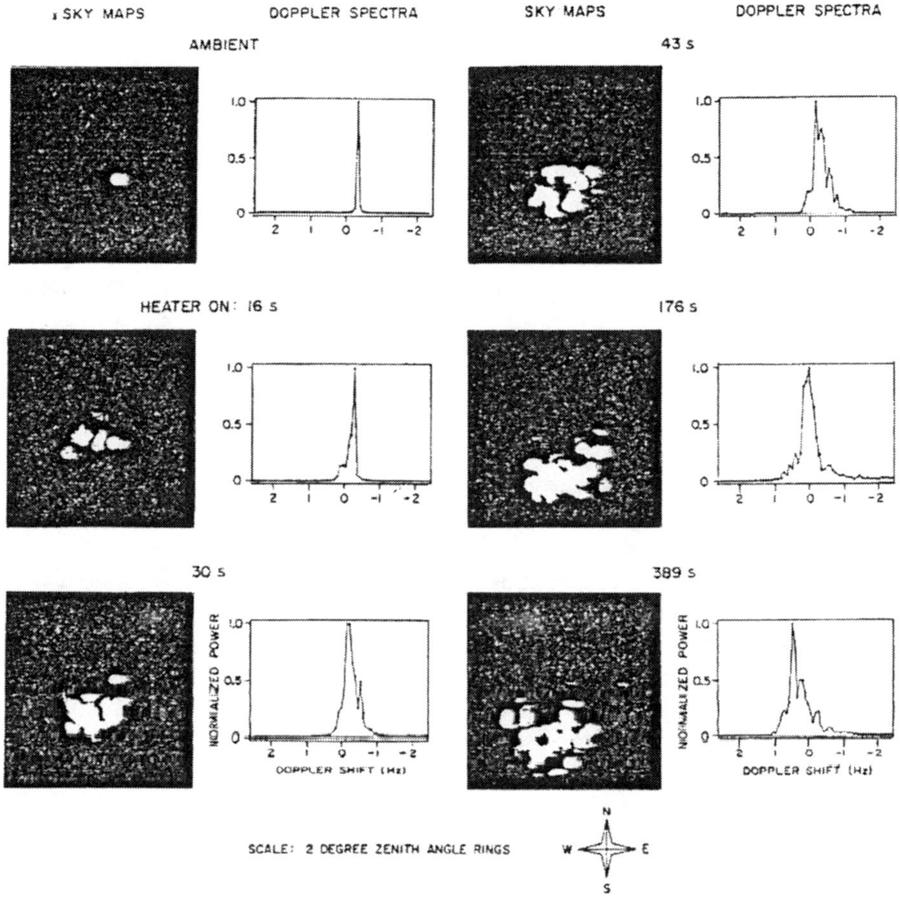

Fig. 14.13 *Time sequence pairs of Doppler spectra and sky maps, obtained during a Platteville heater turn-on (From Allen et al. 1974 © American Geophysical Union)*

14.5.4.3 Plasma line scatter

In field-aligned scatter (FAS) the frequency of the received echo is essentially the same as that transmitted. Another weaker type of scatter has been observed that results in received echoes with two frequencies, one of which is shifted up and the other down from the emitted frequency by an amount nearly equal to the frequency used for heating the ionosphere, that is, the local plasma frequency (see Sec. 14.5.3). This type of scatter is called plasma line scatter (PLS). The differences between FAS and PLS have been discussed by Utlaut (1974) who pointed out that with FAS the strongest returns occur in the perpendicular configuration whereas with PLS the maximum backscattered signal comes from the region of maximum heating at which the ionospheric plasma frequency equals the heater frequency. Recent observations of plasma-line phenomena have been reported by Birkmeyer and Hagfors (1986).

14.5.4.4 Ion–acoustic scatter

A third type of scatter is the result of Bragg-type scattering from ion–acoustic waves that propagate nearly along the geomagnetic field. This is called ion–acoustic scattering (IAS). The received signal differs by only a few kilohertz (i.e. ion–acoustic frequencies) from the transmitter frequency (Utlaut 1975; Beynon and Williams 1978).

14.5.4.5 E-region scatter

Effects in the E region (100–120 km) have been reported from Platteville, Arecibo, and Tromsø, Utlaut and Violette (1972) reported short-lived E echoes on ionograms on frequencies above $f_o F_2$. The echoes appear to be from partially reflecting sporadic E. Similar sporadic-E echoes have been observed at Arecibo (Gordon and Carlson 1974). Wright (1975) attributed E_s echoes to stimulated particle precipitation. Using the STARE radar (see Sec. 4.8.2) Hibberd *et al.* (1983) observed direct backscatter on 150 MHz.

14.5.4.6 ELF/VLF radiation

The conducting E region has been used as a very large antenna to radiate waves in the ELF/VLF range. For example, the Tromsø facility was operated on 2·8 MHz (*x*-wave) with a radiated power of 270 MW (Barr *et al.* 1986). The heater was modulated on frequencies in the range 205 Hz to 9·97 kHz. The heating of the ionosphere modulates the conductivity and, hence, modulates the current in the polar electrojet. The modulation of the current is equivalent to a large oscillating ELF/VLF dipole in the ionosphere, called the polar electrojet (PEJ) antenna, situated at the level of maximum Hall conductivity. ELF/VLF excitations have been reported for Arecibo (Lunnen *et al.* 1985), and ULF excitation has also been reported (Dowden 1985).

14.5.5 The satellite power station

In the 1970s serious consideration was given to generating 5–10 GW of power on a space satellite and transmitting the power to ground, via the ionosphere, on microwaves ($f \approx 2{\cdot}5$ GHz). The power fluxes involved in such an undertaking are very much larger ($\approx 100\,\mathrm{W\,m^{-2}}$) than those in the ground-based heater experiments ($\approx 10^{-4}\,\mathrm{W\,m^{-2}}$) but so are the frequencies; hence, the power absorbed is reduced. Although the heater frequency is too high to produce resonant heating, the net power flux in the microwave beam is still sufficiently large to produce modification by underdense heating. Calculations (e.g. Perkins and Roble 1978) show that pronounced effects can occur in the E region, where a thermal runaway may take place and the electron temperature may increase from about 200 K to more than 1000 K. This effect is enhanced by self-focusing within the microwave beam caused by electron-density depletion along the beam (see Sec. 14.5.3.2). In the lower ionosphere the electron density in the heated region may be enhanced because of the decrease in recombination rate with increase in electron temperature. In the lower E region the electron density may increase by a factor of about 3.

In the F region thermal conductivity plays an important role in redistributing the heated electrons. Temperature increases of 200–500 K will occur along the geomagnetic field lines passing through the radio beam. The electron-density profiles calculated for the undisturbed ionosphere and for the maximum heating case are shown in Fig. 14.14.

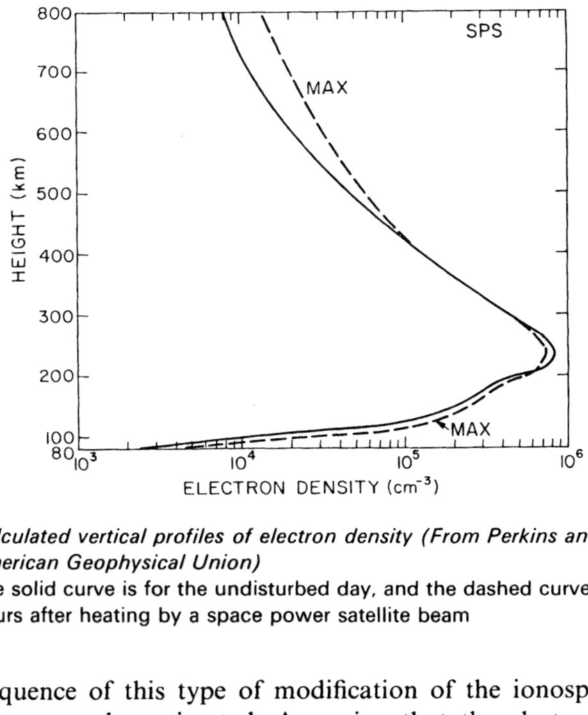

Fig. 14.14 *Calculated vertical profiles of electron density (From Perkins and Roble 1978 © American Geophysical Union)*
The solid curve is for the undisturbed day, and the dashed curve is for the time 3 hours after heating by a space power satellite beam

The consequence of this type of modification of the ionosphere on radio communications can be estimated. Assuming that the electron temperature increases by a factor of 5 and that the D-region electron density increases by a factor of 2, the product $N\nu$ (and hence the nondeviative absorption of a radio signal) will increase by a factor of about 10 (20 dB). Although this is a large increase its extent is limited and its location is well defined, so radio circuits through the modified region can be avoided. Of course, a large number of such power links could affect the power requirements of some HF communication circuits.

An increase in the peak electron density of the E layer by a factor of 2 to 3 will increase the f_oE by about 50%. Such an increase will have some adverse effects on radio propagation, for example, an increase in the lower cut-off frequency for propagation via the F_2 layer.

14.6 Chemical modification of the ionosphere

14.6.1 Types of modification

Impurities are introduced into the ionosphere in several ways. For example:

(i) Deliberate modification
- Provides tracers for the study of ionospheric parameters (e.g. barium releases)
- Induces instabilities into plasma irregularities and thus simulates natural disturbances
- Creates 'holes' in the ionosphere, leading to 'field-aligned' ducts for possible radio communications

(ii) Inadvertent modification can be produced by
- Exhaust gases from rockets
- Chemicals (e.g. chlorofluorocarbons) from the ground.

Chemical seeding of the upper atmosphere can have both beneficial and potentially harmful consequences for the ionosphere.

14.6.2 Effect of chemical impurities on electron loss

We have seen (in Sec. 2.9.2) that electron loss in the F_2 layer is by charge exchange between O^+ and a molecule, followed by dissociative recombination. Hence, chemical modification can affect the electron density in at least two ways:

(i) The introduction of a chemical with a low ionization potential will increase the production rate for a given insolation and thus increase the local electron density.
(ii) The introduction of a molecular gas with a high affinity for electrons, or a high charge-exchange coefficient, will increase the loss rate and hence produce a hole in the plasma.

14.6.3 Inadvertent modification

To illustrate an effect of the inadvertent pollution of the ionosphere we consider the discovery by Mendillo *et al.* (1975) that the exhaust plume of the Saturn V rocket, which launched NASA's Skylab, caused a sudden and large-scale depletion of the electron content of the F region. The observed effect consisted of a large ($\approx 50\%$) decrease in the Faraday electron content, between the ATS–3 satellite and ground, that occurred near 1240, 75°WMT, on May 14, 1973, and lasted for several hours (Fig. 14.15). Observations from several sites revealed that a 'hole' in the F region was created, having a radius of approximately 1000 km. The onset of the depletion occurred within 10 min of the launch of Skylab. The specific mechanism that created the hole was the rapid ion-atom interchange reactions (see Sec. 2.9.2) between the natural O^+ ions and the hydrogen and water-vapor molecules in the plume, from the second-stage engines, followed by dissociative recombination. That is,

$$O^+ + H_2 \rightarrow OH^+ + H \tag{14.4a}$$

$$O^+ + H_2O \rightarrow H_2O^+ + O \tag{14.4b}$$

and

$$OH^+ + e \rightarrow O + H \tag{14.5a}$$

$$H_2O^+ + e \rightarrow H_2 + O \tag{14.5b}$$

$$H_2O^+ + e \rightarrow OH + H \tag{14.5c}$$

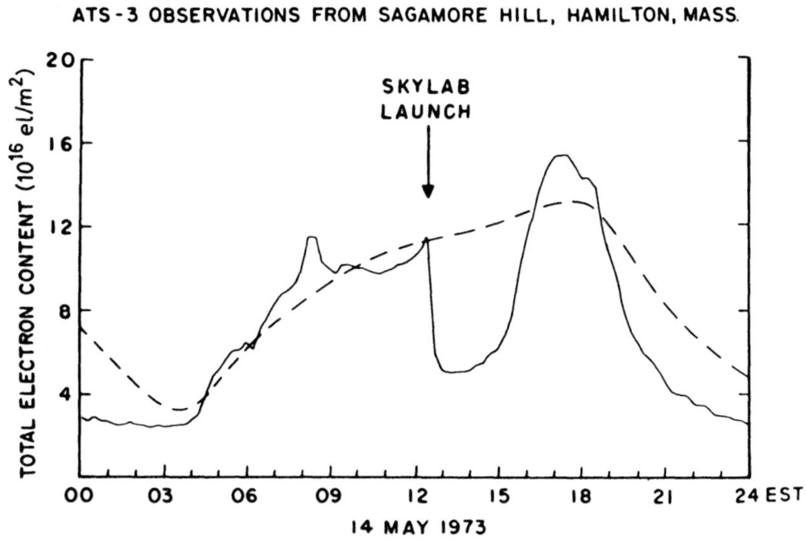

Fig. 14.15 *Variation of total electron content near the time of launch of Skylab by a Saturn V rocket from Kennedy Space Center on May 14, 1973 (From Mendillo et al. 1975 © American Geophysical Union)*
Note the decrease in *TEC* immediately following the launch. The broken line indicates the monthly median.

It should be noted that the dissociative recombinations proceed at a relatively fast rate, so the electron-loss rates are determined by the slower atom-ion reactions. The increase in electron-loss rate results in a decrease of electron content. Side effects of chemically induced depletion of the ionosphere include (*a*) increase of electron temperature, (*b*) decrease of ion temperature, (*c*) enhancement of airglow, (*d*) formation of depleted ducts extending into the magnetosphere that can guide radio waves, and (*e*) production of a focusing lens affecting radio waves traversing the ionosphere.

14.6.4 Deliberate modification

14.6.4.1 Metallic ions

Many experiments have been carried out to test the effect of chemicals on the ionosphere (Haerendel 1977). The most common method of modification is the release of gases in the upper atmosphere. The cheapest way is by chemical reactions such as the barium–copper-oxide reaction, which yields a neutral metal vapor. The ionization provided by EUV illumination is greater than the ambient ionization because of the lower ionization potential of metallic ions. Metal ions above the turbopause ($\approx 120\,\mathrm{km}$) have a much longer lifetime than the times for transport, in contrast to the lifetimes of the dominant ions NO^+ and O_2^+. Because the metallic cloud is only partially ionized we have a diffusing neutral gas that gives information on the properties of the neutral atmosphere, whereas the behavior of the ionized clouds depends also on the interaction with the ambient ionosphere and on the local magnetic and electric fields. To determine these physical properties (e.g. diffusion and E fields), it is necessary to measure the motions of the neutral and ion clouds.

The Ba^+ cloud, with a width of a few kilometers transverse to the geomagnetic field, is generated within about 5 s of release of the barium and expands mainly along the magnetic field lines. The neutral and ionized clouds are observed photographically from two ground stations for triangulation (Föppl *et al.* 1967). After the initial cloud expansion, caused by excess pressure during release, the neutral cloud expands by diffusion. The neutral cloud tends to drift with the ambient neutral wind whereas the motion of the ionized cloud is affected also by $\boldsymbol{E} \times \boldsymbol{B}$ drift; as a result, the two clouds diverge with time (Fig. 14.16). For photographic contrast these seeding experiments must be conducted near sunrise and sunset only. Observations yield neutral horizontal speeds between 60 and $150\,\mathrm{m\,s^{-1}}$. Ion speeds parallel and perpendicular to the magnetic field are approximately $60\,\mathrm{m\,s^{-1}}$ and $40\,\mathrm{m\,s^{-1}}$.

14.6.4.2 Electron attachment

Some chemicals have a high affinity for electrons, so their injection into the ionosphere has relatively little effect except to remove electrons and replace them with heavy negative ions. Such a chemical is sulphur hexafluoride, which was used in Project Firefly (Paul *et al.* 1968) to create an electron hole near the daytime F_2-layer peak. The electron removal is the result of the following reaction:

$$SF_6 + e \rightarrow SF_5^- + F \tag{14.6}$$

The inertia of a heavy negative ion is so great that it does not affect the propagation of HF radio signals. Thus, as far as radio waves are concerned, there is a 'hole' in the F_2 region. An ionogram following a release (Fig. 14.17) illustrates the complicated nature of the echoes. Figure 14.18 gives an example of the radio raypath through the hole.

Auroral zone currents have been modified by the injection of large amounts of water vapor, which modify the electron precipitation from the magnetosphere (Whalen *et al.* 1985).

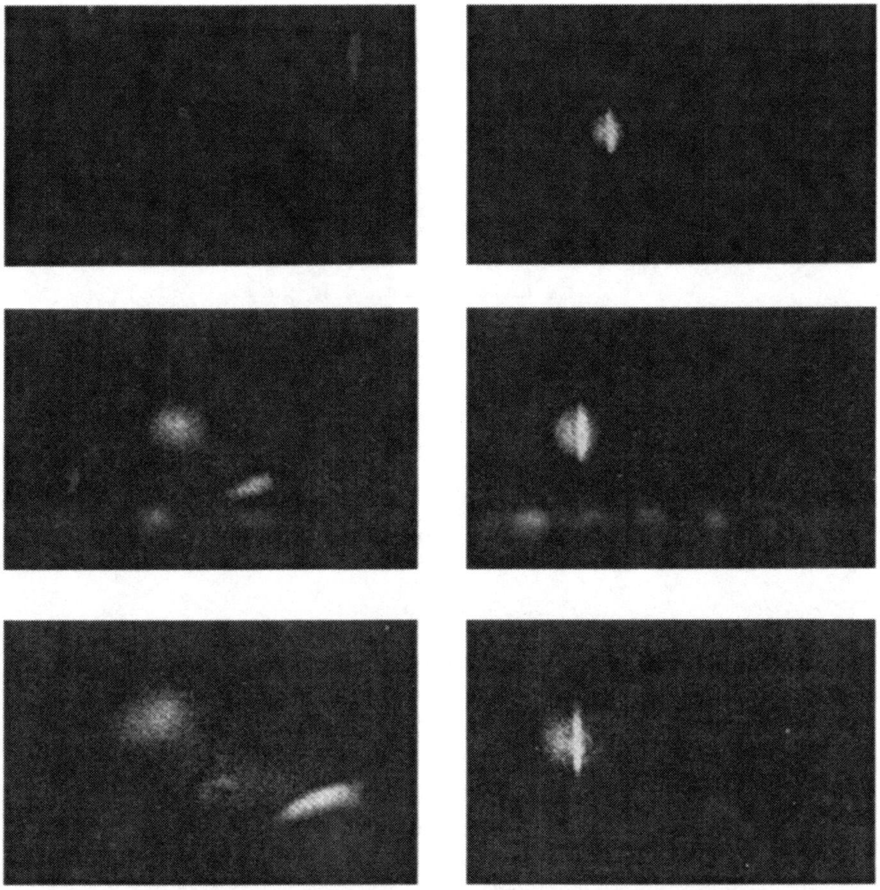

Fig. 14.16 *Barium ion (elongated shapes) and neutral strontium (spherical shapes) clouds at three times subsequent to release of the metal vapor (From Haerendel et al. 1967 © Pergamon Press, Inc.; reprinted with permission)*

14.6.4.3 Shaped charges

Shaped or hollow charges have been effective tools for the injection of ion beams along magnetic lines of force up to high altitudes, i.e. into the magnetosphere (Wescott *et al.* 1974). Forward speeds between 8 and 15 km s^{-1} are characteristic of the fast part of the velocity distribution. The injection of plasma into the ionosphere and/or magnetosphere can set up Alfvén waves and produce charged particle precipitation (Wescott *et al.* 1985).

14.6.5 Effect of hydrogen on the equatorial ionosphere

Under the influence of gravity $\boldsymbol{g}$ and a magnetic field $\boldsymbol{B}_0$ a plasma drifts with velocity

$$\boldsymbol{u} = \frac{\boldsymbol{g} \times \boldsymbol{B}_0}{\omega_H B_0} \tag{14.7}$$

where ω_H is the angular gyrofrequency.

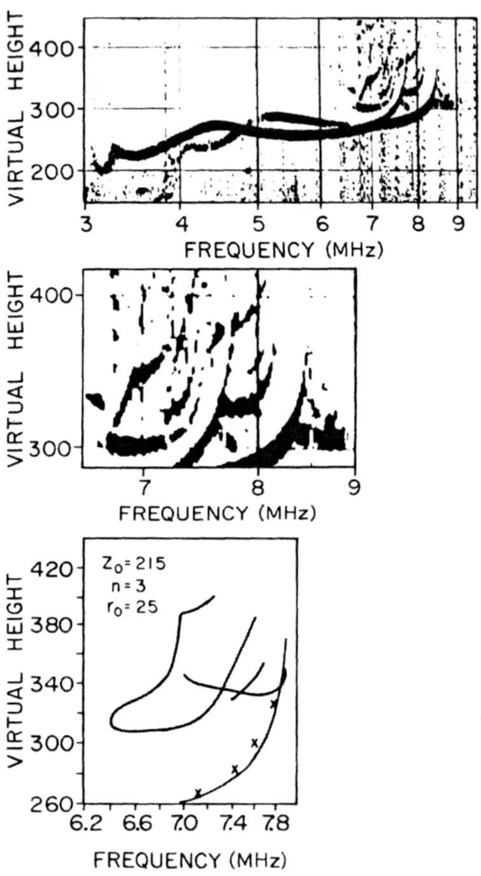

Fig. 14.17 *Ionogram obtained during Project Firefly (From Paul et al. 1968)*

(*a*) Portion between 3 MHz and 10 MHz

(*b*) Enlargement of section between 6·2 MHz and 9 MHz, showing oblique echoes from a 'hole' in the ionosphere

(*c*) Synthetic ionogram obtained by ray tracing. The lowest curve is that calculated for vertical incidence. The x's are the corresponding observed points

The equator is a particularly interesting place to study seeding of the ionosphere because the magnetic field $\boldsymbol{B}_0$ is horizontal and perpendicular to the gravity vector $\boldsymbol{g}$. Although this drift is small ($\approx 1\,\mathrm{cm\,s^{-1}}$) the ion current establishes an electric polarization field at an inhomogeneity. This field is normally short-circuited by currents that flow via the lower F and E regions where the transverse (Pedersen) conductivity is high during the day (Haerendel 1977). At sunset this conductivity is reduced and a situation develops in which a heavy fluid rests on a lighter fluid at the steep ledge of the F_2 layer, which is inherently unstable (the Rayleigh–Taylor instability). It is possible that equatorial spread F is initiated by this process (Anderson and Bernhardt 1978).

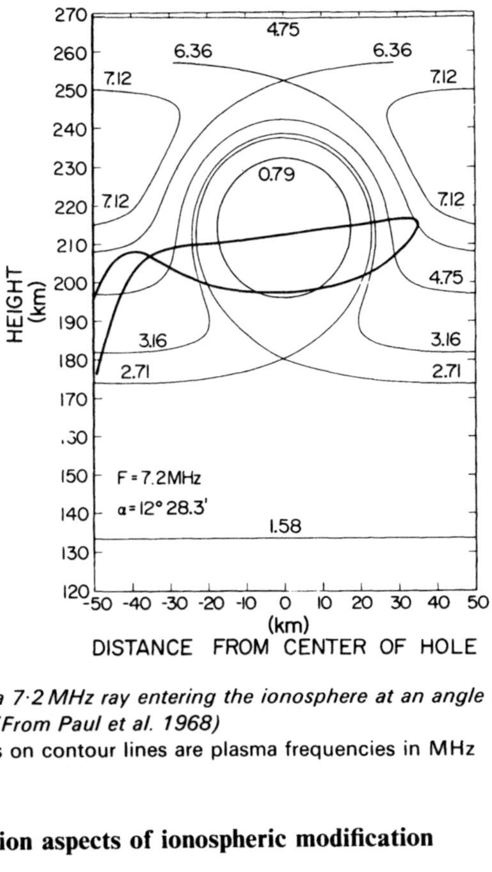

Fig. 14.18 *Path of a 7·2 MHz ray entering the ionosphere at an angle of 12° 28·3′ from the vertical (From Paul et al. 1968)*
Numbers on contour lines are plasma frequencies in MHz

14.7 Communication aspects of ionospheric modification

14.7.1 Radio modification

Possible consequences of radio heating of the ionosphere include the following:

(i) Interference between broadcasting stations caused by wave interaction
(ii) Self-demodulation in the D region, when high powers are employed, which partially offsets the increase in power
(iii) Production of F-region irregularities that allow field-aligned forward scatter and backscatter on frequencies both below and above the penetration frequency

(iv) Creation of irregularities that make possible propagation in ducted modes (see Sec. 6.7.2)
(v) Creation of plasma lenses (Bernhardt and Duncan 1982) produces focusing and/or defocusing of radio waves
(vi) Anomalous absorption of F-layer reflections (Jones *et al.* 1984) produces attenuation of radio waves.

Inadvertent ionospheric modification by high-power (e.g. 10 MW) HF transmitters that already exist could result in undesirable communication effects. Although these transmitters are designed primarily for oblique circuits there may be sufficient vertical sidelobe emission to produce F-region effects and thus degrade communication circuits (Albrecht 1977).

14.7.2 Chemical modification

Chemical modification might be used in several ways to facilitate or hamper ionospheric radio communications:

(i) Plasma seeding in high altitudes may be used to create an artificial duct in the magnetosphere and thus make possible propagation of radio waves on frequencies ranging from VLF through VHF.
(ii) Overdense (e.g. barium) clouds can be used to scatter waves that might be used directly. Alternatively the clouds can be used to modify the angle of incidence of radio signals on a reflecting or ducting layer, thus enhancing the MUF.
(iii) Underdense clouds (e.g. H_2, H_2O, etc.), near the magnetic equator, provide holes that can be used to simulate natural equatorial spread F. Normally this phenomenon has adverse effects on radio communications (e.g. flutter fading, scintillation).

14.7.3 Nuclear explosions

Nuclear explosions are one of the most devastating means of ionospheric modification. They often produce violent chemical and dynamic effects throughout the upper atmosphere, some of which last for several days. Low-altitude explosions produce large acoustic–gravity waves that travel long distances. High-altitude explosions produce X-rays that ionize the D region and result in high absorption of radio signals. They also modify the energetic charged-particle populations in the radiation belts surrounding the Earth. Particle precipitation from these modified belts ionizes the D region and affects the phase of long-distance VLF propagation for many days. The D-region effects are noticed first at two places, one near the explosion and the other at the far end of the geomagnetic field line. With time, the trapped electrons drift in an easterly direction and after about 1 h they envelope the whole Earth.

The main F-layer effect of nuclear explosions is the creation of traveling ionospheric disturbances (Sec. 14.3.2). In the E region, sporadic E is produced (AGARD 1972).

14.8 Concluding remarks

The ionosphere can be altered by radio, mechanical, and chemical means, and such alterations may be advantageous or adverse from the viewpoint of the radio communicator. From the viewpoint of military communications, F-region modification provides a relatively secure means of communication because of the highly aspect-sensitive characteristics of field-aligned scatter.

Several means of modification that have not been discussed are nevertheless of considerable importance. One example is the injection of very-low-frequency waves, which leads to stimulation of VLF radio emissions (Helliwell *et al.* 1977). A number of other aspects of modification are presented in Fejer (1974).

14.9 References

AGARD (1972): 'Effects of atmospheric acoustic gravity waves on electromagnetic wave propagation'. AGARD Conf. Proc. No 115, Advisory Group on Aerospace Research and Development, North Atlantic Treaty Organisation

ALBRECHT, H. (Ed.) (1977): 'Artificial modification of propagation media'. AGARD Conf. Proc. No. 192, Advisory Group on Aerospace Research and Development, North Atlantic Treaty Organisation

ALLEN, E. M., THOME, G. D., and RAO, P. B. (1974): 'HF phased array observations of heater-induced spread-F', *Radio Sci.*, **9**, p. 905

ANDERSON, D. A. and BERNHARDT, P. (1978): 'Modeling the effects of a H_2 gas release on the equatorial ionosphere', *J. Geophys. Res.*, **83**, p. 4777

BAKER, D. M. (1968): 'Observations of acoustic waves in the ionosphere following nuclear explosions'. ESSA Tech. Report ERL 89–SDL 5, Space Disturbances Laboratory, Boulder, CO

BARR, R., STUBBE, P., RIETVELD, M. T., and KOPKA, H. (1986): 'ELF and VLF signals radiated by the 'Polar electrojet antenna': Experimental Results'. *J. Geophys. Res.*, **91**, p. 4451

BARRY, G. H., GRIFFITHS, L. J., and TAENZER, J. C. (1966): 'HF radio measurements of high-altitude acoustic waves from a ground-level explosion', *J. Geophys. Res.*, **71**, p. 4173

BAUER, S. (1958): 'Correlations between tropospheric and ionospheric parameters', *Geofisica Pura e Applicata*, **40**, p. 235

BERNHARDT, P. A. and DUNCAN, L. M. (1982): 'The feedback-diffraction theory of ionospheric heating', *J. Atmos. Terr. Phys.*, **44**, p. 1061

BERTIN, F., TESTUD, J., and KERSLEY, L. (1975): 'Medium scale gravity waves in the ionospheric F-region and their possible origin in weather disturbances', *Planet. Space Sci.*, **23**, p. 493

BEYNON, W. J. G. and WILLIAMS, P. J. S. (1978): 'Incoherent scatter of radio waves from the ionosphere', *Rep. Prog. Phys.*, **41**, p. 909

BIRKMEYER, W. and HAGFORS, T. (1986): 'Observational technique and parameter estimation in plasma line spectrum observations of the ionosphere by chirped incoherent scatter radar', *J. Atmos. Terr. Phys.*, **48**, p. 1009

BROCHE, P., CROCHET, M., and DE MAISTRE, J. C. (1976): 'Gravity waves generated by the 30 June, 1973 solar eclipse in Africa', *J. Atmos. Terr. Phys.*, **38**, p. 1361

CARPENTER, G. B. (1974): 'VHF and UHF bistatic observations of a region of the ionosphere modified by a high power radio transmitter', *Radio Sci.*, **9**, p. 965

CHIMONAS, G. (1970): 'Internal gravity-wave motions induced in the earth's atmosphere by a solar eclipse', *J. Geophys. Res.*, **75**, p. 5545

COHEN, E. A. (1984): 'The study of the effect of solar eclipses on the ionosphere based on satellite beacon observations', *Radio Sci.*, **19**, p. 769

DAVIES, K. (1982): 'Atmospheric gravity waves produced by solar eclipses – A review', *Proc. Indian Nat. Sci. Acad.*, **48A**, Suppl. 3, p. 342

DAVIES, K. and BAKER, D. M. (1965): 'Ionospheric effects observed around the time of the Alaskan earthquake of March 28, 1964', *J. Geophys. Res.*, **70**, p. 2251

DAVIES, K. and JONES, J. E. (1972): 'Ionospheric disturbances produced by severe thunderstorms'. NOAA Professional Paper No. 6, US Govt. Printing Office, Washington, DC 20402

DAVIES, K. and JONES, J. E. (1973): 'Acoustic waves in the ionospheric F_2-region produced by severe thunderstorms', *J. Atmos. Terr. Phys.*, **35**, p. 1737

DAVIS, M. J. and DA ROSA, A. V. (1970): 'Possible detection of atmospheric gravity waves generated by the solar eclipse', *Nature*, **226**, p. 1123

DOWDEN, R. L. (Ed.) (1985): 'Active experiments in space plasmas', Special Issue, *J. Atmos. Terr. Phys.*, **47**, p. 1149

DUNCAN, L. M. (1985): 'The HF-induced plasma line, electron acceleration and resulting airglow', *J. Atmos. Terr. Phys.*, **47**, p. 1267

FEJER, J. A. (Ed.) (1974): 'Non-linear effects in electromagnetic wave propagation'. AGARD Conf. Proc. No. 138, Advisory Group on Aerospace Research and Development, North Atlantic Treaty Organisation

FEJER, J. A., GONZALES, C. A., IERKIC, H. M., SULZER, M. P., TEPLEY, C. A., DUNCAN, L. M., DJUTH, F. T., GANGULY, S., and GORDON, W. E. (1985): 'Ionospheric modification experiments with the Arecibo Heating Facility', *J. Atmos. Terr. Phys.*, **47**, p. 1165

FÖPPL, H., HAERENDEL, G., HASEL, L., LORDL, J., LÜTJENS, P., LÜST, R., MELZNER, F., MEYER, B., NEUSS, H., and RIEGER, E. (1967): 'Artificial strontium and barium clouds in the upper atmosphere', *Planet. Space Sci.*, **15**, p. 357

GEORGES, T. M. (1968): 'HF Doppler studies of traveling ionospheric disturbances', *J. Atmos. Terr. Phys.*, **30**, p. 735

GEORGES, T. M. (1973): 'Infrasound from convective storms: Examining the evidence', *Rev. Geophys. Space Phys.*, **11**, p. 571

GINZBURG, V. L. (1970): 'The propagation of electromagnetic waves in plasmas'. Translators J. B. Sykes and R. J. Taylor (Pergamon Press, Oxford) 2nd edn.

GOODMAN, J. M. (Ed.) (1982): 'Effect of the ionosphere on radiowave systems'. US Govt. Printing Office, Washington, DC 20402

GOODMAN, J. M. (Ed.) (1985): 'Effect of the ionosphere on C^3 I systems'. Nat. Tech. Information Service 5, 285 Port Royal Road, Springfield, VA 22161

GOODMAN, J. M. (Ed.) (1987): 'The effect of the ionosphere on communication, navigation, and surveillance systems'. Naval Research Laboratory, Washington DC

GORDON, W. E. and CARLSON, H. C. (1974): 'Arecibo heating experiments', *Radio Sci.*, **9**, p. 1041

GORDON, W. E. and CARLSON, H. C. (1977): 'The heating experiment at Arecibo' *in* ALBRECHT, H. J. (Ed.): 'Artificial modification of propagation media'. AGARD Conf. Proc. No. 192, Advisory Group on Aerospace Research and Development, North Atlantic Treaty Organisation, p. 7-1

GRAHAM, K. N. and FEJER, J. A. (1976): Anomalous radio wave absorption due to ionospheric heating effects', *Radio Sci.*, **11**, p. 1057

GUREVICH, A. V. (1978): 'Non-linear phenomena in the ionosphere'. Translator J. G. Adashko (Springer-Verlag, New York)

HAERENDEL, G. (1977): 'Modification of ionized media by chemical substances – A review of physical processes' *in* ALBRECHT, H. J. (Ed.): 'Artificial modification of propagation media'. AGARD Conf. Proc. No. 192, Advisory Group on Aerospace Research and Development, North Atlantic Treaty Organisation, p. 13-1

HAERENDEL, G., LÜST, R. and RIEGER, E. (1967): 'Motion of artificial ion clouds in the upper atmosphere', *Planet. Space Sci.*, **15**, p. 1

HANSON, J. D. (1986): 'Techniques for calculating the electric field near the reflection layer in a perturbed ionosphere', *Radio Sci.*, **21**, p. 991

HEDBERG, A., DERBLOM, H., THIDÉ, B., KOPKA, H., and STUBBE, P. (1983): 'Observations of HF backscatter associated with the heating experiment at Tromsø', *Radio Sci.*, **18**, p. 840

HELLIWELL, R. A., KATSUFRAKIS, J. P., and BERNHARDT, P. A. (1977): 'Modification of the propagation characteristics of the ionosphere (and magnetosphere) by injection into the magnetosphere of whistler-mode waves' *in* ALBRECHT, H. J. (Ed.): 'Artificial modification of propagation media'. AGARD Conf. Proc. No. 192, Advisory Group on Aerospace Research and Development, North Atlantic Treaty Organisation, p. 11–1

HIBBERD, F. H., NIELSEN, E., STUBBE, P., KOPKA, H., and RIETVELD, M. T. (1983): 'Production of auroral zone E region irregularities by powerful HF heating', *J. Geophys. Res.*, **88**, p. 6347

HINES, C. (1967): 'On the nature of traveling ionospheric disturbances launched by low-altitude nuclear explosions', *J. Geophys. Res.*, **72**, p. 1877

HUNG, R. J., PHAN, T., and SMITH, R. E. (1979): 'Coupling of ionosphere and troposphere during the occurrence of isolated tornadoes on November 20, 1973', *J. Geophys. Res.*, **84**, p. 1261

HUNTER, A. N., HOLMAN, B. K., FELGATE, D. G., and KELLEHER, R. (1975): 'Faraday rotation studies in Africa during the solar eclipse of June 30, 1973', *Nature*, **250**, p. 205

JATP (1968): 'Special Issue: Symposium on upper atmospheric winds, waves and ionospheric drifts (1967)', *J. Atmos. Terr. Phys.*, **30**, p. 657

JATP (1972): 'Special Eclipse Issue (The eclipse of 7 March 1970)', *J. Atmos. Terr. Phys.*, **34**, p. 559

JONES, T. B. (1973): 'Modification of the lower ionosphere by high power radio waves', *Quart. J. Roy. Astron. Soc.*, **14**, p. 411

JONES, T. B. and SPRACKLEN, C. T. (1974): 'Ionospheric effects of the Flixborough explosion', *Nature*, **250**, p. 719

JONES, T. B., ROBINSON, T., STUBBE, P., and KOPKA, H. (1984): 'Frequency dependence of anomalous absorption caused by high power radio waves', *J. Atmos. Terr. Phys.*, **46**, p. 147

JONES, T. B., ROBINSON, T. R., STUBBE, P., and KOPKA, H. (1986): 'EISCAT observations of the heated ionosphere', *J. Atmos. Terr. Phys.*, **48**, p. 1027

KANELLAKOS, D. P. (1970): 'Spatial and temporal model of the TID from a low-altitude nuclear explosion' *in* DAVIES, K. (Ed.): 'Phase and frequency instabilities in electromagnetic wave propagation'. AGARD Conf. Proc. No. 33, Advisory Group on Aerospace Research and Development, North Atlantic Treaty Organisation, p. 439

KISSICK, W. A. and FERRARO, A. J. (1975): 'Self-heating of the partial reflection experiment', *J. Atmos. Terr. Phys.*, **37**, p. 1365

LEONARD, R. S. and BARNES, R. A., Jr. (1965): 'Observations of ionospheric disturbances following the Alaskan earthquake', *J. Geophys. Res.*, **70**, p. 1250

LERFALD, G. M., JURGENS, R. B., VESECKY, F. J., and WASHBURN, T. W. (1972): 'Traveling ionospheric disturbances observed near the time of the solar eclipse of 7 March 1970', *J. Atmos. Terr. Phys.*, **34**, p. 733

LIU, C. H., KLOSTERMEYER, J., YEH, K. C., JONES, T. B., ROBINSON, T., HOLT, O., LEITINGER, R., OGAWA, T., SINNO, K., KATO, S., OGAWA, T., BEDARD, A. J., and KERSLEY, L. (1982): 'Global dynamic response of the atmosphere to the eruption of Mount St. Helens on May 18, 1980', *J. Geophys. Res.*, **87**, p. 6281

LUNNEN, R. J., FERRARO, A. J., LEE, H. S., ALLSHOUSE, R., CARROLL, K., WERNER, D., and COLLINS, T. W. (1985): 'Detection of local and long-path VLF/ELF radiation from modulated ionospheric current system', *Radio Sci.*, **20**, p. 553

MACDOUGALL, J. (1970): 'Detection of ionospheric gravity waves'. Solar Eclipse Bulletin, National Science Foundation, p. 221

MEGILL, L. R. (1965): 'Self-distortion of radio signals in the D-region', *J. Res. NBS*, **69D**, p. 369

MENDILLO, M., HAWKINS, G. S., and KLOBUCHAR, J. A. (1975): 'A sudden vanishing of the ionospheric F region due to the launch of Skylab', *J. Geophys. Res.*, **80**, p. 2217

MIGULIN, V. V. and GUREVICH, A. V. (1985): 'Investigation in the U.S.S.R. of non-linear phenomena in the ionosphere', *J. Atmos. Terr. Phys.*, **47**, p. 1181

OETZEL, G. N. and CHANG, N. J. F. (1970): 'TID investigation at Stanford, Calif. during the 1970 solar eclipse'. Solar Eclipse Bulletin, National Science Foundation, p. 230

OKUZAWA, T., SHIBATA, T., ICHINOSE, T., TAKAGI, K., NAGASAWA, C., NAGANO, I., MAMBO, M., TSUTSUI, M., and OGAWA, T. (1986): 'Short-period disturbances in the ionosphere observed at the time of typhoons in September, 1982 by a network of HF Doppler receivers', *J. Geomag. Geoelectr.*, **38**, p. 239

PAUL, A. K., SMITH, G. H., and WRIGHT, J. W. (1968): 'Ray-tracing synthesis of ionogram observations of a large local disturbance in the ionosphere', *Radio Sci.*, **3**, p. 15

PAYNE, G. L., NICHOLSON, D. R., DOWNIE, R. M., and SHEERIN, J. P. (1984): 'Modulational instability and soliton formation during ionospheric heating', *J. Geophys. Res.*, **89**, p. 10921

PERKINS, F. W. and ROBLE, R. G. (1978): 'Ionospheric heating by radio waves: Predictions for Arecibo and the satellite power station', *J. Geophys. Res.*, **83**, p. 1611

PRASAD, S. S., SCHNECK, L. J., and DAVIES, K. (1975): 'Ionospheric disturbances by severe tropospheric weather storms', *J. Atmos. Terr. Phys.*, **37**, p. 1357

ROBERTS, D. H., KLOBUCHAR, J. A., FOUGERE, P. F., and HENDRICKSON, D. H. (1982): 'A large-amplitude traveling ionospheric disturbance produced by the May 18, 1980, explosion of Mount St. Helens', *J. Geophys. Res.*, **87**, p. 6291

ROW, R. V. (1966): 'Evidence of long period acoustic gravity waves launched into the F-region by the Alaskan earthquake of March 28, 1964', *J. Geophys. Res.*, **71**, p. 343

ROW, R. V. (1967): 'Acoustic-gravity waves in the upper atmosphere due to a nuclear detonation and an earthquake', *J. Geophys. Res.*, **72**, p. 1599

SCHÖDEL, J. P., KLOSTERMEYER, J., and RÖTTGER, J. (1973): 'Atmospheric gravity wave observations after the solar eclipse of June 30, 1973', *Nature*, **245**, p. 87

SEARS, R. A. (1972): 'Ionospheric HF Doppler dispersion during the eclipse of 7 March 1970 and TID analysis', *J. Atmos. Terr. Phys.*, **34**, p. 727

STUBBE, P., KOPKA, H., RIETVELD, M. T., FREY, A., HØEG, P., KOHL, H., NIELSEN, E., ROSE, G., LAHOZ, C., BARR, R., DERBLOM, H., HEDBERG, A., THIDÉ, B., JONES, T. B., ROBINSON, T., BREKKE, A., HANSEN, T., and HOLT, O. (1985): 'Ionospheric modification experiments with the Tromsø heating facility', *J. Atmos. Terr. Phys.*, **47**, p. 1151

TANAKA, T., ICHINOSE, T., OKUZAWA, T., SHIBATA, T., SATO, Y., NAGASAWA, C., and OGAWA, T. (1984): 'HF-Doppler observations of acoustic waves excited by the Urakawa-Oki earthquake of 21 March 1982', *J. Atmos. Terr. Phys.*, **46**, p. 233

TELLEGEN, B. D. H. (1933): 'Interaction between radio waves', *Nature*, **131**, p. 840

UTLAUT, W. F. (1974): 'A survey of ionospheric modification effects produced by high-power HF radio waves' *in* FEJER, J. A. (Ed.): 'Non-linear effects in electromagnetic wave propagation'. AGARD Conf. Proc. No. 138, Advisory Group on Aerospace Research and Development, North Atlantic Treaty Organisation, p. 3–1

UTLAUT, W. F. (1975): 'Ionospheric modification induced by high-power HF transmitters – A potential for extended range VHF-UHF communications and plasma physics research', *Proc. IEEE*, **63**, p. 1022

UTLAUT, W. F. (1977): 'Ionospheric modification induced by high-power HF transmitters – Potential for communication and plasma physics research' *in* ALBRECHT, H. J. (Ed.): 'Artificial modification of propagation media'. AGARD Conf. Proc. No. 192, Advisory Group on Aerospace Research and Development, North Atlantic Treaty Organisation, p. 6–1

UTLAUT, W. F. and VIOLETTE, E. J. (1972): 'Further ionosonde observations of ionospheric modification by a high-powered HF transmitter', *J. Geophys. Res.*, **77**, p. 6804

WEINSTOCK, J. (1974): 'Enhanced airglow, electron acceleration, and parametric instabilities', *Radio Sci.*, **9**, p. 1085

WESCOTT, E. M., RIEGER, E. P., STENBAEK-NIELSEN, H. C., DAVIS, T. N., PECK, H. M., and BOTTOMS, P. J. (1974): '$L = 1{\cdot}24$ conjugate magnetic field line tracing experiments with barium shaped charges', *J. Geophys. Res.*, **79**, p. 159

WESCOTT, E. M., STENBAEK-NIELSEN, H. C., HALLINAN, T., DEEHR, C., ROMICK, J., OLSON, J., KELLEY, M. C., PFAFF, R., TORBERT, R. B., NEWELL, P., FÖPPL, H., FEDDER, J., and MITCHELL, H. (1985): 'Plasma-depleted holes, waves and energized particles from high-altitude explosive plasma perturbation experiments', *J. Geophys. Res.*, **90**, p. 4281

WHALEN, B. A., YAU, A. W., CREUTZBERG, F., WALLIS, D. D., MCNAMARA, A. G., HARRIS, F. R., PONGRATZ, M. B., BERNHARDT, P. A., KINTNER, P. M., LABELLE, J., SHELDON, W. R., BENBROOK, J. R., BERING III, E. A., FORSYTH, P. A., and KOEHLER, R. A. (1985): 'Waterhole auroral arc modification experiments: Electrodynamic response', *J. Geophys. Res.*, **90**, p. 8377

WRIGHT, J. W. (1975): 'Evidence for precipitation of energetic particles by ionospheric heating transmissions', *J. Geophys. Res.*, **80**, p. 4383

YUEN, P. C., WEAVER, P. F., SUZUKI, R. K., and FURUMOTO, A. S. (1969): 'Continuous, traveling coupling between seismic waves and the ionosphere evident in May 1968 Japan earthquake data', *J. Geophys. Res.*, **74**, p. 2256

Chapter 15

Exercises

Chapter 1

Question 1.1: Discuss the meanings of the following terms: homogeneous medium, inhomogeneous medium, isotropy, anisotropy, nondispersive medium, dispersive medium, absorption. The refractive index of a radio wave of frequency f is given by

$$\mu^2 = 1 - \frac{k}{f^n \cos\theta}$$

where θ is the angle between the direction of phase propagation and a reference direction and k is a constant. Find the value of n for which the group speed v_g and the phase speed v_p are related by

$$v_p v_g = c^2$$

where c is the speed of light. For the case in which $kf^{-n} = 1/2$, determine the angles θ for which propagation is possible and explain the nature of the propagation at other angles.

Question 1.2: A wave propagates through a uniform medium in which the refractive index μ is given by

$$\mu^2 = 1 - 2\cos^2\theta$$

where θ is the angle between the direction of phase propagation and the x axis. Calculate the range of angles for which propagation is possible and calculate the directions of energy flow for

$$90° > \theta > 45°$$

Question 1.3: Given that the strength of the geomagnetic field in the ionosphere is $0{\cdot}4 \times 10^{-4}$ tesla, calculate the thermal speeds of the following ions: electrons, protons, O^+, and NO^+, when the temperature is 1000 K. Find their gyroradii.

Chapter 2

Question 2.1: Show that the equation of a line of force in a magnetic dipole field is

$$r = aL\cos^2\Phi$$

where r is the radial distance to a line of force at a latitude Φ, a is the Earth's radius, and L is a constant. At what latitude does the line of force intersect the Earth's surface? Compare the strengths of the magnetic fields, on a line of force, on the ground, and on the equatorial plane. For a magnetic flux tube of cross-sectional area A, at this latitude, find the cross-section of the flux tube at the equatorial plane.

Question 2.2: Show that the pressure p, mass density ρ, number density n, mean molecular mass $\bar{m}$, and temperature T in a temperature-varying and composition-varying atmosphere are related thus:

$$\frac{dp}{p} = \frac{dn}{n} + \frac{dT}{T} = \frac{d\rho}{\rho} - \frac{d\bar{m}}{\bar{m}} + \frac{dT}{T} = \frac{d\rho}{\rho} + \frac{dH}{H}$$

where H is the pressure scale height. Assume gravity constant. Use this result to discuss the variation of scale height with height in the Earth's atmosphere.

Show that, hence, in a temperature-varying atmosphere, the number density n at a reduced height z above a reference level h_0 at which $n = n_0$, $T = T_0$ is given by

$$nT = n_0 T_0 \exp(-z)$$

where

$$z = \int_{h_0}^{h} \frac{dh}{H}$$

Question 2.3: Consider the case in which electron production is of the Chapman type and in which the electron loss rate varies as the nth power of the electron density (i.e. γN_e^n). Determine the following:

(i) The dependence of the height of maximum electron density on the solar zenith angle χ; (ii) the variation of maximum electron density N_m with χ; (iii) the height profile of electron density. Discuss the case in which γ varies with height [e.g. $\gamma = \gamma_0 \exp(-z)$].

Question 2.4: Near noon at a middle-latitude site, where the magnetic dip is 45°, a neutral wind is observed to blow toward the geographic pole. Explain what effect this will have on the height of maximum density and the magnitude of the critical frequency. What happens at night? Discuss similar effects caused by zonal electric fields.

Question 2.5: Calculate the speed of an electron with kinetic energy (*a*) 1 MeV, and (*b*) 10 MeV. Also calculate the speed of a proton with these kinetic energies.

Question 2.6: Obtain an expression for the electric field in the ionosphere produced by the charge separation between the electrons and positive ions when the temperatures of the electrons and ions differ. Find the contribution to the electric field due to the temperature difference for (*a*) an electron–proton gas, and (*b*) an electron–O^+ gas when $T_e = 1000$ K and $T_i = 500$ K.

Chapter 3

Question 3.1: Show that for a collisionless neutral plasma, in which the positive ions are infinitely massive, the refractive indices are given by the roots of the quartic

$$A\mu^4 - B\mu^2 + C = 0$$

where

$$A = 1 - X - Y^2 + XY_L^2$$

$$B = 2(1 - X)(1 - X - Y^2) - XY_T^2$$

$$C = (1 - X)(1 - X - Y)(1 - X + Y)$$

Question 3.2: Show that in a magnetoplasma in which there are no electron collisions, the angle between the wave normal and the magnetic field θ is given by

$$\tan^2\theta = -\frac{P(\mu^2 - R)(\mu^2 - L)}{(S\mu^2 - RL)(\mu^2 - P)}$$

where

$$P = 1 - X, \qquad R = 1 - \frac{X}{1 - Y}$$

$$L = 1 - \frac{X}{1 + Y}, \qquad S = 1 - \frac{X}{1 - Y^2}$$

Show, further, that cutoff (reflection with normal incidence) occurs when $P = 0$, $R = 0$, and $L = 0$, and that resonance occurs at an angle θ_r given by $\tan\theta_r = -P/S$.

Question 3.3: Show that close to the level of reflection of the ordinary wave with vertical propagation, the refractive index μ_+ of the ordinary wave is approximately given by

$$\mu_+ \approx (1 - X)^{1/2} \operatorname{cosec}\theta$$

Show also that the refractive index of the extraordinary wave μ_- near its level of reflection is

$$\mu_-^2 \approx \frac{2(1 - X - Y)}{(1 - Y)(1 + \cos^2\theta)}$$

The symbols are defined as in the text.

Show that with longitudinal propagation the group refractive index μ' is given by

$$\mu' = \mu + \frac{X}{\mu}\frac{1 \pm \frac{1}{2}Y}{(1 \pm Y)^2}$$

where the + and − signs refer to the ordinary and extraordinary waves respectively. Show that near reflection

$$\mu'_+ \approx (1 - X)^{-1/2} \operatorname{cosec}\theta$$

and

$$\mu'_- \approx (2 - Y)[2(1 - Y)(1 + \cos^2\theta)(1 - X - Y)]^{-1/2}$$

Chapter 4

Question 4.1: Describe the conventional ionosonde technique for sweep frequency sounding of the ionosphere. Discuss the advantages and limitations of the technique and indicate how the limitations have been remedied by modern digital sounding techniques. Describe two ways in which ionosondes have been used to study the ionosphere and illustrate with examples.

Question 4.2: Indicate how ionosonde measurements can be used to measure ionospheric characteristics.

Ionosonde measurements at Boulder, Colorado, show that the F_2-layer penetration frequencies of the ordinary and extraordinary waves are 10·00 MHz and 10·70 MHz respectively. The corresponding critical frequencies of the E layer are 3·00 MHz and 3·76 MHz. The heights of penetration of the E- and F-layers are 120 km and 300 km respectively. What information do these data give about the geomagnetic field? What are the maximum electron densities?

Question 4.3: A conventional ionogram curve is given by

$$h'(f) = \frac{k}{f_c^2 - f^2}$$

for $f < f_c$, where k is a constant. Determine the electron-density profile $N(h)$. Check your result by recalculating $h'(f)$. Neglect collisions and the Earth's magnetic field.

Chapter 5

Question 5.1: Discuss the various factors that can produce transport of plasma in the F_2 layer of the ionosphere. How do these movements affect the plasma densities?

In the absence of winds and electromagnetic drifts a layer is produced with a peak at 300 km. The downward diffusion velocity of positive ions is $10\,\mathrm{m\,s^{-1}}$ at 300 km, and the scale height is 50 km. Discuss the effect on this layer when a west-to-east electric field of $1\,\mathrm{mV\,m^{-1}}$ is present. Estimate the new height of maximum electron density and the fractional increase in the maximum density. Take $B_0 = 0{\cdot}5 \times 10^{-4}$ tesla and $I = 45°$. Assume that the ions and electrons are in thermal equilibrium and that the temperature is independent of altitude.

Make similar estimates in the case of an equatorward meridional neutral wind of $50\,\mathrm{m\,s^{-1}}$.

Question 5.2: Compare/contrast the F_1 layer critical frequencies for Boulder, Colorado, given by eqs. (5.4) and (5.5) at noon in June and December for sunspot numbers of 0, 50, 100, and 150. Assume $IF_2 = R_{12}$ and $\Phi = 48°$.

Question 5.3: Determine the total electron contents (vertical) for the following layers having a critical (*o*) frequency of 10·0 MHz and semi-thicknesses of 100 km: (*a*) parabolic layer, (*b*) quasi-parabolic layer, (*c*) biparabola. Determine the variation of the virtual height with frequency for these layers.

Question 5.4: At a height of 1000 km in the topside of the ionosphere the electron temperature $T_e = 2000\,\mathrm{K}$ and the ion temperature $T_i = 1000\,\mathrm{K}$. Calculate the plasma scale height and the polarization electric field for the cases where the positive ions are (i) oxygen atomic ions, (ii) protons. [*Answer:* (i) 213 km, $0{\cdot}81\,\mu\mathrm{V\,m^{-1}}$; (ii) 3400 km, $0{\cdot}05\,\mu\mathrm{V\,m^{-1}}$.]

Chapter 6

Question 6.1: A radio wave of frequency 2 MHz is incident on a flat non-absorbing ionospheric layer at an angle of 45° in a magnetic meridian. The electron density in the layer increases linearly with height such that the plasma frequency is zero below 200 km and reaches 2 MHz at 300 km. The gyrofrequency is 1 MHz and the dip is 45°. Determine the following parameters for north–south and south–north propagation, with a vertical antenna and with a horizontal east–west antenna: (i) The heights at which the propagation vectors of the ordinary and extraordinary waves are horizontal; (ii) the angles made by the ordinary *ray* and the extraordinary *ray* and the horizontal at these heights; (iii) the fractions of the incident power that go into the ordinary and extraordinary waves; (iv) the polarizations of the ordinary and extraordinary waves at the heights at which the wave normals are horizontal; (v) the polarizations of the emergent waves.

Question 6.2: A radio wave of frequency 3·0 MHz is incident at 45° on a plane parabolic layer, at a magnetic pole where the gyrofrequency is 1·5 MHz. The critical frequency of the layer is 3·0 MHz, the height of the bottom of the layer is 200 km, and the peak is at 300 km. Find the heights of reflection of the ordinary and extraordinary waves and the polarizations of these waves (*a*) at entry, and (*b*) at the levels of reflection. How is the incident power divided between the ordinary and extraordinary waves?

Question 6.3: A radio wave of frequency 3·0 MHz is incident on a plane parabolic layer having a critical frequency of 3·0 MHz, a semi-thickness of 100 km, and a height of peak density of 300 km. The gyrofrequency is 1·5 MHz and the dip is 45°. Determine the heights of reflection of the ordinary rays for angles of incidence of (*a*) 20°, and (*b*) 10°. Sketch the ray paths. What is the wave polarization in the vicinity of the spitze?

Question 6.4: From ionospheric vertical soundings the following data are obtained:

f (MHz)	1	2	5	8	9	9·5	9·8	9·9
h' (km)	201·0	204·1	227·5	287·9	332·5	374·0	425·2	462·0

Neglecting electron collisions and the Earth's magnetic field, determine the electron-density profile. What is the maximum electron density and thickness of the layer?

Question 6.5: A radio wave of frequency 20 MHz is incident on a flat parabolic layer that has a critical frequency of 10 MHz, semi-thickness of 100 km, and height of maximum density of 300 km. Determine the maximum angle of elevation at which rays are reflected. Find the minimum (skip) distance within which signals are not reflected. For this distance what is the maximum usable frequency (MUF), and what is the real height of reflection? Neglect electron collisions and the Earth's magnetic field.

Chapter 7

Question 7.1: Describe the 'phase integral' method for the determination of ionospheric absorption near the level of reflection. A radio wave of frequency f is reflected, with vertical propagation, from a layer in which the electron density increases linearly with height and in which the collision frequency ν is independent of height. Show that the additional absorption L suffered in the process of reflection is

$$L = 2\nu^{3/2}/[3ca(\pi f)^{1/2}]$$

Here a is the height gradient of electron density and c is the speed of light.

Question 7.2: A radio wave is reflected vertically from a sinusoidally corrugated reflection surface moving horizontally with velocity v in a direction normal to the lines of constant amplitude. Neglecting any lateral movement of the point of reflection, derive an expression for the instantaneous (Doppler) frequency shift Δf in terms of the period τ, the amplitude Δh of the height variation, and the radio wavelength λ.

In a 'Doppler' measurement on a 3·0 MHz radio wave reflected from the ionosphere, with vertical propagation the following data are obtained: $\tau = 10$ min; max $\Delta f = 0{\cdot}5$ Hz. Determine (i) the maximum change in reflection height; (ii) the wavelength of the corrugation, assuming that the horizontal speed is 100 m s^{-1}.

Question 7.3: Interference takes place between two waves of wavelength λ, of equal amplitude, reflected simultaneously from two sharply bounded ionospheric layers at different heights moving vertically. Show that the frequency n of the periodic fading is given by

$$n = \frac{2}{\lambda}|v_1 \cos\phi_1 \pm v_2 \cos\phi_2|$$

where ϕ_1 and ϕ_2 are the angles of incidence on the higher and lower layers, and v_1 and v_2 are the vertical speeds of layers 1 and 2 respectively. When is the plus sign used and when is the minus sign used?

Question 7.4: An elliptically polarized radio wave is incident vertically on the ionosphere at the magnetic equator. The ratio of the major axis of the ellipse to the minor axis is R: 1, and the major axis makes an angle γ with the equator. Show that the ratio of the power in the ordinary wave to that in the extraordinary wave is

$$(R^2 \sin^2\gamma + \cos^2\gamma)/(R^2 \cos^2\gamma + \sin^2\gamma).$$

Question 7.5: A reflecting surface is of the form

$$z = H + D \cos kx$$

where $D \ll H$, $k = 2\pi/L$, L is the horizontal wavelength, D is the amplitude, and H is the mean height. Show that the range R to and from the reflection point is given by

$$R = 2(H + D\cos\theta)(1 + k^2D^2 \sin\theta)^{1/2}$$

where $\theta = kx$.

Show that when D exceeds a certain value three reflections are possible. Show that when the surface moves with a horizontal speed V the frequency shift(s) of the echo(es) are given by

$$\Delta f = \frac{2fD\,kV}{c} \cdot \frac{\sin\theta}{(1 + k^2 D^2 \sin^2\theta)^{1/2}}$$

where f is the wave frequency and c is the speed of light.

Question 7.6: Describe the meaning of (*a*) frequency dispersion and (*b*) angular dispersion in the case of propagating waves. Illustrate these phenomena in the case of atmospheric gravity waves of frequency f in an isothermal horizontally stratified atmosphere, in which dissipation and the Earth's rotation are negligible, for which the propagation factor k is given by

$$\frac{C^2 k^2}{\omega^2} = \frac{X - 1}{Y^2 \sin^2\theta - 1}$$

where the symbolism is as follows:

$$C = \text{the speed of sound}$$

$$X = \left(\frac{f_a}{f}\right)^2$$

$$Y = \left(\frac{f_g}{f}\right)$$

$$f_a = \text{the acoustic cutoff frequency}$$

$$f_g = \text{the Brunt-Väisälä frequency}$$

$$\omega = 2\pi f$$

$$\theta = \text{angle between } \boldsymbol{k} \text{ and the vertical}$$

Show that the ray makes an angle β with the vertical, where $\tan\beta = -(Y^2 - 1)\tan\theta$. Show also that the phase velocity v and the group velocity u are related by

$$\frac{C^2}{uv} = 1 + \frac{Y^2 \sin^2\theta(X - 1)}{(Y^2 \sin^2\theta - 1)^2}$$

What are the minimum angles to the vertical at which the phase and energy of gravity waves can propagate? Compare and contrast some of the properties of gravity waves compared with whistlers.

Question 7.7: A radio signal of carrier frequency f and modulation bandwidth f' is reflected vertically from a plane parabolic layer with a critical frequency f_c and a semi-thickness a. Show that the effective bandwidth B of the layer is given by

$$B^{-1} = \frac{a f'}{c f_c}\left(\ln\frac{f_c + f}{f_c - f} + \frac{2ff_c}{f_c^2 - f^2}\right)$$

where c is the free space velocity and $f' \ll f$.

Calculate the variation with carrier frequency of a signal with pulse modulation of duration 0·1 ms, with layer critical frequency of 10 MHz and semithickness of 100 km. What happens when the carrier frequency approaches to within 5 kHz of the critical frequency?

Chapter 8

Question 8.1: A polarimeter at 45° latitude measures the Faraday rotation of a radio signal from a geostationary satellite over the equator in the same meridian plane. The wave frequency is 500 MHz, the vertical electron content is 10^{17} el. m^{-2}, the geomagnetic field is dipolar, and the dipole axis coincides with the geographic axis. The electron gyrofrequency at the Earth's surface on the equator is 0·8 MHz. Calculate approximately the Faraday rotation suffered by the radio wave.

Discuss the limitations of the Faraday technique for the measurement of total electron content and indicate how these can be overcome.

Question 8.2: A radio wave of frequency 40 MHz propagates vertically completely through a Chapman layer having a critical frequency of 4 MHz and a scale height of 50 km. The gyrofrequency is constant at 1·0 MHz and the magnetic dip is 45°. Calculate (i) the polarizations of the *o* and *x* waves; (ii) the fractions of the incident power in the *o* and *x* waves; (iii) the lateral deviations of the *o*-ray and the *x*-ray; (iv) the excess time delays of the waves; (v) the Faraday rotation of the electric field of the wave.

Question 8.3: Show that in the absence of appreciable ray refraction the modulation phase delay ϕ of a radio wave of frequency f MHz produced by an electron content N (el. m^{-2}) along the ray path is given by $\phi \approx 4{\cdot}84 \times 10^{-11} N\Delta f/f^2$ degrees, where f is the modulation frequency in MHz. Discuss the problems that arise in the measurement of phase. Show that when there is small refraction in a flat ionosphere the apparent angle of elevation is greater than the true elevation Δ at a flat Earth by

$$\xi \approx 40{\cdot}3N \cos\Delta/f^2 T \sin^2\Delta$$

where T is the slab thickness. Hence show that when refraction is small $\phi \approx 4{\cdot}84 \times 10^{-11} N\Delta f(1 + QN)/f$ where $Q \approx 40{\cdot}3N \cot^4\Delta/f^2 T$.

Question 8.4: A geostationary satellite beacon with position vector $\boldsymbol{S}$ emits a radio signal from a linear antenna aligned along a vector $\boldsymbol{A}$ $(= A_x, A_y, A_z)$. Show that to an observer on the Earth with position vector $\boldsymbol{R}$ the direction of the antenna with respect to the horizon is given by

$$\alpha = \cos^{-1}\frac{\boldsymbol{A}\cdot(\boldsymbol{R}\times\boldsymbol{S})}{|A_p|\,|\boldsymbol{R}\times\boldsymbol{S}|}$$

where A_p is the magnitude of the projection of A onto a plane perpendicular to the ray path. The x-axis is along the line joining the Earth center and the satellite, the y-axis is tangent to the satellite orbit (east), and the z-axis is north. In terms of h, the height of the satellite, Λ $(=\Lambda_R - \Lambda_S$, the longitude difference between satellite and receiver) and the latitude difference between satellite and receiver, Φ, show that

$$\cos\alpha = \frac{1}{|A_p|}\frac{A_y\sin\Phi - A_z\sin\Lambda\cos\Phi}{(\sin^2\Phi + \cos^2\Phi\sin^2\Lambda)^{1/2}}$$

where

$$A_p =$$

$$\left\{1 - \frac{[A_x(R+h-R\cos\Lambda\cos\Phi) - A_yR\sin\Lambda\cos\Phi - A_zR\sin\Phi]^2}{(R+h-R\cos\Lambda\cos\Phi)^2 + R^2\sin^2\Lambda\cos^2\Phi + R^2\sin^2\Phi}\right\}^{1/2}$$

Calculate the values of α when the emitting antenna lies (i) in the equatorial plane and (ii) perpendicular to the equatorial plane; when (*a*) the satellite and receiver are in the same meridian plane, and (*b*) when the longitude difference is 45°.

Question 8.5: Find the approximate changes in the angle of elevation (=45°) at the ground, produced by a change of electron content of 10^{16} el. m^{-2}, of radio signals from a geostationary satellite at a height of 35 780 km on frequencies of 40 MHz, 140 MHz, and 360 MHz.

Question 8.6: A satellite at a height of 2000 km emits radio signals on very high frequencies f_1 and f_2. The signals traverse a flat ionosphere that has a vertical total electron content N_{Tv}. The true elevation at the (flat) ground is about 45°. Show that the approximate difference S between the paths of the two signals is given by

$$S = 40{\cdot}3N_{Tv}(f_1^{-2} - f_2^{-2})\cos^4\Delta/\sin^5\Delta.$$

Find the path difference for the following values:

$$f_1 = 1228\,\text{MHz}, f_2 = 1575\,\text{MHz}, N_{Tv} = 10^{17}\,\text{el.m}^{-2}, \text{ and } \Delta = 45°$$

Obtain the expression for the approximate difference between the true elevation and the apparent elevation at the receiver on the ground. Find the excess time delays suffered by the two signals in traversing the ionosphere.

Show that in the presence of refraction the Faraday electron content is given by

$$N_{Tv} = A\Omega(1 - Q\Omega)$$

Find values of A and Q for the following conditions: $B_L = 0{\cdot}4 * 10^4\,\text{T}$, $\Delta = 45°$, $f = 40$ MHz, slab thickness = 250 km.

Chapter 9

Question 9.1: Explain the various features in the ionosphere that accompany a burst of solar X-rays in the wavelength range 0·5 Å to 10 Å.

During a sudden cosmic noise absorption the following increases in absorption are observed on a frequency of 30 MHz at an equinox: London (51·48°, 0°), 0·888 dB; Moscow (55·50°, 37·30°), 0·615 dB; Delhi (28·60°, 77·10°), 0·665 dB, Washington (37·9°, 284·00°), 0·661 dB. Determine the dependence of absorption on the solar zenith angle. Calculate the expected absorption on 30 MHz at Accra (05·63°, 359·83°) and at Boulder (40·00°, 254·80°).

Question 9.2: Discuss the effects of solar X-ray bursts (0·5 Å to 10 Å) on the operation of very low frequency systems.

During a sudden phase anomaly over a 200 km path, the phase of an 18 kHz signal is observed to change by 100°. Calculate the change in the effective reflection height of the D region.

Question 9.3: In a sudden frequency deviation, caused by an enhancement of solar ionizing radiation, the frequency shift increases linearly with time from zero to 2 Hz in 2 min. At this time the enhanced solar radiation falls to zero. Assuming that the propagation is vertical and that the carrier frequency is 3·0 MHz, find the total (peak) enhancement of electron content in a vertical column of 1 m^2 cross section. Assume that electron production takes place in a nondeviating region below the level of reflection in which the loss process is linear (that is, attachment) with a relaxation time τ of 30 s. Neglect effects due to electron collisions and the Earth's magnetic field. What is the increase in electron content and the frequency shift of the reflected wave 4 min after the start of the frequency deviation? What would the maximum electron content have been in the absence of all electron losses? [*Answer:* $1{\cdot}33 \times 10^{15}$ electrons m^{-2}, $0{\cdot}244 \times 10^{14}$ electrons m^{-2}, 0·0733 Hz, 3×10^{15} electrons m^{-2} (approximately).]

Question 9.4: Explain the meaning of polar cap absorption and the factors that affect it.

Calculate the rigidities and the cutoff latitudes for protons with energies of 1 MeV, 10 MeV, and 100 MeV.

Question 9.5: Describe the various stages of an ionospheric storm and the effects on radio systems.

Chapter 10

Question 10.1: Show that on frequencies much lower than the electron gyrofrequency, and for infinitely massive positive ions,

$$\mu^2 = \frac{1 - X}{1 - X\cos^2\theta}$$

Show that, under this condition, the electrons are constrained to move along the imposed magnetic field. What is the wave polarization? Determine the forward tilt of the electric field of the wave.

Question 10.2: Show that on a frequency f that is low compared with the gyrofrequency f_H and the plasma frequency f_N, the group speed is

$$V_g = 2c\,\frac{(Y\cos\theta - 1)^{3/2}}{XY\cos\theta}$$

where $X = f_N/f$, $Y = f_H/f$, c is the speed of light, and θ is the angle between the direction of phase propagation and the external magnetic field. Obtain an approximate expression for the velocity of a wave packet.

Question 10.3: A radio wave of frequency 3 kHz enters a dipole tube with $L = 2{\cdot}5$ and is received in the opposite hemisphere. The tube moves inward at a rate of $dL/dt = 0{\cdot}05$ per second. What is the Doppler frequency shift suffered by the wave? Assume that the propagation conditions within the duct remain the same. [*Answer:* 6·8 Hz.]

What magnitude of electric field at the equatorial plane would produce this inward motion? [*Answer:* 1·6 V/m.]

Question 10.4: Discuss the factors that affect the propagation of ELF waves in the Earth–ionosphere waveguide. An extremely low frequency propagates in the Earth–ionosphere waveguide at the upper boundary of which the electron collision frequency is $10^6\ \mathrm{s}^{-1}$ and the electron gyrofrequency is $1{\cdot}0 \times 10^6\ \mathrm{s}^{-1}$. Calculate the ratio of attenuation with magnetic field to that in the absence of the magnetic field. [*Answer:* 1.1]

Question 10.5: Describe the diurnal variations of the effective reflection heights of very low frequency signals (*a*) over short ranges, and (*b*) over long ranges.

Assuming that the height of reflection changes from 70 km by day to 90 km at night, estimate the changes in phase of VLF signals on frequencies of 2 kHz, 5 kHz, 10 kHz, and 20 kHz propagating over ground ranges of 500 km and 5000 km.

Chapter 11

Question 11.1: Discuss qualitatively the variation with azimuth, in the strength of a skywave signal emitted by a vertical antenna of a medium wave transmitter ($f = 800$ kHz) on the dip equator where the electron gyrofrequency is 800 kHz.

Determine (approximately) the polarizations of the ordinary and extraordinary waves, at exit from the ionosphere in the magnetic meridian with south–north propagation, at an angle of 30° with the vertical. The wave frequency is 1·0 MHz, the gyrofrequency is 1·5 MHz, and the value of electron collision frequency at exit is $1{\cdot}9 \times 10^6\,s^{-1}$. Calculate the rotations of the axes of the polarization ellipses.

Question 11.2: A radio wave of frequency 1·0 MHz is incident vertically on the ionosphere at Boulder where the gyrofrequency is 1·48 MHz and the dip is 68·1°. The electron-density distribution is parabolic with base height of 200 km, semi-thickness of 100 km, and critical frequency of 2 MHz. Calculate (i) the heights of reflection of the ordinary and extraordinary waves; (ii) the wave polarizations at the bottom of the layer and at the levels of reflection; (iii) the wave polarization of the extraordinary wave at the level of reflection of the ordinary wave. Ignore effects of electron collisions.

Question 11.3: A radio wave of frequency 500 kHz is incident at an angle ϕ_o in the magnetic meridian where the magnetic dip is 68°. What happens to the wave on entry into the ionosphere? Sketch the ray paths for the cases in which $\phi_o = 10°$ and $\phi_o = 60°$. Take the electron gyrofrequency to be 1·5 MHz. Discuss the behavior of similar waves as the latitude changes from the equator to the poles.

Question 11.4: A radio wave of angular frequency ω propagates in a uniform plasma of density N. The electron collision frequency ν is proportional to electron energy ε (i.e. $\nu = \nu_m \varepsilon$) and the electron energies have a Maxwellian distribution. Use the Boltzmann equation for momentum transfer to show that, for small perturbations of electron energies, the refractive index is given by

$$n^2 = 1 - Xu^2\,\mathscr{C}_{3/2}(u) - i\,\tfrac{5}{2}\,Xu\mathscr{C}_{5/2}(u)$$

where $u = \omega/\nu_m$, $X = Ne^2/\varepsilon_0 m\omega^2$

$$\mathscr{C}_p(u) = \frac{1}{p!}\int_0^\infty \frac{x^p \exp(-x)\,dx}{x^2 + u^2}$$

Question 11.5: Define the term cymomotive force and discuss the temporal and frequency variations of medium frequency waves.

Using the CCIR method (Section 11.6) calculate the field strength received at a circle of radius 500 km from a transmitter near Denver (40°N, 105°W) on a frequency of 1490 kHz with power of 5 kW, at noon and midnight for June and December when the sunspot number is 100. Assume an antenna gain of 3 dB over an isotropic radiator. Compare the diurnal curves of signal strengths at 500 km north and 500 km west of the transmitter for March, under the same sunspot conditions.

Chapter 12

Question 12.1: Using eqn. (12.12) draw diurnal curves of absorption for vertical propagation of waves of frequency 5 MHz at latitudes of 0°, 20°, 40°, and 60°

on the zero meridian for sunspot numbers 0 and 100 in December, March, and June. Assume $\nu = 10^6\ s^{-1}$ and neglect the winter anomaly factor.

Question 12.2: Using Fig. 5.12 calculate the maximum usable frequencies and optimum working frequencies for the following circuits: Washington, DC (38°N, 78°W) to Boulder, Colo. (40°N, 105°W); Washington, DC, to London (52°N, 0°); Washington, DC, to Buenos Aires (34·5°S, 58·7°W).

Question 12.3: Using the procedures outlined in Sec. 12.3.1 find the optimum working frequency for a path from New York to Chicago at 1800 UT in December 1987 when the 12-month smoothed sunspot number was 60 and the *o*-wave critical frequency was 10 MHz. The $M(3000)F_2$ was 3·0. What is the effect of the E layer(s)?

Question 12.4: In the case of Question 12.3 find the transmitter power required to ensure performance of an A3 telephony double sideband system with a 5 kHz bandwidth on the optimum working frequency. The grade of service is marginally commercial for 90% of the hours, so the required signal/noise is 27 dB; the antennas are half-wave horizontal dipoles at a height of 2 m

Question 12.5: Describe the procedures for calculating the optimum working frequencies on paths of length greater than 4000 km. Use Figs 5.12, 5.13, and 6.15 to estimate the FOT at 0600 UT for the path London to Washington, DC, in March 1987 when the 12-month running mean sunspot number was 60 and in March 1979 when the 12-month running mean sunspot number was 158.

Chapter 13

Question 13.1: Discuss the advantages and disadvantages of using D-region forward scatter (*a*) as a communication medium, and (*b*) to study the ionosphere.

Question 13.2: Discuss the advantages and disadvantages of using meteor-scatter (*a*) for communication purposes, and (*b*) for the study of the ionosphere.

Question 13.3: Describe a technique of using meteor radio reflections for the determination of horizontal neutral winds in the ionosphere. At what heights do these apply?

A Doppler meteor radar operating on 50 MHz includes two antennas, both inclined at 10° to the zenith; one is tilted to the north and the other is tilted to the east. In a measurement the Doppler shift is −0·72 Hz on the north-pointing beam and −1·25 Hz on the east-pointing beam. Determine the neutral wind.

Chapter 14

Question 14.1: Discuss the HF Doppler technique for studying the ionosphere. In particular, show how it can be used for the determination of a traveling ionospheric disturbance.

During a severe thunderstorm over the central USA a Doppler sounder operating over a short ground base on 5·0 MHz observes a disturbance with amplitude 1 Hz. The sounder is 100 km north of the storm, and the real height of reflection is 250 km at which the atmospheric density is $10^{-10}\,\mathrm{kg\,m^{-3}}$. Estimate the power flux density at the reflection height of the probing radio signals and the acoustic power radiated from the source. Assume a lossless atmosphere and, if required, use the following data: sound speed at reflection = $800\,\mathrm{m\,s^{-1}}$; dip = 68°.

Question 14.2: Plot (i) the acoustic cutoff period, (ii) the Brunt–Väisälä period, and (iii) the speed of sound, using an atmospheric model of your choice.

Question 14.3: Describe the Luxembourg effect and show how the phenomenon can be used to obtain electron-density profiles. A transmitter radiates 100 kW of power isotropically on 1·0 MHz. Show that the mean oscillatory energy $\bar{U}_0$ imparted to an electron is given by

$$\bar{U}_0 = E^2 e^2/16\pi \mathrm{m} f^2$$

where E is the electric field of the wave in the ionosphere (100 km). Find expressions for the maximum electron displacement and maximum electron velocity. On the assumption that the excess energy cooling factor is approximately $1{\cdot}3 \times 10^{-3}$, show that the mean energy of the electrons exceeds the mean unheated energy by about 0·6%.

Question 14.4: Distinguish between resistive heating and resonance heating. Describe the sequence of processes that usually accompany the onset of high-power heating of the F region.

Subject Index

Name Index

Printed in the United Kingdom
by Lightning Source UK Ltd.
133597UK00001B/100/P

9 780863 411861